Houston Freeways

A Historical and Visual Journey

Erik Slotboom

Published by Oscar F. Slotboom
www.HoustonFreeways.com

Oscar F. Slotboom is solely responsible for all content. This book is not affiliated with the Texas Department of Transportation or any other government agency.

Printed in the United States of America by C. J. Krehbiel, Cincinnati, Ohio

Library of Congress Control Number: 2003094252

ISBN: 0-9741605-3-9

Dedicated to

Everyone who has helped build Houston's freeways

Contents

Preface and Acknowledgements

As I grew up in Houston, I always wondered why no one had written a book about Houston's freeways. I never outgrew my childhood fascination with freeways, and I never stopped thinking about the book. As I drove on Houston's freeways, questions would appear in my mind. What was here before the freeway? Who decided the freeway would be built here? Why was the freeway built with its design? In 2001, I decided to write the book.

As a subject of historical study, the freeway is a unique entity. It's a history we live every day. We see its results. We receive its benefits and suffer from its shortcomings. We build our lives and lifestyles around it. It continues to influence the future, just as much as it has influenced the past.

The influence of the freeway on the development of modern Houston is a central theme of *Houston Freeways*. But behind the freeways are the stories of people—the people who made the freeways happen, and the people who were empowered by the freeways to build something new. Writing *Houston Freeways* has given me a new appreciation for the dedication, vision, and creativity of the many individuals who have shaped Houston through its freeways. Our freedom of mobility is the product of their efforts. The ability to efficiently go where we want, when we want, is something we tend to take for granted, but perhaps we should pause to recognize the efforts of those who made our mobility possible.

Writing a historical book about a recent period of time provides many opportunities. The freeway era has generally been a well-photographed period of history. Many of the key participants in the history are still alive, able to tell their stories. Most documents are still available. But even more important to the writing of this book was the advance of computer and software technology. Just five years ago, it would not have been economically feasible for an individual to write this book. Digital photography with resolution sufficient for document capture has only become available within the last three years. The disk capacity, processor speed, and memory required for processing large digital images has been available at economical prices for less than 5 years. Although production, raster processing, and vector drawing software have been available for some time, recently available powerful packages using the portable document format (PDF) were essential to this project.

Bringing my vision of *Houston Freeways* to reality required the assistance of many. I would like to extend my special thanks to the content reviewers, Chuck Fuhs and Chris Olavson. Both Chuck and Chris have great knowledge of Houston's freeways, and they share my enthusiasm and interest in the subject. Chuck coauthored the Freeway Mass Transit chapter, his area of greatest expertise, and provided many of the photos in the book.

Many others have made important contributions to this book and my research efforts. I won't attempt a comprehensive listing here, but my gratitude goes to everyone who has helped. Two individuals in particular deserve mention: Joel Draut of the Houston Public Library and Anne Cook at the TxDOT library. Without their help, many photos would not have found their way into this book.

The Texas Department of Transportation has been exceptionally helpful in my research for *Houston Freeways*. Every single TxDOT employee I have been in contact with has been cooperative, helpful, and responsive. Many have gone out of their way to provide assistance. The citizens of Texas should be proud of their highway department, both for its accessibility and its work to build Texas' transportation system.

Sometimes I wonder: Hundreds of years from now, how will our freeway era be viewed? Will history say that we lived in an era when cities built great freeway networks to foster an unprecedented level of individual mobility? Will Houston be recognized as the ultimate product of the freeway era, even more so than Los Angeles? As I make a case in the text, I believe that Houston is in fact the world's most freeway-focused city. None of us alive today will know history's judgment of the freeway legacy. But perhaps better yet, we're the ones who are living in it. This important period in Houston's history—and the history of the freeway—has now been documented for the ages. Enjoy your journey through *Houston Freeways*, and enjoy your journeys on Houston's freeways.

Erik Slotboom, July 2003

Abbreviations, Definitions, and Explanatory Notes

Map Legend

(blue line)	Freeway	45	Interstate Highway	(curved arrow)	Indicates direction of aerial view
(green line)	Tollway	59	United States (US) Highway System	(flag arrow)	Indicates direction of ground view
(orange line)	Major Highway	225	State Highway System	(green box)	Park
(black line)	Street				
(dashed black)	Under construction or planned for near-term construction	(logo)	Logo of the Harris County Toll Road Authority		
(dashed pink)	Long term corridor				

Houston and Harris County

Houston was founded on August 30, 1836, by brothers Augustus and John Allen. Nearly all of the city of Houston is contained within Harris County.

	Population	Source	United States Rank
City of Houston	2,009,834	2002 Census estimate	4
Harris County	3,557,055	2002 Census estimate	3
Houston Metropolitan Statistical Area	4,669,571	2000 Census	10
Texas	21,779,893	2002 Census estimate	2

The Texas Department of Transportation (TxDOT) and Texas Transportation Commission

The designations TxDOT and Texas Transportation Commission are used for the present-day entities and all predecessors. The previous and current names for these entities are listed in the table below. The Texas Transportation Commission consists of three members appointed by the governor with the advice and consent of the Texas Senate. Members serve six-year terms which are staggered so that an appointment is generally made every two years. The Texas Transportation Commission is the governing body of TxDOT, responsible for establishing policy and rules. The commission also decides which highway projects receive funding for construction.

1917-1975	Texas Highway Department State Highway Commission
1975-1991	State Department of Highways and Public Transportation State Highway and Public Transportation Commission
1991-current	Texas Department of Transportation Texas Transportation Commission

Organization of TxDOT Houston Offices

TxDOT District 12 includes Houston, Harris County, and five adjacent counties (Fort Bend, Waller, Montgomery, Galveston, and Brazoria). The Houston Urban Project Office existed from 1945 to 1984 and was responsible for Loop 610 and all freeways within Loop 610. The Houston Urban Project Office was sometimes called the Houston Urban Office or the Houston Urban Expressways Office. The administrative head of District 12 is designated the "District Engineer"; the head of the Houston Urban Project Office was designated as the "Engineer-Manager."

District Engineer, District 12	
Jim Douglas	1945-1954
Wiley Carmichael	1955-1973
Omer Poorman	1973-1986
Milton Dietert	1986-1995
Gary Trietsch	1995-present

Engineer-Manager, Houston Urban Project Office	
William J. Van London	1945-1955
Albert C. Kyser	1955-1972
William V. "Bill" Ward	1972-1984

Harris County Government

The county judge is the political and administrative head of county government, chosen by public vote. The county judge is not a judicial position. There are four elected commissioners in Harris County Commissioners Court. Commissioners court handles administrative functions of county government; it is not a judicial body.

Photo Credits

Definitions, Abbreviations, and Acronyms

braided ramp	A design in which entrance and exit ramps are built at the same location, with one ramp passing over the other.
contraflow lane	A lane taken from one traffic direction and temporarily converted for use by vehicles travelling in the opposite direction.
diamond lane	A freeway lane designated for use by buses and high occupancy vehicles, usually marked with a diamond. There is no barrier between the diamond lane and adjacent general-purpose freeway lanes. Also called a concurrent-flow HOV lane.
expressway	A non-access-controlled highway free of traffic lights, but with at-grade intersections.
FM	Farm-to-Market road. Although the areas served by many FM roads have been urbanized, the FM designation has been retained.
freeway	A limited-access highway facility free of intersections.
freeway lanes	The number of general-purpose main lanes in both directions, usually excluding any central transitway lanes. For example, a 10-lane freeway has 5 lanes in each direction.
frontage road	A road on a freeway right-of-way separate from the main lanes but generally parallel to the main lanes, providing access to property along the freeway.
HCTRA	Harris County Toll Road Authority
HGAC	Houston-Galveston Area Council, the regional planning organization
HMRC	Houston Metropolitan Research Center, division of the Houston Public Library
HOV	High occupancy vehicle; a vehicle with more than one occupant.
IH	Interstate Highway. IH is the official designation used by TxDOT, but is not commonly used by the general population. The IH designation is used here to avoid confusion between the numeral 1 and the letter I.
managed-lane freeway	A freeway with certain lanes that are managed to obtain optimal traffic flow. In Houston, managed lanes are planned as toll lanes on the Katy and Northwest Freeways.
Metro	Metropolitan Transit Authority, Houston's public transit provider.
parkway	In the Houston area, the term "parkway" has no meaning. A parkway can be a conventional freeway with frontage roads (e.g. the Tomball Parkway), a tollway (e.g. the Fort Bend Parkway), the frontage roads of a tollway (e.g. the Sam Houston Parkway), a conventional arterial street (e.g. the Lake Houston Parkway), or a parkway in the traditional sense of a limited-access facility in a parklike setting (e.g. Memorial Parkway).
record of decision	An administrative action by the federal government certifying that a project is in compliance with federal regulations and authorizing local agencies to begin construction.
interchange levels	The ground is considered a level. Note that ramps typically do not intersect over a single point, so a level 5 ramp may intersect a level 4 ramp at a point where the level 4 ramp is below its peak.
SH	State Highway, generally a high-volume highway and often a freeway.
spur	A short length of freeway branching from a larger freeway, providing access to a specific area.
stack	An informal name for a multilevel freeway-to-freeway interchange.
tollway	A limited-access highway facility free of intersections and requiring payment of a toll. When referring to the general class of limited-access highway facilities, the term "freeway" is often used to indicate both freeways and tollways. For example, references to regional freeway lane-miles include both freeway and tollway lane-miles.
transitway	A barrier-separated lane, typically in the center of the freeway, for buses and high occupancy vehicles. Also called an HOV lane.
TTA	Texas Turnpike Authority. Created in 1953, it became a division of TxDOT in 1997.
TTI	Texas Transportation Institute, a research organization at Texas A&M University in College Station which performs most highway-related research for TxDOT.
TxDOT	Texas Department of Transportation
US	United States, used to designate a highway in the US highway system

Where it all began: This photograph of the Gulf Freeway overpass at Calhoun shows the exact location of the dedication ceremony for the first freeway segment in Houston, which took place on September 30, 1948. This photograph was taken one month later on November 3, 1948. (Photo: National Archives 30-N-48-1436)

Building the System

It was a warm autumn evening in late September 1948. A crowd of politicians, dignitaries, guests, and curious onlookers began to congregate at an unusual location—a place most had never been before. Many surely sensed it was an important event for Houston, but few realized how dramatically it would transform the way they lived and the way their city would grow and sprawl. It was the introduction of a new concept in transportation that had been tried elsewhere and had now reached Texas, and in the spirit of Texas was declared to be done bigger and better than anywhere else. On that night, the crowd gathering on the freeway overpass would witness the dedication and official opening of the first segment of the Gulf Freeway.

After the usual statements by the officials, Mayor Oscar Holcombe threw a circuit breaker to illuminate the freeway lighting, and a brand-new stretch of freeway emerged from the darkness. The scene was rife with symbolism, as the freeway would light the way to a new future for Houstonians. Just eight years before, the very location of the freeway had been an electric railway providing transit service for Houstonians. Now private automobiles lined up for a half-mile in both directions from the dedication point, awaiting the official green light to proceed on the freeway. Motorists kept the freeway busy into the evening as they took their first drive on the new concrete wonder.

This scene would repeat itself in cities all across the United States. An automobile-hungry society, technological advances, rapidly increasing standards of living, and government policies combined to make the postwar American city a freeway metropolis. But only in a few cities would the freeway metropolis reach the greatest proportions: Los Angeles, New York City, Chicago, Dallas–Fort Worth, and Houston. Among the nation's leading freeway cities, no other city has demonstrated Houston's sustained and ongoing commitment to an ever-expanding and improving freeway system. Houston has built one of the world's most impressive freeway systems. As with any history, it is a drama with leaders and visionaries, victories, defeats, crises, controversies—and continuously evolving visions of the future.

Cruising in style: Car 420 shown above was part of a batch of 20 deluxe streetcars purchased for Houston's electric railways in 1927. The electric railway system reached its peak in 1927 with 90 miles (144 km) of track. This 1928 photo shows a streetcar passing in front of the City Auditorium on Texas Avenue. Jones Hall for the Performing Arts, which opened in 1966, was later constructed at this location. (Photo: HMRC MSS 334-486)

Before the Freeways

On the eve of the freeway era in 1948, Houston was ill-equipped to handle the large number of automobiles that were starting to appear on its roadways. The founders of Houston, the Allen brothers, laid out a high-quality grid system for the city in 1836, but the growth of the city soon exceeded the bounds of the original grid and a "cancerous ring of haphazard growth took shape" around the originally planned city. Houston's streets were extended from the periphery of the city in a generally unplanned manner and efforts to provide a high-quality, planned street system were unsuccessful. Houston's electric streetcar system served the area well and helped fuel suburban growth from 1890 through the 1920s, but the system went into steep decline in the 1930s and was fully replaced by buses in 1940. After World War II, Houston was facing rapidly increasing automobile ownership with a poorly planned street network designed for the pre-automobile era. An extensive freeway system would soon become a necessity.[1]

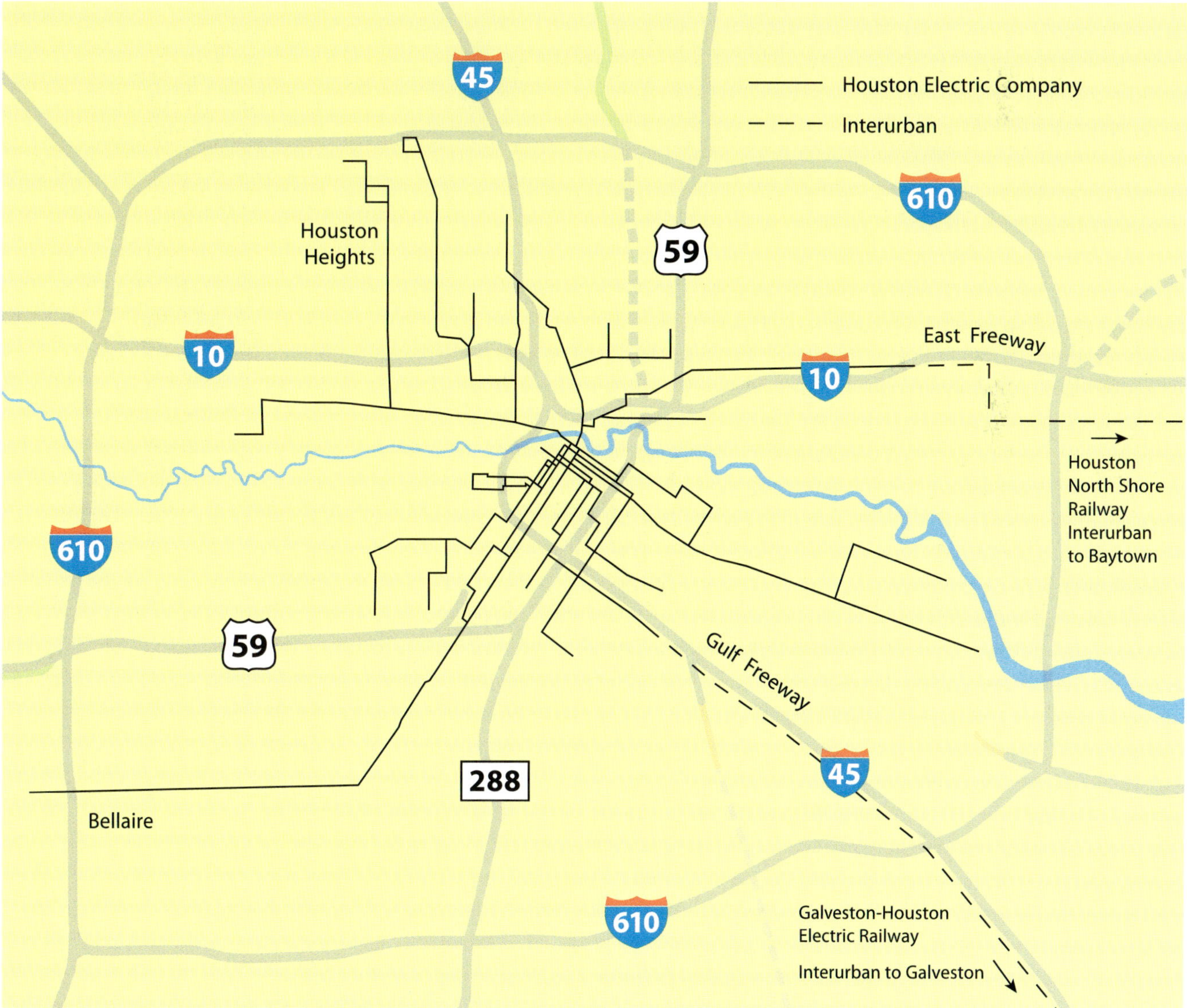

Houston's electric streetcar system at its 1927 peak: The streetcar system was very compact, fitting almost entirely within today's Loop 610. The only routes extending substantially outside Loop 610 were the interurban railways—the Galveston-Houston Electric Railway along the Gulf Freeway and the Houston North Shore Railway near the East Freeway. Suburban track was most prevalent on the northwest side of downtown, serving the Heights neighborhood. (Map adapted from *Houston Electric*[2])

Houston's streets bustle with streetcars: For 49 years electric-powered streetcars navigated the streets of Houston on the tracks of the Houston Electric Company. This 1920s view looks north along Travis Street, near the hub of streetcar activity on the north side of downtown. (Photo: HMRC MSS 200-311)

Streetcars

Houston's first streetcar service was a simple, mule-drawn car that began service in 1868 and by the end of the year had ceased operation. The first sustained service began on May 2, 1874, with a mule-powered system that traveled at less than 10 miles per hour (16 km/h). By 1890 the successful streetcar system had approximately 30 miles (48 km) of track and was spurring rapid residential development in the suburbs along its rail lines. The first electric streetcar was put into service on June 15, 1891, and by 1892 the electric streetcar system had 35 miles (56 km) of track on which 13 routes operated. Suburban real estate development further accelerated, with new subdivisions being built around new rail lines. One of Houston's best-known neighborhoods of the era, the Heights, was built around a streetcar route that opened in 1892. The streetcar system suffered some financial difficulties in the 1890s, but resumed its growth after the turn of the century and expanded to reach its peak of 90 miles (144 km) of track in 1927.

The streetcar first faced competition from rubber-tired vehicles in 1914 when the first "jitney" carriers appeared. A jitney was a private vehicle similar to a taxi carrying passengers for a fee, but operating on a somewhat fixed route. By March 11, 1915, 656 jitneys were operating in Houston. The jitneys were a serious financial threat to the streetcar operator, Houston Electric Company. A city ordinance in 1915 imposed a license fee on jitneys, thinning the ranks to 125 by the end of 1915, much to the relief of Houston Electric. The beginning of the end for streetcars in Houston occurred on April 1, 1924, when the first bus entered service in Houston. The streetcar company, Houston Electric, also operated the buses and found them particularly suited to providing new service in developed sections of the city, express service to outlying residential areas, and feeder routes from areas with low population density. The bus was especially attractive to real estate developers, who could establish service to their new subdivisions at a low cost.

One by one, the streetcar lines were replaced by bus service. The Bellaire streetcar line, which had always been a financial loser and whose track had substantially deteriorated, was the first to be replaced in September 1927. The Great Depression, which began in 1929, reduced streetcar and bus patronage from 42 million fare-paying passengers in 1929 to 26 million in 1932. The weakened financial position of Houston Electric accelerated the transition to less capital-intensive buses. The abandonment of streetcar

The beginning of the end of streetcar service: The first bus was placed into operation in 1924, beginning the conversion to buses that gained momentum in the 1930s and was completed in 1940. This circa-1926 photo shows an early bus model with an electric streetcar passing in the background. (Photo: HMRC MSS 200-358)

Key dates in Houston streetcar history	
1874	First sustained mule-powered streetcar service begins.
1891	First electric streetcars enter service.
1924	First bus service.
1927	Peak extent of streetcar system with 90 miles (144 km) of track.
1936	Widespread conversion to buses begins.
1940	Houston Electric Company ends streetcar service.
2004	Electric streetcars return to Houston with light rail on Main Street.

lines and replacement by buses began moving at full speed in 1936 when six streetcar routes were replaced. The Galveston-Houston Electric Railway ceased service in 1936, although the track within Houston continued to operate as the Park Place streetcar line.

On April 12, 1940, Houston Mayor Oscar Holcombe announced that he had reached an agreement with the Houston Electric Company for the termination of streetcar operations. The Houston Electric Company would be authorized to abandon all streetcar operations, and the city of Houston would be responsible for removing the remaining 42 miles (67 km) of track and refurbishing city streets. In return, Houston would receive a $50,000 payment and, most importantly, the prize that Mayor Holcombe was seeking: the right-of-way of the Park Place streetcar line. Holcombe's pet project was the construction of a new superhighway to Galveston—on the route of the former electric railway.[3]

At 1:32 A.M. on Sunday, June 9, 1940, Houston's streetcar era came to a close as the last regularly scheduled streetcar pulled into the maintenance facility. In August, Houston City Council requested bids for the removal of streetcar rails and the demolition of the Galveston-Houston Electric Railway concrete overpass over railroad tracks at Calhoun Road. The stage was set for the freeway era, but it wouldn't happen right away.[4]

Houston and Los Angeles Compared

Houston's electric railway system was dwarfed by the electric railways of southern California. The Los Angeles Railway Company, which served the city of Los Angeles, had 385 miles (616 km) of track in 1924, far more than

Houston vs. Los Angeles: This view of the Hollywood Freeway through Cahuenga Pass north of downtown Los Angeles illustrates some of the differences between Houston and Los Angeles in the pre-freeway and early freeway eras. The electric railways of southern California dwarfed Houston's streetcar system, both in track mileage and area served. The Hollywood Freeway had a rail corridor in its median until 1952. Freeways and electric railways coexisted in Los Angeles for 22 years. (Photo: National Archives 30-N-44-1296)

Houston's 90-mile peak. Even more impressive was the Pacific Electric Railway, which connected the cities spread out across the Los Angeles Basin and San Fernando Valley. It had more than 1,100 miles (1,760 km) of track in 1925. Prior to the freeway era, electric railways in the Los Angeles region had enabled the region to accommodate a large amount of sprawl. In contrast, Houston's electric railway system was very compact, fitting almost entirely into the 5-mile (8 km) radius of today's Loop 610.[5]

In Houston there was no overlap of the streetcar and freeway eras. Houston's electric streetcar service was shut down in 1940, 8 years before the opening of the first freeway in 1948. In Los Angeles, freeways and streetcars coexisted for 22 years after the opening of the Pasadena Freeway in December 1940, with the last streetcar line remaining in service until March 1963. The Hollywood Freeway at Cahuenga Pass had streetcar tracks in its central median until 1952. Once the freeway era began, Houston and Los Angeles went down similar paths, building extensive freeway networks.

The 1942 Street Plan

Houston in 1942 was not a city designed for large numbers of automobiles. With rapidly increasing private automobile ownership in the United States and Houston, this was recognized as a serious problem and prompted the development of the 1942 *Major Street Plan for Houston and Vicinity*. The first effort to impose a street plan in Houston was part of a general city planning report in 1913. However, the 1942 document reported "no record of consistent efforts to carry out this plan." A second major street plan was published in the comprehensive 1929 *Report of the City Planning Commission*, but fell victim to the financial crisis of the Great Depression. Due to the lack of a City Planning Commission with enforcement power from 1929 to 1935, structures were built and other development occurred in the path of thoroughfares contemplated in the 1929 plan, rendering much

The "deathtrap": This 1920s photo shows Allen Parkway, then called Buffalo Drive, passing underneath a railroad trestle. The 1929 *Report of the City Planning Commission* stated that "this deathtrap on the most heavily travelled drive in the city should be removed without delay." The Allen Parkway rail crossing was typical of the poorly planned street conditions that existed during the early automobile era and persisted until after World War II, when better plans were developed and, more importantly, implemented. (Photo: HMRC MSS 157-111)

of it "impracticable." The 1942 report warned, "Dead-end streets, jogs, offsets, and narrow rights-of-way in many parts of Houston give evidence of what has happened in the past and what will happen in the future unless there shall be adherence to a plan for the general framework—a Major Street Plan."[6]

The 1942 plan presented a comprehensive, new thoroughfare plan for Houston, but it did not include freeways. However, the report did recognize that freeways would be necessary in the future and identified three corridors where freeways would be feasible. Limited-access highway facilities were later built on all three locations. The corridor of the former Galveston-Houston Electric Railway would become the Gulf Freeway, the corridor along Buffalo Bayou west of downtown would become Memorial Parkway, and the corridor along White Oak Bayou north of downtown would become the North Freeway. The 1942 report seemed to suggest that freeways could be constructed only where wide, structure-free corridors with few intersecting streets already existed. Most likely for financial reasons, the authors of the report did not consider the clearance of urbanized areas for freeways to be a viable option.

Key dates in Houston pre-freeway road planning

1836	The Allen brothers found the city of Houston and lay out a grid system for central Houston.
1913	The first development plan is published, but its road recommendations are largely ignored.
1929	The City Planning Commission releases a comprehensive report which is not implemented because of the Great Depression.
1940	The City Planning Commission is reestablished and strengthened after three previous dissolutions.
1942	The *Major Street Plan for Houston and Vicinity* is published, recognizing the need for freeways but not officially designating any freeways.
1943-1946	TxDOT designates freeways in Houston.
1955	The city of Houston's *Major Thoroughfare and Freeway Plan* becomes the first official local plan to include the freeway system.

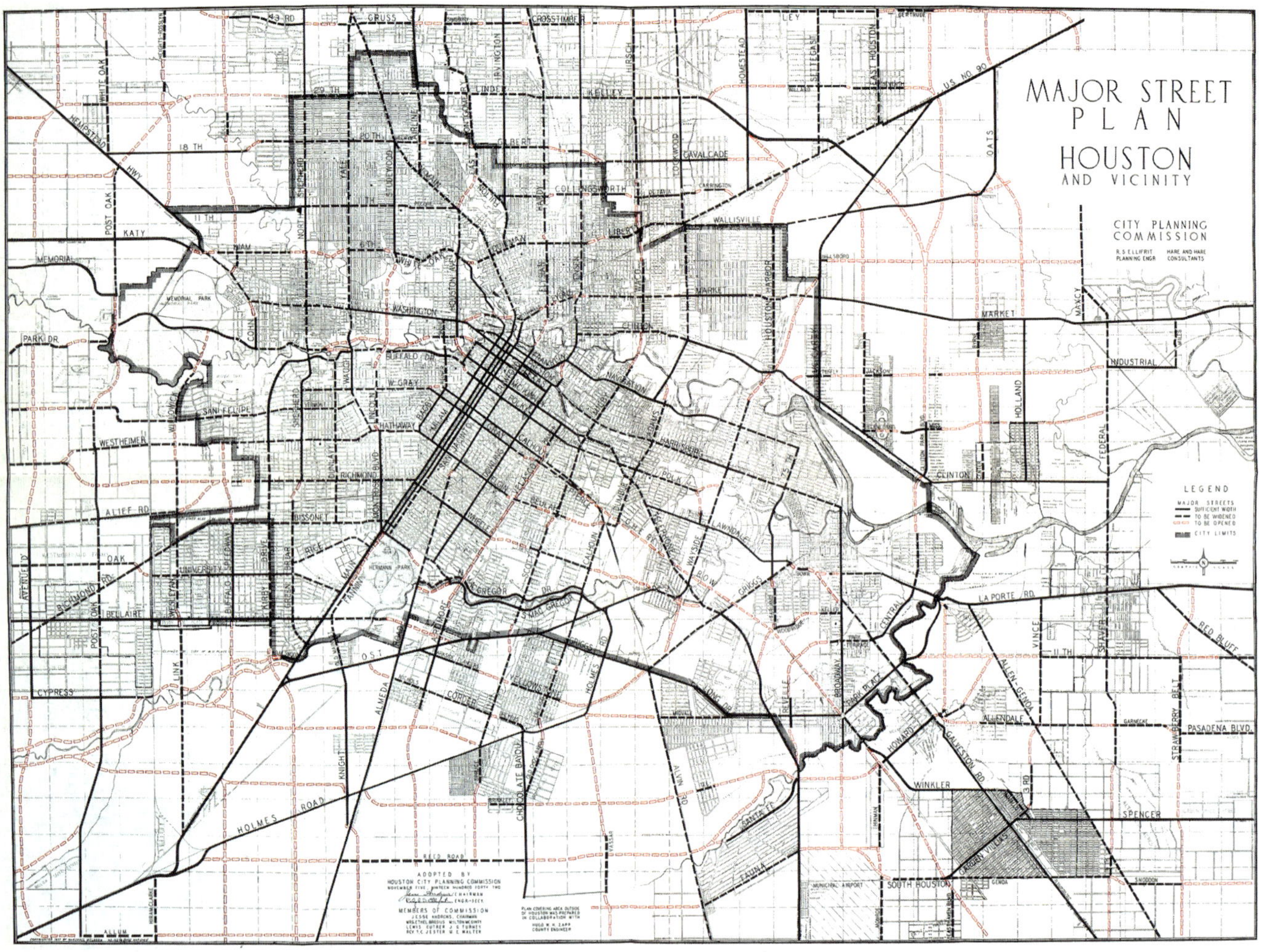

Freeway-less plan: The *Major Street Plan for Houston and Vicinity* published in 1942 was the first major update to Houston's street plan since 1929. The planned street network for Houston is shown above. The plan attempted to correct the largely haphazard and unplanned growth in the street network that had occurred in the previous decades. The 1942 plan did not include freeways, but recognized the need and identified three corridors that were good candidates for freeways.

Ideal freeway corridors: The three freeway corridors discussed in the 1942 plan are superimposed on today's freeways in this map. Two of the corridors became freeways, and the third became Memorial Parkway, a limited-access facility.

Streetcars return to Houston: After a 64-year absence, streetcars will return to Houston streets in 2004 on the $325 million, 7.5-mile (12 km) light rail line. Construction began in 2001 after substantial controversy and litigation. This photo shows tracks being laid south of downtown on Main Street. (Photo: November 2002)

The ribbon cutting: As Houston's freeway system took shape in the 1950s and 1960s, ribbon cuttings were a frequent event as officials celebrated one new freeway opening after another. This ribbon cutting took place on the Galveston Causeway on July 26, 1961. (Photo: TxDOT)

Putting the Freeway System on the Map

The era of the freeway was waiting for its time. In the 1930s, plans for superhighways had been dashed by the Great Depression. In the early 1940s, World War II precluded freeway construction not directly related to the war effort. Houston's 1942 *Major Street Plan* was very modest in terms of freeways, recognizing the future need but not putting any routes on the official planning maps.

But local officials had seen the future in the 1930s, and it was a freeway future. Houstonian Ross Sterling, chairman of the Texas Transportation Commission from 1927 to 1930, initiated efforts to build the Gulf Freeway in 1930. Mayor Oscar Holcombe took the lead in promoting the Gulf Freeway in the late 1930s, and planning for the Gulf Freeway continued during World War II. But it would be the 10-year period after World War II that would define Houston's freeways, and Houston itself.

Early Efforts

August 14, 1945: V-J Day. Japan had surrendered, and World War II was over. The nation's resources could be directed toward peacetime activities, and highway construction was near the top of the list. The freeway routes that officials in Houston had contemplated during the wartime years would now begin to develop into an overall system.

The planning for Houston's freeway system was somewhat behind the efforts of other cities. The first comprehensive freeway plan for Los Angeles had been proposed in 1937, and Los Angeles completed its first freeway, the Pasadena Freeway, in December 1940. In New York, Robert Moses had already constructed an extensive parkway system in the greater New York City area prior to World War II. Detroit, Michigan, was another early leader. However, there seemed to be very little influence from other cities on the development of Houston's plans. None of the cities in the vanguard of the freeway movement were as geographically suited for the new type of plan that would take shape in Houston: the loop and radial system.[7]

The key players in the early development of Houston's freeway system were the city of Houston, the Houston office of TxDOT,* the Houston Chamber of Commerce, and to a lesser extent, Harris County. Each of the players took specific roles in the process. This freeway construction partnership built Houston's original freeway system and would generally stay intact over the following decades, although the influence of certain players would wax and wane over time.

The first task was to define the needed freeways and their general alignments. The city of Houston and the TxDOT Houston office worked together to perform this task. In the early years, the city of Houston took charge of the process with Planning Director Ralph Ellifrit, Mayor Oscar Holcombe, and the Houston Planning Commission taking the lead. TxDOT acted in more of a consultative role.

When the needed routes were identified, the Houston Chamber of Commerce Highway Committee took responsibility for the crucial political tasks in the process. This included building public support, leading the pro-freeway efforts of the business community, obtaining state approval of the proposed freeway routes, and securing funding.

Once the freeway routes were adopted into the state highway system, exact alignments were approved by the Houston office of TxDOT, often in coordination with the city of Houston and Harris County. The city of Houston and Harris County then established "building lines,"† which prevented construction on property in the path of proposed freeways. Until 1956 the city of Houston and Harris County were responsible for all right-of-way acquisition costs, a burden which was substantially shifted to state and federal agencies in 1956 and 1957. TxDOT was fully responsible for the final step in the process, detailed design and construction.

In October 1943, the Texas Transportation Commission* officially designated Houston's first freeway—US 75 south of downtown, now the Gulf Freeway. In May 1945, the commission approved the US 75 route (now IH 45) along the western and southern edges of downtown, the US 59 Eastex Freeway from downtown to Loop 137 (now Loop 610), and a short section of SH 225, the La Porte Freeway, east of the Gulf Freeway. In October 1946, the commission approved the addition of the US 90 Freeway (now IH 10) across Houston, and the northward extension of the US 75 Freeway from downtown to the city limit.[8]

At this point, Houston's freeway system consisted of five spokes converging on downtown, plus the SH 225

* The present-day acronym for the Texas Department of Transportation, TxDOT, is used for all references to the highway-building agency of the state of Texas, which was the Texas Highway Department from 1917 to 1975 and the State Department of Highways and Public Transportation from 1975 to 1991. Similarly, the Texas Transportation Commission refers to both the present-day commission and its predecessors.

† "Building lines" defined the right-of-way for planned freeways in the Houston region and were established by the city of Houston and Harris County. If property was within the building lines for a freeway, the property owners generally could not receive permits to improve the property, and if improvements were made after building lines were established, the improvements would not be entitled to compensation.

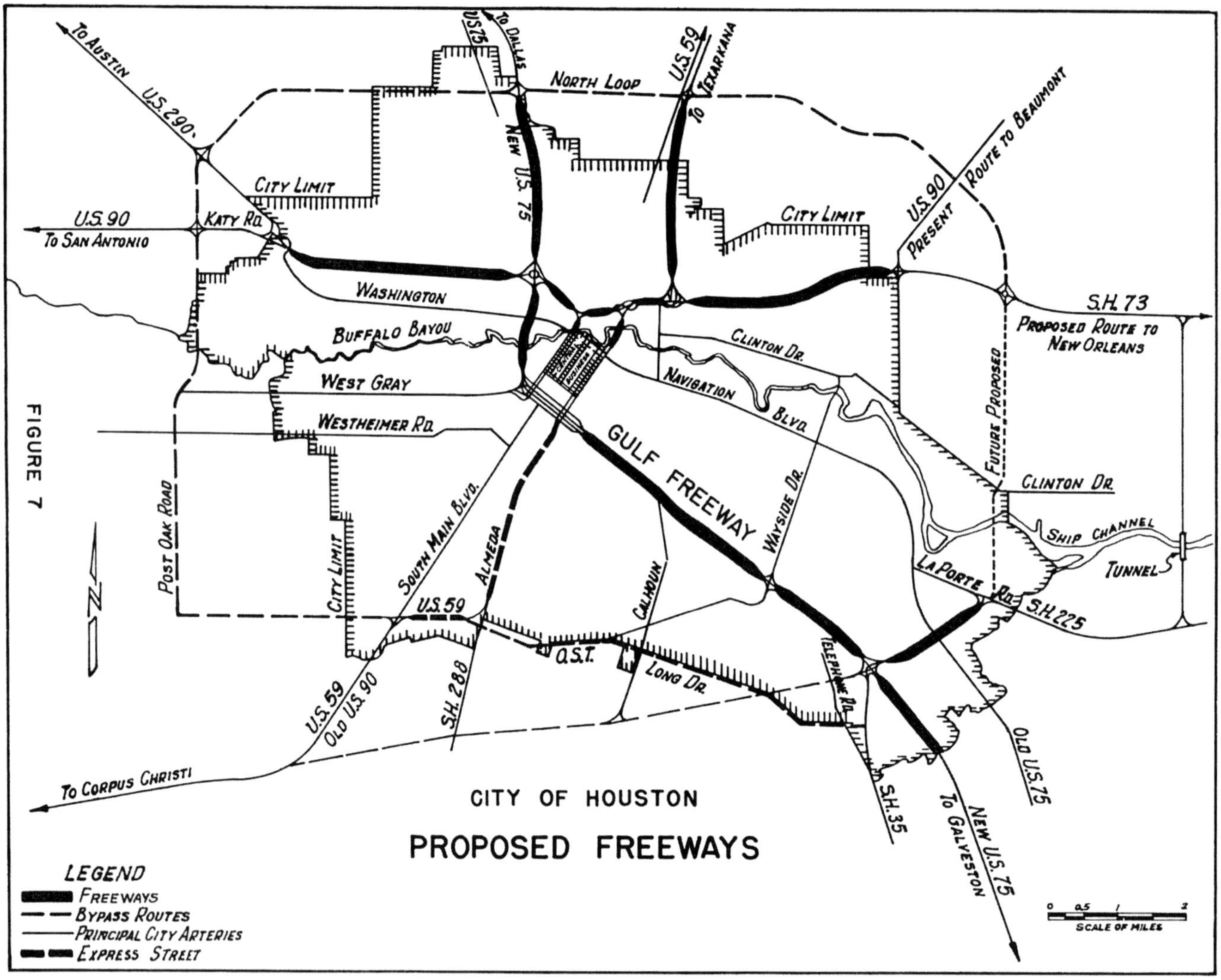

The first freeway plan, 1947: The first plan was actually more of an ad-hoc collection of the freeway routes that had been approved by the Texas Transportation Commission. The plan consisted of five spokes converging on downtown Houston and a spur on the Gulf Freeway that connected to SH 225 in southeast Houston. The Houston office of TxDOT developed a plan for constructing these routes in late 1947, and this map was drawn in 1949.[9]

spur route off the Gulf Freeway. The North Loop, approved as the US 90 bypass route in June 1942, was still planned as an arterial street. These approved routes served as the basis of a highway construction plan formulated in late 1947. Most of this work came under the jurisdiction of the TxDOT Houston Urban Project Office, which was headed by William James Van London, generally known by his initials W. J. The Urban Project Office was established in 1946 and was responsible for TxDOT projects generally within the city limits of Houston, which at the time were located near present-day Loop 610. The TxDOT Houston District Office was responsible for projects outside Houston. Van London managed the design of Houston's first freeway, the Gulf Freeway, and was the first person to have a big influence on the design and construction of freeways in Houston.[10]

The Master Plan

In the early 1950s local officials began to formulate a more comprehensive freeway plan for Houston. This effort resulted in a plan that was presented to the Texas Transportation Commission in July 1953. The freeway system was taking shape, but it was still a work in progress, with final route determinations to take place in 1954. The 1953 plan showed a freeway along the full length of the Westpark-Alief Road corridor and a freeway along US 90A. These two routes were later replaced by the Southwest Freeway. The 1953 proposal showed the South Loop along Holmes Road. Both the Northeast and Northwest Freeways were shown on preliminary alignments. Near Hobby Airport in southeast Houston, the plan included a freeway which was removed from the final version. In September 1953, the Texas Transportation Commission officially approved freeway status for US 59 (the Southwest Freeway), SH 288 (the South Freeway), and the eastward extension of SH 225 toward La Porte. In December 1953, the commission officially approved US 290 (the Northwest Freeway) into the freeway plan.[11]

In late 1953 the City of Houston Planning Department staff was busy refining the freeway plan, including the

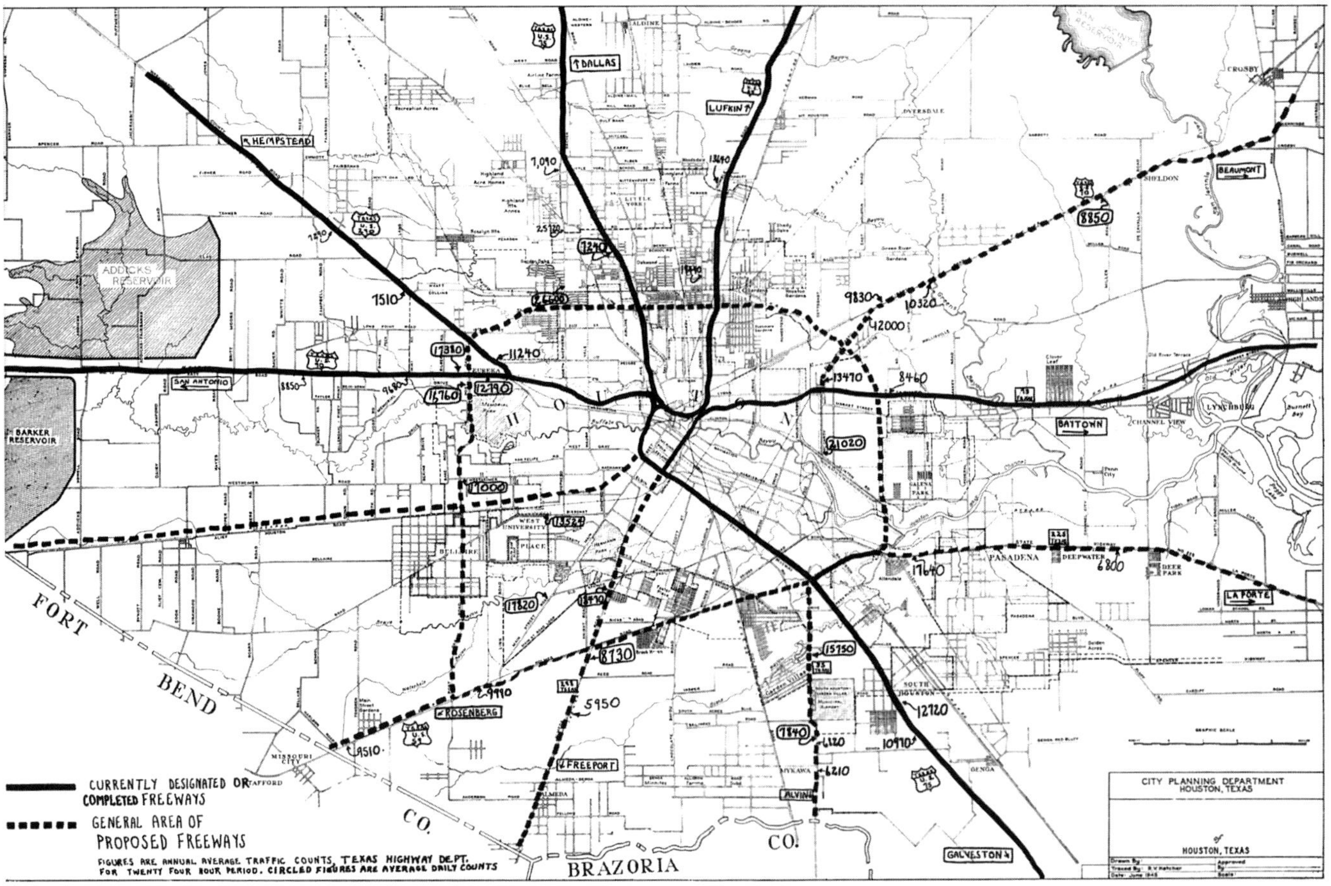

The developing master plan, 1953: This map was included in a booklet summarizing the presentation Houston officials made to the Texas Transportation Commission in July 1953. Houston officials were requesting the commission to adopt the proposed new routes shown by dashed lines into the state highway system. The Southwest, South, and La Porte Freeways were approved in September 1953. This original plan included many features that were dropped from the final plan.

alignment of the Southwest Freeway and the South Loop. A revised alignment for Loop 610 was presented to the Texas Transportation Commission in September 1954 and officially accepted in October. By the end of 1954 Houston's core freeway system had been defined. A few adjustments were made to the routes in the plan, but for the most part the plan was complete. The 1955 edition of the city of Houston's *Major Thoroughfare and Freeway Plan* was the first to show the full freeway system.[12]

The 1954 Annual Report of the Houston City Planning Commission, published in early 1955, described the cooperation that produced the freeway plan: "Success in connection with the planning and development of the freeway system has been largely due to the splendid coordination of all agencies involved. From the beginning, the State Highway Department, the City and County governments, and the Highway Committee of the Chamber of Commerce have worked closely and unselfishly. The Highway Committee has worked as a coordinating agency and has organized the important presentations which have been made to the Highway Commission regarding expansion and development of the freeway system."

Probably the most influential individual in the development of the freeway plan was Ralph Ellifrit, who served as director of Houston's Planning Department from 1940 until 1963, except for a three-year absence for military service during World War II. In a 1979 interview looking back on his career, Ellifrit stated, "We actually had more influence on the location of our freeways than almost any other city, as opposed to the highway department. In fact, we were way out ahead of them and had the whole system planned before [the highway department began its efforts]." Some in TxDOT may not fully agree with Ellifrit's recollection, but a review of project files shows Ellifrit taking the lead in defining the freeway system again and again, especially in regard to alignment decisions undertaken in cooperation with the Houston office of TxDOT. Harris County played a role in freeway alignment decisions outside the Houston city limits, but was a secondary player to the city of Houston in the early years of the development of the freeway system.[13]

Getting the Construction Machine into Gear

The plan was made. The routes were approved. Now it was up to the city of Houston to acquire the right-of-way so TxDOT could begin construction. State law in 1955 required local entities to pay 100% of the right-of-way cost for freeways and other state highways, but the city of Houston lacked sufficient financial resources to pay for the cost on its own. Houston had reached the limit of

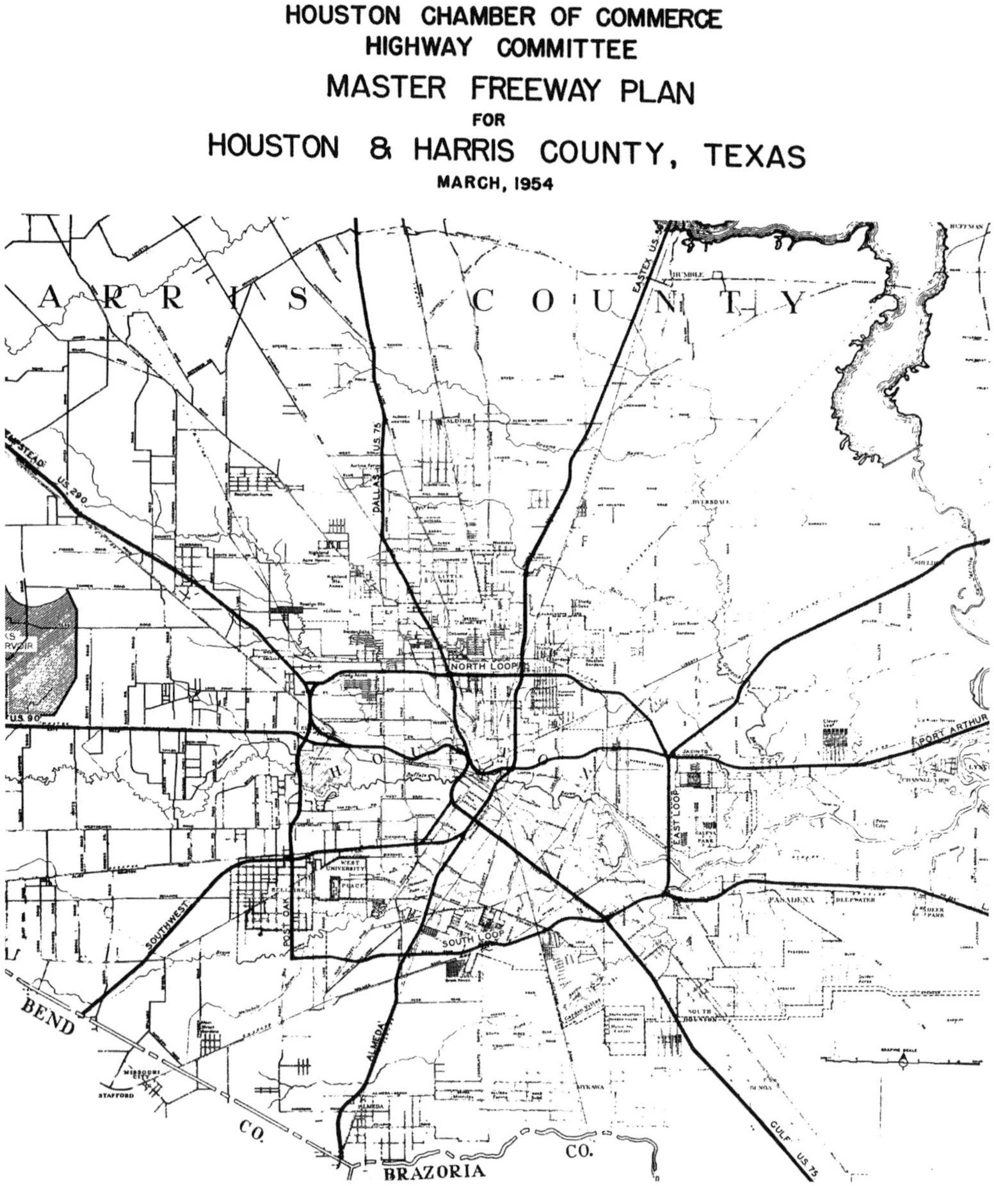

The final plan, March 1954: The general alignments of the core freeways in Houston's freeway system were finalized in late 1953 and early 1954. Ralph Ellifrit, director of the City of Houston Planning Department, led efforts to define the alignment of the South Loop and the Southwest Freeway during this period. Only the South Freeway would sustain a major realignment after 1954. (Source: Greater Houston Partnership)

the bonds it could issue and simply could not obtain more money for right-of-way acquisition.

Right-of-way acquisition came to a stop around April 1955, and Houston Mayor Oscar Holcombe explained that it occurred "because the city just hasn't had the money." Harris County stepped forward to take over the responsibility for freeway right-of-way acquisition in 1955. Harris County formulated a plan to pay for right-of-way with a special auto license fee—an extra $4 on the existing $12 fee. Voters approved the "wheel tax" on September 17, 1955, and it appeared that the freeway construction program would get back on track. However, the state attorney general ruled the wheel tax unconstitutional three months later, before any bonds backed by the tax could be issued. On May 2, 1956, the Supreme Court of Texas upheld the attorney general's decision. Without money for right-of-way acquisition, construction funds designated for Houston were at risk of being shifted elsewhere.[14]

The crisis deepened in July 1956 when a $15 million county bond issue requiring two-thirds voter approval failed by a slim margin. Local officials went to Austin in late July to request one more chance to come up with the money for right-of-way. A second election for $15 million in county bonds was called for September 8, 1956. The bond issue passed easily. The crisis was over, but the freeway construction program had sustained a setback. Texas Transportation Commission chairman Marshall Formby remarked that there had been a "crippling of the overall highway program" during the period.[15]

Fortunately major relief was on the way to cash-strapped

localities in Texas and nationwide. The Federal-Aid Highway Act of 1956 was signed into law by President Dwight Eisenhower June 29, 1956, and became effective July 1. The act provided for 90% federal payment of right-of-way costs on the federal Interstate Highway System, with the states paying the remaining 10%. Harris County would no longer need to make any contributions to right-of-way costs for Interstate 45, Interstate 10, and Loop 610. However, the full right-of-way cost for the extensive US and state route freeways would still be substantial. Relief from that burden would come from the Texas legislature in May 1957 with bill HB 620, which required the state of Texas to pay for no less than 50% of the right-of-way cost on non-interstate freeways. The TxDOT Houston Urban Project Office immediately set up a new department to purchase right-of-way. In September 1957, the Texas Transportation Commission authorized the expenditure of $43 million on the interstate system in Houston, with $19 million allocated to right-of-way acquisition and $24 million allocated to construction.[17]

> "We actually had more influence on the location of our freeways than almost any other city, as opposed to the highway department. In fact, we were way out ahead of them … ."
>
> Ralph Ellifrit, director of the City of Houston Planning Department, 1940–1963[16]

While agencies were struggling to acquire right-of-way, the freeway construction program was sputtering along, prompting criticism about the slow progress. The *Houston Chronicle* published a multipart series on the problems of Houston's freeway and road construction programs in September 1957. The interagency cooperation that existed in the early 1950s had diminished and a communications breakdown between TxDOT, the city of Houston and Harris County was slowing progress. In what could be described as a Houston freeway summit meeting, Houston Mayor Oscar Holcombe and his key staff went to Austin in October 1957 to meet with TxDOT head Dewitt Greer to discuss the problems and find a solution. All agreed that better communication and coordination were needed. A month later top Harris County officials met with TxDOT to iron out the county-state problems. Slowly the bureaucratic dysfunction was corrected and the Houston freeway construction program started to gain momentum.[18]

Right-of-way crises were not yet over, however. By September 1958, Harris County had used up just about all of the $15 million in right-of-way bonds approved in September 1956. Officials called another $15 million bond election for November 4, 1958. The bond issue fell short of the two-thirds majority needed for passage. Officials quickly called a special election for December 4, 1958, for another vote on the bonds. Without passage of the bonds, construction money earmarked for Houston would have been diverted elsewhere, due to lack of available right-of-way. The bonds passed with a better than 5-to-1 margin on the second try. With funding difficulties overcome, the only thing holding back the freeway program in the late 1950s was the time required to purchase the large number of right-of-way parcels needed for the freeways. County and state officials strengthened their right-of-way acquisition departments to handle the huge job, and by 1959 acquisition was proceeding much more quickly. As the calendar reached 1960, all hurdles had been cleared. The stage was set for the great freeway construction era of the 1960s.[19]

The freeway system would be an expensive undertaking, local authorities were soon to find out. In 1957 the freeway system in Harris County was estimated to cost $435 million, which included $88 million for right-of-way. To put those numbers into perspective, total expenditures for the budget of the state of Texas for fiscal year 1958 were $1.02 billion. Federal funding for highways nationwide under the 1956 Federal-Aid Highway Act totaled $1.518 billion for fiscal year 1958. The freeway system would be built as funding became available. A 20-year construction period was expected.[20]

The Visionary and the Pragmatist

Just as the freeway era was beginning to ramp up in the mid-1950s, the two men who would manage Houston's program through the golden age of freeway construction were appointed to their positions. In March 1955, Albert C. Kyser became head of TxDOT's Houston Urban Project Office. By that time, the scope of the Urban Project Office had been redefined to cover Loop 610 and the freeways inside the loop, including the extensive downtown interchange complex. Kyser filled the position previously occupied by the retiring W. J. Van London, who had presided over the construction of Houston's first freeways, the Gulf and Eastex Freeways. Also in 1955, Wiley E. Carmichael became head of TxDOT District 12, which covered all highways in Houston and the eight surrounding counties, but normally excluded the area under the jurisdiction of the Urban Project Office.* Kyser was the visionary who built Houston's freeways to be among the best in the United States and the world. Carmichael, in contrast, was a pragmatist who built minimal freeways—the smallest and most basic designs he could get by with.

Design deficiencies of the first generation of Houston's freeways were certainly well known by the mid-1950s. Kyser's approach to the next generation of freeways was not incremental improvement. His freeway designs were a huge leap forward to designs that are still modern by today's standards. Kyser first demonstrated his forward thinking in January 1956 when he announced that all future Houston freeways would have interior emergency shoulders and median barriers. Freeways under Kyser's juris-

---

*The limits of the Urban Project Office and District 12 were not strict, with each office occasionally participating in projects in the other's territory.

The Freeway Builders

The development and implementation of Houston's freeway plan from the early 1940s through the 1960s was a large and sustained effort, with many business leaders and city, county, and state officials all making important contributions. Three individuals stand out as the most influential.

William James Van London, 1893-1957

William James Van London, known as "Van" to his colleagues and W. J. to the general public, presided over the early development of the Houston freeway system. Van London joined TxDOT in 1921 and started work in Houston in 1938. Soon after arriving in Houston, Van London was developing preliminary plans for the Gulf Freeway, the first freeway in Houston and Texas. In 1946 Van London became head of the Houston Urban Project Office, which had jurisdiction over Houston's urban freeways. Van London served as chief engineer for the Gulf Freeway, which opened in October 1948, and also contributed the development of Houston's freeway master plan in the early 1950s. He retired from TxDOT in March 1955. (Photo: TxDOT)

Ralph Ellifrit, 1909-1999

Ralph Ellifrit served as the director of planning for the city of Houston from 1940 until 1963, except for a three-year absence from 1943 to 1946 for service in the army. Ellifrit and the City Planning Commission took a leading role in defining the alignments of most of Houston's freeways and identifying new freeways to be added to the system. Houston has never had any zoning or land-use planning, but the well planned freeway system stands out as the greatest accomplishment of Houston's planning efforts in the post-World War II era. (Photo: HMRC RG-D5-1182)

Albert C. Kyser, 1907-1984

Albert C. Kyser was the visionary and mastermind behind the design and construction of Houston's freeway system. Kyser, who was generally known by his initials A. C., joined the highway department in 1934 and the Houston district in 1938, working under W. J. Van London. Upon Van London's retirement in March 1955, Kyser was appointed head of the Houston Urban Project Office, which was responsible for Loop 610 and all the freeways inside Loop 610. Kyser's freeway designs propelled Houston into the ranks of the nation's top freeway cities. Among his design accomplishments were freeways with larger traffic-carrying capacity, four-level freeway-to-freeway interchanges, the downtown Houston central interchange complex, the incorporation of safety features into freeways, and the South Freeway dual-freeway design. Kyser retired in July 1972. (Photo: TxDOT)

Inspecting Houston's freeway system:
A delegation from Australia visited Texas in 1965 as part of a worldwide highway study. Pictured at right is Dewitt Greer (1902-1986), who is regarded as the father of the modern Texas highway system and served as head of TxDOT from 1940 to 1968. At left is J. A. L. Shaw, commissioner of the Department of Main Roads in Sydney, New South Wales, Australia. The Australians looked to California and Texas for design standards for highway and bridge construction. In New South Wales, rural interests had a disproportionately large influence in highway funding decisions, making it difficult to obtain funding for needed freeways in Sydney. Shaw must have been a little envious of the impressive urban freeway systems taking shape in Houston and cities across the United States. (Photo: TxDOT)

> "Creative, progressive, innovative, forward-looking, one of the best, not afraid to change, and an engineer's engineer."
>
> Dewitt Greer, father of the modern Texas highway system, describing A. C. Kyser in 1972.[23]

diction were designed to include the latest highway safety features, many of which were developed at the Texas Transportation Institute. The freeways under design at the time, notably the West Loop and the Southwest Freeway, would have 8 to 10 main lanes, continuous frontage roads, and modern geometrics. Four-level stack interchanges became the standard at freeway-to-freeway intersections, with cloverleaf interchanges banished to the scrap heap of obsolete freeway designs. Freeways were designed to minimize neighborhood impacts by being depressed below grade or elevated. In the early 1960s Kyser developed the "dual freeway" design* for the South Freeway, a design which has few equals in North America. Design work for the downtown interchange complex was finalized by Kyser's design team, including the sprawling US 59–SH 288 interchange south of downtown.[21]

It was Kyser's relentless pursuit of bigger, better, and more modern freeway designs that set him apart from other engineers and brought Houston one of the nation's better-designed freeway systems. Kyser's visionary designs often "brought him into conflict with some of the divisions in Austin," according to TxDOT head Dewitt Greer, referring to the more conservative design groups in Austin who were hesitant to approve Kyser's plans. Kyser would then go straight to Greer's office to get approval. "Every time he showed up, I knew he was going to ask for another million," Greer recollected in 1972. "And, although I was prepared to resist, what he had in hand was almost irresistible—and nine times out of ten he got what he wanted. A. C. Kyser is extremely creative and ingenious and he is always looking for innovation to inject into his work." Greer made these remarks at a freeway dedication in May 1972, just prior to Kyser's retirement in July.[22]

During this period Wiley Carmichael (1912-1996) was building the freeways outside of Loop 610, as well as all the roadways on the state highway system in the counties surrounding Houston. Carmichael was generally a by-the-book engineer, building the minimal facilities to get the job done. Just outside Loop 610, Houston's radial freeways were generally reduced to six main lanes and frontage roads. Although it was probably not anticipated in the 1950s, the largest traffic volumes in Houston's freeway system would occur on the radial freeways just outside Loop 610—exactly where Carmichael built the minimal six-lane facilities. In retrospect, Carmichael can legitimately be faulted for not acquiring more right-of-way on critical urban freeway segments to enable future expansion. The issue of right-of-way width for the Katy Freeway outside Loop 610 was a point of dispute between Kyser and Carmichael, with Kyser urging the acquisition of more right-of-way and Carmichael advocating a minimal right-of-way width. Carmichael would get his way, but west Houston would pay a heavy price for his lack of foresight.[24]

To be fair, there was a limited amount of resources

---

*The dual freeway consists of an outer freeway for local traffic and an inner freeway for express traffic. The South Freeway (SH 288) was designed as a dual freeway, but only the outer freeway was constructed. A wide inner median is reserved for a potential future 6-lane express freeway. For more information, see the South Freeway history on page 184.

Transportation Planning in Houston

The history of transportation planning in Houston descends deep into the underworld of cryptic acronyms, mind-numbing regulatory procedures, and bureaucracy that seems to evolve and mutate faster than a cold virus. But the ultimate product of the process is one of the most important elements of building a freeway system: the official local transportation plan. For readers ready to delve into this arcane world, the history of transportation planning in Houston follows.

The formulation of Houston's core freeway network in the early 1950s was a quick and decisive process. Engineers and planners identified the needed freeways and the best freeway alignments. Political officials worked to secure funding, and the freeways moved forward to construction as soon as money became available. But the relative simplicity of early freeway planning soon gave way to more formal planning procedures.

The 1953 Central Business District Parking Survey and accompanying origin-destination survey were the first studies geared toward better transportation planning. In 1959 TxDOT provided the leadership in the creation of the first regional transportation planning agency, the Houston Metropolitan Area Transportation Study (HMATS). Starting in 1959, HMATS performed Houston's first comprehensive traffic studies. The studies were the first in Houston to use digital computers for analysis. HMATS published several reports, including *Freeway Phase* in 1961 which recommended the addition of new freeways.[25]

Federal regulation of the transportation planning process first began with the Federal-Aid Highway Act of 1962. The act required all urban areas with a population over 50,000 to have a continuing and comprehensive transportation planning process, carried on cooperatively by states and local communities—the "3C" process. HMATS was reorganized in 1964 to meet federal requirements and was renamed the Houston-Harris County Transportation Study. H-HCTS was officially launched on February 1, 1965. The Galveston County Urban Transportation Study was created in 1964. These two organizations produced major planning reports through the rest of the 1960s up to 1970. The two planning organizations combined in 1970 to form the Houston-Galveston Regional Transportation Study (H-GRTS). TxDOT remained the lead agency in H-GRTS, with local governments acting in a supporting role. Decision-making authority was assigned to a committee of elected and government agency officials.[26]

The 1973 Federal-Aid Highway Act strengthened local planning by introducing the concept of the Metropolitan Planning Organization. The MPO was responsible for regional planning decisions and received funding from the Federal Highway Trust Fund to conduct its operations. The Houston-Galveston Area Council (HGAC) was designated as the MPO for the Houston-Galveston urbanized area. The H-GRTS structure was adapted to meet the new federal requirements, and H-GRTS operated as a partner with HGAC. HGAC has served as the Houston region's MPO since 1973. HGAC provides guidance on a wide range of regional issues, not just transportation.

Transportation decision-making authority has resided with a group of elected and government officials in a committee structure that has evolved over time. The technical work involved in the planning process has been performed by H-GRTS and the transportation planning department of HGAC. HGAC updates the long-term transportation plan on a regular basis, with major reviews occurring every three to five years. By the 1990s, federal regulation and increased interest in regional cooperation shifted more authority to HGAC. HGAC's public visibility increased during the 1990s with its public involvement efforts.

The complexity of transportation planning increased dramatically with the 1990 revisions to the Clean Air Act and the 1991 Intermodal Surface Transportation Efficiency Act. Today, the transportation planning process requires sophisticated computer programs for traffic and air-quality modeling. Meeting federal requirements requires a large staff, and the official long-term plan is subject to continuous review and revision. It generally takes about 5 years for a major highway construction project to work its way through the planning and approval process, and then another 5 to 20 years for construction.

Key dates in Houston transportation planning

1954	The core freeway network is defined with a minimum of formal study.
1959	Houston's first comprehensive and formal transportation study is launched.
1962	The Federal-Aid Highway Act of 1962 mandates an ongoing planning process.
1964	Houston-Harris County Transportation Study and Galveston County Urban Transportation Study are formed.
1970	Houston and Galveston planning agencies merge.
1973	The Federal-Aid Highway Act of 1973 strengthens local planning efforts. The HGAC becomes the region's MPO.
1990, 1991	The Clean Air Act amendments of 1990 and the Intermodal Surface Transportation Efficiency Act of 1991 increase the complexity of transportation planning.

available for highway construction, and there was room for only one A. C. Kyser in Houston. Carmichael took the expedient course and stretched his money as far as he could. Whereas Kyser's job was focused almost entirely on urban freeways, allowing him to take a large role in design, Carmichael was responsible for a wide range of farm-to-market roads and lower-tier highways in the large eight-county area around Houston. Carmichael retired in 1973, one year after Kyser.

Growth and Expansion

Houston in the 1950s was a boom town. Vehicle registration in Houston increased from 290,000 in 1949 to 607,000 in 1959. Officials were soon looking to expand the freeway system to accommodate anticipated future growth and suburban sprawl.[27]

The first substantial change to the freeway plan occurred in late 1959, when the South Freeway was realigned to a new location east of the originally planned route along Almeda Road. The first major addition to the freeway plan came on July 11, 1960, when Harris County Commissioners Court voted to designate the entire 87.5-mile (140 km) Outer Belt (now the Beltway 8-Sam Houston Parkway) as a full freeway. Harris County was responsible for the construction, and adoption into the state highway system was not pursued at the time. In 1961 two new freeways were added to the plan: the Harrisburg Freeway and the West Loop Extension. The Harrisburg Freeway was an extension of the La Porte Freeway, SH 225, into downtown Houston. The West Loop Extension was a southward extension of the West Loop which later became known as the Bay City Freeway and the Fort Bend Parkway. Officials began efforts to obtain state approval of the two freeways, but initial efforts were unsuccessful.[28]

The Grand Parkway first began to receive attention from planners in 1961. The Grand Parkway, a mega-loop around the metropolitan area estimated to be 130 miles (208 km) long at the time, was officially added to planning documents in 1965. A 1968 planning map published by the city of Houston showed the Grand Parkway as not only a freeway, but a broad corridor of parks and lakes around the periphery of Houston.[29]

The September 1961 announcement that the Manned Spacecraft Center (now the Johnson Space Center) would be located southeast of Houston prompted the Houston Planning Commission to initiate a study of the transportation needs of the southeast region. The report, completed in August 1963, recommended a new freeway along Red Bluff Road and a short section of Fairmont Parkway between Red Bluff Road and Beltway 8. The Red Bluff Freeway was in the city of Houston's *Major Thoroughfare and Freeway Plan* by 1966. The Red Bluff Freeway was planned as a Harris County project, and approval into the state highway system was not sought. Also in 1966, the Texas Transportation Commission designated US 90 in northeast Houston as a controlled access freeway.[30]

While all the previous 1960s freeway additions were conceived and promoted by local interests, TxDOT took the lead in the Alvin Freeway, SH 35. In May 1964 the Texas Transportation Commission authorized a study which returned its findings in late 1965. A full freeway was recommended on a new location from Alvin, 21 miles (34 km) south of Houston, to the Gulf Freeway near downtown Houston. The freeway was adopted into the state highway system by the commission in November 1967.[31]

Key dates in the history of the freeway plan, 1940-1970

1943	The first freeway, the Gulf Freeway, is officially approved by TxDOT.
1946	Five freeways into downtown Houston are approved by 1946. A construction plan is formulated in 1947.
1953	The first comprehensive freeway plan is presented to the Texas Transportation Commission.
1954	The core Houston freeway system is finalized.
1955	The 1955 edition of the city of Houston's *Major Thoroughfare and Freeway Plan* is the first to include the full freeway plan.
1960	Beltway 8 is designated as a full freeway.
1961-1967	The West Loop Extension, Harrisburg Freeway, Red Bluff Freeway, Grand Parkway, and Alvin Freeway are officially added to long-term plans.

With the addition of the Alvin Freeway, Houston's freeway plan reached its pre-1970s peak, about the same time freeway plans across the nation reached their peak extents. It turned out that adding new freeway routes to the master plan would be the easy part. Getting the new routes adopted into the state highway system would be more difficult. After ongoing efforts throughout the 1960s, the La Porte Freeway extension was finally accepted into the state highway system in 1969. The Outer Belt (now Beltway 8) was also accepted into the state highway system in 1969 when it became clear that Harris County would not have the resources to build the freeway. The Grand Parkway and West Loop extension would not be adopted into the state highway system in the 1960s.[32]

Actual construction was even more difficult than getting the freeways into the state highway system. The 1970s would not be kind to any of the freeways added to the master plan during the 1960s. The 1970s freeway construction crisis was about to begin, and by 1975 it appeared that none of the 1960s additions would ever be constructed. But the story has a happy ending for all the 1960s additions except the La Porte Freeway extension.

Name That Freeway

When Houston's first section of freeway opened in 1948, it was nameless. This was immediately recognized as a problem, and Mayor Oscar Holcombe promptly called for a contest to name the freeway. After sifting through 13,000 entries, the judges named it the Gulf Freeway. Houston's second freeway, the Eastex Freeway, was named in a contest coinciding with the freeway opening in 1953. The master plan for the freeway system

Route number	Name (year officially adopted)	Original, informal, or planning names	Previous names, proposed names (prop.), comments
Interstate Highway System			
IH 45 South	Gulf Freeway (1948)	Interurban Expressway	Named by contest
IH 45 North	North Freeway (1956)		Dallas Freeway (prop. 1965)
IH 10 East	East Freeway (1956) Baytown-East Freeway (1990) (in Baytown)	Port Arthur Short Line	Beaumont Freeway (prop. 1965). East Freeway is more commonly used than Baytown-East Freeway.
IH 10 West	Katy Freeway (1965)		West Freeway (1956-1965)
IH 610	the Loop (1956)	Defense Loop	Loop 137 (1942-1956). 610 officially designated in 1959.
US Highway System			
US 59 North	Eastex Freeway (1953)	Jensen Drive Expressway North Side Expressway Humble Expressway	Named by contest
US 59 South	Southwest Freeway (1956)		Also the Lloyd Bentsen Freeway, but this designation is never used.
US 90 East	Crosby Freeway (1988)		Northeast Freeway (1956-1988)
US 290	Northwest Freeway (1956)	Hempstead Freeway	Hempstead Freeway (prop. 1965)
State Highway System			
BW 8	Beltway 8 (1969) Sam Houston Parkway (1986)		Outer Belt (1952-1969). Tollway main lanes are called the Sam Houston Tollway.
SH 35	Alvin Freeway (unofficial)		
SH 99	Grand Parkway (1965)	Outer-outer loop	
SH 122	Fort Bend Parkway (1988)	West Loop Extension Bay City Freeway Blue Ridge Highway	
SH 146	Lanier Freeway (Baytown only)		Named after Bob Lanier, chairman of the Texas Transportation Commission 1983–1987 and mayor of Houston 1992–1998
SH 225	La Porte Freeway (1956)	Sterling Highway Harrisburg Freeway (cancelled section)	Also called Pasadena Freeway inside the city of Pasadena
SH 249	Tomball Parkway (1988)		
SH 288	South Freeway (1956)	Almeda Freeway	Brazosport Freeway (prop. 1965). Named the Nolan Ryan Expressway in Brazoria County south of Houston.
Tollways and Other Freeways			
None	Hardy Toll Road (c. 1982)		Named after previously existing road
None	Westpark Toll Road (c. 1999)		Named after previously existing road
None	Red Bluff Freeway (1963)		Named after previously existing road. Freeway main lanes not built, now called Red Bluff Road.
FM 1764	Emmet Lowry Expressway (1989)		Named after Texas City mayor who promoted the freeway

was finalized in 1954, and the planned freeways were given informal, unofficial names. It soon became apparent that a formal freeway naming plan would be needed, rather than the existing practice of naming freeways as an afterthought or by their planning names. The so-called compass-point formal naming scheme was approved by Houston City Council in November 1956, based on the recommendation of City Planning Director Ralph Ellifrit. It assigned most freeways a name based on the geographic region it served, for example, the Southwest, Northwest, and South Freeways. In addition, each of the four sections of the loop was split into two sections; for example, the West Loop was split into the West Loop North and West Loop South. There was some discussion of the potential

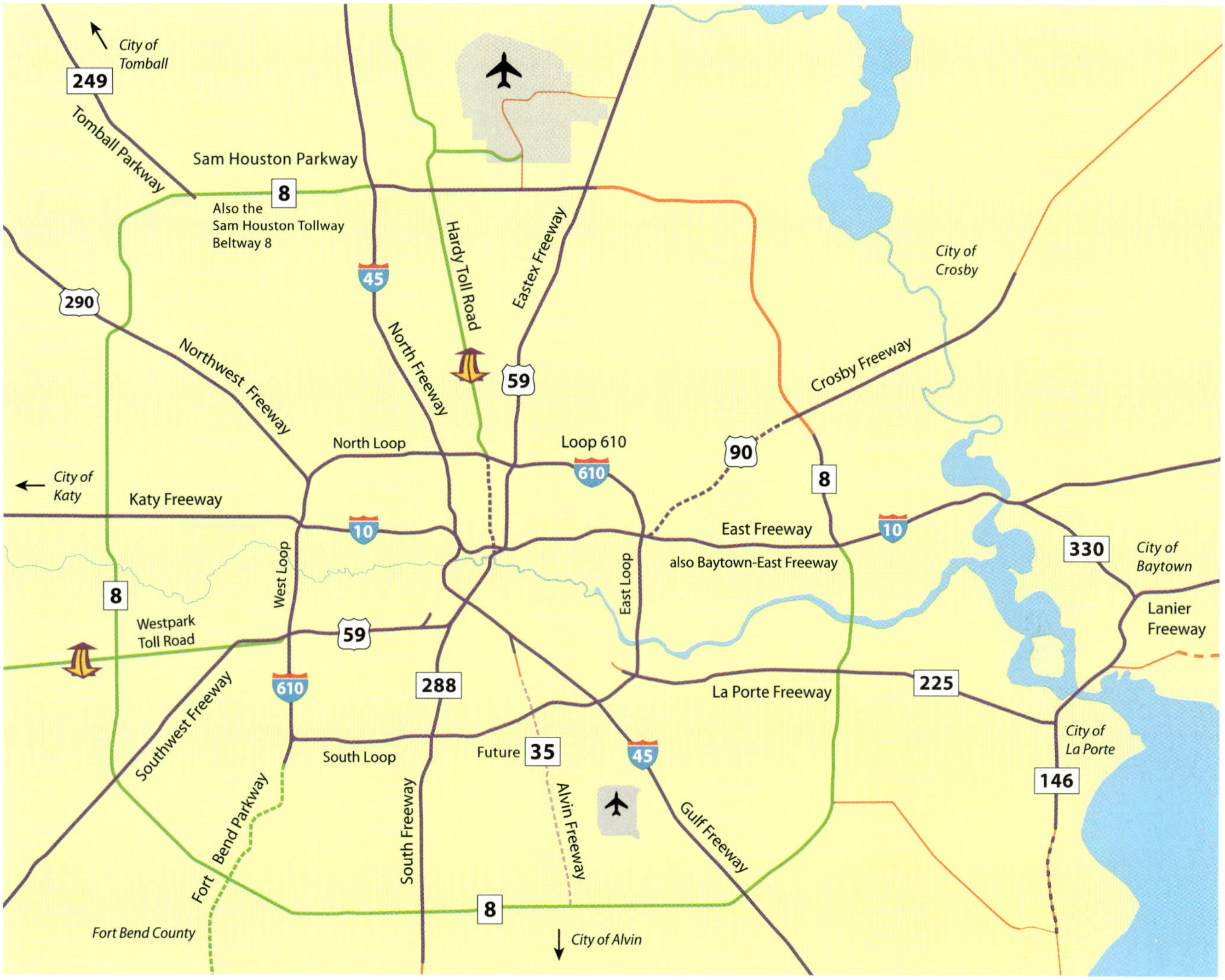

Freeways around the compass: Houston's freeways are generally named after their geographical direction or the suburban city they serve. On its original formulation in 1956, the naming scheme was called the "compass-point" system since several freeways were named after compass directions. Only one freeway in the immediate Houston area is named after an individual—the Sam Houston Parkway, also called the Sam Houston Tollway.

for motorist confusion in certain situations, such as going west on the East Freeway or north on the West Loop South.[33]

In 1965, TxDOT asked Houston City Council to rename five freeways to eliminate the compass-point names and instead adopt geographic names to prevent motorist confusion. The TxDOT proposal changed the North Freeway to the Dallas Freeway, the Northwest Freeway to the Hempstead Freeway, the West Freeway to the Katy Freeway, the South Freeway to the Brazosport Freeway, and the East Freeway to the Beaumont Freeway. City Council rejected all the proposals except the Katy Freeway proposal. Interstate 10 West was officially renamed the Katy Freeway, mainly because the original road along the corridor was known as Katy Road.[34]

After 1965 the compass-point naming system remained intact with only one exception: the Northeast Freeway was renamed the Crosby Freeway by Harris County in 1988. New freeways and tollways added to Houston's system since the 1960s have been named after the existing corridor roadway in the cases of the Westpark and Hardy Toll Roads, or by the geographic destination in the cases of the Tomball Parkway and Fort Bend Parkway. Only one freeway in the immediate Houston area has been named after an individual. In 1986 Houston City Council passed a resolution to name Beltway 8 the Sam Houston Parkway. Sam Houston served as the commander of the Republic of Texas Army in its 1836 victory over Mexico and as first president of the Republic of Texas. The city of Houston is named after him.[35]

Outside of Houston, the SH 201 freeway in Baytown was named the Lanier Freeway in honor of native Baytonian Bob Lanier, who served as chairman of the Texas Transportation Commission in the 1980s and mayor of Houston in the 1990s. In Texas City the FM 1764 freeway was named the Emmett Lowry Expressway in honor of the Texas City mayor who served from 1964 to 1989 and was instrumental in getting the freeway built.

The Galveston County master plan, 1969: Galveston County had an ambitious freeway plan as of 1969. This map shows two planned freeways that were later cancelled, the West Bay Freeway and the SH 146 freeway. The plan also shows three planned new crossings between Galveston Island and the mainland. Only one of the crossings, the Galveston to Bolivar crossing at the eastern end of the island, may still be built. (Source: *Galveston County Transportation Plan Volume 3*, 1969)

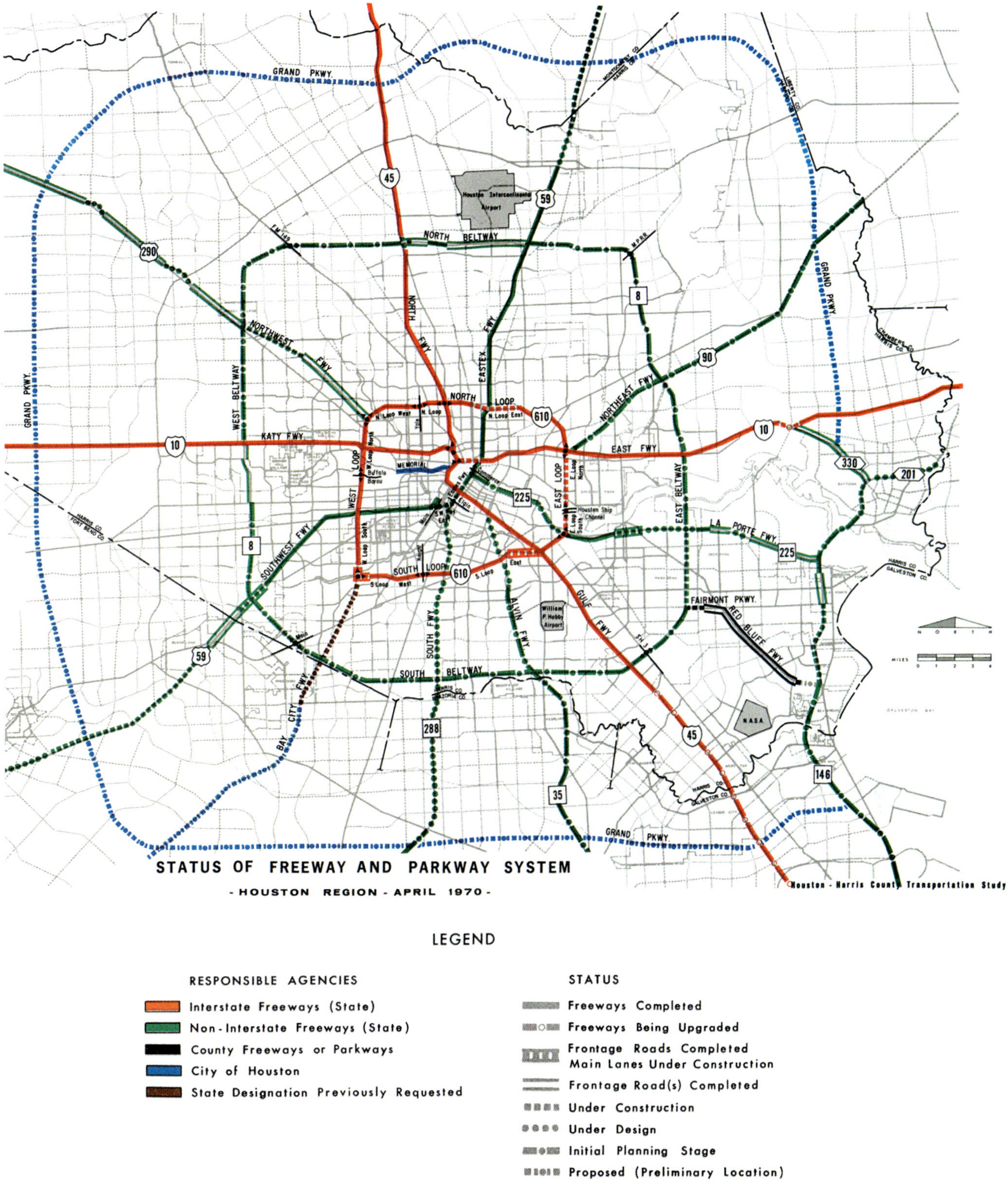

The Houston freeway master plan, 1970: The plan reached its peak extent in 1967 after the addition of the SH 35 Alvin Freeway. (Source: Houston-Harris County Transportation Study)

Freeway parade: After the narrow defeat of a critical bond issue for freeway right-of-way in July 1956, the Houston Chamber of Commerce launched an all-out campaign to ensure approval on a second vote on September 8, 1956. The Junior Chamber of Commerce organized an antique car parade to emphasize the need for "modern freeways." This photo was taken at the 1400 block of Main Street. The bond issue passed easily on the second try with 80% of the vote. (Photo: Houston Photographic and Architectural Foundation Trust)

Building Local Support

Planning Houston's freeway system wasn't just about transportation. It was also about the emergence of Houston as a major city—a city that had the scale and economic strength to build an extensive freeway system to match the nation's largest and richest cities. Houston's population base, economic output, and business infrastructure were far smaller than many other cities planning large freeway systems just after World War II. Houston's regional population was only about 20% as large as metropolitan Los Angeles, but the plans for Houston's freeways were big—big enough, business leaders hoped, to be a source of prestige and propel Houston into the ranks of the nation's top cities.

Building a big freeway system was expensive, and Houston had nowhere near the wealth of cities like New York, Los Angeles, Chicago, or San Francisco. It would take a large commitment of resources to build the freeways, and a substantial percentage of that money would need to be generated locally. Building local support for the extensive freeway program—and the taxes needed to fund it—became a top priority of political leaders and civic organizations in the 1950s.

Prior to 1956 the local role in freeway construction was substantial. There was no strong national program to build freeways. The state of Texas had an active highway construction program, but it did not proactively plan new urban freeways. The first task facing local authorities was to obtain state approval of the needed urban freeway routes. Once approval was obtained, the local money would need to start flowing. Local entities in Texas were responsible for 100% of the right-of-way cost for all new freeways prior to 1956 and 50% of the right-of-way cost on non-interstate routes after 1957. Often the right-of-way cost was higher than the construction cost due to expensive right-of-way clearances through urbanized areas. Getting the money to acquire the right-of-way was the key task for local authorities. Without available right-of-way, the federal and state construction money designated for Houston would go elsewhere. The Houston Chamber of Commerce led efforts to secure state approval of freeway routes and then generate the local funds for right-of-way acquisition.

The Houston Chamber of Commerce

In the years after World War II, Houston's future was shaped by a group of businessmen, political leaders, and philanthropists who were determined to build Houston into a large, prosperous, and respected city. The main vehicle for building their vision of Houston was the Houston Chamber of Commerce.* The Chamber of Commerce was a very large organization, actively guiding and influencing nearly every aspect of Houston's development and growth. The 1950 annual report of the Chamber of Commerce included summaries from 22 major committees whose responsibilities encompassed nearly every aspect of business and civic affairs. The 1950 report stated that there had been 1,375 committee and subcommittee meetings during the year with more than 15,000 in attendance. One of the chamber's most active committees in the 1950s was the Highway Committee.

The Chamber of Commerce was very influential because its numerous committees were headed by some of the most prominent, well-connected individuals in Houston. The Highway Committee in the crucial years of the development of the freeway plan was headed by two very prominent Houstonians. William P. Hobby, former Texas governor and owner of the *Houston Post-Dispatch* newspaper, headed the Highway Committee from 1938 to 1953. The Highway Committee would maintain its prestige and influence with the February 1953 appointment of *Houston Chronicle* president John T. Jones, nephew of legendary Houstonian Jesse Jones, as committee chairman. The leadership of newspapermen in the Highway Committee ensured a steady flow of freeway-friendly press in the local newspapers. In the early 1980s another

Population	1950	2000
Los Angeles County	4,151,687	9,519,338
Los Angeles Metro Area	4,819,599	15,620,448
Harris County	806,701	3,400,578
Houston Metro Area	919,767	4,427,804

Data: U.S. Census. 2000 metro area population excludes Ventura for Los Angeles and Brazoria for Houston.

Keeping up with the big boys: Houston's population is only a small fraction of the Los Angeles region. In 1950, when freeway plans were being made, Houston was about 19% as large as Los Angeles. Even in 2000, the Houston metro area population had not reached the Los Angeles 1950 metro population. With a smaller population base and no wealth-producing industries comparable to Los Angeles' entertainment and aerospace industries, building Houston's freeway system required a strong local commitment.

* The Houston Chamber of Commerce joined with the Houston World Trade Association and the Houston Economic Development Council to form the Greater Houston Partnership in 1989.

Highway appreciation: Highway Week 1952 was a statewide event in Texas and, up to that time, was the largest effort to educate the public about the need for better highways and increased highway funding. In Houston, the Chamber of Commerce organized an extensive program for Highway Week, including the overhead banner on Washington Street shown above and the billboard on Navigation Boulevard shown at left. (Photos: TxDOT)

very influential Houstonian, John B. Turner Jr., became chairman of the Highway Committee and provided key leadership to launch Houston's second wave of freeway construction. Turner was president of Friendswood Development Company, a land development subsidiary of Exxon Corporation that was active in developing master-planned communities around Houston, including Clear Lake City and Kingwood.[36]

The Highway Committee of the Houston Chamber of Commerce performed the all-important political role in the development, implementation, and funding of the freeway plan. Up until about 1970, these efforts covered a wide range of political territory, from building public support at the grassroots level to working on national funding issues. At the public level, the Highway Committee focused its efforts on securing voter approval of freeway bond issues. Events such as Highway Appreciation Week were designed to educate the public about the need for better roads. The Chamber of Commerce was funded by the business community, and freeways were good for business, especially in the construction, real estate, and land development industries. The Highway Committee coordinated the efforts of the business community to convey the pro-freeway message to political officials. At the local political level, the Highway Committee coordinated the efforts of the numerous agencies involved in developing the freeway plan, including the city of Houston, Harris County, TxDOT, and municipalities in the Houston region. The Highway Committee's efforts were crucial in the formulation of the original freeway plan in the early 1950s and in the formulation of the Regional Mobility Plan in 1982. At the state level, the Highway Committee was responsible for presentations to the powerful three-person Texas Transportation Commission. The Transportation Commission determines which freeway routes become part of the state highway system and which highway projects receive construction funding. The Highway Committee was also involved in legislative efforts relating to highway funding at the state and national levels. By the early 1990s the political role of the Chamber of Commerce in freeway development had greatly diminished.

The Freeway as a Money-Maker

Starting in the late 1940s, road advocacy groups were analyzing data from the first freeways to measure the benefits and show that the benefits exceeded the costs. One of the first reports to appear was *Economic Evaluation of the Gulf Freeway*, published by the city of Houston in July 1949, less than a year after the opening of the first segment of the freeway. It listed many of the benefits of the freeway and provided detailed calculations on the time savings for motorists. It was calculated that the annual value of the time savings to motorists was $2,668,664. Based on the $11,000,000 cost for construction and right-of-way, this suggested that the freeway would pay for itself in time savings alone in about five years. It was reported that the total savings to motorists had reached $21,060,000 by March 1954. The 1949 study couldn't make firm conclusions about property value increases or reduced accident rates due to the short operational period of the freeway, but these benefits would be revealed by subsequent studies.[37]

A February 1954 report, *Texas Expressways*, published by the Texas Highway Department, reported that between 1945 and 1950 the value of land within the Gulf Freeway "zone of influence" increased by 103%, while land values outside the zone increased by only 50%. It was also recognized that freeways leading to downtown would enhance property values in the central business district, but this was a difficult value to quantify. Improved safety was also a key selling point. In a report prepared for the observance of the 500 millionth mile of vehicle travel on the Gulf Freeway on February 18, 1954, TxDOT reported that the Gulf Freeway sustained only 2 fatal accidents per 100 million vehicle-miles of travel. This compared to the national rate of 6.9 for all traffic arteries, 5.7 on the Pennsylvania Turnpike, and 7.0 on the New Jersey Turnpike.[38]

Armed with the statistics for time savings, property value increases, and improved safety, Chamber of Commerce Highway Committee chairman John T. Jones led the campaign to promote support for the freeway system and the needed funding. The campaign slogan: Freeways don't cost—they pay!

Highway Week 1952

Promoters of a modern highway system realized that a large and expensive public works program was needed to build the system, and strong public support was crucial for funding and implementing the program. Highway Week 1952 was the first large-scale public information event in Texas to promote a major investment in the highway system. The statewide event was conceived and promoted by the Texas Good Roads Association, and Texas Governor Allan Shivers signed a proclamation designating January 27 to February 2, 1952, as Highway Week. The publication of the Texas Good Roads Association, *Texas Parade* magazine, reported that road boosters were "chomping at the bit … to let loose the biggest statewide road celebration in Texas history to observe the 35th anniversary of the founding of the Texas Highway Department."[39]

Houston would lead Texas with the most extensive observance of Highway Week. The Houston Chamber of Commerce formed a special subcommittee for the observance and enlisted the participation of 35 civic clubs. The Chamber of Commerce installed street banners and billboards emphasizing the need for more and better highways. Movies depicting highway problems were shown at local civic club meetings and broadcast on local television. Displays of large road-building machinery were set up around the city. The major Houston newspapers featured a special highway section on January 28. For the Houston Chamber of Commerce, it set the stage for the many public support campaigns that would take place during the 1950s to promote the passage of bond referendums and highway-related fees.

Paying the Bills

After the right-of-way funding crisis during 1955 and 1956, Harris County was practically solely responsible for local freeway right-of-way acquisition costs. Harris County depended exclusively on bond issues for raising money. There was no question that Houston had a pro-freeway electorate in the 1950s, but there was one complication. According to the Texas constitution, county bond issues for roads and highways required a two-thirds majority for passage. Passage of county bond issues would have been a near-certainty had only a simple majority been required, but the two-thirds requirement made the local funding process an ongoing challenge.

Assuring passage of a critical Harris County bond issue in 1956 required an all-out effort by the Chamber of Commerce. The bond issue failed by a slim margin, but was quickly brought back for a revote because state highway construction funds were about to be diverted elsewhere. Executives of Houston's 200 largest businesses

The wheel tax: In 1955 Harris County proposed an automobile registration fee to pay for freeway right-of-way. The "wheel tax" was placed before voters on September 17, 1955, and the images above show the first and last pages of a brochure promoting passage of the fee. Voters approved the fee, but the fee was ruled unconstitutional a few months later, setting the stage for bond elections that would require extensive pro-freeway campaigns in the following years. (Source: HMRC)

and industries were contacted and asked to brief their employees on the importance of the bonds. The Junior Chamber of Commerce organized antique automobile parades for downtown and the suburbs to draw attention to the antiquated state of Houston's roads. In the days preceding the election, "freeway caravans" with scores of workers fanned out across the city to distribute literature. The bonds passed with 80% of the vote on the second try.[40]

In 1958 another critical county bond issue for freeway right-of-way failed by a slim margin. A group called the Citizens Freeway Bond Committee worked with the Chamber of Commerce to coordinate grassroots efforts to pass a second vote on the issue. Efforts were successful, with 84% of the voters approving the bonds on the second try.[41]

With those two bond issues, the freeway construction program was able to proceed at full speed. But bond issues would be needed regularly, and in 1963 a major bond issue was once again rejected, this time by an overwhelming margin, and was not brought back for a revote. A key bond vote occurred in January 1966 to authorize funds for the South Freeway, the Southwest Freeway connection to

Bottleneck busters: The Houston Junior Chamber of Commerce placed this portable billboard on the Southwest Freeway just before the downtown exit to remind motorists to vote for road improvement bonds on January 29, 1966. The bond issue included funds to pay for the local portion of right-of-way acquisition costs for the extension of the Southwest Freeway to connect with the Eastex Freeway. The bond issue passed. (Photo: *Houston Chronicle*, January 26, 1966)

the Eastex Freeway, and the first major work on Beltway 8. The Chamber of Commerce led the pro-bond campaign and the bond issue was approved. In 1970 the state constitution was changed to require only a simple majority for county road bond approval. Since achieving a simple majority was a near certainty in Houston, the era of extensive bond election campaigns was brought to an end. Of course, there were new challenges awaiting the freeway system in the 1970s.

Creating the toll road authority: The approval of $900 million in bonds to create the Harris County Toll Road Authority on September 13, 1983, was the beginning of a seismic shift in transportation planning in Houston. Tollways would steadily gain importance over the next 20 years. The image above came from a brochure printed to promote passage of the bonds. Houston was near the peak of its transportation crisis in 1983, and the possibility of restoring mobility resonated with voters, with the bond issue receiving 69.7% of the vote. (Source: HMRC)

Washington, D.C., 1965: In the mid-1960s anti-freeway protests were becoming more prevalent and vocal across the United States. Plans for a central city freeway network in Washington, D.C. made the nation's capital a hotbed of anti-freeway protest. In response to the growing protest, the 1968 Federal-Aid Highway Act imposed the first comprehensive regulations to govern the planning of new freeways. Anti-freeway protest would continue to increase into the 1970s, affecting just about every city in the United States, including Houston. The anti-freeway movement reached Houston in 1970 when residents in the path of the planned Harrisburg Freeway organized to stop the freeway. Three other projects in the early 1970s also ignited some controversy, but overall Houston had relatively little anti-freeway protest. Houston never experienced an anti-freeway street demonstration, such as the Washington, D.C. protest shown above. Protest in Houston was contained within the public hearings, which often became quite heated. (Photo: Washington, D.C. Public Library, Washingtonian Collection)

Crisis

December 2, 1968: The builders of the nation's highways gathered in a gray, chilly Minneapolis, Minnesota, for the annual convention of the American Association of State Highway Officials (AASHO).* Delegates were greeted by a harsh editorial in the *Minneapolis Tribune* accusing state highway departments of being poor local planning agencies and urging them to move decision-making authority to the urban level. The state highway officials didn't need another unfriendly press report to get them stirred up, however. Many were already incensed about the new wave of regulations that were coming out of Washington in response to the Federal-Aid Highway Act of 1968. State highway officials felt that the Federal Highway Administration's new rules went way beyond the intent of Congress. In fact, the federal government was now telling the nation's highway builders what they could and couldn't do. Federal interference was increasing dramatically, and the state highway departments were becoming compliance agencies. Truculent delegates lashed out at the congressional leaders responsible for the new legislation, but congressional leaders and officials from the Federal Highway Administration held their ground, telling the fuming state highway officials that the world was changing, and state highway departments would need to change too.

For the previous nine years, state highway departments had been living in the golden decade of freeway construction. But the clock was ticking ever so quickly to January 1, 1970, a date which would mark the beginning of a very difficult period for freeway construction. Numerous factors were converging to decimate the nation's freeway construction program, particularly in urban areas. On one front, environmental, neighborhood preservation, minority representation, and energy consumption issues were poised to stop freeway after freeway. On another front, spiraling inflation and stagnant fuel tax receipts were beginning to choke off funding for highway construction.

While most large cities across the United States were dealing with these two challenges, Houston had a third factor that escalated its crisis to a level beyond anything other cities were facing: an unprecedented population, construction, and commercial boom fueled by the energy shortages of the 1970s. This three-pronged attack on Houston's transportation system would plunge Houston into possibly the worst transportation crisis any city in the United States has experienced in the post-World War II era.

The Changing Regulatory Climate for Urban Freeways

Federal regulation of transportation planning first began with the Federal-Aid Highway Act of 1962, which mandated a comprehensive and ongoing local planning process. Requirements for public involvement in the planning process were expanded by the Federal-Aid Highway Act of 1968, which was finalized in August. On October 25, 1968, the Federal Highway Administration first published a preliminary set of new regulations that provided the public with a much stronger voice in the location and design of highways. The new regulations required state highway departments, in public hearings, to give full consideration to a wide range of factors, including social, economic, and environmental effects. At least two public hearings would need to be held during project development, rather than the existing practice of having one hearing, and any individual dissatisfied with a state decision could appeal the decision to the Federal Highway Administration, delaying project approval. The new regulations were prompted by widespread complaints from civic, conservation, and other groups that highways had disrupted communities or disregarded scenic and other environmental considerations. State highway officials would have to dedicate substantial effort to comply with federal rules, and would be forced to spend more time and effort holding public hearings to ensure a continued flow of federal money.[42]

Anger at the state level over the federal government's intervention in their highway construction affairs initially boiled over at the AASHO meeting in Minneapolis. In December 1968 and January 1969, states continued their efforts to have the regulations withdrawn or delayed. The Texas Transportation Commission was vocal in its opposition and commission chairman Jack Kultgen testified at hearings in Washington, D.C. Texas officials felt that restrictive regulations were a response to abuses that had occurred in other states and situations that existed in the urban northeast. "We do not believe that an all-encompassing set of regulations which would provide solutions for a minority of cases should become an obstacle which

* The organization was renamed the American Association of State Highway and Transportation Officials (AASHTO) in 1971.

The regulatory maze: TxDOT officials unroll the freeway approval flowchart—15 feet of twists and turns through the oppressive federal bureaucracy. This view starts at month 31 on the flowchart. The flowchart was formulated in 1971 to illustrate the huge difficulty that freeway projects faced in trying to negotiate the myriad of regulations imposed by the federal government, especially the National Environmental Policy Act of 1969 and the Federal-Aid Highway Acts of 1968 and 1970. The flowchart was featured in a congressional hearing in 1971 and received widespread press coverage. At left, Bannister L. DeBerry, TxDOT head 1973-1980; at right, James C. Dingwall, TxDOT head 1968-1973. (Photo: TxDOT)

could substantially delay, if not completely curtail, further construction and reconstruction of highways in our state." But if state highway officials were upset about an extra public hearing and the potential for delay, they were totally unprepared for the regulatory bomb that would hit them in less than a year.[43]

On January 1, 1970, President Nixon signed the National Environmental Policy Act of 1969 (NEPA) into law. Included in the broad reach of NEPA were requirements for environmental impact statements for major federal actions having a significant effect on the environment. This included federally funded highway projects. Environmental impact statements were comprehensive, costly studies that could take years to complete, and the vagueness of the early regulations ensured that many disputes would end up in court. The Federal-Aid Highway Act of 1970 further reinforced the requirements for a full consideration of the social, economic, and environmental effects of federally funded highways. Numerous other federal laws added to the complexity of highway construction.[44]

Perhaps even more disturbing to the nation's road builders were the profoundly anti-automobile statements and policy directives coming out of the federal Department of Transportation, which had been formed in 1966. Secretary of Transportation John Volpe took the lead in vilifying the automobile. "More and more, the hallowed right to jump into our cars and drive anywhere we please is being tallied against other community and individual values—the need for elbow room, clean air, stable neighborhoods, more park land, and many others." "America must now accept the fact that the private automobile will not forever be the absolute monarch of our core cities." Other spokesmen for the U.S. Department of Transportation were making similar statements: Dependence on the automobile had created a host of social problems in the nation's cities. Highways had gone just about as far as they could go in central city areas, and alternative means of transportation had to be found. With approximately 12,000 new motor vehicles being added in the United States every day, the nation was at risk of turning into a giant parking lot. There was discussion of halting all construction of major highways in urban areas so cities could step back and implement full-scale land-use and transportation planning.[45]

By late 1970 the U.S. Department of Transportation had announced a new emphasis in its work: moving people—rather than vehicles—on urban freeways and streets. Volpe stated that efforts would be made to "encourage the greatest use of buses in preference to individual automobiles." With the new directives from federal legislation and department leadership, the bureaucrats at the Federal

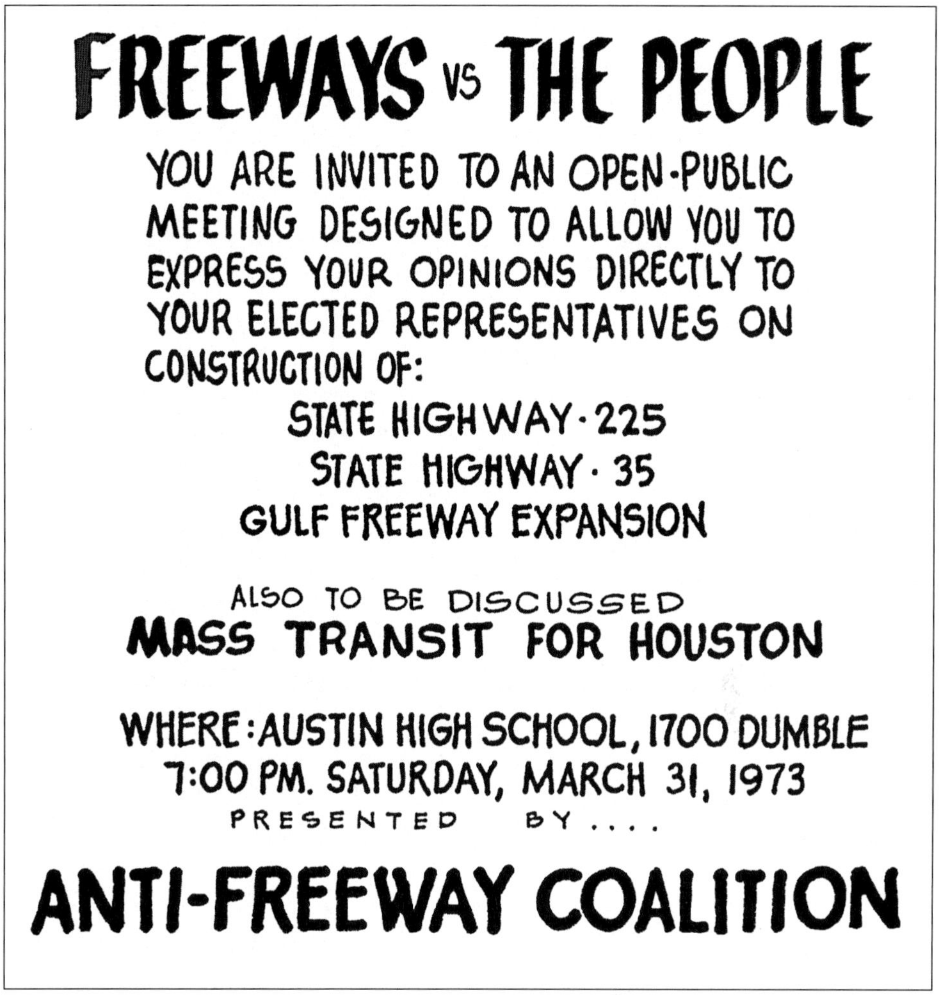

The freeway as the enemy: Anti-freeway protest in the early 1970s in Houston focused on three projects: the SH 225 La Porte Freeway extension into downtown (known as the Harrisburg Freeway), SH 35 (the Alvin Freeway), and the expansion of the Gulf Freeway. The Anti-Freeway Coalition was a rather informal group that formed in February 1973 after the first public meeting on the Gulf Freeway expansion. It appears to have disbanded nearly as quickly as it formed. (Source: TxDOT documents for public hearing held on February 12, 1973)

Highway Administration in Washington, D.C. responded as they knew best: with more regulation.[46]

Freeway construction would never be the same. In fact, the new regulations forever doomed most urban freeway construction projects requiring extensive right-of-way clearance. A few cities in the nation would have the political climate to continue ongoing, large-scale freeway construction and expansions—Houston would be one. But doing so would not be an easy task.

In 1971 top officials at TxDOT headquarters in Austin charted the tortuous route a planned highway project would need to take through the federal bureaucracy. The 15-foot long (4.6 m) chart showed a nearly incomprehensible maze with all forms of twists and turns through the bureaucracy—and plenty of opportunities for detours, blind alleys, and dead ends. The chart, it was reported, "resembled a complex set of hieroglyphics or a medieval tapestry." It received national attention and was even featured in congressional hearings in Washington, D.C. In 1956, when large-scale construction on the Interstate Highway System began, a 34-month delay from the start of preliminary planning to the start of construction was standard. By 1961 the delay had reached 44 months. In 1972 the delay was 78 months—six and a half years. Albert C. Kyser, who managed Houston's urban freeway construction from 1955 to 1972, commented in May 1972, "It will be increasingly difficult for the state to build new freeways. That is, freeways where we have not already acquired the right-of-way. Public hearings are fine, but the procedures and red tape are too cumbersome. It used to be that if we had a problem we could resolve it here or in Austin—now it has to go all the way to Washington." James C. Dingwall, head of TxDOT in 1972, spoke about the impact of the federal regulations in a 1972 address, "We in the highway field find every bulletin, policy, and procedure memorandum, or what-have-you stemming from the federal level only to add to the proliferation of red tape, compound the already abundant confusion and, incidentally, escalate the cost."[47]

The Anti-Freeway Movement

Nationally, protests against freeways had been seen since the beginning of large-scale freeway construction in the 1950s. One of the better-known controversies of the early freeway era concerned the construction of the Cross-Bronx Expressway in New York City. In spite of intense opposition and huge engineering challenges, New York City's master builder and power broker Robert Moses was able to push the freeway through one of the nation's most densely populated urban areas in the period from 1948 to 1963. The city of San Francisco had its "freeway revolt" in 1959, banning the construction of additional freeways within the city. But for the most part, large urban freeway projects requiring extensive displacements advanced unimpeded until the mid-1960s.[48]

By 1964, however, there were signs of trouble in freeway paradise—Los Angeles. A member of the California State Highway Commission resigned over the approval of the Long Beach Freeway* (Interstate 710) extension through South Pasadena, stating, "You must be aware of how freeways are going to affect people." The *Los Angeles Times* published an editorial on December 8, 1964, about the future of freeways, saying, "There is still time to plan the freeways of the future, but we are running out of community and aesthetic resources." The Houston newspapers took note of the trouble in paradise; it was an omen of things to come. The city of Houston's wavering commitment to freeways was already beginning to become evident in the mid-1960s. In January 1965 Roscoe Jones, City of Houston Planning Director, expressed concerns about freeway construction, saying, "We shouldn't slow down our freeway building program. But there's a limit to how many we can build. Eventually, vital parts of the city will have to be destroyed for more freeways." The departure of Ralph Ellifrit as City of Houston Planning Director in December 1963 marked the beginning of the decline in the city of Houston's role in freeway and thoroughfare planning. Whereas the city of Houston and its City Planning Commission had once been the most proactive government agency in promoting freeways, it took on a more passive role under the mayoral tenure of Louie Welch from 1964 to 1974. Although Welch was generally supportive of freeways, a top TxDOT official from the era remarked that he "wouldn't stick his neck out for a freeway." In the 1970s the city of Houston also reduced its commitment to the planned major thoroughfare system for the Houston region. The lack of a well-planned, high-capacity thoroughfare system would put an increased traffic load on the freeways.[49]

> "We shouldn't slow down our freeway building program. But there's a limit to how many we can build. Eventually, vital parts of the city will have to be destroyed for more freeways."
>
> Roscoe Jones, City of Houston Planning Director, January 12, 1965

By the late 1960s the protest era had arrived and freeways were one of the targets. The broad public support the highway construction program had enjoyed was now being fractured by vocal anti-freeway efforts. While previous anti-freeway protests were more isolated and focused on specific projects, a new national trend was now emerging—the anti-freeway movement. Three groups were particularly influential in the anti-freeway movement: neighborhood preservation groups, environmentalists, and minority groups. These groups had been empowered by new legislation and regulations in the late 1960s and early 1970s, and now had legal and regulatory tools that allowed a vocal minority to have a decisive impact on freeway-related decisions. The anti-freeway movement was active nationwide by the early 1970s, and its wide reach was described in the 1971 book, *Superhighway-Superhoax*, written by Washington, D.C. resident Helen Leavitt. Several other scathing anti-freeway books followed.[50]

By the early 1970s urban freeway projects across the nation were falling like a row of dominoes, and by the mid-1970s urban highway planners were sifting through the wreckage of their freeway plans. Boston's inner loop was cancelled in December 1971. Washington, D.C.'s inner city freeway network was largely wiped off the map and replaced by a mass transit system. Numerous planned expressways in New York City were cancelled, including the Lower Manhattan Expressway. Closer to Houston, a dense inner-city freeway network was cancelled in Dallas. The planned Riverfront Expressway was deleted in New Orleans. The list went on and on. Practically every large and medium-size city was affected. But perhaps no city was hit as hard as Los Angeles. The freeway capital of the world would see its grand plans for an unbelievably dense freeway grid largely reduced to what had been completed by 1970. Only a few new freeways would move forward in Los Angeles after 1970. Funding issues also played a very large role in the collapse of the Los Angeles freeway plan.

The Anti-Freeway Movement in Houston

Houston's freeway construction program was generally free of controversy until 1970. Major right-of-way clearances through urbanized areas began in the early 1950s and continued through the late 1960s with minimal or negligible opposition. The only rumblings came from residents in the city of Bellaire in 1954 in opposition to the alignment of the West Loop 610, and from residents of the Memorial Bend subdivision in 1962 in opposition to the alignment of the West Belt. In both cases, the freeways moved forward on the originally planned alignments, and Bellaire, in particular, went on to become one of the more desirable neighborhoods in Houston.

Almost as soon as the calendar reached 1970, the climate began to change. Four freeway projects became the focus of opposition: the planned extension of the La Porte Freeway into downtown Houston, an expansion of the Gulf Freeway, the planned alignment of Beltway 8 through Jersey Village, and a section of the planned

* The Long Beach Freeway extension remained highly controversial and was put on hold in the 1970s due to opposition from the city of South Pasadena. The project came back to life in the 1990s due to the critical need to close the gap in the Los Angeles freeway system. A record of decision was issued by the federal government in April 1998, authorizing the California Department of Transportation to proceed with the project. An unfavorable court ruling in July 1999 put the project in limbo again. As of 2003, the prospects for constructing the freeway were highly uncertain.

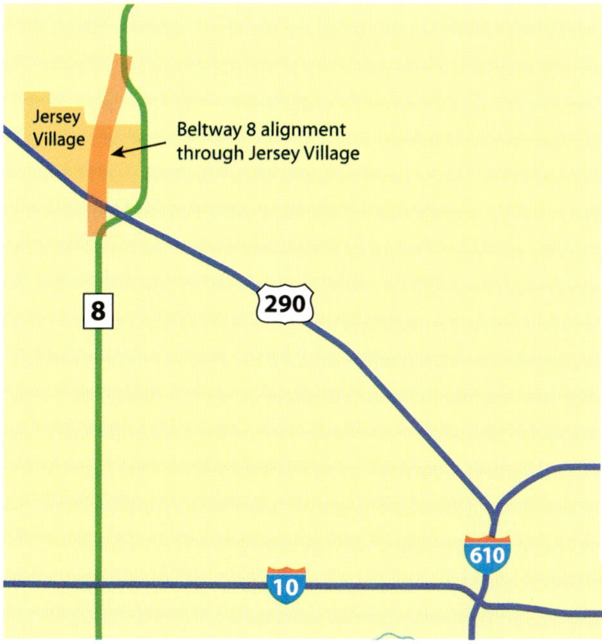

Alvin Freeway. In all four cases, localized groups were protesting the impact of the freeway projects on their neighborhoods.

In May 1970 the Houston TxDOT office received its first taste of well-organized freeway opposition. A group of predominantly Hispanic residents along the corridor of the La Porte Freeway extension, more commonly called the Harrisburg Freeway, organized a meeting to launch their opposition effort. What set this group apart was the dedication and commitment of its leader, area resident Richard Holgin. Holgin became the symbolic David in a classic David versus Goliath battle. For the following four years Holgin relentlessly opposed TxDOT's efforts to get the freeway approved and built. Prospects for the freeway became tenuous as the 1970s financial crisis hit TxDOT, but Holgin's efforts were probably a decisive factor in the ultimate fate of the freeway. At the end of the battle Holgin had succeeded in doing what no one else had ever done in Houston, and no one has done since: prompting the cancellation of a planned urban freeway. The story of the Harrisburg Freeway is covered in full detail starting on page 198.

In 1973 controversy arose over the three other projects. Opposition to the Gulf Freeway expansion focused on the displacement of homes in the predominantly black, low-income neighborhood adjacent to the freeway. The opposition was vocal but did not succeed in stopping the Gulf Freeway expansion, which moved forward to construction approximately 10 years later due to funding shortfalls during the 1970s. Controversy over the alignment of Beltway 8 through Jersey Village began in 1973 and continued through most of the 1970s. Residents were ultimately successful in rerouting Beltway 8 around their neighborhood, and the tollway with frontage roads was completed in 1990. The Alvin Freeway sustained considerably less opposition than the other projects, with controversy focusing on the alignment through the site of Law Park, at that time a largely undeveloped tract of land. The freeway alignment through the park site was approved with mitigation measures, but the budgetary crisis of the 1970s caused the freeway to be placed on long-term hold. As of 2003 the freeway remains an item for long-range planning, with no construction planned for the near future.[51]

Overall, the anti-freeway movement in Houston was weak compared to most cities in the United States. It consisted of localized efforts against freeway projects rather than a widespread anti-freeway sentiment, and Houston remained one of the most consistently pro-freeway cities in the nation. Nevertheless, the early 1970s anti-freeway movement did result in the cancellation of the Harrisburg Freeway. Houston's greatest setback in terms of freeway planning would not occur until 1992, however, when anti-freeway forces stopped a planned expansion of the West Loop.

Funding: The Devastating Blow

Even as citizen opposition had cancelled or delayed projects in Houston, the vast majority of planned freeways and improvement projects faced no opposition. But the 1970s brought an even more potent freeway-killer to Houston which had a much greater impact on plans for the system than citizen opposition. Money for new freeways was shrinking. Rapidly escalating highway construction costs and stagnant revenue from fuel taxes combined to dramatically reduce the purchasing power of available highway construction funds.

Revenue from the motor fuel excise taxes on gasoline and diesel fuel is the principal source of funds for highway construction. The federal motor fuel excise tax was created in 1932 to provide funds to help offset the federal budget deficit that had developed during the Depression. The revenue from the original 1-cent tax was not earmarked for highway construction. After a period in 1933 during

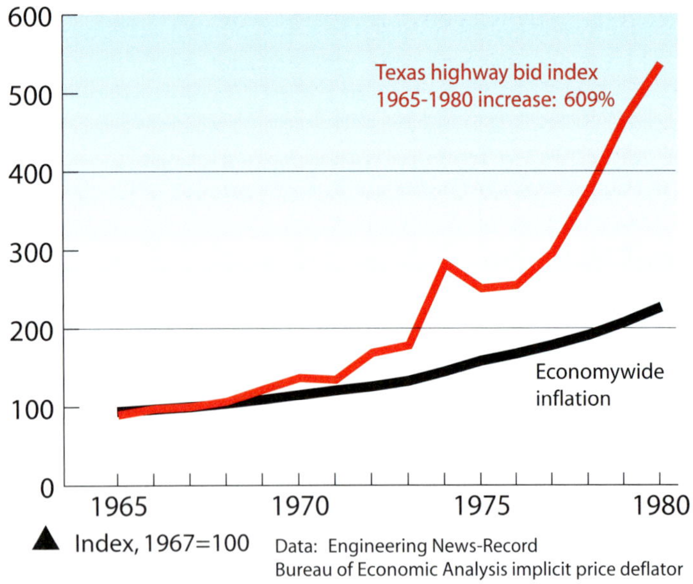

Highway construction costs were skyrocketing, far exceeding the overall inflation in the national economy...

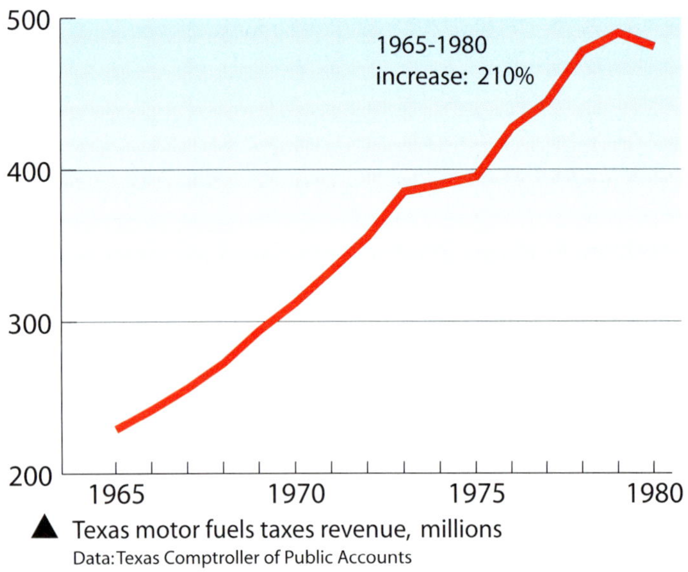

...while revenues from Texas motor fuels taxes were growing about one-third as quickly...

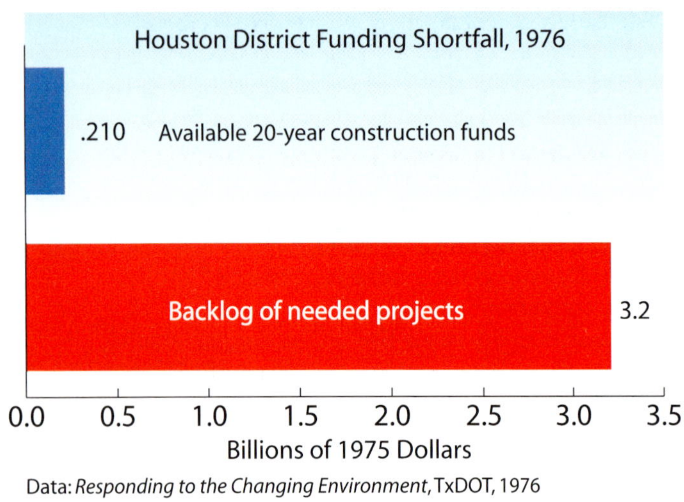

...leading to a huge shortfall in funding.

which the tax was 1.5 cents per gallon, the tax returned to 1 cent per gallon on January 1, 1934. On July 1, 1940, the fuel tax rate was increased to 1.5 cents per gallon and then to 2 cents per gallon on November 1, 1951.[52]

The year 1956 brought the single most important event in the history of highway construction in the United States: the Federal-Aid Highway Act of 1956 and the accompanying Highway Revenue Act of 1956, signed into law by President Dwight Eisenhower on June 29. The National System of Interstate Highways had been created in the Federal-Aid Highway Act of 1944 and a route map was approved in 1947, but legislation prior to 1956 did not provide a federal commitment to build the system. The 1956 act designated the highway network as the National System of Interstate and Defense Highways and provided financing for the construction of the entire highway network, with planned completion by June 30, 1972. The federal fuel tax was increased to 3 cents per gallon to provide funding for construction. Most significant, however, was the establishment of the Highway Trust Fund. A percentage of the proceeds from various highway-user taxes, including 100% of the gasoline tax, was to be deposited into the Highway Trust Fund. The funds were dedicated solely to financing the federal-aid highway program. In 1959 President Eisenhower sought a 1.5-cent increase in the fuel tax, but settled for a 1-cent increase which took effect on October 1, 1959, raising the total federal tax to 4 cents.[53]

In Texas the state gasoline tax began with a 1 cent-per-gallon tax in 1923. Seventy-five percent of the revenue from the tax was dedicated to highways and 25% was earmarked for public schools. This ratio of revenue distribution became part of the Texas Constitution in 1946. The state fuel tax was increased to 4 cents per gallon by 1929 and reached 5 cents per gallon in 1955. In 1959 the combination of federal and state fuel taxes in Texas was 9 cents per gallon. From that point on, the real value of the fuel tax would be continuously eroded by inflation, especially in the 1970s when inflation began to run rampant.[54]

Starting in 1965, inflation accelerated throughout the United States economy. The gross domestic product deflator, the widest measure of inflation, began its upward climb in 1966 after remaining below 2% annually from 1959 to 1965. From 1966 to 1975, economywide inflation rose steadily and reached an annual rate of 9.3% in 1975, with a 68% rise in the 10-year period from 1965 to 1975. Similarly, the consumer price index rose 70.6% from 1966 to 1975, reaching a peak of 11% in 1974. Highway construction cost, as measured by the Federal Highway Administration's highway bid price index, surged upward much more quickly than the overall inflation rate in the period, increasing by 125% nationally and by a devastating 279% in Texas. In 1974 alone, the Texas highway bid price index increased an astonishing 57.9%. This staggering increase was partially offset by a decrease of 11% in 1975. But the construction cost index didn't tell the full story. The time and expense to bring projects to the construction phase had increased dramatically as well. Highway departments were now required to perform comprehensive environmental studies and public involvement programs. Large staffs were needed to guide projects through the new regulatory bureaucracies.[55]

Inflation in itself is not necessarily a problem if funding keeps pace, but the receipts from the federal and state fuel taxes were increasing at a much slower rate than the increase in construction costs. In the same 1965–1975 period, receipts from the four-cent federal fuel tax increased by only 69.1%.[56]

In Texas, receipts from the state motor fuel taxes increased by only 72.5%. Some of the growth was due to a strong population growth of 21.1% during the period. Until 1973 gasoline consumption and associated fuel tax revenue had been increasing at the rate of 6 to 7% per year. However, revenue growth slowed dramatically with the energy crises of the 1970s. Because of the Arab oil embargo in 1973, state gasoline tax revenue actually fell 10% in the first two months of 1974 as compared to 1973, and the annual increase from 1973 to 1974 was 1.2%. As Americans purchased smaller, more fuel-efficient automobiles, the growth rate of fuel tax receipts dropped even more. Federal regulations requiring automakers to increase the fuel efficiency of their vehicles and the newly enacted 55-mile-per-hour (88 km/hr) national speed limit further reduced gasoline consumption. The Texas state gasoline tax of 5 cents per gallon was among the lowest in the nation and had not been raised since 1955.[57]

In 1975 the funding crisis became critical and drastic action was necessary. TxDOT had reached its peak employment level of about 20,500 employees in 1972. Employment had dropped to 19,500 by 1975, but far greater reductions were needed to correct the budgetary imbalance. The department head count was reduced 28% to 14,000, with about 3,000 employees being laid off. "It was painful," recalled Bannister L. DeBerry, TxDOT head at the time who authorized the mass work force reductions with a letter to district managers. "That was the hardest thing I had ever done, to sign that letter." In a November 1975 presentation to the Texas House Ways and Means Committee, TxDOT representatives warned that the future of the state's highway construction program was in "serious jeopardy" and that by 1980 the department would have no funds for new highway construction. As a further blow to the morale of the state's highway builders, in 1975 the name of the Texas Highway Department was officially changed to the State Department of Highways and Public Transportation, suggesting a possible shift of focus toward mass transit.[58]

In order to shore up the deteriorating finances of TxDOT, the Texas legislature began to supplement the TxDOT budget with revenue from the general fund in 1977. This was possible during the period of budgetary surplus resulting from the 1970s oil boom, but was not a long-term solution.

Shared Company—the Mighty Succumb

The funding crisis was a nationwide phenomenon and no states were spared. The mightiest freeway construction machine the world has ever known, the California Division of Highways of the 1960s, was decimated by funding shortfalls and a changing political climate. Governor Jerry Brown entered office in January 1975 and set out to dismantle the world-leading highway construction program that his father, Governor Pat Brown, had built during his years in office between 1959 and 1967. In response to "critical financial problems" in July 1975, the California Department of Transportation (successor to the Division of Highways, known as Caltrans) placed a freeze on all new highway construction and halted nearly all land acquisition for freeway right-of-way. Plans were announced to eliminate 3,300 department positions. California could not even afford to provide its 10% share of construction costs for the federal Interstate Highway System. It was estimated that by 1978 or 1979 Caltrans would not even be able to meet its basic maintenance needs and obligated financial commitments. Caltrans weathered the crisis, but for practical purposes it abdicated its role in building new freeways. In the future, it would be up to localities to raise their own money to build new freeways. Several areas of California launched their own programs in the 1980s, including Orange County, San Diego County, Riverside County, and to a lesser extent Santa Clara County.[59]

The world's freeway capital, Los Angeles, was hit especially hard. In November 1975 the California Highway Commission approved an $843.5 million budget for the 1976-1977 budget year, including only $240 million for new construction—the smallest new construction outlay since 1958. Only $58.1 million—about 152 million in 2003 dollars—was allocated for the huge Los Angeles region. There was a spate of official freeway cancellations in the Los Angeles area in the period. The halt of new freeway construction in 1975 left numerous missing links, gaps, and unusable sections of completed freeway. On September 14, 1975, the *Los Angeles Times* reported on the uncertain future of Los Angeles freeways. 680 miles (1088 km) of the planned 1,540-mile (2,464 km) freeway system had been completed, but it was "increasingly doubtful that the system would ever get much beyond 700 miles." A Caltrans official stated, "This is it. There no longer is any money. Sure, there will be a few more miles built, just to fill in the gaps, but our freeway system is here ... well, it's seen its heyday." For the most part, the Caltrans official was correct. Los Angeles would recover from the crisis and close most of the gaps in its freeway system over the following years. The embattled Interstate 105, now known as the Century Freeway, was even built. But the era of new freeway construction in the immediate Los Angeles area was over. In the future Los Angeles would focus its efforts on retrofitting a carpool lane system onto the median shoulders of its existing freeways.[60]

Standing By While the Fuel Tax Shrank

An increase in the fuel tax would have been the most effective way of obtaining additional revenue for highway construction to help keep up with inflation. But there would be no fuel tax increase at the national level or in Texas for the entire 23 year period from 1959 to 1982. This was somewhat surprising since an increase in the federal fuel tax to encourage conservation was discussed and actively promoted by various interests, including President Jimmy Carter, for most of the 1970s. It would take Mr. Anti-tax himself, President Ronald Reagan, to finally sign into law an increase in the fuel tax in 1982.

Soon after the Arab oil embargo of 1973 there were growing calls for an increase in the gasoline tax as a measure to promote conservation. In 1974 Congress' Joint Economic Committee was recommending an increase of 30 cents per gallon—about 91 cents in 2003 dollars—and President Ford's top advisers were recommending an increase of at least 10 cents per gallon. In November 1974 President Ford acted against the wishes of his advisors and publicly announced his opposition to any fuel tax increase, stating, "We are not going to have [a higher gasoline tax] as long as I am in the White House." In terms of highway construction funding, the proposed taxes would not have provided much relief from the highway construction cost inflationary spiral, since proposals would have diverted the proceeds to various nonhighway uses, including mass transit, conservation subsidies, alternative energy development, and national debt reduction.[61]

In 1975 the leadership of the Democrat-controlled Congress pressed ahead with plans for a large increase in the fuel tax—up to 23 cents per gallon, depending on national consumption levels. The proceeds of the proposed 3-cent base increase were to be dedicated to the development of alternative energy sources. An additional tax up to 20 cents would be levied based on the level of gasoline consumption. Both tax increases were overwhelmingly defeated by the House in June 1975 with a 345-72 vote against the 20-cent tax, handing a major setback to House leadership.[62]

President Jimmy Carter entered office in January 1977 and began to formulate an energy policy. In June 1977 he proposed a 50-cent-per-gallon standby gasoline tax that would be triggered if consumption in a given year increased over consumption in the previous year. The proposal was rejected by the House Ways and Means Committee. An amendment to increase the fuel tax by 5 cents per gallon, with the proceeds equally divided between highways and mass transit, was brought to a full vote on the House floor and was rejected by a margin of 339-82 in August 1977. Clearly, there was no willingness in Congress to raise the already high fuel prices Americans were paying. Carter would make one last effort to push through a gasoline tax increase in 1980. Calling it an import fee, he sought to impose a 10-cent-per-gallon gasoline tax. Congress passed a bill prohibiting Carter from imposing the fee. Carter subsequently vetoed the bill which prohibited the fee. Congress then proceeded to override President Carter's veto by a vote of 335-34 in the House and 68-10 in the Senate. It was the first time since 1952 that Congress had overridden a Democratic president's veto.[63]

At the end of 1980, the federal gasoline tax stood at 4 cents per gallon, exactly where it stood in 1959. In the meantime, nationwide highway construction inflation had reduced the value of the 1959 dollar to just 23 cents.[64]

In Texas, a 2 cent increase in the state gasoline tax was included in a tax bill passed in the 1971 legislative session. Governor Preston Smith objected to the size of the tax bill and was particularly critical of the gasoline tax increase. In order to ensure the governor's approval of the tax bill, the Texas House and Senate promptly voted to remove the gasoline tax increase from the bill. The Texas gasoline tax would remain at 5 cents per gallon until 1984.[65]

Overall, political support for the highway system started its decline in the 1960s and continued its downward trend through the 1970s. Bill Ward, head of the TxDOT Houston Urban Project Office from 1972 until 1984, described the decline of political support in the Houston area during the period as consistent and steady—a "linear" decline, he called it. It seemed that several factors were coming into play in the political landscape. Highway construction was a long-term investment, requiring a substantial investment of money in the near term to provide benefits and economic development over the long-term horizon. In the short term, however, there would be disruptions and possible public protest, often receiving disproportionately large press coverage.

Of course, politicians hate negative press coverage and being thrust into controversial situations. A more expedient route for politicians was to focus limited resources on less controversial programs that would provide instant gratification for constituencies, such as housing subsidies, social programs, or increased social security benefits. Some politicians were sympathetic to environmental causes, and choking off funding was a very effective way to stop highway construction. Other politicians were focused on preventing tax increases in the inflationary environment of the 1970s, especially in consideration of the rapidly rising cost of gasoline. The constituencies that had strongly supported freeway construction in the 1950s and 1960s had, for the most part, achieved their objectives. They had their freeways, so they turned their attention to other issues. Anti-freeway interests stepped forward to fill the void left by the departed pro-freeway constituencies and very effectively disseminated their messages through the press, which was often very sympathetic to protest causes. The underinvestment in the nation's highway infrastructure would eventually produce another crisis, but it was something that the next generation of politicians and voters would have to deal with. Because of Houston's astronomical growth during the 1970s, the transportation crisis would occur very quickly.

Taking a Slice Out of the Trust Fund

The federal Highway Trust Fund was the recipient of a small river of money generated by the various motor vehicle-related excise taxes. The highway construction program was one of the few federal programs to have its own dedicated revenue source. But in Washington, D.C., no money is safe forever. Other political interests with their own agendas would cast an envious and covetous eye on the money in the Highway Trust Fund. The highway lobby and its supporters in the United States Congress had been successful in protecting the Highway Trust Fund from any major threats through the 1960s. In 1968 President Johnson withheld payments from the fund as an anti-inflationary measure, but normal spending was soon resumed. The real threat to the Highway Trust Fund was diversion—taking the money and spending it on something other than highways. And the threat would be the proverbial enemy from within.

Public transportation in the United States was in a period of transition. Historically, public transit operators had been private firms that relied on fare box revenue to sustain operations, make capital investments, and earn a profit. But as Americans switched to private automobiles after World War II, the demand for public transportation declined sharply, leaving the private transit operators with a shrinking revenue stream—one that could no longer sustain capital investments and, by the 1960s, often could not even sustain operations. By the late 1960s practically all private transit firms had become bankrupt, were taken over by government agencies, or were certain to go out of business in the near future. It was clear that public transportation could be viable only as a taxpayer-subsidized, government operation.

In 1970 Houston Mayor Louie Welch realized that a city takeover of Houston's private bus operator, Rapid Transit Lines, would soon be needed. Welch appealed for federal mass transit assistance in front of a House Banking Subcommittee in Washington considering mass transit legislation. Mayor Welch testified, "The problems of congestion in cities of urban sprawl like Houston are not the same as those in more densely populated cities, but those problems are just as acute in our cities. The revenue from the fare box will not be sufficient to provide the essential capital, and financial assistance must be received from other sources." Like municipalities all over the nation, Houston was facing a new and expensive public service obligation. Even Houston wouldn't be shy about lining up for federal transit subsidies. Of course, the demands from older, financially strapped cities in the Northeast would be even stronger. Cities were desperate for money to help pay for the new transit obligation, and they didn't care where the money came from—even if it meant raiding the Highway Trust Fund.[66]

In 1972 Welch began to push for the city of Houston to take ownership of Rapid Transit Lines. On April 17, 1974, the city of Houston finalized the purchase. The city found itself with a barely functioning operation with aging equipment, maintenance facilities dating back to the streetcar era, and an unhappy work force that would soon go on strike.[67]

Throughout the 1960s interest in urban mass transit had been steadily increasing. By 1963 virtually all major cities in the United States were studying rapid transit requirements for the future. The Houston Chamber of Commerce, which had been a key driving force in the development and construction of Houston's freeway system, stated that its top transportation priority for 1967 was to "encourage immediate steps by public officials to determine the need for a rapid transit system for the area." In 1968 the Chamber of Commerce established a new committee, the Urban Transit Committee, which

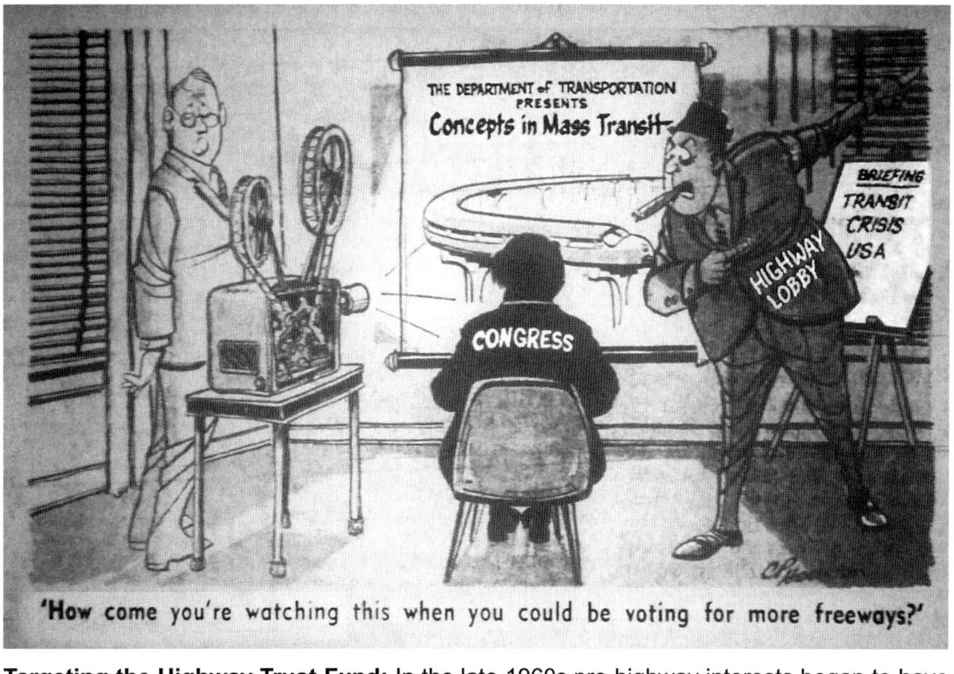

Targeting the Highway Trust Fund: In the late 1960s pro-highway interests began to have to defend the Highway Trust Fund. Opposing interests, notably mass-transit proponents and environmental groups, felt that the nation's transportation policies placed too much emphasis on highways and were seeking to divert trust fund money to mass transit. This 1968 cartoon illustrates the view that the highway lobby wielded excessive influence in national transportation policy. The federal Department of Transportation under John Volpe supported and promoted trust fund diversion. In 1973 pro-transit interests achieved their objective when federal highway legislation authorized the diversion of Highway Trust Fund money to mass transit starting in 1975. The trust fund diversion further exacerbated the nation's highway construction funding crisis. (Source: *Houston Chronicle*, September 22, 1968)

The Mass Transit Crisis

The highway construction crisis developed very quickly, with highway construction going from a relatively strong position in the late 1960s to a crisis situation by 1975. But urban mass transit had been in decline since the end of World War II, and by 1970 the condition of urban transit had deteriorated to a crisis point. The rising use of the private automobile caused a huge reduction in the market for transit services. Transit patronage in 1970 was one-third the level of 1945.[68] Privately owned transit operators could not justify investments in transit services. Equipment became obsolete, service deteriorated, and transit fares went up, further accelerating the switch to private automobiles and the vicious cycle of decline in mass transit. The last year transit operations in the United States showed a profit was in 1962. Deficits of urban transit systems grew dramatically in the late 1960s, reaching \$332 million in 1970—1.26 billion in 2003 dollars.[69] The bankruptcy of the Penn Central Railroad in June 1970 was indicative of the direction of the private passenger rail and transit industry in the United States. Government agencies took over big-city commuter rail lines abandoned by private firms, but small towns weren't so lucky. About 300 urban transit companies operating in 1950 had gone out of business by 1970.[70] The major national news magazines—*Newsweek*, *U.S. News and World Report*, and *Businessweek*—all featured the mass transit crisis on their covers in 1971 or 1972.

> "City rapid transit systems are in trouble nearly everywhere."
>
> *U.S. News and World Report*, January 19, 1970

Many large urban cities such as New York, Chicago, and San Francisco were already subsidizing urban transit with state and local funds by 1970, but as the cost of building and operating transit systems escalated, cities all across the country were looking for help. Federal assistance for urban mass transit began with the Urban Mass Transit Act of 1964, but funding had been relatively low. The first legislation providing large-scale federal assistance was the Urban Mass Transportation Assistance Act of 1970, which authorized \$10 billion (38 billion in 2003 dollars) in federal funding over 12 years, with funding coming from general federal government revenue. But cities wanted more. The federal Highway Trust Fund became the target for obtaining additional funding. The issue of using Highway Trust Fund money was so contentious that it prevented agreement on a new federal-aid highway bill in 1972. The Arab oil embargo of 1973 made mass transit look more attractive, and after a long stalemate in congressional conference committee, the Federal-Aid Highway Act of 1973 authorized the use of Highway Trust Fund money for mass transit. The spigot had been opened for federal government transit assistance and subsidies, which in the long run would be funded almost exclusively by money from the Highway Trust Fund.

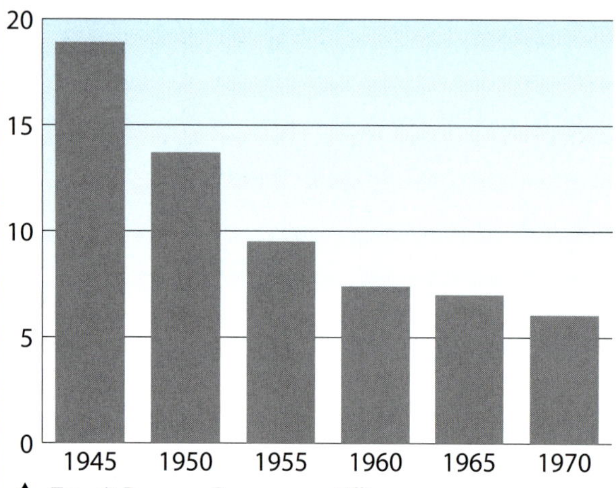

▲ Transit Revenue Passengers, Billions

Data: *Business Week*, June 3, 1972; original data, American Transit Association

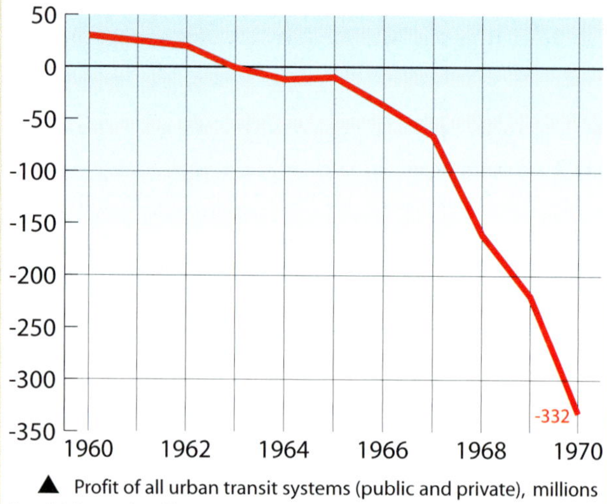

▲ Profit of all urban transit systems (public and private), millions

Data: *U.S. News and World Report*, April 26, 1971; original data, American Transit Association

would "participate in studies of the movement of people in an urban environment by modes other than roads and highways." In a dramatic change from past policy, Big Oil also started to publicly state its support for diversion of Highway Trust Fund money to mass transit. In February 1972 Thomas D. Barrow, President of Humble Oil, the predecessor to today's ExxonMobil, expressed Humble's support for transforming the Highway Trust Fund into a transportation trust fund and allowing local entities to use the money for roads or transit.[71]

In 1972 Secretary of Transportation John Volpe made his move to bust the Highway Trust Fund and dedicate a portion of it to mass transit. Under the plan, federal mass transit funding would begin in 1974 and reach \$2.25 billion in 1976. For comparison, total federal aid for highway construction in 1976 was approximately \$6.27 billion. The United States Senate easily approved an amendment authorizing the transfer of \$800 million in funds to transit,

Houston area freeway cancellations (also see map on page 66)		
1974	Harrisburg Freeway (SH 225 La Porte Freeway extension inside Loop 610)	Permanently cancelled; deleted from all planning documents by 1992.
1975	West Loop Extension, also called the Bay City Freeway (now called the Fort Bend Parkway Toll Road)	Reinstated in 1988; construction underway in 2003 as a tollway.
1976	Grand Parkway	Reinstated in 1984; first section opened in 1994.
1983	SH 146 in Galveston County	Freeway permanently cancelled; a 2002 study recommended highway upgrades.
Late 1980s	West Bay Freeway in Brazoria and Galveston Counties (also called the GAP Parkway)	Appears to be permanently cancelled in 2003, but could be revived as a toll road.

Freeways placed on long-term hold in the 1970s		
1976	Alvin Freeway, SH 35	Short section opened in 2000; entire corridor under study in 2003.
1976	Crosby/Northeast Freeway, US 90	Freeway outside Beltway 8 opened in 1991; section inside Beltway 8 scheduled to be built prior to 2010.
–	Red Bluff Freeway main lanes	Main lanes never formally planned; could be built as a tollway if demand warrants.

which was about 13% of the $6.26 billion in Highway Trust Fund receipts for 1974. Through intense effort and political procedural maneuvering, the highway lobby was able to prevent trust fund diversion in the House version of the highway bill. A conference committee could not formulate a compromise bill and deferred action until 1973. That year, President Nixon and the new Secretary of Transportation Claude Brinegar both stated their support for diverting highway funds to mass transit. The Senate version of the new federal highway legislation once again included provisions to use Highway Trust Fund money for mass transit, and just as in 1972 the House version kept the Highway Trust Fund intact. A conference committee attempted to resolve the differences between the Senate and House versions of the federal transportation legislation, but the committee remained in a stalemate, meeting after meeting. Pressure was building to reach a compromise to avoid a major disruption of federal funding for highways.[72]

> "[In 1977], not a single mile of freeway or expressway was opened for traffic."
>
> Houston-Galveston Regional Transportation Study newsletter, January 1978

Finally, after 14 meetings and more than two months of negotiation, a compromise was reached, allowing final approval of the Federal-Aid Highway Act of 1973. The Highway Trust Fund would remain intact in 1974. But starting in 1975, highway funds would be diverted to mass transit and in 1976 cities would have the authority to cancel planned interstate highway construction and divert the money to rail transit facilities. The Highway Trust Fund had been busted. In Houston there were no major diversions of funds previously set aside for freeways, but the loss of funds to transit further shrunk the federal highway funding pie, exacerbating the crisis that existed nationwide.

Houston Feels the Effects

Like all other places in the United States, Houston felt the effects of the highway construction funding crisis. On May 5, 1974, the *Houston Chronicle* reported how the inflationary spiral was impacting Houston's freeway construction program. The cost of six major freeway projects, valued at $91.3 million in 1968, had increased an average of 60% between 1968 and 1974 to $145.8 million. Another set of four projects had risen from $11.8 to $23.8 million, a 101% increase. The article also reported that a 25% cost increase had occurred in 1973 alone. These increases mirrored the escalation of the Texas highway bid price index.

New freeway construction in Houston slowed to a trickle in the 1970s. Progress on the major unfinished radial freeways, the Northwest and South Freeways, was very slow or nonexistent. The Houston-Galveston Regional Transportation Study (H-GRTS) reported in its January 1976 newsletter, "During 1975, a total of 6.5 miles of new freeway facilities were opened to traffic, about par with the average rate during the last 5 years." However, an average of only 2.9 miles (4.6 km) per year were being opened in the urban areas of Harris County. The freeway completion rate was "a far cry from the construction rate experienced during the sixties, which saw an average of 18 miles of freeways and expressways being completed annually."

In 1976 TxDOT head Bannister DeBerry warned, "The crisis is upon us. It is real and it is serious." In response to the developing crisis, TxDOT hired the management consulting firm McKinsey & Company to analyze statewide revenue and needs, and make a recommendation on allocation of dwindling resources. The TxDOT Houston Urban Project Office under Bill Ward played a supporting

Traffic jam capital of the United States: In the early 1980s Houston's traffic congestion reached crisis proportions and Houston was generally recognized as having the most congested freeway system in the United States. This 1982 photograph looks east along the Katy Freeway (IH 10) with the Loop 610 interchange in the background. (Photo: Texas Transportation Institute)

role in the development of the report. The report began with a summary of the grim funding situation. In the Houston region, available funding over the following 20 years would cover only 7% of the region's backlog of needed highway projects. The report provided a blueprint for sustaining TxDOT and improving its operational efficiency in the changing environment, but its recommendations gutted the Houston freeway program. Recommended for cancellation were the Harrisburg Freeway, Alvin Freeway, and Beltway 8. The Northwest and South Freeways were downsized, to be built as highways rather than freeways for most of their length. Other long-term freeway projects, such as the widening of the North and Eastex Freeways, were put on hold indefinitely. In general, the report urged TxDOT to focus on smaller, less expensive projects which still provided substantial transportation benefits. McKinsey also recommended lowering TxDOT's freeway design standards, turning the clock backward to an era of smaller, less modern freeways. McKinsey had really just told TxDOT that it was going to have to live within its means. The McKinsey report was adopted, but it would not be the final word.[73]

Official planning documents began to reflect the harsh new reality of lowered transportation funding. The Harrisburg Freeway had already been put on hold in 1974, partly in response to neighborhood opposition, but budgetary issues would ultimately kill the freeway by 1976. The Grand Parkway was deleted from long-term plans in 1976. The January 1977 newsletter of H-GRTS stated that, at the current rate of construction, "it would take 45 years to complete the system as planned." Progress on Houston's freeway system hit rock bottom in January 1978 when the H-GRTS reported that in 1977 "not a single mile of freeway or expressway was opened for traffic." The West Loop extension, also known as the Bay City Freeway and later reincarnated as the Fort Bend Parkway, was deleted in 1978. Two other freeways in Galveston County, the SH 146 Freeway and the proposed West Bay Freeway, were not officially cancelled in the 1970s, but it became doubtful that the freeways could ever be built. The SH 146 Freeway in Galveston County was officially cancelled in 1983, and the West Bay Freeway, which was never more than a line on planning maps, disappeared from long-term planning documents by the late 1980s.

A potentially greater risk for the long-term viability of Houston's freeway system was the skyrocketing cost of right-of-way and the intense development occurring around Houston. In order to preserve critical right-of-way for future needs, TxDOT had to make large expenditures to purchase land. This further cut into construction funds. The high cost of right-of-way for Beltway 8 in the booming suburban areas around Houston was putting the entire project at risk for cancellation.

> "Immobility is a vexing fact of Houston life. It now figures heavily in business and personal decisions, the large and small. Everyone is seeking a way to cope with the need to get from point to point."
>
> *Houston Downtown* magazine, January 4, 1982

Most congested cities in the United States, 1984

1. Houston
2. New Orleans
3. New York
4. Detroit
5. San Francisco
6. Seattle
7. Los Angeles
8. Boston
9. Charlotte
10. Atlanta

Data: *U.S. News and World Report*, September 7, 1987

Skyrocketing Demand for Freeways

When the energy supply shocks of the 1970s occurred, the national economy was hit hard. But there was one big winner: the energy industry. Houston in the 1970s was a petroleum city, so Houston's economy boomed as the high price of oil fueled an unprecedented expansion of drilling in the United States. The count of drilling rigs in the United States, which hovered around 1,000 in the early 1970s, began its upward trend in the mid-1970s and reached a record of 4,530 drilling rigs the last week of 1981. Population in the Houston metropolitan statistical area increased from 1.9 million in 1970 to 2.8 million in 1980. The downtown skyline was transformed. "Edge cities" sprouted around the periphery of Houston, providing impressive suburban skylines. Demand for transportation facilities in Houston was skyrocketing.

All the new arrivals in Houston brought automobiles—lots of them. The number of registered vehicles in Harris County increased from 1.1 million to 1.9 million during the 1970s, a 71% increase that added an average of 220 new vehicles to Houston's roads daily. In 1979, near the growth peak, it was estimated that 375 new vehicles were hitting Houston's roads daily. While all the new autos were pouring into Houston, capacity of the freeway system was barely growing. A mismatch of supply and demand was rapidly developing.[74]

Houston achieved its pre-crisis peak mobility in 1970 with the relatively new freeway network providing a high level of congestion-free mobility. But the 1960s freeway system and the few enhancements that were added in the 1970s were no match for the additional demand that came in the 1970s. Whereas Los Angeles had built most of its freeways with 8 or 10 traffic lanes in the 1960s, Houston's planners had not been as forward-looking. Surely they never anticipated the phenomenal growth that Houston would experience, and as always, financial resources for building freeways were limited. As of 1970 Houston had only one sustained length of 10-lane freeway, the Katy Freeway (IH 10) between downtown and the West Loop. Some freeways had 8 main lanes, including the Southwest Freeway inside Loop 610, the East Freeway inside Loop 610, and Loop 610. All other freeways generally had 6 main lanes, and the Eastex and East Freeways outside Loop 610 had only 4 main lanes.[75]

As traffic congestion became worse, even more disturbing was the prediction of a 1979 Rice Center study on Harris County mobility and congestion issues. Based on funding and population trends at the time, it concluded that traffic congestion would be a permanent and chronic

The 1970s boom: The number of active drilling rigs in the United States is a good indicator of the level of activity in the energy industry and the health of Houston's economy. The steady climb in the rig count through the 1970s and into 1981 was mirrored by the boom in Houston's economy. The boom brought an influx of people and automobiles to Houston, an estimated 375 new cars per day in the late 1970s, overwhelming Houston's freeway system. The subsequent bust in the energy industry was even more dramatic than the boom.

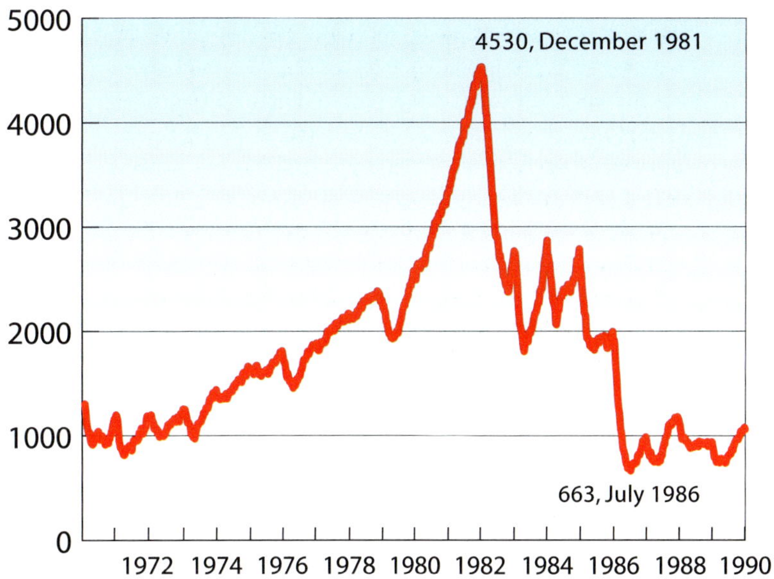

▲ Active Drilling Rigs

Data: Baker-Hughes

Houston Post, May 30, 1984

condition in Houston. The vice president of the Rice Center stated, "Experience has shown us that, once the major freeways in a city become congested, they never become uncongested. That's what Houston faces. There will be congestion in Houston through the turn of the century and beyond."[76]

Houston had been working its way up the list of most traffic-congested cities in the United States through the 1970s. Ranking cities for traffic congestion is somewhat of an art, dependent on the criteria used and subject to available data, which usually varies from city to city. The Texas Transportation Institute produces one of the better-known reports for urban congestion, the *Urban Mobility Report*, which was first published in 1981. Rankings are provided in 14 categories. Los Angeles is the long-running traffic congestion champion, earning the number one ranking for all years since 1981 in 4 categories. In 7 categories, Los Angeles has been number one for all years except from 1983 to 1985 when Houston displaced Los Angeles in one or all of the years. In the remaining 3 categories Los Angeles has shared number one rankings with Houston and other cities.

But Houston was ranked as the undisputed most congested city on other lists. A comprehensive report by *U.S. News and World Report* ranked Houston as number one in traffic congestion in 1984. And given the inherent imperfection in any congestion measure, Houston was clearly among the most congested cities in United States and almost certainly the worst congested during the early 1980s.[77]

Moving the Gridlock to the Airways

As Houston's freeways and streets were approaching gridlock, another transportation crisis began to develop—in the air. Houston's business executives were finding that the best way to beat Houston's traffic was to avoid it entirely by flying above it in helicopters. The river of petro-dollars fueling Houston's boom provided plenty of money to pay for the helicopters, and the Gulf of Mexico offshore drilling industry required constant personnel shuttles to and from the offshore drilling rigs. The number of helicopters based in the Houston region increased from 103 in 1977 to 315 in 1982. In 1981 it was estimated that 780 helicopters were operating within a 300-mile (480 km) radius of Houston. In 1982 the *Houston Post* reported, "A spokesman for the Federal Aviation Administration said the Houston area, with its 800 or so based helicopters, is easily the nation's center for helicopter activity." Armadillo Airways began a helicopter shuttle service between west Houston and the city's airports in 1980.[78]

In addition to backed-up streets around their neighborhoods, Houstonians now faced the relentless thunder of the nation's largest fleet of helicopters crisscrossing the city at low altitudes. It became a serious quality-of-life issue, especially for neighborhoods near the hub of helicopter activity in the Galleria/Uptown commercial district along the West Loop. As one frustrated Houston resident explained at a 1981 public hearing, "What little quality of life is left in this city will be totally destroyed without a noise abatement rule." In March 1981, Houston Mayor Jim McConn formed a committee to study the helicopter problem and recommend regulations to mini-

A helicopter on every roof: The severe traffic congestion in the late 1970s and early 1980s helped contribute to an unprecedented helicopter boom in Houston. Allison/Walker Interests, a developer of suburban office buildings, used a helicopter to visit project sites around the Houston area. "If we had to drive to every one of our projects, we'd spend all day on the road," remarked the company's chief pilot in 1982.[79] (Photo: HMRC RGD6)

mize the impact on neighborhoods and reduce the risk of accidents. Houston was entering uncharted territory when it attempted to regulate helicopter activity in December 1981. The Federal Aviation Administration (FAA) rejected Houston's plans in June 1982, saying that the city of Houston had overstepped its authority into areas of regulation that belonged to the federal sector. Houston was the nation's first city to attempt extensive regulation of helicopter operations. The FAA wanted to ensure that Houston would become a prototype for other cities that were considering helicopter regulation. Fine-tuning of the regulations continued, with the FAA, helicopter interests, and neighborhood groups all participating in the process. In December 1983 Houston City Council adopted a revised ordinance. However, all the fuss over the helicopter invasion would soon be forgotten as the imminent oil bust would eliminate the demand for helicopter service.[80]

Key dates in the 1970s crisis

1968	Federal regulations dramatically increase and give expanded powers to small but vocal groups opposing freeways, making it more difficult to build new freeways.
1970	The National Environmental Policy Act of 1969 is signed into law, requiring environmental impact statements for federally funded highway projects. Building new freeways becomes even more difficult.
1971	The Texas State Legislature rescinds a planned increase in the fuel tax.
Early to mid 1970s	Highway construction costs escalate dramatically, increasing 58% in Texas in 1974 alone. Fuel tax revenues stagnate.
1975	TxDOT initiates large-scale workforce reduction, including 3,000 layoffs.
1976	A comprehensive study recommends a downsizing of new freeway plans in Texas.
1974–1978	The Harrisburg Freeway, West Loop Extension, and Grand Parkway are cancelled. Other freeways are put on hold. New freeway construction is reduced to a trickle.
Late 1970s –early 1980s	Houston's population boom brings the city's freeway system to near gridlock.

EXIT 757
Gessner Rd
1 MILE

The Second Wave

Getting politicians to take decisive action on an issue that doesn't involve national security or public safety isn't easy. And when new taxes are involved, it can be nearly impossible. But politicians always have their noses in the wind, sniffing for paths of political gain or expediency. By the early 1980s the air around Houston was thick with discontent. Traffic congestion had reached crisis proportions, and politicians would need to take action if they wanted to keep their jobs. Reaching a consensus wouldn't be easy and required another bruising round of the freeway versus rail debate, but in the end the freeway prevailed and the confrontation of the crisis would begin.

Houston's response to the traffic crisis was strong and decisive. If it was a war to be fought, Houston was bringing overwhelming force into battle. Some of Houston's brightest minds in business, transportation, and government would formulate a unique solution emphasizing cost-effectiveness and results. It was a solution that would meet demand by increasing supply of both highways and transit. The second wave would establish Houston as one of the nation's most aggressive builders and expanders of freeways and would propel Houston into the elite of the world's freeway cities.

The First Steps

Houston's freeway construction program had reached its low point in 1976 when the McKinsey report recommended a dismantling of the ambitious plans for the freeway system. The following year, state political leaders in Austin took steps to ease the immediate crisis and delay a day of reckoning. The first substantial action to reverse the decline took place in early 1977.

At legislative hearings in November 1975, a top TxDOT official testified that TxDOT would completely run out of funds for new highway construction by 1980 if the department's finances were not strengthened. In February 1976 Governor Dolph Briscoe and State Comptroller Bob Bullock were publicly stating that additional funds would be needed for the highway program. That task became a top priority in the 1977 legislative session. A bill to pump an additional $528 million (1.3 billion in 2003 dollars) into the highway program over two years cleared the legislature by early April. It was an unusually swift passage through the legislature, which usually completes most of its work in the days just before the end of the session in early June. The bill was designed to keep TxDOT even with inflation by funneling general tax revenue into TxDOT's budget when dedicated highway funds could not keep up with highway construction inflation. Governor Briscoe signed the bill into law April 12, 1977. The new funding reversed a steep decline in TxDOT's budget, which had been $832, $731, and $662 million in the years 1975–1977, respectively. For fiscal years 1978 and 1979 TxDOT funding was $922 million and $1.02 billion. The legislation would continue to pump between $130 and $150 million per year in general revenues into the TxDOT budget until its repeal in 1984. The injection of new funds brought TxDOT back to life, allowing the formulation of a new 20-year plan in December 1977. Planning resumed on previously moribund projects, including Houston's Beltway 8.[81]

Help was also on the way from the federal government. By the early 1980s, the high price for underinvestment in the nation's highway infrastructure was becoming evident. In August 1982, *Newsweek* featured the state of the nation's crumbling infrastructure on its cover, calling it "The Decaying of America." The report stated, "Aging and neglected, the nation's network of roads, bridges, sewers, and rails is nearing collapse." It was more than an embarrassment—it was a risk to the national economy. President Ronald Reagan had entered office in January 1981 with the intent to cut taxes and reduce federal aid to states, including cutting aid for highways and mass transit. But the ailing economy was the main issue of the day, perhaps giving President Reagan cover to call for a five-cent increase in the gasoline tax in November 1982. Senate and House leadership were immediately supportive, and Congress approved the Surface Transportation Assistance Act of 1983 on January 6, 1983, increasing the fuel tax from four to nine cents per gallon. It was the first increase in the federal fuel tax since 1959. The tax increase provided $295 million in new federal funds to Texas in 1983, which was about 22% of the 1982 TxDOT expenditures of $1.35 billion.[82]

(Opposite page) **Houston's first five-level stack interchange:** One of the key projects of Houston's second wave of freeway construction was the Beltway 8-Sam Houston Tollway—Houston's second loop. The Sam Houston Tollway features one of the nation's most impressive collections of modern stack interchanges, most of them with five levels. This April 1988 photo shows construction on Houston's first five-level stack interchange at the Katy Freeway and the west Sam Houston Tollway. This interchange will be Houston's most short-lived, as it will be demolished and reconstructed during the Katy Freeway expansion project, scheduled for 2003-2008. (Photo: *Houston Chronicle*)

The Birth of a Toll Road Authority

Houston's effort to solve the traffic crisis began in the mid-1970s. One of the first local officials to make transportation a top priority was Harris County Judge Jon Lindsay. In 1974, at the age of 38 with no previous political experience, Lindsay stunned the local political scene by winning the election for the powerful position of county judge, defeating the incumbent Bill Elliot. Lindsay was somewhat of a rarity among elected officials—he was a civil engineer. Harris County had been one of the key driving forces in the development and construction of Houston's freeway system in the 1950s and 1960s. An engineer was now at the helm of the region's most aggressive road-building entity. Lindsay was positioning Harris County to step forward and once again take a leadership role in freeway construction.

"Harris County is in the toll road business and County Judge Jon Lindsay is the man of the hour."

Houston Chronicle, September 14, 1983, one day after the successful bond election to create the Harris County Toll Road Authority

Starting in 1975, Lindsay focused his efforts on rescuing Beltway 8. In 1975 and 1976 Lindsay led Harris County in requesting the Texas Turnpike Authority (TTA) to study parts of the Beltway as toll road projects. Both studies declared the projects infeasible, but Lindsay continued his efforts to save the project. By 1977 improved financial conditions at TxDOT allowed planning to resume on small portions of the Beltway. However, Lindsay was already beginning to realize that the county would need to take matters into its own hands if it wanted to see the Beltway constructed any time in the near future. He stated that Harris County should be responsible for the Beltway toll road rather than the TTA. In 1978 Harris County resumed right-of-way acquisition and corridor preservation efforts for the Beltway.[83]

In 1979 the Texas Turnpike Authority identified the Hardy Road corridor as a potential candidate for a toll road. The issue of further study of the Hardy corridor was considered by Harris County Commissioners Court in August 1980. With Lindsay absent, Harris County commissioners voted 4-0 to oppose the toll road due to large and vocal community opposition. When Lindsay returned to Houston, he put the toll road back on the agenda for another vote and brought two commissioners to his side, resulting in a 3-2 vote in favor of continued studies. With that vote Lindsay first took ownership of the Hardy Toll Road. As the project became increasingly embattled in the following years, Lindsay would almost single-handedly push the project forward to construction.

By mid-1982 the TTA was backing away from plans to construct the Hardy Toll Road due to opposition from the city of Houston, the Metropolitan Transit Authority, and elected officials. Lindsay was now the only person who could save the Hardy Toll Road. In August 1982, Lindsay instructed the county attorney to investigate if the county could create its own toll road authority. It turned out that special legislation would be needed. The legal authority for Harris County to form a toll road authority came with Texas legislative bill SB970, signed by Governor Mark White in June 1983.

Lindsay arranged a $900-million bond election for September 13, 1983, to launch the toll road authority. Winning the election was by no means assured. On June 11 Houston voters had soundly defeated the Metropolitan Transit Authority's plans for a billion-dollar heavy rail line. Opposition to the toll road authority was strong in predominantly minority, low-income areas in the vicinity of the controversial Hardy Toll Road on the north side of Houston. Support was strongest in the northern and western suburbs. The election hinged on which group could get its voters to the polls.

Lindsay's forces set up a sophisticated direct mail and phone bank operation to get their targeted voters to the voting booths. Opinion polls were used to identify pro-tollway voters. It was a grassroots effort, designed to get frustrated motorists to vote, and it worked. A strong turnout in suburban west Houston drove the toll road bonds to victory with 69.7% of the vote. For a bond issue, the election had an unusually high 13% voter turnout. A post-election analysis by the *Houston Chronicle* summarized the vote by saying, "Jon Lindsay is the man of the hour."[84]

The Regional Mobility Plan

In the early 1950s the Houston Chamber of Commerce played a key role in gaining state approval and securing funding for Houston's freeway system. After a long period of relative dormancy, the transportation committee of the Chamber of Commerce came back to life to address the transportation crisis. In 1980 Roger Hord, director of the chamber's transportation department, initiated efforts in conjunction with the transportation committee. The committee was led by John B. Turner Jr., president of Friendswood Development, a land development company active in building master-planned communities around Houston. Turner was the strongest and most influential leader of the transportation committee since the 1950s. As a land developer, he was well connected with local business interests and government agencies. He focused his civic service on solving the transportation problem that was threatening the future of Houston.[85]

Hord and Turner began their efforts by developing a comprehensive list of transportation projects needed in the Houston area and obtaining endorsements for the projects from agencies in the region. As they identified projects, it became apparent that a more systematic approach would be necessary—one that included growth estimates, traffic projections, and cost-effectiveness measures. In 1981 Hord and Turner assembled a group of leading transportation professionals in the Houston area and developed a plan to solve Houston's mobility problem. In February 1982 the group released *A Regional Mobility Plan for the Houston Area*, the most comprehensive and sweeping proposal for updating Houston's transportation system since the freeway system was first proposed in the early 1950s. The Regional Mobility Plan would form the backbone of Houston's response to the traffic crisis.

The plan produced by the task force was intended to deliver the maximum amount of traffic relief for the money spent. The plan was a unique solution, tailored for Houston's low density, sprawl, and decentralized travel patterns. It emphasized increasing the supply of transportation facilities to meet increasing demand. The report limited the use of expensive options, like rail transit, only to situations where no other method could get the job done. It didn't fall victim to the transportation fad of the day, "heavy rail" mass transit, an extremely expensive system used in Atlanta's MARTA, Washington, D.C.'s Metro, San Francisco's BART, and many others.* More than anything else, the Regional Mobility Plan was a freeway plan. The plan had three key recommendations: expand existing freeways, build a comprehensive system of dedicated bus lanes into the freeway system, and build new facilities, both freeways and tollways.

The report recommended major expansion of nearly all the radial freeways. Elevated express lanes were recommended for the Katy, Southwest, North, West Loop, and Northwest Freeways. The Eastex, Gulf, and North Loop Freeways were also slated for expansions. In all, 170 miles (272 km) of freeways were recommended for widening. In addition, 184 miles (294 km) of new freeways and tollways were recommended for construction during the 15-year implementation period, and an additional 100 miles (160 km) of new freeways or tollways were proposed to serve anticipated future growth. The substantial mileage of new freeways in the plan was really just a build-out of the planned freeway network that existed in 1970, before cancellations and delays began. The only new facility not previously planned was the Hardy Toll Road.

The distinguishing characteristic of the Regional Mobility Plan was the comprehensive use of freeway transitways—barrier-separated lanes designed for buses and high-occupancy vehicles (HOV). The extensive use of transitways was a solution unique to Houston, bucking the national trend of building rail systems. In all, 128 miles (205 km) of transitway lanes were recommended. The plan included provisions for 30 miles (48 km) of "high-capacity transitway" but did not specify a technology, recommending that "more costly transportation technologies are added only when travel demands cannot be met by lower cost options."

Reaction to the plan was good. A representative of the Federal Highway Administration remarked, "It's a massive undertaking, but we have a massive problem." John Butler of the Texas Transportation Commission called the plan "the finest thing the Chamber [of Commerce]

REGIONAL MOBILITY PLAN

For The Houston Area.

February, 1982

TRANSPORTATION COMMITTEE
John B. Turner, Jr.
Chairman

Houston
CHAMBER OF COMMERCE

No flash, all substance: The 1982 Houston Regional Mobility Plan was not flashy in its appearance or presentation, but its recommendations formed the blueprint for the major freeway expansion and construction program that would take place over the next 20 years. (Photo: Chuck Fuhs)

* "Heavy rail" transit typically uses a third rail for electrical power. The third rail is at ground level, next to one of the regular rails that the rail cars ride on. The electrified third rail necessitates a secure right-of-way for the track with no crossings. Light rail, in contrast, typically uses overhead wires for electrical power, allowing the tracks to be placed on city streets.

Leaders of the Second Wave

Like the original construction of the freeway system, orchestrating the second wave of Houston's freeway construction was a broad, coordinated effort. Two individuals stand out as the most influential in Houston's second wave.

Robert C. "Bob" Lanier, born 1925

A native of Baytown just east of Houston, Bob Lanier became a successful banker and real estate developer after practicing law early in his career. His active role in politics led to his appointment as commissioner on the three-member Texas Transportation Commission in February 1983, and he became commission chairman in June 1983. Lanier's influence on transportation in Houston and in Texas is probably greater than that of any other individual in the post-1970s era. He played a key role in securing 5-cent increases in the state fuel tax in 1984 and 1986, tripling the tax from 5 cents to 15 cents and providing critically needed funding to modernize the Texas highway system. He directed increased resources to Houston to address the transportation crisis and compensate for prior underfunding of the region. After his service on the transportation commission, Lanier played an active role in transportation planning in Houston as chairman of the Metropolitan Transit Authority board. Lanier's emphasis on cost-effective transportation solutions led to his opposition to rail systems and the cancellation of plans for monorail when he became mayor of Houston in 1992. Lanier then turned his attention to non-transportation issues, including public safety and downtown revitalization. In 2002, Lanier was inducted into the Texas Transportation Hall of Honor. (Photo: TxDOT)

Jon Lindsay, born 1935

Jon Lindsay won the election for the powerful position of Harris County Judge at the age of 38 and began service in January 1975. As a civil engineer, Lindsay was a rarity in the political world. Through the remainder of the 1970s, Lindsay led efforts to keep Beltway 8 alive. In response to the inability of state resources to solve Houston's transportation problems, Lindsay worked to create the Harris County Toll Road Authority, which was officially launched with voter approval in 1983. Lindsay brought the Hardy Toll Road to construction almost single-handedly, and his toll road authority would go on to become one of the most important factors in the second wave of construction in Houston. Lindsay did not seek reelection in 1994 but was elected as state senator serving the northwest Harris County district in 1996. (Photo: office of Senator Jon Lindsay)

has embarked upon in some time." Other local officials and agencies were soon throwing their support behind the plan, even while there was some apprehension over the tax increases that would be needed. For the Chamber of Commerce, the plan served as an effective rallying tool to focus the efforts of local governments and business interests. The united, focused effort was particularly important for obtaining political support at the statewide level.[86]

The plan's recommendation for the 15-year implementation period did not include the Grand Parkway or the Fort Bend Parkway, two freeways cancelled in the 1970s. But those two freeways were not forgotten. Suburban interests worked to revive the cancelled freeways, and in 1984 the Grand Parkway was reinstated to Houston's long-term freeway plan. Also in 1984, Fort Bend County southwest of Houston began its efforts to revive the Bay City Freeway, which was renamed the Fort Bend Parkway. Years of study followed, and in 1988 the Fort Bend Parkway outside Beltway 8 was officially adopted into the state highway system. In the late 1990s plans were formulated to construct the Fort Bend Parkway between Loop 610 and Beltway 8. The restoration of Houston's freeway plan was complete. Houston would realize nearly its entire pre-1970s freeway plan and would even go on to add new freeways and tollways. Only one section of freeway in the immediate Houston area, the La Porte Freeway inside Loop 610, remained cancelled. Two other freeways outside Houston in Galveston County, SH 146 and the West Bay Freeway, were deleted from long-term plans in the 1980s.

Even with its focus on cost-effective solutions, the Regional Mobility Plan wouldn't be cheap. It was projected to cost $16.2 billion over 15 years, approximately 27 billion in 2003 dollars. Only $6.9 billion was anticipated to be available during this period, leaving a gap of $9.3 billion. The plan was very direct in its financial approach, recommending a range of tax and fee increases to put the funding in place. To solve the problem, it was going to take money—and lots of it.[87]

Now Houston needed a political money man, someone who could push through tax increases and get money flowing into Houston. Fortunately, Houston's transportation messiah was just about to step forward.

Needed: One Master Freeway Builder for Houston

When the time comes for decisive action, there's always the temptation to look back in history. When road-building is the subject, one name stands alone among the great urban freeway builders of the United States: Robert Moses, the master builder of New York. Starting in the 1920s, Moses (1888-1981) developed a political empire that enabled him to build most of the New York City and Long Island freeway system and the bridges connecting the city, as well as Long Island's park system and other large projects in New York state. With his political power, Moses had an extraordinary ability to get things done. He also had a disregard for the negative consequences of his projects, which would ultimately tarnish his standing in history. Moses' overuse of power would contribute to the implementation of mechanisms ensuring that no one would ever again have his power to build at will. But perhaps there was a new incarnation of Robert Moses, refined for the modern era and its limitations, waiting in the wings to step forward for another great building era. Would Houston have the good fortune of the next Robert Moses?

Robert Clayton "Bob" Lanier was born in 1925 in Baytown, about 30 miles (48 km) east of Houston, the third of three children. His father was a Methodist minister and later a refinery worker. The Lanier family struggled through the Depression in a small house with no indoor plumbing. Lanier wrote for the Baytown newspaper while in high school and planned to become a reporter. He studied English at the University of New Mexico and was a sportswriter for the college paper. After World War II service in the Navy, he put himself through the University of Texas law school by writing for the *Austin American-Statesman*. After graduation he joined the Baker & Botts law firm in Houston, one of the state's most influential law firms. After three years at the firm, Lanier started his own practice and began dealing in real estate.[88]

It was fitting and perhaps prophetic that Lanier's first real estate venture was the purchase of an apartment complex on the Gulf Freeway in the early 1960s. Shortly afterwards Lanier focused his efforts on real estate and rescuing distressed banks, selling the banks after returning them to profitability. Lanier was a fanatic about information, details, and number crunching. His ability to absorb information and use it to his advantage is credited as one of the major reasons for his success. At various times in his career, Lanier would take a strong interest in a certain issue and then learn everything he could about it. If he couldn't learn enough from books, Lanier would often bring world-renowned specialists and authors to Houston to discuss their works with him. Scientist Linus Pauling and economist Milton Friedman were among the visitors to his River Oaks home.

While Robert Moses had dedicated his life to building highways, infrastructure, and parks, Bob Lanier had spent his time building personal wealth. Having achieved business and financial success, Lanier started to take more interest in civic issues. By the early 1980s transportation was clearly the biggest challenge facing Houston. Perhaps it was the size and difficulty of the problem that attracted Lanier, or perhaps it was his real estate development background and his knowledge of the importance of transportation and freeways to Houston. Lanier was well connected in Democratic Party circles, making frequent donations to candidates and often holding fund raisers at his home. Lanier's support of winning Democratic gubernatorial candidate Mark White in the 1982 Texas governor's election proved to be a key event in the transportation history of both Houston and the state of Texas, launching Lanier into his first position of great influence in transportation.

Mr. Chairman

As the calendar reached 1980, Houston recorded a full half-century of a transportation drought. For 50 years Houston did not have a representative on the three-person Texas Transportation Commission, the powerful commission that determines the allocation of highway funds in Texas. Houston's last representative, Ross Sterling, left the commission in 1930 when he became governor. The long wait for the next Houston commissioner ended in 1981 when Republican Governor Bill Clements appointed Houstonian John Butler to the commission. In 1982 Democrat Mark White defeated Clements and became governor. In February 1983 he appointed Bob Lanier to the Texas Transportation Commission, and in May 1983 Lanier became chairman. Houston had gone from zero representation to a two-thirds majority. Of course, the increased presence didn't translate directly into money for Houston, but the Lanier era of the Texas Transportation Commission would prove to be the most influential for

Houston's freeways turn into an obstacle course: The reconstruction and expansion program got underway in the early 1980s and launched the seemingly never-ending era of construction detours on Houston's freeways. This July 1985 view shows the Gulf Freeway at Lockwood, with northbound traffic diverted to an adjacent structure. On the other side of this overpass was the notorious "S" curve detour. The Gulf Freeway was Houston's first freeway to undergo a comprehensive rehabilitation with a fully integrated bus transitway. (Photo: Chuck Fuhs)

highway construction funding in Texas since the 1950s.

Lanier now turned his attention to transportation and began to learn about it with the same zeal he had applied to his other endeavors. Lanier would bring a bottom-line, results-oriented approach to transportation planning. The more Lanier looked at the numbers, the more he realized that highway construction was the best—and probably only—way to rescue Houston from its crisis.

First, Lanier collected data showing that Houston had been short-changed over the 66-year history of TxDOT. Houston had contributed 22 to 25% of state-generated revenue, but had received only 9 to 11% of the highway construction contracts. These statistics gave Lanier immediate justification to dramatically increase Houston's share of the pie. Houston was ready to build, with project plans complete and ready for immediate contract award. In 1983 approximately 33% of statewide construction funding was allocated to Houston. During the 1984–1993 period Houston received 26.4% of state highway construction funding.[89]

However, if Houston was going to get the needed support from the state for its Regional Mobility Plan, the highway funding pie would have to be made much larger. That meant an increase in the fuel tax, definitely not a popular idea among politicians, including Governor Mark White. The Texas energy economy was in a steep decline during this period, putting severe strain on the state budget. Many politicians were even looking to take money out of the TxDOT budget to cover shortfalls in other areas. During this crucial period Lanier worked to protect and strengthen the highway program. He managed to convince key political leaders, including Governor Mark White, that Texas needed an increase in highway funding—and that a decrease in funding could be devastating to the long-term economic future of the state. Lanier focused his efforts on building support for an increase in the fuel tax, not an easy task in an antitax state like Texas.

Transportation wasn't the only crisis brewing in Texas. Education reform was also in need of attention. The 1983 Texas Legislature failed to take action on the top two issues facing Texas, so Governor Mark White was forced to call a special session* in 1984 to deal with education and transportation. It was the critical moment for transportation funding in Texas. If a fuel tax increase could

*The Texas Legislature meets in its regular session for the first five months of every odd-numbered year. Special sessions are often called to complete business not finished in the regular sessions.

not be pushed through at the time of crisis, then maybe it could never be done. At the end of the session the Texas Legislature responded with money. In fact, it was a record tax increase for Texas. Included was a doubling of the fuel tax from 5 cents per gallon to 10 cents per gallon, the first increase in 29 years. 75% of the new revenue from the fuel tax would go to highways, with the remaining 25% constitutionally dedicated to education. In addition, motor vehicle registration fees were raised by $25. Bob Lanier played a key role in the enactment of the revenue increases.

On July 11, 1984, at the construction zone for the North Belt Freeway in Houston, Governor Mark White signed the highway funding bill into law. All the pieces were now in place for the second wave of Houston's freeway construction.[90]

Lanier continued his efforts to strengthen funding for highway construction in Texas. In 1986 a temporary additional 5 cent increase in the fuel tax was enacted by the Texas Legislature and Governor Mark White, who subsequently lost his bid for reelection in November 1986. In July 1987, after substantial political maneuvering on tax issues, Republican Governor William Clements approved a permanent extension of the fuel tax increase. Lanier's six-year term on the Transportation Commission ended on July 28, 1987. In just three years, the state fuel tax had been raised from 5 cents per gallon to 15 cents per gallon. 75% of the revenue generated by the fuel tax was constitutionally allocated to the state highway fund. The funding improvement would be critical for meeting the transportation needs of not only Houston, but the entire state of Texas.

The Transportation Improvement Program

The second wave of construction was a three-pronged attack on Houston's congestion. First, the unfinished freeways, then consisting of only frontage roads in the suburban areas, would be completed. This included the La Porte Freeway, the Northwest Freeway, and the North Belt. Second, new facilities would be constructed, most of which had been on long-term planning maps since the 1960s. This included the Hardy Toll Road, nearly all of the Beltway 8/Sam Houston Tollway, the Crosby Freeway, the SH 249 Tomball Parkway, and the western section of the Grand Parkway. Two other all-new facilities, the Westpark Tollway and the NASA 1 bypass freeway, would later be added. Third, most existing freeways would be reconstructed and expanded, usually with the inclusion of a transitway. This included the North Freeway from downtown to Conroe, the Gulf Freeway from downtown to outside Beltway 8, the Southwest Freeway from downtown to Sugar Land, the Eastex Freeway from downtown to Kingwood, and sections of the East Freeway and Loop 610. Large-scale construction began in the mid-1980s, with intensive work continuing through the late 1990s.

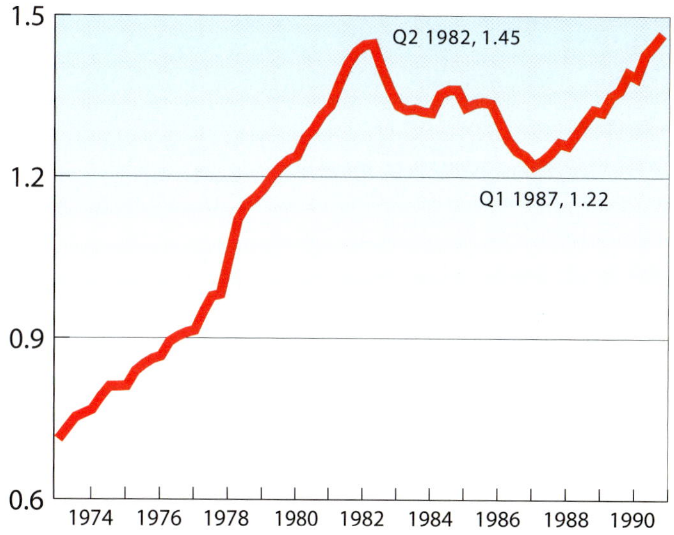

▲ Harris County Employment, millions

Data: Texas Workforce Commission

The bust: This plot of total employment in Harris County in the 1980s shows the impact of the energy industry collapse. The energy boom reached its peak in 1982 when employment topped out at 1.45 million. Houston's economy then went into decline, culminating with a sharp drop in oil prices in 1986 and the collapse of Houston's economy. Houston's employment reached its low point in the first quarter of 1987, and it would take until 1990 to return to the pre-bust employment level. For the entire 1980s, employment in Harris County grew by only 18%, providing critically needed breathing room for the freeway system and time for capacity improvements to catch up to demand.

Breathing Room

While political officials engineered a multipronged attack on Houston's transportation problem, increasing the supply of transportation resources, the other factor in the equation—demand—was providing relief. The mid-1980s would bring a devastating collapse to the energy industry, and oil-dependent Houston was hit particularly hard. The energy boom reached its peak in 1981 and 1982. In 1981 the price of a barrel of oil as measured by the average crude oil refiner acquisition cost reached $35.24—about $62 per barrel when measured in 2003 dollars. Prices stayed high in 1982, averaging $31.87 per barrel, or $53 per barrel in 2003 dollars. The number of drilling rigs operating in the United States, a good indicator of the amount of activity in the energy industry, soared to unprecedented levels in the early 1980s, reaching an all-time record of 4,530 active rigs the last week of 1981. The boom began a downward trend around 1983 and culminated in the collapse of oil prices in 1986, with average prices for the year dropping to $14.55 per barrel, $21 per barrel in 2003 dollars. The effect on the Houston economy was dramatic and devastating. Huge layoffs at energy-related firms were part of the daily news. When the bottom was hit in the first quarter of 1987, Harris County had lost 225,000 jobs since the peak employment of 1.45 million in the second quarter of 1982. Employment recovered in the following years, reaching 1.47 million jobs by the fourth quarter of

Key Dates in the Second Wave

1982	The Houston Chamber of Commerce releases the Regional Mobility Plan.
1983	*February:* Bob Lanier is appointed to the Texas Transportation Commission. *September:* Voters approve the creation of the Harris County Toll Road Authority.
1984	On June 11 Governor Mark White signs a transportation funding bill into law, raising the fuel tax from 5 to 10 cents per gallon.
1986	Fuel tax is increased to 15 cents per gallon. Houston's economy collapses as oil prices drop, reducing the growth of freeway demand.

1989, but for the entire decade of the 1980s jobs in Harris County increased by only 18%. This slow growth provided the freeway system much-needed breathing room, time to catch up to demand that had so drastically exceeded supply in the 1970s and early 1980s.[91]

By the late 1980s Houston was dropping like a rock down the list of the nation's most congested cities. The combination of the freeway expansion program, new tollways, better mass transit, and moderated growth in demand provided a huge improvement in Houston's mobility.

Houston's decision to focus on buses and freeway transitway lanes was also making a strong contribution to improved mobility. In November 1993 the Federal Highway Administration reported, "Of the thirty-three metropolitan areas [studied], all except for Houston showed declines in the share of bus commuters between 1980 and 1990." Although the 3.65% of Houston workers traveling to work by bus was still below the 1990 national average of 4.89%, the increasing trend was unique in the United States.[92]

Build Your Own Freeway

Another important factor in the 1980s freeway resurrection was the increased role of local real estate and land development interests in promoting the construction of new freeways. At the June 1984 special session of the Texas Legislature, a new state law was passed authorizing the formation of nonprofit transportation corporations. A transportation corporation was an entity funded by independent sources outside of TxDOT. It was responsible for performing all work up to the point of construction, including right-of-way acquisition, environmental studies, and engineering. TxDOT would then pay for the actual construction of the highway or freeway. In October 1984 Houston's first transportation corporation, the Grand Parkway Association, was authorized by the Texas Transportation Commission. The Grand Parkway Association was initially funded by west Houston real estate interests who were pushing for the construction of the western segment of the Grand Parkway. The Galveston-Alvin-Pearland transportation corporation was formed soon afterwards to promote the Galveston-Alvin-Pearland Parkway (the "GAP Parkway"), a resurrection of the West Bay Freeway. The commission authorized the Galveston-Alvin-Pearland transportation corporation to perform feasibility studies for the GAP Parkway in November 1985. In 1990 the Fort Bend Parkway Association was designated as a transportation corporation by the commission.[93]

Bob Lanier was a strong supporter of the transportation corporation concept. After all, if real estate and land development interests were going to be big beneficiaries of the new transportation facilities, it was certainly reasonable to expect them to provide financial assistance for their construction. Lanier viewed transportation corporations as a way to stretch available funding as far as possible by increasing private sector participation. The Grand Parkway Association was the most active transportation corporation in the Houston area, and probably in the state of Texas. In the 1980s the Texas Transportation Commission entered into several agreements with the Grand Parkway Association for construction of Grand Parkway segments in west and north Houston.[94]

The concept of the transportation corporation was well received by the interests who were seeking transportation facilities, and those interests were initially willing to provide funding to get the projects back into the development process. But as these projects progressed, the complexity and high cost of bringing a major transportation project to the point of construction became a problem. Securing right-of-way donations was the easy part, although it was generally not possible to obtain 100% of the required right-of-way through donations. Then, millions of dollars would be needed for environmental studies and engineering. Increasingly stringent environmental regulations were adding to the cost, and by the time the United States Congress approved the Intermodal Surface Transportation Efficiency Act of 1991, the costs of environmental studies had become so high that it was impossible to obtain the needed funds through donations. The Grand Parkway Association was able to move one segment of the Grand Parkway to construction, completing it in 1994, but the hopes of constructing additional segments had faded by the late 1980s. The association received large cash infusions from local government agencies in the 1990s to sustain its operations. The GAP Parkway faded quickly after a feasibility study was completed. That project had little merit from a transportation perspective and would have promoted development in environmentally sensitive and hurricane-susceptible areas. The driving force behind the Fort Bend Parkway was Fort Bend County Commissioner's Court, not the transportation corporation.

The transportation corporation era had largely run its course by the 1990s, but it did produce one big accomplishment: the Grand Parkway. Although only one segment was constructed, the Grand Parkway Association kept the project alive and ultimately, with government agency and TxDOT financing, would continue to move the entire project forward.

Behind the Scenes

While Bob Lanier and Jon Lindsay publicly led the charge for an improved transportation system, many others worked behind the scenes to make the second wave happen. United States Congressional Representative Tom Delay, from Sugar Land southwest of Houston, provided key support at the federal level for Houston's highway construction program. The Harris County Toll Road Authority (HCTRA) was the pet project of Harris County Judge Jon Lindsay, who almost single-handedly brought the authority into existence with voter approval in 1983. However, HCTRA owes much of its operational success over the following 20 years to Wesley Freise, who served as executive director from 1985 to 2000. Freise negotiated agreements to provide innovative financing for the expansion of the toll road system, and he was instrumental in moving new projects forward, especially the Westpark Tollway. Under his management the authority consistently completed new construction projects on time and on budget. In recognition of his efforts, the Texas Transportation Institute inducted him into its Hall of Honor in April 2002.

Tollway dealmaker: Wesley Freise served as executive director of the Harris County Toll Road Authority (HCTRA) from 1985 to 2000. Freise is credited with the operational success of HCTRA and the negotiation of agreements that facilitated the expansion of the toll road system. His biggest deal was the transfer of ownership of the Beltway 8 Ship Channel Bridge to HCTRA in exchange for funding to help complete the south and southeast sections of the Sam Houston Tollway. (Photo: Tom Bailey Photography via HCTRA)

Further out of the public view are Houston's engineering consultant and construction firms. They have played a key political role in building Houston's infrastructure, particularly in the post-1980 era. Engineering firms doing business with Harris County are some of the largest contributors to political campaigns, particularly in county elections. The *Houston Chronicle* has reported on the close relationship between contractors and politicians, and critics' charges that the system lacks accountability. The issue of business contributions to politicians extended to all aspects of county contracting and was not limited to road and tollway construction. The subject has even arisen as a potential item for campaign reform, receiving the attention of county judge and road-building advocate Robert Eckels in 1999 when he publicly supported a law that would have banned contributions from architectural, engineering, legal, and other professional firms doing business with the county.[95]

In spite of critics' complaints, the system has successfully delivered needed transportation infrastructure to Harris County, and the outstanding record of the Harris County Toll Road Authority stands as a testament that the system has worked in the public interest.

The Unseen Power

Perhaps no good story is complete without an inconspicuous, intensely private figure who works behind the scenes, pulling strings and calling shots. Houston's freeways have such a character: James D. "Doug" Pitcock, the highly influential owner and president of Houston's leading highway contractor, Williams Brothers Construction.

Inspiring adulation, admiration, respect, fear, and scorn in Houston's construction industry, Pitcock (born 1928) has been one of the key players in Houston's freeway construction machine. Pitcock has been an active participant in building Houston's freeways longer than anyone else, joining Houston's construction industry in 1950 when plans for Houston's freeways were just tentative lines on a map. With the influence he gained through his connections and political contributions, he has been a valuable advocate for highway funding in Houston, statewide, and nationwide. He has served in dozens of high-level positions in industry associations at the local, state, and national level, including president of the Associated General Contractors of America in 1984. And perhaps most importantly, his construction firm has lowered the cost of freeway construction in Houston, allowing Houston to build more freeways with available funding.

After receiving a degree in civil engineering from Texas A&M in 1950, Pitcock went to work in Houston's construction industry and joined Williams Brothers when it was founded in 1955. Pitcock has run Williams Brothers since its founding and obtained full ownership of the firm in 1991. His legendary work habits—he worked 11.5-hour workdays plus weekends at the age of 70 in 1998—helped propel Williams Brothers to the dominant position in Houston's highway construction industry. The year 1991 also marked the beginning of a difficult period for Williams Brothers, with controversies and long delays engulfing the firm's projects on the Fred Hartman Bridge and the Eastex Freeway. But by the late 1990s, Williams Brothers had corrected its long-running tendency for com-

Freeway construction magnate: James D. "Doug" Pitcock, shown in a 1991 photo, has been practically invisible to the public but has been very influential in developing and building Houston's freeway system since the 1950s. His construction firm, Williams Brothers, dominates highway contracting in Houston and has lowered the cost of freeway construction, enabling Houston to build more freeways with available funds. (Photo: Williams Brothers Construction)

pleting jobs behind schedule. Williams Brothers became more dominant than ever, winning approximately 38% of Houston-area construction contracts greater than $10 million completed between 1987 and 2002. Pitcock took the contracts he wanted and let the rest of the industry divide what remained. In 2001 Williams Brothers received numerous awards for its rapid completion of an emergency bridge repair of the Queen Isabella Causeway in south Texas near Brownsville (see page 350), and in 2003 the firm was gaining a reputation as being Houston's fastest highway contractor. Officials uniformly praise the quality of Williams Brothers' work. In 1998 the first public light was shed on Pitcock in a comprehensive article written by the *Houston Chronicle*.[96]

How is Doug Pitcock able to consistently bid so low on Houston's freeway construction jobs? The 1998 article provides some answers. Pitcock maintains very tight control of his company. Some people say he is "strict"; others might say he rules with an iron fist. He maintains a small central office staff and keeps overhead low. His volume provides economies of scale, allowing Williams Brothers to own concrete, asphalt, and prefabrication plants around the city. Pitcock strictly enforces safety policies, keeping workman's compensation costs well below industry averages. He is very aggressive in front-loading his bids; that is, earning a huge profit on tasks early in a project while sustaining losses on work near the end of the project. This front-loading brings in cash flow during the early phases of the project cycle. Perhaps most importantly, Pitcock answers to no one. He can bid aggressively without having to worry about shareholders scrutinizing profit margins.

Pitcock is perhaps one of the last of a certain breed—some would say a dying breed—of Houstonians. They are the singularly focused, relentlessly dedicated individuals who are intent on achieving success and building something great in Houston. These individuals are the essence of the Houston spirit—a form of raw entrepreneurship and business savvy that found a home in a city where the deal was king and the future was limitless. A little cowboy blood seems to flow through the veins of these Houstonians. Houston's freeway system has had the good fortune of receiving the benefits of the hard work of many great Houstonians, such as Oscar Holcombe and Bob Lanier. Doug Pitcock, in his own quiet way, has helped build Houston. Although few Houstonians outside of the construction industry have ever heard of Doug Pitcock, they drive on his completed projects every day. The mystery man behind Houston's freeway system has been a big asset to Houston's freeway construction program.

The 1990s

During the 1990s the implementation of the freeway and tollway construction program continued. By 2003 the intent of the 1982 Regional Mobility Plan had largely been realized, although many details of the actual implementation differed from the 1982 plan. For example, no double-deck freeways were constructed.

With the completion of so many large expansion and construction projects in the 1980s and 1990s, and the substantial improvements to mobility in Houston, the early 1980s transportation crisis was a distant memory. But the crisis was the key factor that helped bring more state money to Houston. With the crisis over, the Texas Transportation Commission turned its attention to other pressing needs across the state of Texas. Most other regions of Texas could legitimately claim to have urgent, unmet transportation needs. The term of Houston advocate Bob Lanier on the Texas Transportation Commission ended in July 1987. History shows that regional funding in Texas follows cycles, so there was really only one direction for the funding to go. The result was a steady downward trend in Houston's share of state highway construction funding through the 1990s.

After a peak of 33% in 1983, Houston's share of construction funding went into a steady decline, falling to 13.3% in 2001. Houston's "fair share" of state transportation funding is generally regarded to be at least 22%, based on population, vehicle-miles travelled, and motor vehicle registration fees. Reversing the downward trend became one of the top priorities of local political officials. They presented Houston's case to the Texas Transportation Commission in April 2001 and September 2002, and the entire political establishment in Houston was organizing to restore Houston's funding. The downward trend in funding was reversed in 2003 in conjunction with the huge Katy Freeway expansion project.[97]

In the 1990s, the role of the Houston Chamber of Commerce as a lobbying organization for Houston-area transportation improvements was greatly diminished. The chamber's influence and power had always been directly related to the strength and influence of the business community. With the collapse of the Houston economy between 1986 and 1989, Houston business interests were severely weakened. Many of the key supporters of the Chamber of Commerce—real estate, land development, and banking interests—were forced into bankruptcy or disappeared entirely. The business community that had so strongly promoted Houston for the previous 50 years was struggling for its own survival. In 1989 the Chamber of Commerce joined with the Houston World Trade Association and the Houston Economic Development Council to form the Greater Houston Partnership. The Clean Air Act revisions of 1990 and the Intermodal Surface Transportation Efficiency Act of 1991 made transportation planning a highly complex and bureaucratic task. A detailed vision of Houston's transportation future like the 1982 Regional Mobility Plan was no longer within the scope of the Greater Houston Partnership. Transportation planning responsibility was more firmly entrenched in the transportation department of the regional metropolitan planning organization—the Houston-Galveston Area Council (HGAC). Political leaders ensured a pro-mobility, pro-freeway policy for the HGAC and its key policy-making committee, the Transportation Policy Council (TPC). Jimmy Schindewolf, an advisor of Houston mayor and highway advocate Bob Lanier, served as chairman of the TPC from 1994 to 1998. Harris County Judge Robert Eckels took the chairman position in 1998. The responsibility for lobbying shifted to individual political leaders. In the 1990s, Eckels was the most vocal and influential advocate of transportation investments in Houston.

The 1990s saw the emergence of the Harris County Toll Road Authority (HCTRA) as a major force in Houston-area highway construction. Rapidly increasing patronage of HCTRA's tollways by Houston motorists provided a strong revenue stream that easily covered bond payments and provided financing for tollway expansions and new facilities. Through the 1990s tollway patronage doubled every five years, reaching 792,000 vehicle-transactions per day in 2002. The financial strength of HCTRA in the mid to late 1990s allowed it to move forward with advance planning and construction for several projects. The South Sam Houston Tollway was completed in 1997, the Hardy Toll Road Airport Connector opened in 2000, and capacity improvements to the Hardy Toll Road were underway by the late 1990s. The Westpark Tollway began construction in 2001, and work was underway in 2002 to expand the Sam Houston Tollway to eight lanes from the Southwest Freeway to the North Freeway. Other projects in the advanced planning phase included the Harris County section of the Fort Bend Parkway, the Hardy Toll Road extension into downtown Houston, and a tollway on the Katy Freeway corridor. In June 2001, HCTRA released a

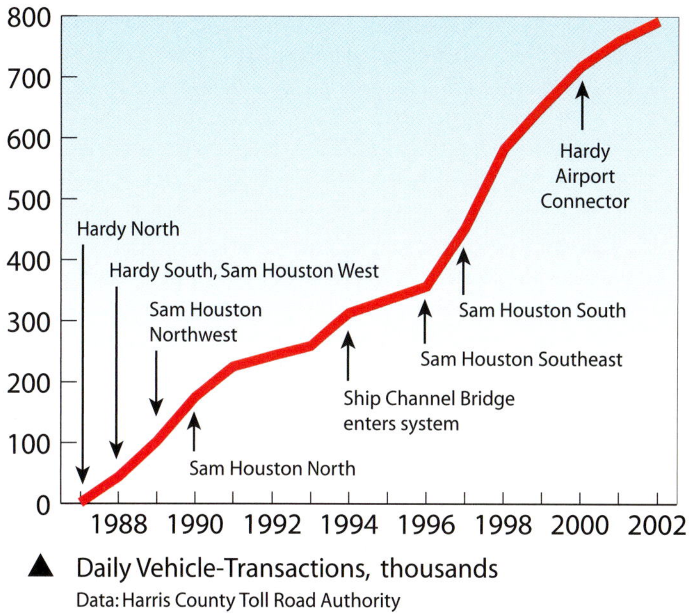

Phenomenal success of Houston's tollways: This plot shows the increase in vehicle-transactions on the Harris County toll road system since the first tollway opened in 1987. This success has positioned the Harris County Toll Road Authority to be a leader in building the next generation of Houston's transportation infrastructure.

map showing additional potential tollway projects for the long-term future—the "pooled projects" map.[98]

The 1990s brought continued improvement and expansion of the transitway system on Houston's freeways, generally in conjunction with major freeway expansion and reconstruction projects. Houston Mayor Lee Brown, successor to Mayor Bob Lanier, was able to shift the priorities of the Metropolitan Transit Authority toward a light rail construction program. Houston's first light rail line on Main Street began construction in 2001 and is scheduled for operation in 2004. Future political events will determine if light rail will become the future focus of the Metropolitan Transit Authority, rather than the transitway system.

Assessing Houston's Freeway System: A Critique

As the calendar reached 2003, it marked the 50th anniversary of the unveiling of Houston's first comprehensive freeway plan. The group of civic leaders who traveled to Austin in July 1953 to present the plan to the Texas Transportation Commission brought with them a blueprint for what would become the best loop and radial freeway system in the United States. But it was only the beginning of planning efforts that would continue for the next 50 years. So some questions naturally arise: How good is Houston's freeway system? What was done right? Where did Houston's freeways go wrong? How could Houston's freeways have been better?

The Houston freeway system report card will answer those questions. Giving a grade to a freeway system is no simple task. During the past 50 years, the American city has continuously evolved, transportation patterns have shifted, political standing for highway construction has had its ups and downs, and the American economy has been transformed. Houston experienced one of the nation's greatest booms of the 20th century—and a severe bust. All the while, the freeway system was struggling to make progress. Tracing the progress over this long period of time is a lot like following a student from preschool through college. So the report card will focus on key categories and distinct periods of time, tally up the good and the bad, and assign a grade. It is a subjective approach, taking into account the ability of the freeway system to meet Houston's needs, the vision and innovation of Houston's efforts, and the quality of Houston's efforts in comparison to other cities. In the end, a final grade is assigned.

Report Card Methodology

Rating a freeway system can be a highly complex task if formal, computed measures are used to compare freeway systems in different cities and assess the ability of a freeway system to meet a region's needs. The Texas Transportation Institute's annual *Urban Mobility Report* uses a methodology of computed measures to rate urban traffic congestion, but the multitude of complex metrics can cause a casual reader to lose sight of the big picture.

Instead, a subjective approach will be used. In assigning the grades, three factors are considered: the ability of Houston's freeway system to meet the region's needs, the amount of vision and innovation demonstrated in the particular category, and the performance of Houston in comparison to other cities. The grades are nonscientific and are the opinion of the author. The following symbols are used to rate the performance in specific categories.

✓	Strength
✘	Weakness
↔	Mixed result

Planning

Over the last half century Houston's planning efforts are rated to be just slightly above average—a grade of C+. There have been periods of both strong and weak performance. The original formulation of the freeway plan was a period of strength for freeway planning. The City of Houston Planning Department under Ralph Ellifrit and the City Planning Commission deserve most of the credit for the strong planning during this critical early period.

Below-average planning performance occurred from 1955 to 1970 when Houston's freeway planning was poorly adapted to trends that became evident, especially the shift in growth to west and north Houston and the need for a more gridlike freeway pattern to serve suburb-to-suburb commute patterns. The city of Houston abdicated its leadership in freeway planning in the mid-1960s and would go on to practically abandon its freeway advocacy. Many major activity centers in Houston, such as the Texas Medical Center and the Uptown Houston District along the West Loop, do not have well-planned access to freeways. Improved access to major activity centers could have been accomplished with better interchanges and short spur freeways, but this kind of connectivity has historically not been part of Houston's freeway planning. During this period other cities, such as Los Angeles and Dallas-Fort Worth, were developing more aggressive plans.

Houston exhibited above-average planning through the 1970s crisis and the 1980s freeway resurgence. The freeway plan was only minimally damaged by the 1970s crisis, and in the 1980s new freeway routes were added. The transitway system was born and greatly expanded during this period. These strengths were offset by a continued adherence to the outdated loop and radial model, insufficient planning to relieve downtown bottlenecks, and the greatest planning failure in the history of Houston's freeway system—the cancellation of plans to widen the West Loop in 1992.

Houston finished the most recent period in freeway planning on a high note with one of the best planning efforts in the United States. With TxDOT in the lead, a series of corridor major investment studies defined the future of numerous Houston freeways, with managed-lane freeway designs being planned for two corridors. HCTRA has also become a regional planning leader with its toll road planning efforts.

The 1996 to 2005 era will likely mark the high point of TxDOT influence in planning. In 2003 the political leadership of Texas is attempting to shift both planning and financing responsibility for urban transportation infrastructure to the local level. In the future, planning leadership will need to come from Harris County, HCTRA, adjacent counties, and suburban areas.

Over the entire 50-year period there has been a trend in transportation planning in Houston: an increasing reliance on freeways. For the most part, this trend has benefits for an urban setting like Houston. In the case of mass transit, it was a logical and economical strategy to incorporate transit services into freeways. Freeways are efficient movers of traffic, and keeping vehicles on freeways and off local streets can be viewed as a plus. Freeways are very cost-effective in terms of the number of daily trips they serve.

Still, the focus on freeways has caused one element of Houston's transportation planning to be neglected: the regional arterial street network. Getting to or from the freeway is often a large part of any travel trip in Houston. In addition, as Houston continues to grow, new development will be located further from freeways. Houston's arterial street network is at best mediocre, and little progress is being made to improve it. Travel through Houston on arterial streets is generally not efficient. Houston can look to numerous other cities, such as Dallas-Fort Worth, Phoenix (Ariz.), and South Orange County (Calif.), to see how arterial street networks can be better integrated into a freeway system. The lack of a well-planned regional arterial street network has forced more traffic onto Houston's freeways, increasing the reliance on freeways.

To a certain extent, local governments have been able to shirk their arterial street responsibility because of the

Report Card: Freeway Planning in Houston

Grade	Period		Assessment
B	1940-1955 The Original Plan	✓	The loop and radial system was a logical choice based on downtown-focused urban development patterns in Houston at the time.
		↔	Development of the freeway plan was somewhat behind other major cities, but the quality of the final plan was not affected.
		↔	The original plan included an extensive downtown interchange complex with excellent access to downtown. In the long run, the downtown interchange would need to predominantly serve through-traffic, not traffic originating or terminating downtown.
D+	1955-1970 Refinement	✓	Recognition of suburb-to-suburb commuting patterns with the addition of the Beltway as a freeway in 1960 and the Grand Parkway in 1965.
		↔	Houston's plan was not as aggressive as other cities such as Dallas-Fort Worth and Los Angeles, but it remained a plan that could be reasonably constructed with available resources.
		✗	During this period the need for a more gridlike freeway pattern for suburban areas should have been recognized. Suburban freeway planning should have been adapted to a grid pattern, including the addition of a freeway on or near SH 6-FM 1960.
		✗	New freeways were added to south and southeast Houston rather than to the areas where growth was shifting to—west and north Houston.
		✗	Additional radial freeways should have been added to serve future growth areas: Westpark corridor, the present-day SH 249 corridor, US 90A corridor, and FM 529.
		✗	Spur routes should have been planned to provide connectivity to major activity centers such as the Texas Medical Center, Uptown-West Loop, and NASA.
B	1970-1980 Crisis Management	✓	Beltway 8 is saved, primarily by Harris County.
		✓	Transitway development begins.
		↔	Damage to the freeway master plan is not as extensive as in other cities.
		✗	The Grand Parkway, Fort Bend Parkway, and Harrisburg Freeway are cancelled. The future need for the Grand Parkway and Fort Bend Parkway remained, and those routes should have been retained.
		✗	The city of Houston abandons freeway advocacy and reduces its commitment to the regional arterial street plan.
B	1980-1995 Recovery	✓	One of the strongest recoveries in terms of freeway and tollway planning in the U.S.
		✓	The Grand Parkway and Fort Bend Parkway are restored to long-term plans.
		✓	New facilities are added: Hardy Toll Road and Tomball Parkway (SH 249).
		↔	Although some progress was made, there is still no regional commitment to a high-quality arterial system to complement the freeway system.
		↔	Regional planning moves toward a greater reliance on freeway frontage roads.
		✗	The last chance to adopt a more gridlike pattern for suburban areas is lost.
		✗	Cancellation of plans to expand the West Loop in 1992 is the greatest planning setback in the history of Houston's freeway program.
		✗	Insufficient planning to relieve downtown bottlenecks.
A-	1996-2005 Planning for Growth	✓	TxDOT leads planning efforts with major investment studies of several freeway corridors. Plans include enhanced integration of tollways and transit into freeways.
		✓	HCTRA plans future tollways to meet regional needs.
		✓	New tollways are added: Westpark Tollway, Fort Bend Parkway, Hardy Toll Road airport connector and downtown extension.
		✗	Lack of a strong regional arterial plan causes increased reliance on freeways. Still no planning to reduce adherence to the loop and radial system configuration.
C+	Overall Grade for Houston's Freeway Planning		

The Loop and Radial System: Did Houston Take the Wrong Fork in the Road?

When the creators of Houston's freeway system were putting routes on the freeway planning maps in the 1940s and early 1950s, the American city was a very different place than it is today. Cities were focused on a vibrant core downtown, and suburban employment centers as we know them today didn't exist. This was particularly true in Houston since there were no large centers of activity competing with downtown. Galveston, 47 miles (75 km) to the south, did not influence freeway plans in the Houston area. A freeway system focused on downtown seemed logical for Houston, and planners developed a loop and radial freeway system for Houston. The radial freeways were designed primarily to serve commuter movements to and from downtown, and the loops served traffic that was crossing or going through the city.

Around the United States, many other cities—especially those with no geographic constraints—also planned freeway systems using the loop and radial concept. Two cities similar to Houston in population and geography, Atlanta and Washington, D.C., both adopted loop and radial systems. Many other cities adopted modified forms of the loop and radial system. In 1965 the *Houston Chronicle* reported on Houston's freeway plan, "Houston's freeways are of the loop and radial pattern [which are] considered by experts to be ideal."

The loop and radial system may have been ideal in the early freeway era, but it would spawn new development patterns and a transportation demand it could not efficiently serve: the suburb-to-suburb commute. Job migration to the suburbs was already underway by 1965 and would accelerate greatly afterwards. This was particularly true in Houston, with the development of large employment centers outside downtown getting underway in the late 1960s. These suburban employment centers would later become known as "edge cities," a term popularized by the 1992 book, *Edge City: Life on the New Frontier*. With the decentralization of employment, fewer and fewer trips originated and terminated downtown. Increasingly, commuters passed through downtown on their way from one suburb to another, or they used the loops to make the suburb-to-suburb commute.

The loop and radial system was having to serve transportation patterns it wasn't designed for, and in most cases, it broke down. Downtown interchange complexes couldn't handle the through-traffic load and became traffic choke points. Houston and Dallas both exemplified this problem. The loops and connections to the loops were receiving far more traffic than they were designed for. Sections of loops in Houston, Atlanta, Washington, D.C., and Dallas became legendary for traffic congestion. The loop and radial system was no longer ideal.

But the loop and radial system wasn't the only game in town. Another freeway system design that appeared in several cities in the early freeway era, the grid system, would prove to be far more ideal in serving decentralized travel patterns that would ultimately develop in nearly all cities. Grid systems provide a more direct travel route for the suburb-to-suburb commute, reducing the total vehicle miles travelled in a region. The system is less susceptible to choke points downtown and along loops. Grid systems also provide the key characteristic of network redundancy. If one link goes out due to an accident or construction, traffic can easily shift to efficient alternate routes. The cities that selected the grid system, or modified forms of it, generally already had a certain amount of decentralization in the 1950s. In metropolitan Los Angeles, cities were spread out across the L. A. basin and San Fernando Valley, largely the result of the Pacific Electric Railway and beach communities. Many twin-city metropolitan areas, such as Dallas-Fort Worth, Minneapolis-St. Paul, and New York-New Jersey, were more logically suited for a grid system or modified form thereof. Houston had no such luck.

Did Houston take the wrong fork in the road by selecting the loop and radial system? Yes. Was it reasonable or politically feasible for a city like Houston to adopt a grid system in the 1950s? No. After all, when Houston's core freeway network was defined in the early 1950s, the urbanized area of Houston was generally contained within the present-day Loop 610. The idea of building a grid system in the vast areas around Houston to serve edge cities of the future would have been a tough sell. Houston's influential, downtown-based business establishment surely wanted the freeway system to be focused on their real estate.

But when the impact of decentralized transportation patterns became evident in the 1960s, Houston could have adapted its system to be more gridlike in suburban areas. Yet Houston continued to adhere to the loop and radial model and did almost nothing to evolve the system toward a more gridlike pattern. The addition of a freeway on the SH 6-FM 1960 corridor would have greatly contributed to gridlike performance in suburban areas, but was never seriously considered. As time passed, increased suburban development closed the window of opportunity for constructing a suburban grid. Even in 2003, Houston's freeway planning still clings to the outdated loop and radial model.

Of course, Houston has responded to the need for suburb-to-suburb transportation capacity by the construction of two loops and active planning for a third. In terms of loop development, Houston is the national leader and is far ahead of similar cities like Atlanta and Washington, D.C. But the inherent flaws and inefficiency of the loop and radial system cannot be erased. The end result is that Houston will need more, wider freeways to meet its transportation needs.

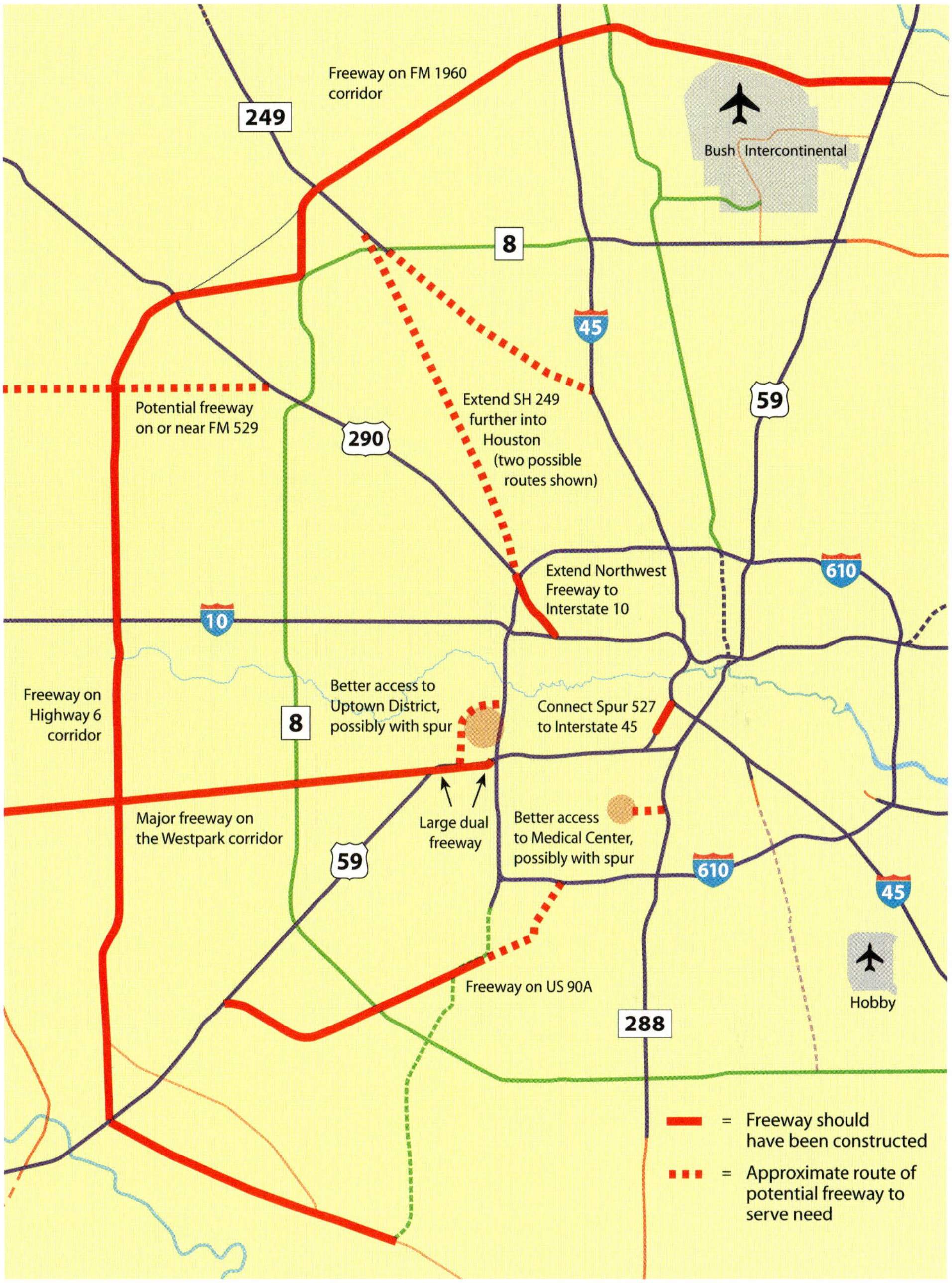

How Houston's freeway system could have been better: This map shows some ways Houston's freeway system could have been better planned to meet modern-day needs. All these potential improvements were realistic and feasible if they had been planned at the proper time. Some of these routes were included in preliminary plans for Houston's freeway system in the early 1950s but were dropped from the final plan. Other enhancements likely were considered by planning officials. Financial limitations surely played a large role in curtailing plans for Houston's freeways. Also, freeways are often added to the regional plan as a result of lobbying by politically powerful groups. If a freeway did not have a powerful government or well-connected private entity behind it, it had less chance of becoming reality.

widespread presence of frontage roads. It is easier and less expensive for localities to rely on frontage roads than to fund and build high-quality arterial streets. Localities also want the numerous other benefits of frontage roads, including better freeway accessibility for motorists, business accessibility, and commercial buffers between freeways and residential areas. The net result is that Houston freeways have become not only transportation corridors, but also commercial strips, local traffic collector-distributors, and economic development tools. In Houston the freeway is required to do more than probably any other place in the world.

Freeway Design

Once the planning is done, design comes next. Houston earns a B for its freeway design efforts over the last 50 years. Houston's pre-1960 freeways exhibited many design deficiencies but were typical of all freeways in the United States, so Houston is rated an average performer in freeway design in the 1945–1960 period. The adoption of a de facto policy for the pervasive use of frontage roads was the most significant development during this era. Frontage road development is counted as a plus for Houston's freeway design due to increased access and connectivity, although it would go on to have unintended consequences in urban growth and freeway performance.

For the era of freeway construction that built most of Houston's core freeway system from 1960 to 1985, Houston receives a C+. There are both great strengths and great weaknesses during this period. Some of Houston's freeways exhibit design characteristics that are among the best in the United States. The South Freeway corridor is among the best-designed freeways in the nation; the downtown interchange complex is one of the most

Report Card: Designing Houston's Freeways

Grade	Period	Mark	Assessment
C+	1945-1960 Early Freeways	✓	Frontage roads become a standard design feature of Houston's freeways.
		↔	Houston's first freeways have many shortcomings, but so do freeways everywhere else. Houston's freeways are typical for the era.
C+	1960-1985 Construction of the Original System	✓	Many freeway segments are built to design standards that are comparable to the best freeways in the nation. The South Freeway is one of the best-designed in the U.S.
		✓	Houston is a leader in the implementation of highway safety features.
		✓	Multilevel stack interchange design is standard at freeway-to-freeway interchanges.
		↔	One of the best downtown interchange complexes in the United States in terms of traffic originating or terminating downtown, the downtown interchange complex would later serve mostly through-traffic and was not adequately designed for that task.
		↔	Frontage roads become pervasive, promoting greater access but also having unintended consequences, such as heavy commercial development along freeways.
		✗	Insufficient capacity on most freeways, including severe underdesign of the radial freeways just outside Loop 610 and insufficient capacity on the Pierce Elevated (Interstate 45) downtown.
		✗	Insufficient right-of-way is acquired on most radial freeways outside Loop 610.
		✗	Poor lane balance on freeways, causing traffic-flow and safety problems.
		✗	Frequent overpasses on main lanes on many freeways cause a roller-coaster effect, lowering traffic-flow potential and reducing driver sight lines.
		✗	Poorly designed interchanges at key freeway-to-street intersections, including West Loop intersections and South Loop near the stadium complex.
A-	1985-2003 Reconstruction and Expansion	✓	The extensive, successful, and cost-effective transitway system is integrated into the freeway system.
		✓	Most of Houston's radial freeways are modernized and expanded to high design standards.
		✓	The five-level stack becomes standard at freeway-to-freeway interchanges.
		✓	Faults in original construction, such as poor lane balance, are corrected. Houston becomes the leader in high mast freeway lighting.
		✓	The managed-lane freeway becomes part of freeway design in the late 1990s.
		✓	Freeway durability improved, with 15-inch-thick (38 cm) concrete the standard.
		↔	Standards for frontage roads are improved, but reliance on frontage roads for local distribution continues.
		↔	Frequent overpasses with up-and-down grades remain typical of Houston's freeways, although geometrics are improved over original construction.
		✗	Some reconstructed freeways still have insufficient capacity.
B	Overall Grade for Houston's Freeway Design		

extensive and modern in the United States. Multilevel stack interchanges are standard at freeway-to-freeway intersections, and Houston has always been a leader in the implementation of safety features.

But there was a dichotomy in Houston's freeway design largely resulting from the division of responsibility for freeway design. The Houston Urban Project Office, which existed from 1945 to 1984, was responsible for Loop 610 and the freeways inside the loop. This office typically produced top-quality freeway designs, often among the best in the United States. Outside Loop 610, TxDOT District 12 was responsible for freeway design. The original construction of freeways outside Loop 610 exhibited below-average design characteristics, especially in terms of freeway traffic-volume capacity and right-of-way planning. Poor lane balance was another characteristic of freeways constructed in this period, particularly at intersections with Loop 610. Lane balance is the use of auxiliary lanes and merging lanes to allow an orderly flow of traffic in transition zones. In practical terms, it means that when lanes of traffic enter a freeway, there should be plenty of distance to allow traffic in the new lane to merge

You Can Land a Jumbo Jet on It

As Houston's freeway system has expanded over the years, the growth has been paralleled by the increase in freeway pavement thickness. Houston's original freeways were typically constructed with 8-inch-thick (20 cm) concrete. But a lot has changed since the 1960s. Traffic volumes are much larger and include more trucks, and trucks have become heavier with tires inflated to higher pressures. In addition, the ground beneath Houston has some of the worst soil for serving as a base for freeways—"expansive" clays that shrink, expand, and shift depending on moisture conditions. Ongoing research at the Texas Transportation Institute has identified strategies for improving the longevity of Houston freeway pavement.

In 2003 Houston freeways are built with 15-inch-thick (38 cm) concrete. How thick is 15 inches? Heavy duty runways at the busiest hub airports in the United States generally have concrete thickness in the range of 15 to 18 inches (38 to 46 cm). Depending on conditions at particular locations, thicknesses can exceed 20 inches (51 cm) on some runways. A new 9,400-foot (2,865 m) runway at Houston's Bush Intercontinental Airport is paved with 17-inch-thick (43 cm) concrete, and other runways at Bush Airport range between 17 and 19 inches (43 to 48 cm) thick. Runways at the Dallas-Fort Worth International Airport are typically 17 to 18 inches (43 to 46 cm) thick.

with the existing lanes. Poor lane balance was the cause of many congestion and safety problems, and remains a problem on freeways that have not been modernized. Design standards for Houston's freeways built up to the 1970s didn't call for lane balance, but a little common sense would have gone a long way in improving freeway performance.

In terms of freeway design, the benchmark for high quality in the 1960–1985 era is California. California's urban freeways typically have long sight lines due to smooth grades, frequent use of depressed and elevated sections through urban areas, and high traffic volume capacity. Some Houston freeways—especially inside Loop 610—met California standards, but most did not. Most Houston freeways were constructed at ground level with elevated overpasses at intersections. This contributes to a "roller-coaster" effect, causing poor sight lines and reduced traffic-carrying potential. Houston's geography has made below-grade freeway construction less of an option than it was in California or other regions. The perfectly flat, low-lying coastal plain of the Houston region makes depressed freeways particularly susceptible to flooding. Large pump stations are required to drain depressed freeways, and flooding is an all-too-frequent occurrence when heavy rainfall occurs. Still, sight lines could have been improved by raising intersecting streets over freeways more frequently.

Houston adopted new design standards with the freeway expansion and reconstruction program that began in the 1980s. The new design standards corrected most of the shortcomings of the original construction of Houston's freeways. Capacity was added, lane balance and geometrics were improved, freeway-to-freeway interchange design was advanced to the five-level stack, and transitways were incorporated into freeway design. Frontage road design, capacity, and service levels were improved, generally a plus but further contributing to a heavy reliance on frontage roads. With the ever-present financial, regulatory, and political limits to freeway construction, Houston did an excellent job of building modern freeways with high design standards. The 1980s freeway standard does not meet standards of the top quality designs such as the South Freeway, but some freeways, such as the reconstructed Eastex Freeway, rank among the better-designed in the United States.

In the late 1990s Houston began planning for its first managed-lane freeways. Managed lanes are special freeway lanes for use by transit-oriented vehicles (buses, vanpools, and carpools) and toll-paying single-occupant vehicles. The lanes are managed to maintain free-flow traffic conditions. The future use of managed-lane freeways will keep Houston in the forefront of freeway design.

Freeway Political Leadership

Pro-freeway political leadership is critical to the implementation of a freeway construction program. Houston has seen periods of strong and weak leadership. Over the last 50 years Houston receives a B for the efforts of its political leadership in building the freeway system.

The city of Houston was a very strong leader in the early freeway era from 1940 to 1960. Mayor Oscar Holcombe, the City Planning Department, and the City Planning Commission propelled the city of Houston to its leading role in promoting the freeway system. The

Report Card: Freeway Leadership and Politics

Grade	Period		Assessment
B+	1940-1960 Building the System	✓	Mayor Oscar Holcombe was a strong freeway advocate.
		✓	The Houston Chamber of Commerce provides key leadership.
		✓	Strong political partnerships exist to pass freeway bond issues.
		✓	The city of Houston shows strong leadership. The City Planning Department under Ralph Ellifrit and the City Planning Commission are most influential in the early development of the freeway system.
D+	1960-1975 Decline	✓	TxDOT remains proactive in regional transportation planning.
		✓	Political leaders support freeway construction through the 1960s and continue to seek TxDOT acceptance of new routes.
		✗	The city of Houston abdicates its leadership role in 1964 and by 1975 it abandons advocacy of freeways and reduces its commitment to the arterial street system.
		✗	Starting in the late 1960s, real (inflation-adjusted) funding for highway construction drops dramatically at local, state, and federal levels of government.
		✗	Many politicians, particularly those representing inner cities and environmental interests, are hostile or indifferent to freeways in the early 1970s.
A	1975-1995 Strong Resurgence	✓	An era of great leadership, led by Bob Lanier and Jon Lindsay.
		✓	Other leaders work in a less visible role, including Congressman Tom Delay and Doug Pitcock of Williams Brothers Construction.
		✓	Jon Lindsay creates the Harris County Toll Road Authority.
		✓	Houston Chamber of Commerce develops the Regional Mobility Plan in 1982 and leads the implementation of the plan.
		✓	Dramatic funding increases at local and state levels
		✓	Strong regional interagency cooperation began with the North Freeway contraflow lane in the 1970s and continued through the entire period.
B	1996-2003 Steady Leadership	✓	Harris County Judge Robert Eckels leads regional transportation political efforts.
		✓	Interagency cooperation continues and expands with the participation of HCTRA.
		↔	The city of Houston remains indifferent on highway leadership issues during the tenure of Mayor Lee Brown but is supportive of regional planning efforts.
B	Overall Grade for Houston's Freeway Political Leadership		

Houston Chamber of Commerce also provided key leadership in the 1950s. The role of the city of Houston began to decline around 1964 with the administration of Mayor Louie Welch. The period from about 1965 to 1975 was an era of below-average leadership. No person or agency stepped forward to fill the vacuum left by the city of Houston.

Starting in 1975, a period of great freeway leadership began. Harris County Judge Jon Lindsay entered office in January 1975 and began to reverse the decline in Houston's freeway system. His creation of the Harris County Toll Road Authority in 1983 was an event of monumental importance for the future transportation infrastructure of Houston. Houstonian Bob Lanier was appointed to the Texas Transportation Commission in 1983 and led efforts during the most influential era for highway funding in Texas since the 1950s. The Houston Chamber of Commerce played a key role in the formulation and implementation of the 1982 Regional Mobility Plan. Regional cooperation among government agencies helped get things done, especially in regard to the regional transitway system.

The era of great leadership came to an end around 1995. In the following years some decline in Houston's freeway program occurred, but the efforts of Harris County Judge Robert Eckels and the proactive planning efforts of the TxDOT Houston office kept the period above-average in terms of leadership. In 2002 and 2003 there were signs of a potential strengthening in freeway leadership due to the increasing traffic problems in the Houston region. Still, it will be difficult, if not impossible, to repeat the leadership of the 1975-1995 period.

Other Criteria for Grading Houston's Freeways

One area in which Houston has been a national leader is in the implementation of transportation plans. Virtually all of Houston's originally planned freeways will be

Report Card: Getting the Freeways Built

Grade		
A	✓	Houston will construct nearly all of its originally planned freeways, something few cities have been able to achieve.
	✓	Regional interagency cooperation has been key to implementation, particularly in the post-1975 era.
	✓	Innovative financing and toll financing have played a key role in building out the freeway system.
	✓	Houston freeways have been relatively inexpensive to construct, mainly due to a favorable construction environment in the state of Texas and also due to a competitive contractor environment in Houston.
	↔	Long delays have occurred in the construction and expansion of needed facilities. However, long delays are typical everywhere in the United States.

constructed, an accomplishment that few large cities have been able to achieve. Houston's recovery from the 1970s highway construction crisis was one of the strongest in the nation. The impact of setbacks which occurred in the 1970s was reversed in the 1980s and 1990s with the restoration of cancelled routes to the master plan, increased funding, and tollway construction. The only major setback of the 1970s that was not reversed—the cancellation of the Harrisburg Freeway—did not have a big impact on the performance of the freeway system. The greatest setback to Houston's freeway system was the failure of plans to expand the West Loop in 1992. It is indicative of the success of Houston's implementation that the inability to expand an existing eight-lane freeway is the greatest setback to the system.

Houston has also excelled in the economical construction of freeways, providing an outstanding value for taxpayers. The low cost of freeway construction is largely a statewide phenomenon in Texas but has been especially true in Houston. Several factors contribute to the good value for taxpayers. Texas has a large highway construction program that supports a large and competitive contractor environment. For example, the large demand for precast bridge beams and retaining wall segments keeps multiple casting yards in business, fostering price competition. Houston's Williams Brothers Construction has been one of the most competitive construction bidders in Texas. Houston and Texas can use economical construction methods with precast components for bridges. California, for example, must use expensive cast-in-place concrete for its bridges due to the risk of earthquakes. Texas is a right-to-work (non-union) state, which helps keep labor costs reasonable. TxDOT staff and researchers at the Texas Transportation Institute are continuously working to get better material performance and improved safety for less cost. Property values in Texas are generally lower than national averages, making the acquisition of right-of-way more affordable than most places in the United States. Houston and Texas have always had a low to moderate level of taxation in comparison to most of the United States, so the low cost of construction has been especially crucial for the implementation of the freeway construction program.

In terms of operation and intelligent transportation systems, Houston has lagged somewhat behind other cities in the development of a traffic management system. After a closed-circuit television research program on the Gulf Freeway in the 1960s, there was little progress until the Houston Transtar Operations Center took shape in the 1990s. By 2003 Houston Transtar was recognized by the Federal Highway Administration as one of the best in the nation, receiving national awards in 2001 and 2002 for having one of the top traveler information web sites.

The Final Grade

When the final report card is opened, Houston receives a B for its freeway system. Houston has a good system, but it could have been better.

It should be noted that probably no metropolitan area in the United States would receive an A by the criteria used to rate Houston. Other regions have excelled during certain periods in the last 50 years, but practically no region got everything right for the entire period. Perhaps Detroit, Michigan, is the best candidate for an A, but as one of the slowest-growing metropolitan areas in the United States in the last 50 years, Detroit's transportation system has been minimally challenged and really cannot be compared to the high growth metropolitan areas in the South and West.

The Final Grade

B

The greatest setback to Houston's freeway system:
Cancellation of plans to expand the West Loop (1992)

A Tale of Three Cities

Houston was not the only city developing big freeway plans up until the first freeway planning peak around 1970. Two urban areas comparable to Houston, Los Angeles and Dallas-Fort Worth, were also thinking big, aggressively planning dense freeway networks that included far more centerline miles than Houston. This was to be expected for Los Angeles, since its metropolitan area has about four times as many people as Houston. But the populations of Houston and Dallas-Fort Worth are very similar, so Dallas-Fort Worth could be credited with winning the battle

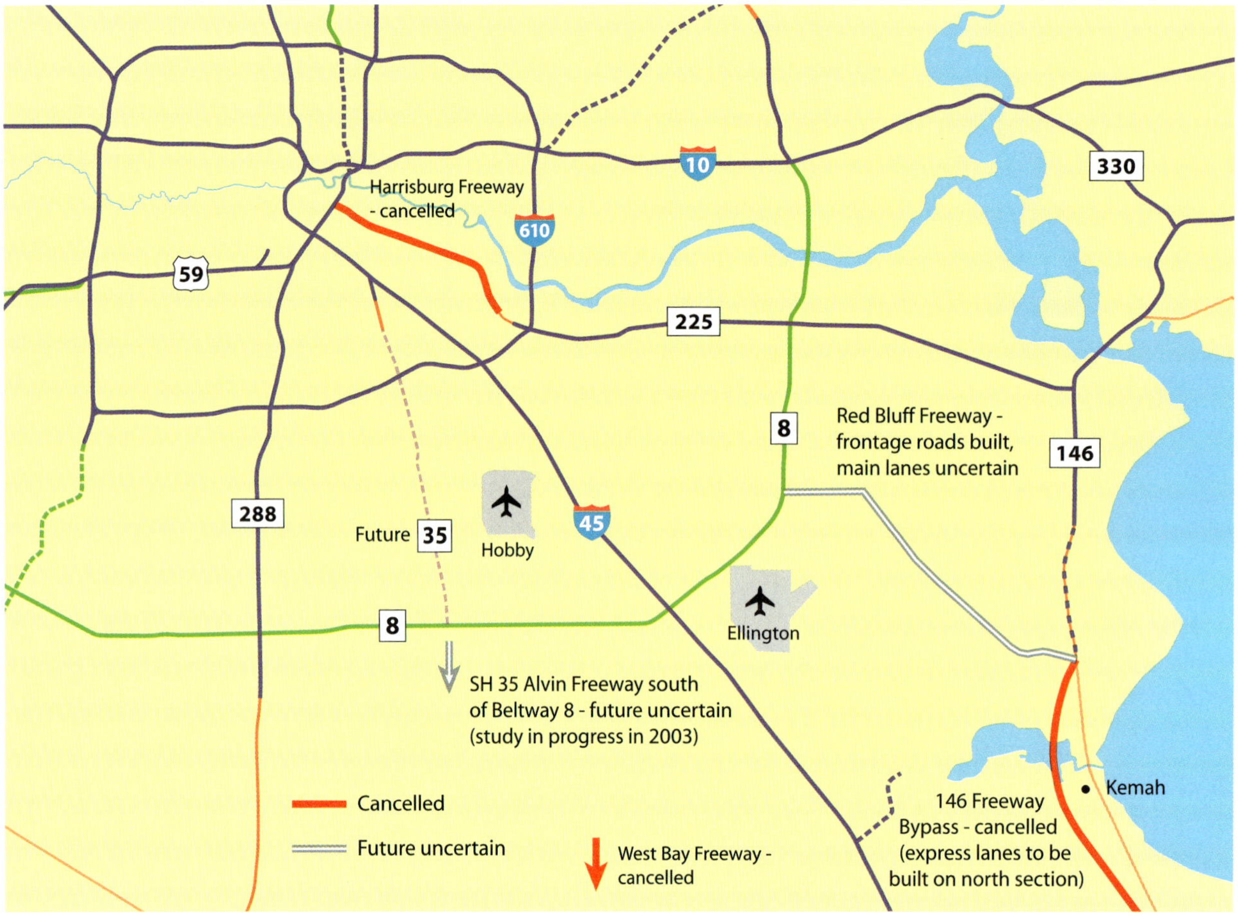

Houston's cancelled freeways: Houston's freeway plan has survived nearly entirely intact with only one cancelled section of freeway in the immediate Houston area, the Harrisburg Freeway. Southeast of Houston, the SH 146 Freeway Kemah bypass and West Bay Freeways were cancelled. Probably all remaining unbuilt routes will be constructed in the future.

of long-range freeway planning as of 1970.

Houston, however, distinguished itself in another way. When all is said and done, Houston will actually construct nearly all of its originally contemplated freeway network. Based on 2003 plans, Houston will sustain only one partial freeway cancellation within the immediate Houston vicinity: the Harrisburg Freeway section of the La Porte Freeway. The freeway plans of Los Angeles and Dallas, in contrast, didn't fare so well after 1970. Both plans were decimated by the 1970s antifreeway backlash and funding crisis, with planned freeways getting wiped off the map like a wholesale liquidation. Houston's plan was realistic—meeting the region's needs without being overly ambitious or disruptive. It stood the test of time, perhaps the greatest test of all.

A Good Deal for the Texas Taxpayer

Houston's freeways have been relatively inexpensive to construct, especially compared to California. Here's why:

- A large highway program in Texas, supporting a broad and competitive contractor and supplier base
- Construction methods with heavy use of precast components, especially for bridges
- Labor laws in Texas—a right-to-work (non-union) state with reasonable labor costs
- Ongoing research at TxDOT and the Texas Transportation Institute for better value in freeway construction
- Relatively low property values in Texas, making right-of-way acquisition affordable

Dallas-Fort Worth

250 miles (400 km) north of Houston is Houston's cross-state rival, the Dallas-Fort Worth metroplex.* In some ways, Houston and Dallas are like close Texas siblings. In other ways, they may as well be on different planets. As one adage says, Dallas is champagne, caviar, and BMWs, while Houston is beer, barbeque, and pickup trucks. But when it came to freeways, Houston and Dallas certainly kept an eye on each other, especially in the 1950s. In 1953 Dallas interests became keenly aware that Houston was pulling ahead in terms of freeway planning.

* The term metroplex refers to the entire Dallas-Fort Worth metropolitan area.

Dallas-Fort Worth Cancelled Freeways

A	Fort Worth, south east-west freeway
B	SH 199 Freeway (later reinstated but cancelled again in 2000)
C	Northside freeway (later reinstated but cancelled again in 2000)
D	River Freeway (now known as the Trinity Parkway) connection to Interstate 30
E	Oak Cliff east-west freeway
F	Oak Cliff north-south freeway
G	Garland north-south freeway
H	North Dallas east-west freeway
I	East Dallas north-south freeway
J	Interstate 635 to Loop 9 connection
K	Interstate 635 extension east freeway
L	Plano Loop

Dallas-Fort Worth freeway casualties: The Dallas-Fort Worth regional freeway plan was updated in 1967 to include a dense inner-city grid to accommodate forecasts that future travel patterns would largely bypass the central business district of Dallas. As of the late 1960s, Dallas-Fort Worth had a more ambitious freeway plan than Houston with more centerline miles. This freeway plan was reaffirmed in 1971, but there was really no chance of ever constructing the inner-city routes due to the highway funding crisis of the 1970s and increased opposition to urban freeways. (Source: map adapted from *Dallas-Fort Worth Regional Transportation Study, Interim Report*, 1971)

Both the *Dallas Morning News* and the *Dallas Times Herald* published editorials in June 1953 stating that Harris County's ambitious plan should serve as a challenge to Dallas and the rest of the state. The *Times Herald* went on to say, "Harris County leaders are planning soundly for the future. For some reason, Dallas County cannot seem to get up the steam necessary to execute a master plan for handling its ever-mounting vehicle traffic. Harris County is stepping out ahead of us." The risk of falling behind Houston was enough to spur Dallas-Fort Worth into action. Dallas County would go on to implement a loop and radial plan very similar to Houston's. With Dallas' addition of its second freeway loop in 1964, both Houston and Dallas had radial, double-loop systems which focused on the central business district.[99]

In the mid-1960s, Dallas began to pull ahead of Houston in terms of freeway planning. The Federal-Aid Highway Act of 1962 required all cities with a 1960 population exceeding 50,000 to have a comprehensive, cooperative, and continuing transportation plan—the so-called "3C" process. Local authorities launched a transportation study in 1964 to develop plans for projected 1985 traffic volumes. For the next three years the study team collected and analyzed data, then devised a plan. The final report was published in 1967.[100]

The report recommended a dramatic transformation of Dallas' freeway system from a radial system focused on the central business district to a grid system. There was nothing incremental or evolutionary about the new plan. It was a fundamental change in thinking about how to move traffic around Dallas. The planners formulating the recommendation had found that very few future trips had origins or destinations in the central business district, and the existing radial system could not funnel all the projected traffic through or around downtown. In order to implement the new grid system, the plan recommended five major new freeways for Dallas and numerous other connecting links in both Dallas and Fort Worth. Most of the proposed freeways cut through older residential areas near downtown. At the time new freeways could be built through established areas with little or no opposition, and the study team concluded that it was less costly to build new routes than to expand existing routes. The recommendations of the 1967 report were adopted into the region's long-term plan.

But 1967 was not a good time to launch such an ambitious transformation of the freeway system. The acceptance of new freeways in established urban areas was diminishing, and the protest era was gaining momentum. The new political environment made it very difficult to build new freeways. The 1967 plan was reaffirmed by the regional planning organization in 1971, but all the factors that worked against freeways nationwide—urban issues, environmental issues, and cost issues—would also be felt in Dallas. In 1973 the planned grid system was wiped off the planning map. The original radial system, most of which had already been constructed, remained. In the end, it was politically easier to improve existing freeways than to construct new freeways.[101]

Time would prove that the authors of the 1967 report were remarkably clairvoyant. Transportation patterns developed just as they predicted, with traffic largely passing through the central business district, rather than originating or terminating in it. The Dallas downtown freeway interchange complex, known as the Mixmaster, would become totally incapable of handling the through-traffic load. In 2003 studies are in progress to develop a plan to rebuild the Mixmaster to meet modern needs.

Los Angeles

Perhaps the most interesting story of the rise and fall of a planned freeway network is Los Angeles. Los Angeles distinguished itself as the freeway capital of the world before its freeway construction program was curtailed in the 1970s. But what is most mind-boggling is the freeway system that could have been if a full build-out of the planned network had occurred.

The greater Los Angeles freeway system had its roots in four reports issued between 1937 and 1943. Although the planned freeway networks of the four reports would need to be reconciled, it was clear by 1943 that Los Angeles was going to have a large and sprawling freeway system. Los Angeles was 10 years ahead of Houston in preliminary planning, since Houston's first large-scale freeway plan was not formulated until 1953.[102]

After World War II, events would work in favor of expanding the planned freeway network for the Los Angeles region. In 1946 the Metropolitan Parkway Engineering Committee, an ad-hoc committee of various city and planning engineers from the area, released *Interregional, Regional, and Metropolitan Parkways*, the first unified plan that was officially endorsed by top officials. In 1947 the California State Legislature approved the Collier-Burns Highway Act, raising fuel taxes and fees to provide a reliable funding source for highway construction and allowing large-scale urban freeway construction to begin. The regional freeway plan was adjusted and expanded during the 1950s, reaching its peak around 1965. Freeways began to disappear from planning maps in the late 1960s. First, freeway cancellations were a mere trickle. But by the mid-1970s, especially 1975, freeways were being removed from the state highway system on a regular basis. By the late 1970s any thought of building new freeways in the immediate Los Angeles area was gone. Instead, officials focused efforts on completing the missing links in the system. Only one new freeway would be constructed, the Century Freeway, which was completed in 1993. It was able to move forward mainly because its right-of-way had already been cleared by the early 1970s. In 2003 it appeared unlikely that a critically needed missing link in the Los Angeles freeway system, the Long Beach Freeway (Interstate 710) extension, would ever be built. With one possible exception, a potential extension of the Route 57 Orange Freeway, every cancelled freeway from Los Angeles' original plan is permanently dead. By the 1990s freeway construction activity in the greater Los Angeles area had shifted to Orange and Riverside Counties, where officials were able to build new tollways and freeways through non-urbanized areas.

Similar to Dallas-Fort Worth, Los Angeles officials had correctly predicted future needs and planned a freeway network that would meet those needs. With the cancellation of approximately half of Los Angeles' planned freeways, the region was left with one of the nation's most inadequate freeway systems on a per-capita basis. Data from the Texas Transportation Institute's *Urban Mobility Report* for 2002 show just how heavily the freeways of the Los Angeles region are loaded with traffic. Based on data for the year 2000, Los Angeles averaged 23,400 vehicle miles traveled per day per lane-mile of freeway (DVMT/LM). Los Angeles had the highest freeway traffic load in the United States and was far ahead of the freeway loadings of 15,460 DVMT/LM in Dallas-Fort Worth and 15,310 DVMT/LM in Houston.

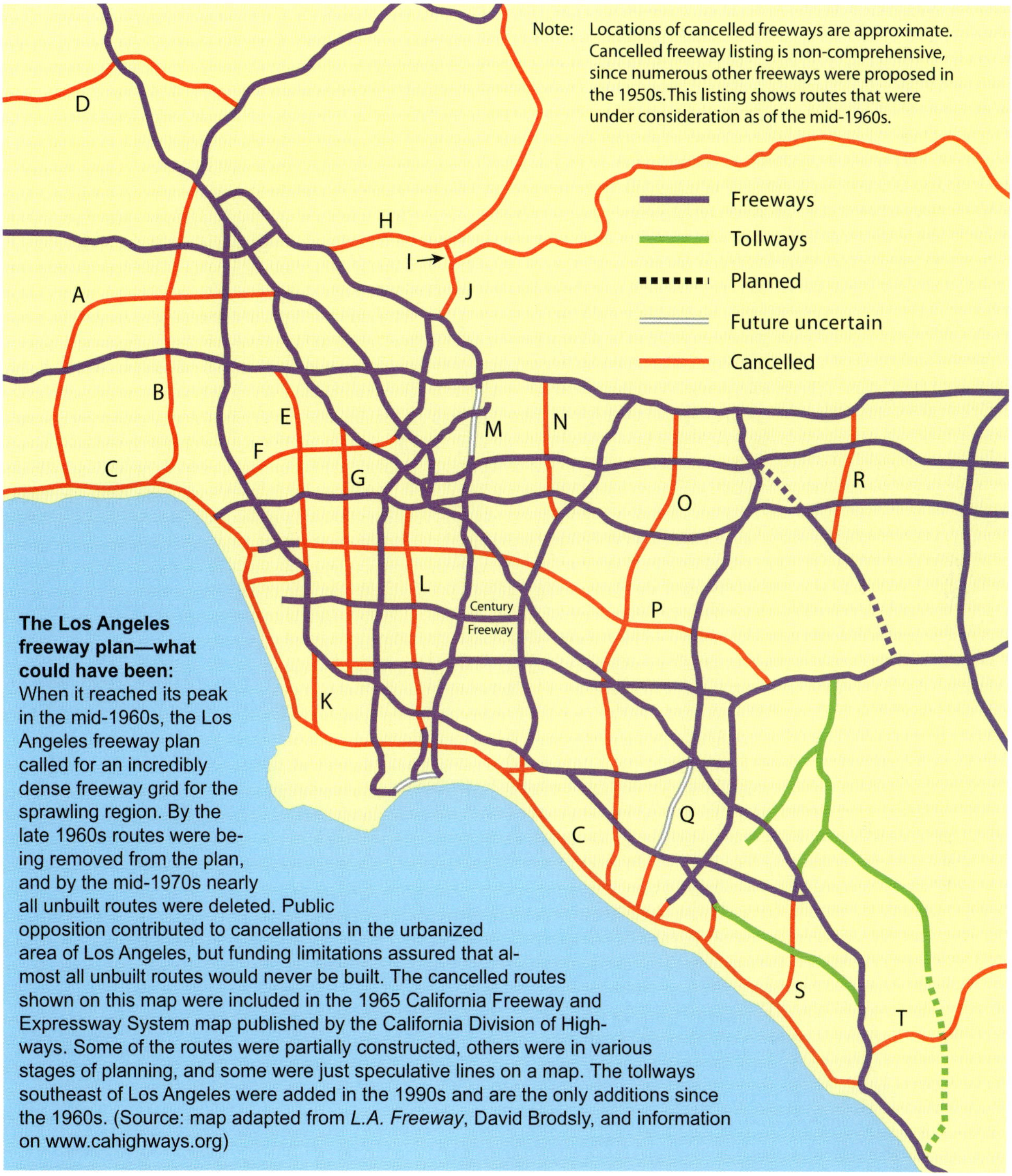

The Los Angeles freeway plan—what could have been: When it reached its peak in the mid-1960s, the Los Angeles freeway plan called for an incredibly dense freeway grid for the sprawling region. By the late 1960s routes were being removed from the plan, and by the mid-1970s nearly all unbuilt routes were deleted. Public opposition contributed to cancellations in the urbanized area of Los Angeles, but funding limitations assured that almost all unbuilt routes would never be built. The cancelled routes shown on this map were included in the 1965 California Freeway and Expressway System map published by the California Division of Highways. Some of the routes were partially constructed, others were in various stages of planning, and some were just speculative lines on a map. The tollways southeast of Los Angeles were added in the 1990s and are the only additions since the 1960s. (Source: map adapted from *L.A. Freeway*, David Brodsly, and information on www.cahighways.org)

Listing of Cancelled and Uncertain Freeways in Los Angeles

A	Whitnall Freeway, CA 64	K	Hawthorne Freeway, CA 107
B	Reseda Freeway, CA 14	L	Industrial Freeway, CA 47
C	Pacific/Ocean Freeway, CA 1	M	Long Beach Freeway extension, IH 710
D	Santa Paula Freeway, CA 126	N	Rio Hondo Freeway, CA 164
E	Laurel Canyon Freeway, CA 170	O	Huntington Beach Freeway, CA 39
F	Beverly Hills Freeway, CA 2	P	Slauson Freeway, CA 90
G	CA 258, part of the Whitnall Freeway	Q	Orange Freeway extension, CA 57
H	Route 118 Freeway	R	CA 142 Freeway
I	CA 249 Freeway	S	Laguna Freeway, CA 133
J	CA 2 Freeway	T	CA 74 Freeway

Houston—high mast illumination capital of America: This view of high mast light fixtures lined up along the West Loop in Bellaire is typical of most Houston freeways. Houston makes extensive use of high mast illumination along linear sections of freeway—probably more than any other place in the world. (Photo: November 2002)

Building Better Freeways

One day in March 1968 researcher Ted Hirsch at the Texas Transportation Institute at Texas A&M University gathered a group of onlookers for the first test of a new highway safety device, the vehicle impact attenuator. The attenuator consisted of a bank of energy-absorbing barrels that was designed to be positioned in front of fixed objects along highways, reducing the force of impact on the vehicle and its occupants in the event of a collision.

One of the observers was Wiley Carmichael, head of the TxDOT Houston District. Carmichael was particularly interested in the results of the test. There had been a rash of fatalities involving collisions with concrete abutments on Houston's freeways, especially on the West Loop. Of the 99 fatalities on Houston's freeways between 1961 and 1968, 27 had occurred at concrete abutments. Between September 1965 and October 1968, three abutments on the West Loop were responsible for 8 fatalities.

The group watched as a test vehicle slammed directly into the attenuator at high speed. The vehicle was brought to a halt with practically no damage, never reaching the concrete abutment just behind the energy-absorbing barrels. As soon as he saw the test result, Carmichael told Hirsch, "I'm putting those on the West Loop." The Houston Urban Project Office went to work immediately to finalize the attenuator design, and in October 1968 the first attenuator assemblies were installed on the West Loop. Within weeks, the impact attenuator had received its first hit, preventing serious injury or fatalities.[103]

The crash impact attenuator consisting of 55-gallon drums was a very simple idea. But as inventors often say, the best inventions are the ones that seem obvious once you see them. Others, including researchers in California, had previously tried more sophisticated impact attenuation schemes that turned out to be impractical. The "Texas crash cushion," as it was called, was the first practical impact attenuator in the United States, and the West Loop installation in October 1968 was believed to be the first deployment of such a device on a highway. The Texas crash cushion or similar designs were adopted by most states shortly afterwards. Numerous other attenuator designs appeared on the nation's highways in the following years.[104]

The impact attenuators were just one of many safety improvements that were implemented on Houston freeways starting in the 1950s, mirroring the increasing national interest in safer highways and automobiles. Other important safety advances developed in Texas in the 1960s were a breakaway light pole and a national-standard breakaway sign. In one important element of safety, freeway lighting, Texas was among the research leaders in high mast illumination. Houston's freeway lighting program would go on to use high mast illumination more extensively than any other city in the United States, and probably the world.[105]

Five Generations of Freeways

Just like anything else that is new, building the first freeways was a learning experience. Houston's first two freeways, Gulf and Eastex Freeways, were classified by TxDOT as first generation freeways. Although Houston's first generation freeways improved on the existing parkways of New York City and the Pasadena Freeway in Los Angeles, they still had their share of shortcomings. There were many lessons to be learned.

The first generation freeway didn't have a median barrier. The Gulf Freeway had a four-foot-wide median with curbs only, and first generation freeways often also had curbs on the right side of the roadway. Emergency shoulders were intermittent and almost never extended along bridges. Merging distances on entrance and exit ramps were short or practically non-existent. Geometrics were poor, with poor sight lines and roller-coaster grades.[106]

To be fair to the designers of the first generation freeway, nobody anticipated the huge traffic volumes, heavy trucks, and increasing speeds that would soon become prevalent on freeways. Engineers had no accurate traffic projections to work with, and the still-lingering Depression-era mindset had engineers trying to save money wherever they could. At the time, the first generation freeway was a huge improvement over what was previously available—congested city streets.

Following the national trend, freeway design improved rapidly in the 1950s. The second generation freeway came into being, correcting most of the problems of the first generation freeway and adding capacity to meet expanding freeway demand. In January 1956, the newly appointed head of the Houston Urban Project Office, Albert C. Kyser, announced that all future Houston freeways would have interior shoulders and a center guard rail. Later that year a median barrier was installed on the Gulf Freeway. Starting in the late 1950s, new freeways became substantially more modern and much safer. Geometrics were greatly improved, especially in terms of sight lines and ramp merging distances. Freeway corridors became much

Deadly gore point: Concrete abutments at exit ramp gore points were some of the most deadly hazards on Houston's early freeways. This abutment on the West Loop at the Katy Freeway was one of three West Loop gore points that were particularly hazardous, causing eight fatalities between 1965 and 1968. Between 1961 and 1968, 27% of the fatalities on Houston's freeways occurred at abutments such as this one. Eliminating these hazards became a top priority of the TxDOT Houston offices. In October 1968, the West Loop abutments were the location of the nation's first installation of practical impact attenuators—the "Texas crash cushion." The crash cushion was a bank of modified 55-gallon drums that was placed in front of the hazard. (Photo: *Houston Chronicle*, 1968)

Vehicle collisions occurred soon after the installation of the first crash cushions. During the weekend of October 11, 1968, two attenuators were hit. In one incident, the driver sustained a broken nose and the passenger sustained a broken collar bone. In the other incident, the vehicle was able to drive away from the scene. In December 1968 two vehicles struck an attenuator, and the only injury was a scratched knee. Collisions with attenuators became so frequent in the following months that officials felt further investigation was warranted. A remotely activated surveillance camera was installed at the attenuator on the exit ramp on the northbound West Loop at the Southwest Freeway. A driveway signal bell hose similar to the hoses used to alert attendants at full-service gasoline stations was placed in front of the attenuator to activate the camera, as shown in the lower left photo. The surveillance camera soon captured a collision. The lower right photo shows a vehicle moments after it hit the attenuator at an estimated speed of 70 miles per hour (112 km/h). Injuries to the vehicle occupants were minor. (Photos: Texas Transportation Institute)

The first generation freeway: This view of the Gulf Freeway shortly after its completion demonstrates two of the shortcomings of the first generation freeway: lack of a central guardrail and light fixtures placed close to the main lanes with no protective guardrail. Other shortcomings were short merging distances at on-ramps, poor geometrics, roller-coaster grades, and traffic-carrying capacity designed to meet forecasts that were much too low. (Photo: TxDOT)

wider, frontage roads became standard, and freeway-to-freeway interchanges used multilevel stack interchanges rather than cloverleafs.[107]

However, it would be many years before all traces of the first generation freeway would be erased from Houston. The Gulf Freeway was expanded and modernized in the 1980s, and the Eastex Freeway underwent a huge modernization and expansion in the 1990s. Interestingly, Los Angeles' original freeway, the Pasadena Freeway, remains in its near-original condition 63 years after it was opened in December 1940. No significant improvements are planned for the Pasadena Freeway as of 2003, and it is destined to become a historic monument to the birth of freeways in the world's freeway capital.

Whereas the transition from the first to second generation freeway was a dramatic improvement, future changes to freeways were more gradual and evolutionary. In 1966 engineers at TxDOT were contemplating the future of freeway design, prompting one researcher to state, "I believe we are actually in the transition period between the second and third generation freeways." The third generation freeway included wider rights-of-way and extra-wide medians to allow for future expansion, facilities large enough to accommodate peak demand at high service levels, improved geometrics and sight lines, more extensive use of structures such as braided ramps, and more durable pavement. Two new freeways under development in Houston in the 1960s, the South and Crosby/Northeast Freeways, were designed to the standards proposed for the third generation freeway. The most distinctive feature of the South and Northeast Freeways was their extra-wide right-of-way corridors, 400 to 500 feet (122 to 152 m). But the third generation freeway was emerging just as a very challenging era for highways in the United States was about to begin—the 1970s. The third generation freeway was more expensive than its predecessors and also had the potential for larger environmental impacts due to its larger size. The highway funding crisis of the 1970s resulted in the cancellation of numerous freeways, and the two surviving third generation freeways moved forward in slow motion. The completion of the South Freeway was delayed until 1984 and the first section of the Crosby/Northeast Freeway did not open until 1994. In the 1970s many standards for new freeways, especially right-of-way requirements, were rolled back to second generation standards, mainly due to financial constraints.[108]

> "Gradually, experience is teaching lessons in freeway building. Our newest freeways are far superior to the earlier ones."
>
> *Houston Chronicle*, September 26, 1962

A new design standard for Houston's freeways emerged

The second generation freeway: The second generation freeway began to take shape in the mid-1950s as engineers identified the shortcomings of the first generation freeway. Around 1960, the first freeways designed to second generation standards began to open in Houston. The second generation freeway was a huge leap forward, featuring more lanes, better geometrics, and improved safety. Most of Houston's freeway system was constructed to second generation standards, and the basic design characteristics continue to be used for new construction even in 2003 for facilities with low traffic volume, such as new tollways. The above view looks south over the West Loop at Westheimer in December 1964. The freeway featured eight main lanes, full inner and outer shoulders, and three-lane frontage roads in each direction. This view also shows the U-turn from the frontage road, which allows vehicles to bypass the traffic light and continue on the frontage road in the opposite direction. U-turns were standard on Houston's freeways beginning with the second generation. (Photo: TxDOT)

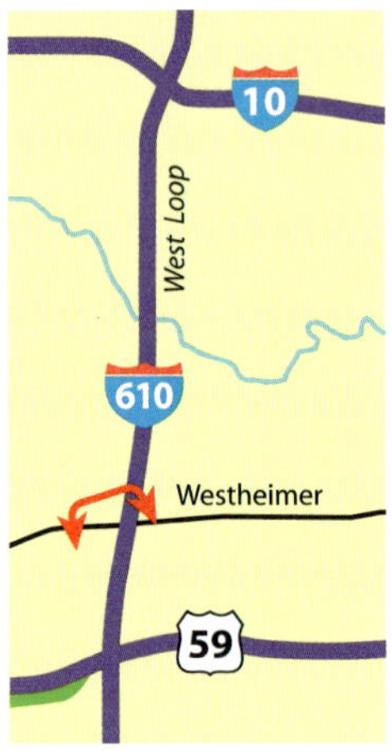

in the 1980s. The 1980s standard was not as grandiose as the third generation freeway, but it was larger than the second generation freeway and incorporated many new distinctive design features. The new standard was applied to existing freeways which were part of the comprehensive reconstruction and expansion program that began in the 1980s. The 1980s design template fully incorporated a reversible, barrier-separated transitway in the center of the freeway for buses and high occupancy vehicles. It featured 8 to 10 continuous general-purpose main lanes, continuous high mast illumination, five-level stack interchanges at freeway intersections, fully continuous frontage roads with improved service levels, more extensive use of braided entrance/exit ramps, carefully planned lane balance, and increased attention to environmental factors such as noise abatement. The preferred right-of-way width was 350 to 400 feet (107-122 m), which was achieved on the Eastex and suburban North Freeways after extensive right-of-way acquisition. In cases where right-of-way acquisition was too costly or infeasible, the 1980s template was applied to freeways on 300-foot-wide (91 m) rights-of-way, such as the Gulf and Southwest Freeways. The more cost-sensitive tollway projects constructed in the 1980s and 1990s were generally constructed to second generation freeway standards.

In the late 1990s, the next phase in the evolution of the Houston freeway began to take shape. Plans for the Katy Freeway expansion were initially formulated in 1998 and included four "special-use" lanes that were separate from

The third generation freeway: The ideas behind the third generation freeway began to take shape in the early to mid-1960s. It featured further improvements to geometrics and safety, more durable pavement, wider rights-of-way, and space reserved for future expansion. However, its high financial cost prohibited widespread implementation, particularly as the financial crisis of the 1970s arrived. Only two Houston freeways would be built to third generation standards: the South and Crosby/Northeast Freeways. The photo above shows the South Freeway, which features most of the third generation freeway design elements. This section of freeway opened in 1983. (Photo: September 2002)

Generation	Period Constructed	Examples	Characteristics
First	1946-1959	None remaining; original Gulf and Eastex Freeways.	4-6 main lanes, poor geometrics, short merging distances, roller-coaster grades, poor safety design
Second	1960-present	Most of Houston's freeway system was constructed to this standard. Remaining examples include the eastern half of Loop 610, the La Porte Freeway, the East Freeway, and the Sam Houston Tollway. Many second generation freeways have been upgraded to the 1980s design standard.	6-10 main lanes, modern geometrics and grades, modern safety design, four-level freeway-to-freeway interchanges
Third	1975-present	South and Crosby/Northeast Freeways. The remaining unbuilt section of the Crosby Freeway is the only third generation freeway that will be built in the future.	6-10 main lanes, very wide right-of-way, wide median for expansion, outstanding geometrics
1980s Design Standard	1980-present	Eastex, Southwest, North, and Gulf Freeways. Other freeways adhere to the general design characteristics but may not include all features, for example, the SH 249 Tomball Parkway.	8-10 main lanes, central transitway, improved frontage roads, improved lane balance, 300-400 foot (91-122 m) wide right-of-way, high mast illumination, five-level freeway-to-freeway interchanges
Managed-lane freeway	2004-future	Katy and Northwest freeway designs as planned.	8-10 main lanes, 4 managed lanes, transit provisions

The 1980s freeway design standard: This view of the North Freeway just north of Beltway 8 is an excellent example of the freeway design that became standard for freeway reconstruction starting in the 1980s. The freeway has 10 main lanes, auxiliary lanes for lane balance, a central transitway lane with connections to adjacent park-and-ride lots, continuous frontage roads generally having 3 lanes in each direction, and continuous high mast illumination. This section of freeway was reconstructed to the 1980s design standard in 1997. (Photo: James Lyle, TTI, June 2001)

the main lanes. The special-use lanes became known by the more generic term of "managed lanes" as the freeway design evolved during the next five years. When a final design was presented to the public in February 2003, it included a four-lane tollway in the center of the freeway. The use of "managed lanes"—lanes for buses, high occupancy vehicles, and toll-paying single-occupant vehicles—appeared to be a trend for the future. In September 2002 a major study of the Northwest Freeway recommended the addition of four managed lanes to the freeway corridor, in addition to expansion of the main lanes. The future of Houston freeways will not adhere to any fixed standard, however. Designs optimized for specific needs will prevail. Major freeways will be expanded to the 1980s standard and the emerging managed-lane design. Low volume freeways, such as the NASA 1 bypass freeway, will be much smaller and more consistent with the second generation freeway. Tollways will generally be smaller than major freeways and will also adhere to second generation freeway standards.

Higher and Brighter

In 1959 the Houston office of TxDOT was about to begin construction on the IH 45 downtown interchange complex. Project engineer Dexter Jones was responsible for developing the lighting system for the new freeway. Until the early 1960s, the nationwide standard for highway lighting was the use of 30-foot-tall (9 m) lighting poles with 400 watt mercury vapor lamps. Jones worked through the design and soon discovered that the conventional fixtures wouldn't work well on the freeway, which featured 80-foot-wide (24 m) elevated structures in each direction. A large number of light fixtures would be needed, illumination uniformity would be poor, and cost would be high. The solution for a better lighting system was to use taller fixtures with more powerful lamps and space the lighting fixtures further apart. The IH 45 central interchange was successfully lighted in 1963 with 1,000 watt lamps mounted on 40-foot-tall (12 m) fixtures. It became clear to Jones that the future of highway lighting would be increased mounting heights with more powerful lamps: higher and brighter. The ultimate form of this concept—high mast illumination—would take shape during

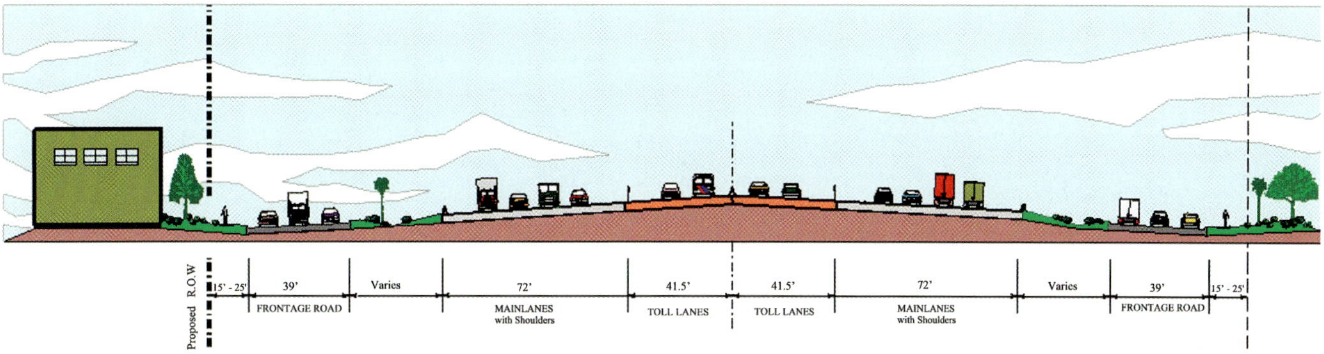

The managed-lane freeway: This cross-section view shows a minimal implementation of the managed-lane freeway. The reconstructed Katy Freeway, scheduled for completion in 2008, will be Houston's first freeway to include managed lanes. Managed lanes are open to transit-oriented vehicles—buses, vanpools, and carpools—and to single-occupant vehicles that pay a toll. In the above view, the managed lanes are identified as toll lanes. Managed lanes are operated to maintain free flow of traffic during peak commute periods.

the 1960s in Texas and elsewhere. Although Houston would not be the first to implement high mast illumination, it would go on to use high mast illumination more extensively than any other place in the United States and probably the world.[109]

In 1965 Jones and the Houston Urban Project Office began a program to develop high mast illumination for Houston. Other research efforts were simultaneously underway at the Texas Transportation Institute, as well as other highway departments in the United States and Europe. Early work focused on highway interchanges. In September 1966, the Texas Transportation Institute installed a temporary telescoping high mast tower at the interchange of IH 35E and IH 820 in north Fort Worth. The tower was able to achieve a height of 120 feet (37 m). The ability of the high mast to light the interchange far exceeded researchers' expectations.[110]

By 1967 the Bureau of Public Roads no longer considered high mast illumination experimental. High mast illumination was reportedly already in use in Europe at the time. The first installation of high masts for highway lighting in the United States was a project with twenty-four 100-foot (30 m) tapered steel poles along IH 5 in Auburn, Washington, just south of Seattle, in 1968. Texas also completed its first two high lighting projects in 1968. Both projects used truss-assembled towers with fixed (non-lowerable) lighting assemblies. A project on IH 410 in east San Antonio featured twenty 100-foot (30 m) towers, and a project on IH 30 at the Texas-Arkansas border featured a 150-foot-tall (46 m) tower on the state line and four 100-foot-tall towers, two on the Texas side and two on the Arkansas side. In 1969, what was billed as the most advanced highway lighting system in the United States was completed on IH 635 in north Dallas. The installation featured thirteen 150-foot-tall masts with a winch system to raise and lower the light assembly, reportedly the first such winch system placed into service in the United States.[111]

Jones and the Houston office of TxDOT were holding back on Houston's first installation of high mast illumination. A little extra time was needed to perfect the winch system which raised and lowered the lights, and new types of lamps well suited for high masts would soon become available. In 1970 all the elements were in place for Houston's first high mast installation. Four 175-foot-tall (53 m) towers were installed at the interchange located at the southwest corner of Loop 610, where the South and

Jersey barrier installation: Houston's freeways were originally constructed with metal central guardrails, officially called the "non-yielding median rail" since they were designed to deflect vehicles rather than absorb an impact. The metal guardrail design also included a fence to screen headlight glare of oncoming traffic. The metal guardrails required excessive maintenance and were replaced by concrete "jersey" barriers starting around 1980. The view above shows installation of jersey barriers on the Katy Freeway just west of downtown in 1982. The construction crew installed the barrier and worked on their sun tans at the same time. (Photo: Chuck Fuhs)

Types of highway lighting	
Mercury vapor	Blue-white light, used extensively in the United States on streets and highways from the 1940s until the 1970s.
High-pressure sodium	Amber light, used almost exclusively to light highways in the United States, including high mast illumination.
Low-pressure sodium	Yellow light, has lowest glare and highest efficiency and is the predominant lighting used on European highways. Extreme yellow color of light provides poor color rendition.
Metal halide	Modern version of mercury vapor provides white light with excellent color rendition and may become more prevalent on highways as efficiency and longevity improves.

Moonlight tower, Austin, Texas: The last remaining example of the nation's first wave of high mast illumination in the 19th century can be found in Austin, Texas. Moonlight towers were built in cities around the United States and were first installed in Austin in 1895. Seventeen remain today. Moonlight towers are 165-foot-tall (50 m) vertical truss assemblies braced by guy wires with a ring of lights at the top. This moonlight tower is located a few blocks away from the TxDOT headquarters. It served as an inspiration for project engineer Dexter Jones, who led Houston's high mast illumination efforts in the 1960s and 1970s.[112] (Photo: June 2002)

West Loops meet. An installation with six 200-foot-tall (61 m) towers followed soon afterwards at the intersection of US 59 and IH 10 in downtown Houston. High mast lighting was gradually installed at interchanges all around Houston in the following years.[113]

Moving Ahead of the Pack: High Mast Lighting of Linear Sections of Freeway

High mast illumination became common at interchanges in the United States in the 1970s. But conventional 40 and 50-foot (12 and 15 m) fixtures still ruled the linear sections of freeways between interchanges. In 1973 Houston first began to plan the construction of reversible, barrier-separated transitway lanes in the medians of freeways for use by buses and high occupancy vehicles. This would necessitate the removal of the metal guardrails in the freeway medians where the freeway lighting fixtures were located. Studies of lighting options for the transitway-equipped freeways showed that conventional fixtures wouldn't do a good job, especially with the continuous frontage roads along Houston's freeways and the usual corridor width of 300 to 400 feet (91–122 m). Continuous high mast illumination of linear freeway sections would be the solution.[114]

In April 1974, Jones and the Houston Urban Project Office launched a joint research program with the Texas Transportation Institute at Texas A&M University to determine the optimal configuration for high mast illumination of linear sections of freeway. The objective was to develop a glare-free system of lighting, something which could be called the holy grail of highway lighting. By the end of 1976, the team had found the optimal solution: the Z-pattern high mast freeway illumination system. The Z-pattern system provided light rays that always came from above or behind the motorist; at no time would the motorist be confronted with light shining into the line of sight. The standard mast height was 125 feet (38 m) but could reach 150 feet (46 m) in certain instances. The linear spacing was typically 700 feet (213 m). The lamp unit featured two large and four small floodlights, all carefully aimed to reduce glare.

Around 1979, Houston's first installation of high mast illumination on a linear freeway section was completed on the Gulf Freeway from downtown to Loop 610. The North Freeway followed soon afterwards. During the 1980s and 1990s, high mast illumination was added to most of Houston's freeways as the freeway expansion and reconstruction program reached full speed.

Sky Glow

Not everyone was pleased with the proliferation of powerful floodlights along Houston's freeways. In particular, astronomers weren't happy. Light fixtures can send light rays upward into the atmosphere, where the light can be scattered by airborne particles and cause an effect called sky glow. The glow reduces the visibility of stars in the sky, effectively drowning out their dim light, and can adversely affect astronomers' nighttime observa-

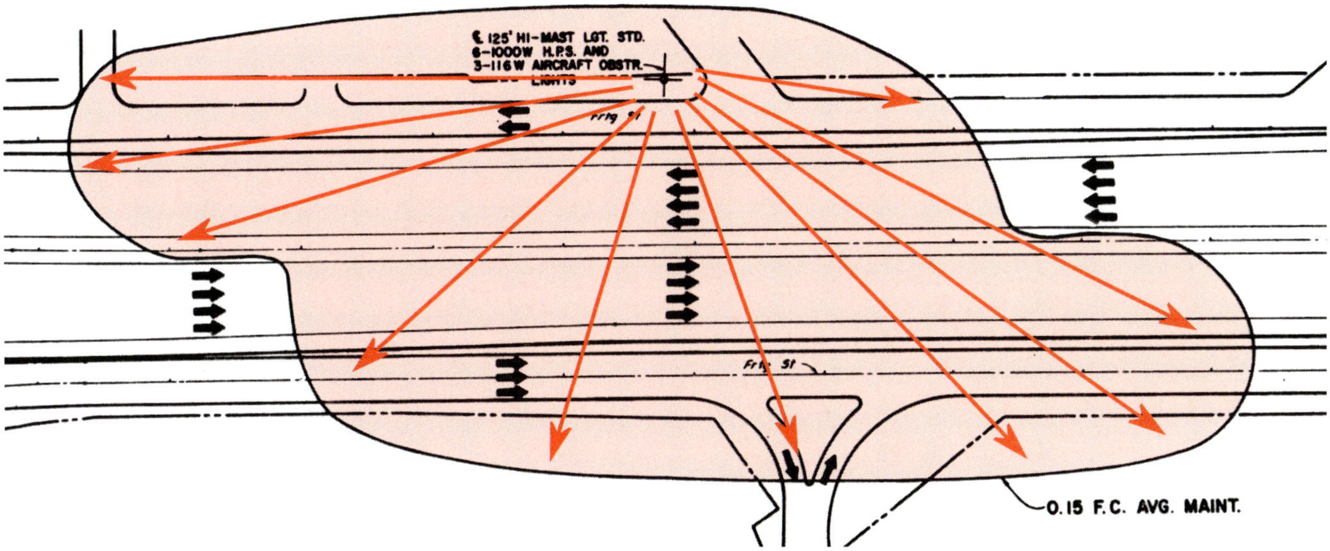

The Z-pattern high mast configuration: The first generation of high mast lighting on linear sections of Houston's freeways, installed from 1979 to the early 1990s, featured the Z-pattern lighting configuration. The Z-pattern was designed to provide light from above and behind the driver so the driver would never look directly into a floodlight. (Graphic: *Z-Pattern High Mast Freeway Illumination*[115])

tions. Astronomers call sky glow "light pollution," a term that conveys a negative spin and aids their political efforts to enact legislation to curtail sky glow. The International Dark-Sky Association was founded in 1988 and has led efforts to curtail sky glow. By the early 1990s, TxDOT started to modify its lighting standards to reduce the contribution highway lighting makes to sky glow. To solve the problem, "cutoff" light fixtures are used on new lighting installations in Texas. Cutoff light fixtures greatly reduce the amount of light that is directed upwards into the atmosphere. This is normally accomplished by recessing the light source inside a metal housing and using a flat-bottomed transparent surface to direct the light towards the ground. In contrast, the hemispherical-shaped covers commonly found on street lights are not cutoff devices, since the hemispherical shape sends a certain percentage of light upwards. Cutoff fixtures have the added benefit of reducing glare since the light source is well shielded.[116]

The adoption of cut-off lighting marked the end of new Z-pattern high mast illumination installations in Houston. The floodlights used on the Z-pattern assemblies would send some of their light upwards into the atmosphere. In addition, properly aiming the Z-pattern floodlights could be difficult, and misalignments could exacerbate loss of light into the atmosphere. The new standard design for high mast illumination uses a ring of cutoff light fixtures, all pointed towards the ground. In the late 1990s foes of light pollution mounted an effort to pass a state law to mandate the use of cutoff lighting fixtures in state-funded installations. Their efforts paid off in 1999 when Governor George W. Bush signed a law regulating state-funded outdoor lighting and mandating the use of cutoff fixtures.[117]

The switch to cutoff lighting was generally accompanied by an increase in mast height. New installations of high mast illumination on linear sections of freeway in Houston in 2003 use 175-foot-tall (53 m) masts. The biggest and best implementation of high mast illumination is planned for the Katy Freeway expansion, scheduled for 2003-2008. Whereas all previous installations on linear sections of freeway have used a single row of masts, the Katy Freeway will have an unprecedented dual mast configuration with 175-foot masts along both sides of the freeway.

As lighting technology advances, Houston's freeway illumination will continue to improve. Perhaps in the future, metal halide lighting technology will become viable for high masts, replacing the high pressure sodium lights that are used today. Metal halide produces a white light which shows the lighted area in its natural color, in contrast to the yellowish cast imparted by high pressure sodium lamps. Metal halide lamps are used for indoor lighting because of their excellent color rendition.

Terms relating to control of artificial light

sky glow - the scattering of artificial light by particles in the atmosphere, resulting in a luminance of the atmosphere at night.

cutoff light fixture - a light fixture that directs a very small percentage of its total light upward into the atmosphere. In Texas no more than 2.5% of the fixture light can be emitted above a plane at the lowest point of the light assembly.

full cutoff light fixture - a light fixture that directs no light upward into the atmosphere.

The Leader

If high mast illumination on linear sections of freeway is so great, why don't highway departments across the United States and the world use it more extensively? "It's a lot more expensive," says Mike Strech, executive

High mast illumination:
The above photo looking west along the Katy Freeway near Kirkwood shows the usual view that Houston motorists see at night—a long line of high masts to the distant horizon. This mast installation was part of the first generation of high masts in Houston and used the Z-pattern design with 125-foot-tall (38 m) masts. (Photo: November 2002)

Since the first installation of high mast lighting on a linear section of freeway in 1979, high mast illumination has been added to most of Houston's freeways. In 2003 new installations use 175-foot-tall (53 m) masts. High mast lighting and improved reflectivity of signs have eliminated the need to illuminate overhead signs on Houston's freeways.

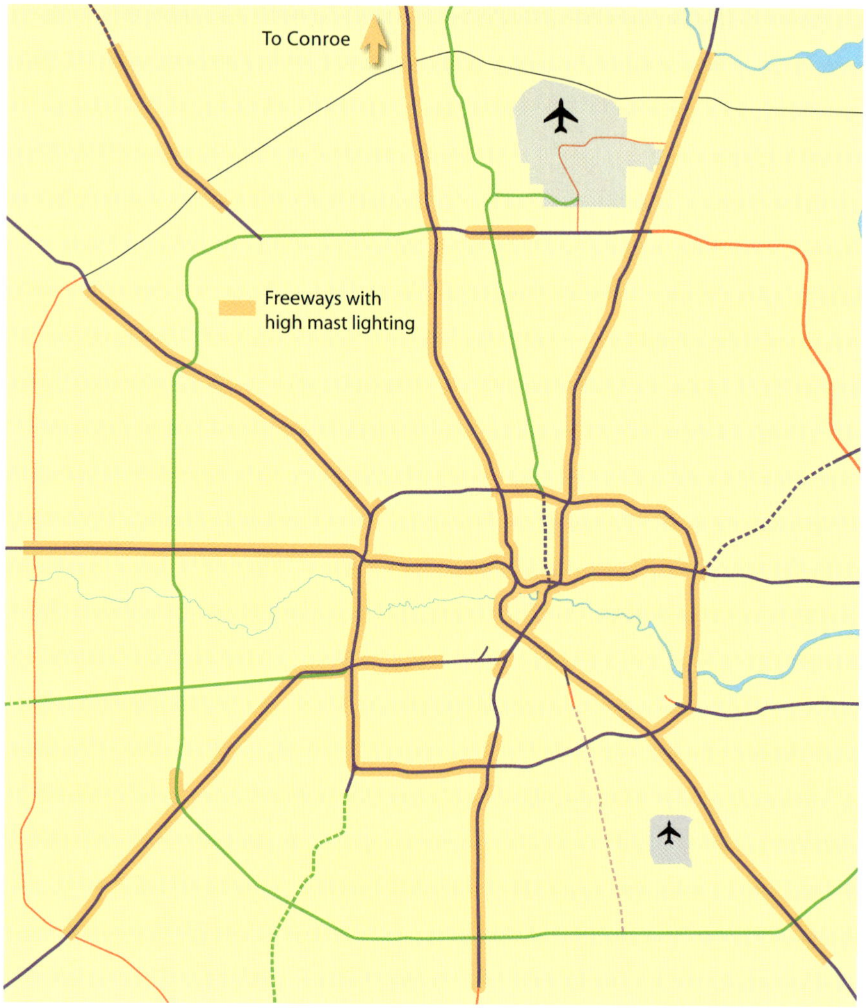

Cutoff high mast lighting: These masts on the Southwest Freeway north of Sugar Land use cutoff light fixtures. (Photo: June 2002)

director of the Harris County Toll Road Authority (HCTRA). For HCTRA the most economical solution for lighting Houston's tollways is conventional 40 and 50-foot-tall (12 and 15 m) fixtures mounted on the center barrier of the tollway main lanes. HCTRA's situation is similar to most highway departments in the United States. It is responsible for lighting only its tollway main lanes, even when the tollways have state-owned frontage roads. Of course, most urban freeways outside Texas do not have frontage roads, so highway departments tend to use lighting designs similar to HCTRA's. On Houston's freeways, where frontage roads are widespread, TxDOT needs to find the most economical solution for lighting both the main lanes and the frontage roads. Lighting Houston's wide freeway corridors, typically 300 to 400 feet (91 to 122 m) wide, is an ideal task for high mast illumination.[118]

In fact, a unique set of conditions is needed to support widespread use of high mast illumination on linear sections of freeway. In the United States these conditions exist on a large scale only in Texas, and they exist to the greatest extent in Houston. First, wide freeway corridors with frontage roads make high mast illumination a competitive engineering solution. Second, a commitment to highway safety and well-lit freeways is needed. Third, the region must have the financial strength to afford the added cost of high mast illumination. And fourth, the region must have public support for high mast lighting. In Houston there is generally good support for powerful lighting, especially among the commercial interests that line Houston's freeways. The result: Houston is widely believed to be the world leader in the use of high mast illumination. Aside from the safety benefits of well-lit freeways, driving Houston's freeways at night can be an impressive experience. Looking forward, the driver sees a long line of 125-foot to 175-foot-tall (38-53 m) masts receding to the distant horizon. Freeway lighting is uniform and consistent. Anyone with a view above ground level can see the paths of distant freeways by their high masts. The seemingly ubiquitous presence of masts along freeways is a distinctive feature of Houston's freeway system.

Why Houston is the leader in high mast illumination

- Wide freeway rights-of-way with main lanes and frontage roads which are more effectively lighted with high masts
- A commitment to safety and well-lighted highways
- Financial strength to afford high mast lighting
- Public and business support

Lighting the freeways, not the sky: The left photo shows a typical high mast equipped with directed floodlights. Floodlights were used for the Z-pattern high mast illumination that was standard in Houston's first generation of high mast lighting. The floodlights were difficult to aim and had the potential to send light upwards into the atmosphere, contributing to the sky glow effect. Starting in the early 1990s, TxDOT began to use a ring of cutoff light fixtures that point toward the ground. The right image shows the arrangement of lights that is now standard for new installations of high mast lighting on Houston freeways. (Photos: July 2002)

Key dates in high mast illumination

Mid 1960s	Research efforts begin in Texas and elsewhere. Reported installations in Europe.
1968	First high mast installation in the United States in Auburn, Washington. First high light installations in Texas using truss towers.
1969	First reported use of winch-controlled light assembly on IH 635 in Dallas.
1970	First high mast installation in Houston.
1979	First high mast installation on a linear freeway segment in Houston—the Gulf Freeway.
1980s	Widespread implementation of high mast illumination on linear freeway sections in Houston.
Early 1990s	New installations use cutoff light fixtures rather than directed floodlights.
2003	New mast installations use 175-foot-tall (53 m) masts. Katy Freeway expansion will have dual rows of 175-foot masts.

It should be pointed out, however, that Houston's freeways are not the world's best lit. That honor is generally given to the freeways of Belgium, which are continuously lighted through both urban and rural areas with higher intensity light than is typically used in United States. High masts are used at interchanges only, however. Other regions in the world, particularly in Europe, also have strong freeway lighting programs.

Freeway lighting in Belgium: Belgium is generally credited with having the world's best-lit freeways. Approximately 90% of Belgium's freeways, both urban and rural, are fully and continuously lighted. Belgium formally adopted its policy of continuously lighting all freeways in the early 1960s. The photo above of the E40 freeway just east of Brussels, dating from the early 1970s, shows the lighting design that is most common on Belgian freeways: central fixtures with low-pressure sodium lamps. Low-pressure sodium lamps impart a strong yellow cast to the light. The more recent image at right provides a closer view of the typical lamp assembly. High mast lighting is used only at large interchanges. The above photo shows that Belgium's continuous lighting policy was well ahead of other needed highway safety improvements. Notice that this freeway has a very narrow median with no guardrail and the lighting fixtures are of a non-breakaway design with no collision protection. (Photos: upper courtesy of Dexter Jones; right, RTech, Belgium)

Locally Preferred Alternative

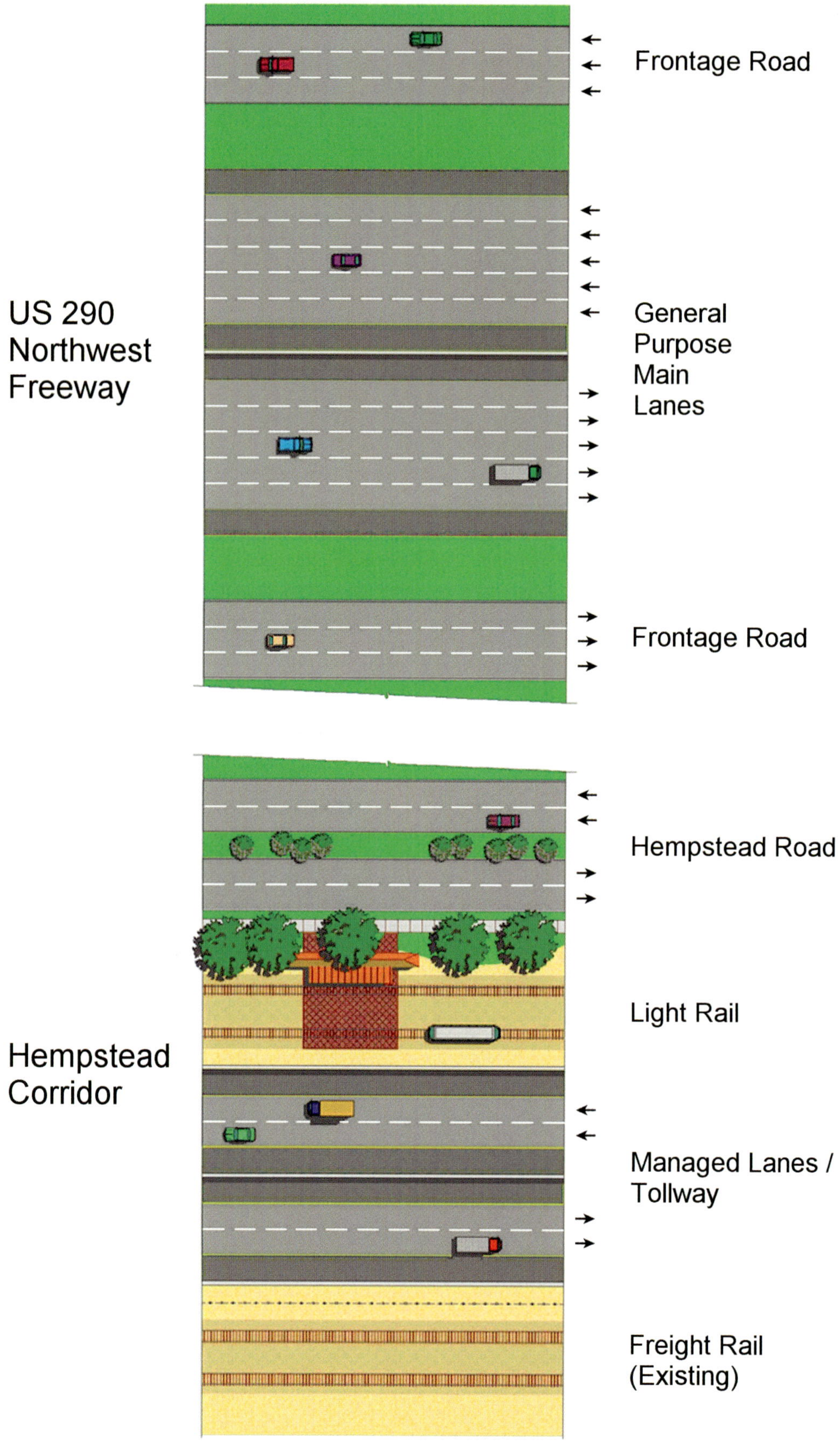

Freeway corridor of the future: The Northwest Freeway reconstruction plan was presented to the public in October 2002. The plan includes a moderate expansion of the freeway main lanes and frontage roads, but more significantly includes a parallel toll road and a high-capacity transit corridor. Houston's freeways will continue to be expanded and improved, but toll roads and transit will play an increasingly larger role in overall corridor planning.

The Future of Houston's Freeways

Nothing is ever certain in the world of freeway planning and development. Turning a line on a planning map into a freeway or tollway is a very difficult task. Numerous factors can send even the best-made plans to the back burner or the scrap heap. And while the challenges of funding and special interest group opposition still remain potent, today's freeway builders face a new challenge that has the potential to be a show-stopper: increasingly stringent federal air quality regulations and antifreeway organizations seeking to use those regulations to stop freeways.

But one thing is certain. If Houston is able to proceed with plans that are in development in 2003, Houston will likely establish itself as the nation's leading city for building and expanding freeways and tollways in the 2003–2020 period. Perhaps only Dallas-Fort Worth and Phoenix will be investing in freeway and tollway transportation facilities at a rate comparable to Houston. Houston will strengthen its position among the top freeway cities in the United States and the world. The next chapter in the story of Houston's freeway system holds the promise of remarkable achievement, but getting to the planned freeway network of 2020 will be a challenge.

The Traffic Cycle Returns

Houston's aggressive action in the 1980s to address the transportation crisis almost wiped traffic congestion off the local political radar screen. The Houston Area Survey, conducted by Rice University, is a comprehensive annual survey of the opinion of Houstonians on a wide range of issues. In 1982, the first year of the survey, 51% of those surveyed rated traffic congestion as the biggest problem facing Houston with crime a distant second at 26%. By 1986 traffic congestion had dropped to second place with 36%, and economic conditions moved up to first with 39%. By 1992 only 3% of survey respondents rated traffic congestion as the biggest issue facing Houston, while crime moved to first place with 65%. Traffic congestion began an upward trend in 1995 with strong economic and population growth fueling an increase in demand for transportation resources that outstripped the increase in supply. In 2000 traffic congestion once again became the most serious problem facing Houston, with 31% of respondents in comparison to 24% for the second item, crime. In 2003 traffic congestion held steady at the top position with 33%. Transportation was back on the political radar screen.

There was a noticeable increase in political efforts to secure highway construction funding starting in 2001, particularly in regard to Houston's share of state highway funding, which had dropped to 13% in 2001. City and county bond issues for road construction totaling $949 million passed with overwhelming margins in November 2001. As an issue at the top of the public's agenda, transportation was once again poised to receive the attention of political leaders. But just as it had in the past, the issue of transportation and how best to solve the problem would become the subject of controversy. It was time for another round in the battle of freeways vs. rail. This time, however, it appears that both freeways and rail will prevail, but tollways will emerge as the biggest winner.[119]

Filling the Pipeline

Houston's previous waves of freeway construction had been guided by single, comprehensive documents that arrived with a bang and served as a rallying point for freeway-building efforts. Houston's first major freeway construction wave in the 1950s and 1960s was launched by the original 1954 freeway master plan. The second wave in the 1980s was launched by the 1982 Regional Mobility Plan. But the next wave will be different. The blueprint for the next generation of freeway improvements is the product of a new era—an era in which transportation projects on individual freeways are subject to years of analysis

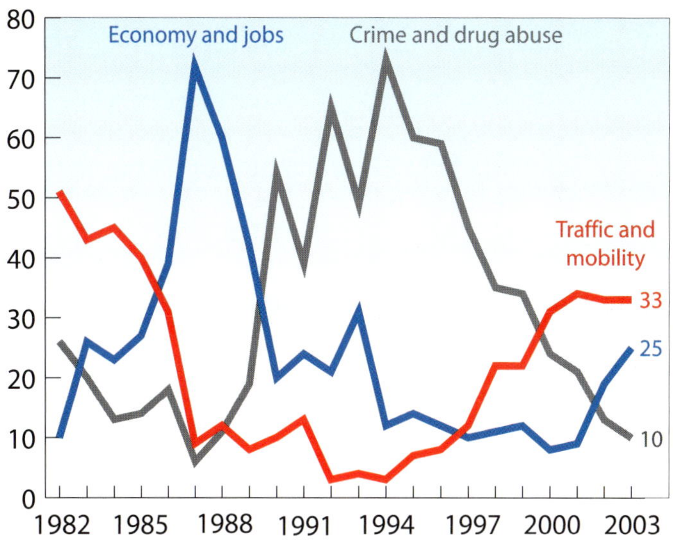

▲ Biggest Problem Facing Houston Data: Houston Area Survey

Percent who identified the issue as the biggest problem facing Houston

At the top of the agenda: In 2000 traffic congestion and mobility returned to the top of the list as the biggest problem facing Houston. The importance of transportation should ensure large investments in both freeways and transit in the near and intermediate future.

Recent Major Investment Studies in Houston Managed by TxDOT

Year	Freeway	Limits	Recommendation
1997	IH 10 West Katy Freeway	Downtown Houston to Brazos River	Expand freeway; add managed lanes from Loop 610 to SH 6
1999	IH 45 South Gulf Freeway	Beltway 8 to Galveston	Expand freeway; add non-barrier-separated HOV lanes
1999	US 59 South Southwest Freeway	SH 6 to Fort Bend-Wharton County line	Expand freeway; add frontage roads
2001	SH 249 Tomball Parkway	Northwest of Houston, Pinehurst to Todd Mission	New freeway alignment north of Tomball
2001	SH 146	Fairmont Parkway in La Porte to IH 45 at Texas City	Extend freeway; add express lanes; upgrade highway
2002	US 290 Northwest Freeway	Loop 610 to FM 2920	Expand freeway; build new toll road on Hempstead Road corridor; preserve corridor for high-capacity transit
2000-2004	SH 99 Grand Parkway	Five sections of the planned loop	Define alignment for future freeway or tollway
Begins 2003	SH 35	IH 45 in Houston to SH 288 in Angleton	
Begins 2003	SH 225	Loop 610 to SH 146	
Begins 2003	SH 288	US 59 (downtown Houston) to Angleton	

and public input. The process is called a major investment study or corridor feasibility analysis and ultimately leads to a locally preferred alternative for transportation improvements. The recommended improvements are then adopted into the region's official long-range transportation plan.

Starting in the mid-1990s, a series of major studies on many of Houston's freeways began. The period from 1997 to 2003 was very influential in defining the future of Houston's freeways. One by one, the recommendations of these studies formed a new plan for Houston's freeways. In all cases, freeway expansions or new freeways were recommended. In a piecemeal fashion, Houston was getting a freeway plan that would be nearly as aggressive as the 1954 and 1982 plans (see table).

Plans of the Harris County Toll Road Authority (HCTRA) and Fort Bend County Toll Road Authority are not subject to the lengthy planning processes that are required for projects receiving federal money. The 1997–2003 period was equally influential for defining the future of Houston's tollway network. Several key projects moved forward, including the Westpark Tollway, Fort Bend Parkway Tollway, Hardy Airport connector, and Hardy Toll Road downtown extension. In 2001 HCTRA released a list of potential future toll road projects—the so-called "pooled projects." The list of projects will be considered on a case-by-case basis as demand develops, but nevertheless it forms a blueprint for an aggressive program over the next 20 years.

Since 1997 there has been an evolution in the approach to freeway corridor planning. With each passing year there seemed to be a movement to more diversified transportation corridors with both toll and transit facilities, in addition to expansions of general-purpose main lanes. The recommendation for the Northwest Freeway in 2002 exemplified this trend, with TxDOT recommending both a tollway and high-capacity transit facility—even without any commitment from agencies which would build those facilities. The North Freeway study, in progress in 2003, considered transit and toll road options before attempting to identify needed highway improvements.

It was almost as if TxDOT was transitioning into an empowerment agency—one that would empower local entities to take a larger share of the responsibility in future transportation corridors. In reality, TxDOT was saying that it couldn't do it alone. Building more mega-corridors around Houston like the $1.7 billion Katy Freeway project was not financially or politically feasible. TxDOT had shown strong leadership by launching the wave of studies for future improvements, but in 2003 it appeared to be repositioning itself as just one player on the team. Both HCTRA and the Metropolitan Transit Authority seem ready to step forward and assume larger roles.

The Plan

The official transportation plan for the Houston region is maintained by the Houston-Galveston Area Council (HGAC), the regional planning agency. Long-term plans are developed 20 years into the future. In mid-2003 the 2022 plan was the most recently approved plan, and the 2025 plan was in development. Any freeway project to be constructed with federal money must first be included in the official regional plan.

The 2022 plan includes the addition of 1,046 new free-

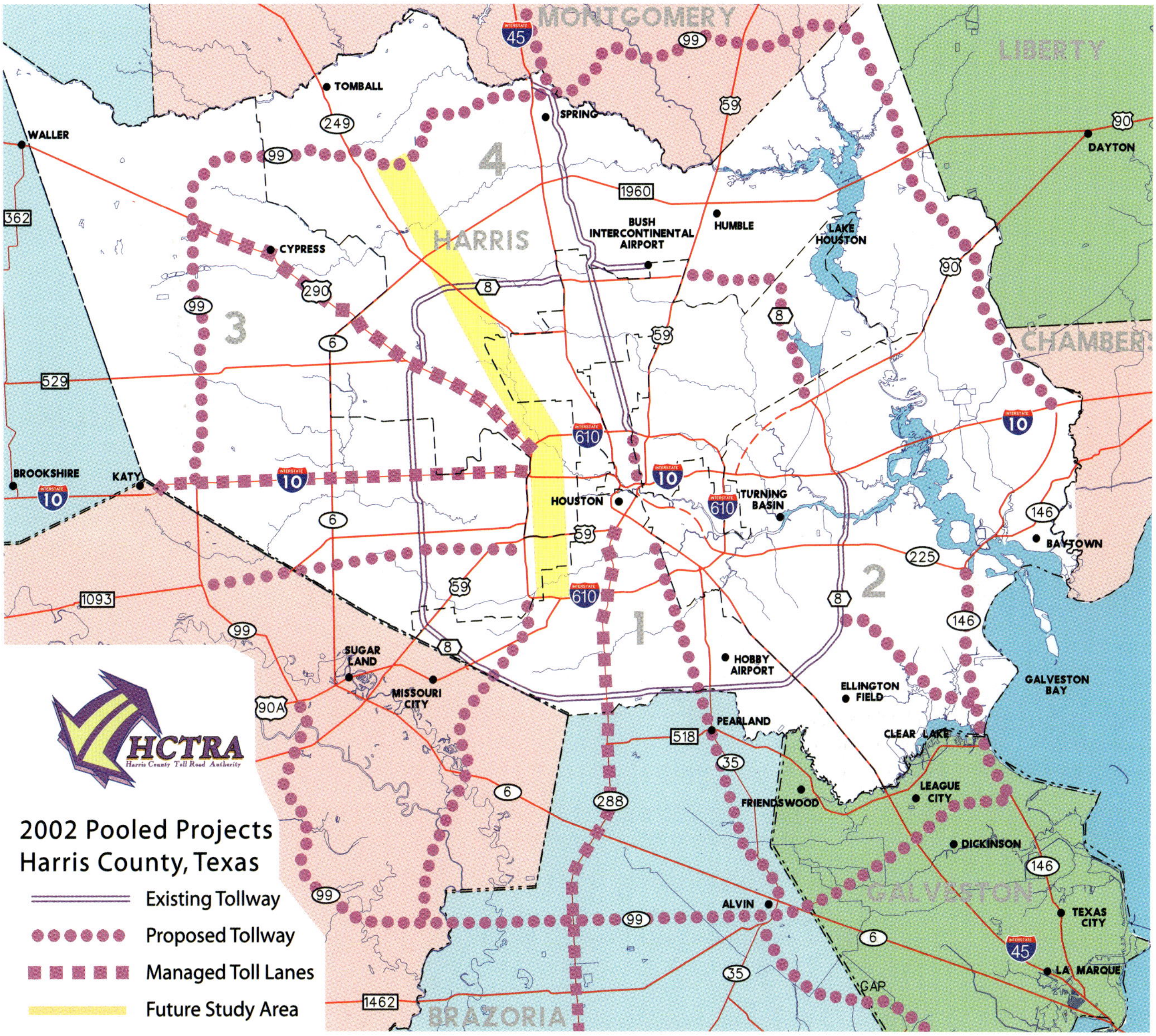

Tollway planning: In 2001 the Harris County Toll Road Authority (HCTRA) first released its "pooled projects" map, showing potential future toll road corridors in the Houston region. Three of the projects shown, the Westpark Toll Road, Katy Freeway toll facility, and Fort Bend Parkway Toll Road, are either under construction or about to begin in 2003. The northeast section of Beltway 8 will be constructed, but the other projects are speculative and may or may not be built. (Source: HCTRA)

way lane-miles (1,674 lane-km), 234 new tollway lane-miles (374 lane-km), and 226 new transitway lane-miles (362 lane-km). The numbers are certainly impressive, but perhaps the greatest achievement will be in the individual projects that make up the plan. The Katy Freeway expansion, if implemented as planned in mid-2003, will give Houston one of the nation's most impressive urban freeway corridors for a sustained distance. The Grand Parkway will establish Houston as the only major city to build most or all of a wide outer loop—Houston's third loop. Several new facilities are planned, mostly toll roads. The toll road system will be among the nation's more extensive, supplementing the large freeway system. Houston's downtown interchange complex will likely be improved, solidifying its position as one of the most extensive and modern in the United States.

If fully implemented, the long-range plan will solidify Houston's position among the world's top freeway cities. Although Houston will never equal the sheer number of freeway and tollway lane-miles in much larger cities such as New York and Los Angeles, Houston's freeway system will be among the most modern and up-to-date.

Making the Plan Happen

Building and expanding a freeway system isn't something that just happens. It takes a sustained and dedicated effort by political leadership. Some of Houston's most influential political officials of the post–World War II era made freeways a top priority. In the 1980s and 1990s, Houston benefited greatly from its pro-mobility

Major upcoming freeway construction, 2003-2008: Houston's freeway construction program will gain momentum in the 2000s after a slowdown in the late 1990s. The above map shows future projects and projects that began in 2002. Ongoing construction that began prior to 2002 is excluded. The projects shown in the above map are all funded as of early 2003 and should be constructed in the 2003-2008 time period barring any funding or approval disruptions. Other new projects may move forward in this period, particularly new toll road projects. The centerpiece of the mobility plan is the expansion of the Katy Freeway, IH 10 West, with an estimated construction cost of $1.2 billion.

leadership, particularly Harris County Judge Jon Lindsay and Texas Transportation Commission Chairman and Houston Mayor Bob Lanier. Nonelected officials, especially Doug Pitcock of Williams Brothers Construction, have also played key roles.

But no one's influence lasts forever. In 2003 the era that had been so strongly influenced by Lanier, Lindsay, and Pitcock was in its twilight. If Houston is to fully implement its next wave of freeway construction, a new generation of pro-mobility leadership will be needed. It will be difficult, and perhaps impossible, for Houston to repeat the strong leadership it had in the past. In 2003 Harris County Judge Robert Eckels is the region's leading advocate for tollways and freeways. Eckels has provided excellent pro-mobility leadership for the region, but one person alone cannot repeat the legacy of the 1980s and

1990s. Without Eckels, Houston's mobility program could have been seriously impaired between 1997 and 2003. A pro-mobility political environment is not something that should be taken for granted.

To be sure, Houston's freeways have a tremendous amount of political inertia behind them. The county governments of Harris County, Fort Bend County southwest of Houston, and Montgomery County north of Houston have been essential in maintaining the strength of Houston's pro-mobility political environment. Pro-mobility suburban communities, the Houston business establishment, and the political influence of the construction and engineering consulting industries have also played key roles in sustaining the strength of Houston's freeway and tollway program.

And then there's the subject of money. The ups and downs of Houston's freeway program have been closely tied to funding. The 1970s highway construction bust and subsequent traffic congestion crisis were largely the result of greatly diminished funding. The second wave of construction in the 1980s was the result of increased funding. Implementing the planned program for the next 20 years will require additional sources of revenue. The first step being pursued by local officials is to raise Houston's share of state funding to its "fair" level, which is generally regarded to be about 22%. That step is relatively painless since it does not involve new taxation. Revenue from the Harris County toll road system will help cover the shortfall and enable the construction of new tollways, possibly as a substitute for new freeways. Fully implementing the plan, however, will require an enlargement of the highway construction funding pie. This could be accomplished with increases in federal funding, but most likely it will be achieved only by an increase in the state fuel tax or a new source of locally generated revenue. The political climate in Texas in 2003 does not appear to permit any tax increases, and an increase in the fuel tax seems highly unlikely in the near future. But as the existing 20-cent state tax is continuously eroded by inflation, at some point an increase may become necessary and politically feasible.

In 2003 the political leadership of Texas is attempting to steer Texas down the path of toll funding for its future highway construction program. There is also an effort to shift financial responsibility for urban transportation facilities to the local level through regional mobility authorities. Constitutional amendments passed by voters in November 2001 provided the state of Texas with new options in toll road and bond funding of highways. It appears that these programs will have little impact in Houston, since the Harris County Toll Road Authority is already aggressively building and planning new toll facilities where they are financially and politically feasible. Still, financial contributions from TxDOT to potential future toll roads, particularly the Grand Parkway, could make some marginal projects viable.

History shows that transportation crises often provide the impetus for new highway construction funding. It happened in the 1950s when the rapidly expanding use of private automobiles prompted new federal funding. It happened in the 1980s when Houston's traffic congestion crisis led to increases in state taxes dedicated to highway funding and the establishment of the Harris County Toll Road Authority. Will it take another crisis to bring funding up to adequate levels? The answer to that question is probably yes, but there's always hope that political leaders will be proactive in preventing a future traffic congestion crisis before it occurs.

One thing is clear for the future. The challenges to freeway construction will be greater than ever. The risk of a major disruption resulting from federal regulations is an ongoing threat. When a political environment has been so strongly pro-freeway, the chances of political weakening are greater than the chances of strengthening. Awaiting the next generation of pro-mobility leadership are the most difficult regulatory environments, funding challenges, and sophisticated antifreeway efforts that Houston's freeways have ever faced.

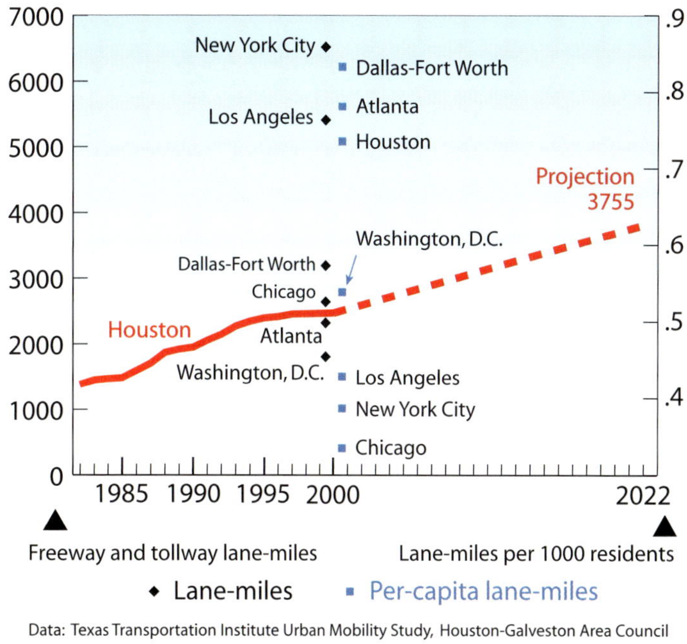

Freeway cities: This plot shows the projected increase in lane-miles on the Houston freeway and tollway system. Houston ranked fifth in lane-miles in 2000, the latest data available from the Texas Transportation Institute (TTI). The TTI lane-mile data tends to lag behind new additions, and the large additions of lane miles in Houston in the late 1990s on the Eastex Freeway, North Freeway, and Sam Houston Tollway are not reflected in the data. By 2003 Houston has surely overtaken Chicago and will continue a strong upward trend in lane-miles with ongoing freeway expansions and tollway construction. On a per-capita basis, Houston is in the top three with Dallas-Fort Worth and Atlanta.

The Anti-highway Regulatory Environment

The anti-highway and anti-freeway mechanisms built into federal regulations are substantial, but perhaps none looms as large as clean air regulations imposed under the Clean Air Act of 1970 and its 1990 revisions. The federal regulations that implement the Clean Air Act are highly complex, and the threat imposed by noncompliance is systemwide, potentially affecting nearly all projects in the area, not just individual projects. If compliance with federal standards is not achieved, federal transportation funds can be withheld, striking a severe blow to the region's freeway construction program. Atlanta, Georgia, was declared to be in a "conformity lapse" with federal air quality regulations in January 1998 and its highway development program was placed under sanctions. Federal transportation funds were frozen, although many grandfathered and exempt projects still continued. Atlanta's highway construction program was subsequently curtailed and most of its planned outer loop, the Outer Perimeter, was cancelled, leaving only the uncertain northern arc section in long-term plans.[120]

Air quality is a highly complex issue involving a large array of factors. This includes the wide variety of pollution sources, six major categories of pollution, weather conditions, and chemical processes that are subject to ongoing research. For Houston, the problem is ground-level ozone. Houston was declared to be a "nonattainment" region for ozone based on data collected from 1987 to 1989, indicating that Houston was not in compliance with federal regulations. Houston was one of five regions in severe nonattainment of ozone standards and had the highest ozone levels among these regions. Only Los Angeles, classified as an extreme nonattainment region, had higher ozone levels during the benchmark three-year period.[121]

Houston and the four other severe nonattainment regions were given 17 years to meet federal standards, with a deadline of 2007. The federal government applies a uniform standard to all regions without regard to the particular characteristics of each region. Houston is in an especially difficult position because of its extensive refining and petrochemical industries, ongoing population and economic growth, and hot summer weather. Fair or not, Houston must comply to ensure an uninterrupted flow of federal transportation funds.

Reducing ozone is a difficult task. Ground-level ozone is formed by a series of complex atmospheric chemical reactions primarily involving sunlight and the ozone precursors: nitrogen oxides and volatile organic compounds. Ozone typically forms during periods of elevated temperatures, bright sunlight (with minimal cloud cover), low wind speeds, and moderate-to-high concentrations of nitrogen oxides and volatile organic compounds. Nitrogen oxides are formed almost entirely by high-temperature combustion. Vehicle engines are a source of nitrogen oxides, with cars and trucks accounting for 30% of the nitrogen oxides in the Houston area. Volatile organic compounds are chemicals that vaporize easily, such as the components of gasoline and solvents. Cars and trucks account for 26% of Houston's volatile organic compound emissions.

Ozone-reduction efforts through the 1990s resulted in a gradual downward trend in ozone levels, in spite of substantial population and economic growth during the period. However, the trend was not sufficient to bring Houston into compliance by 2007. In 1998 it was determined that a 65% to 85% reduction in nitrogen oxide emissions was needed. Drastic action became necessary. In 2002 speed limits on Houston freeways were reduced to 55 miles per hour (88 km/h) and a vehicle emissions testing program was implemented. The speed limits were soon increased due to public opposition to the low speed limits and minimal contribution to air quality improvement. The ozone reduction program will be carefully monitored and refined as the deadline nears. Compliance with federal air quality regulations is the biggest challenge facing Houston, and Houston's political leadership has worked hard to find a solution.

Other federal regulations relating to the environmental impact of freeways have generally slowed Houston's freeway projects but have not stopped them. With enough persistence and political support, the projects can usually move forward. Perhaps the best example of a freeway which struggled to overcome the anti-freeway mechanisms of the regulatory environment is the Fort Bend Parkway Tollway. Local officials tried to build the facility for 40 years, and during the 1990s federal environmental regulations slowed the progress of the project. But persistence paid off, and the tollway had its ground breaking in May 2003.

The Opposition

Federal regulations also serve as an effective tool for anti-freeway forces. Environmental groups tend to oppose all new highway construction, wherever it may be. They have attempted to use federal air quality regulations to pursue their larger agenda of curbing urban sprawl and reducing use of automobiles. In 2001 the Sierra Club and Environmental Defense filed a lawsuit to force adherence to strict limits for nitrogen oxide emissions and remove certain highway construction projects from the area's plan, including the Grand Parkway, Fort Bend Parkway, Westpark Tollway, and Katy Freeway expansion. The Environmental Protection Agency ultimately approved the region's smog reduction plan in October 2001, preserving the region's highway construction program. But the challenges of federal air-quality regulations to Houston's freeway system are far from over in spite of the near-unanimous political support and strong public support for ongoing infrastructure improvements, as evidenced by road bond election results.[122]

A prominent figure in anti-freeway efforts in Houston is environmental attorney James B. "Jim" Blackburn. When environmental groups or localized groups opposing highway projects need legal assistance, they turn to Blackburn. In many cases, the threat of litigation has altered plans or prompted additional studies, satisfying the concerns of the opposition. Blackburn has not yet scored a legal victory that seriously impacts Houston's freeway system, but he will no doubt continue his efforts. In 2003 he represented interests opposing the Katy Freeway expansion, initiating litigation to stop or alter the project. Other groups working to promote new forms of development, such as the new urbanism style of development which gained much publicity during the 1990s, oppose Houston's freeway construction program.*

The battle between roads and rail once again came to life in 1998 when anti-rail Mayor Bob Lanier left office and was succeeded by pro-rail Mayor Lee Brown. Houston voters rejected rail in 1973 and 1983, and Lanier nixed a planned monorail when he entered office in 1992. Mayor Brown put rail on the agenda once again, and this time the outcome of the battle was different—rail won. It was a small victory, but a hard-fought one. Brown was able to move the seven-mile (11 km), $325 million light rail project on Main Street forward without a public vote by using available cash funds. Rail opponents filed lawsuits to stop the project and obtained a restraining order preventing construction, but a court ruling on March 8, 2001, cleared all legal hurdles and allowed construction to begin. By a 54-46% margin in November 2001, voters rejected a proposition that would have forced a referendum on the project, which was about 25% complete. In 2001–2003 the Metropolitan Transit Authority conducted studies to define the alignment of potential future light rail lines.[123]

The future of rail in Houston will depend on a planned public vote in November 2003 and the priorities of future mayors. Still, the ridership of "successful" rail lines is so low that even the extensive system advocated by rail proponents will have a negligible effect on demand for Houston's freeway system. Dallas's 44-mile (70 km) light rail system, touted as a huge success, was carrying 59,360 passenger-trips per weekday in March 2003. Since most major urban freeways serve between 200,000 and 300,000 vehicles per day along their busiest sections, 59,000 passenger-trips per day on an entire system represents a negligible transportation impact. The high cost of rail systems, however, consumes a large amount of financial resources available to a region. Los Angeles invested approximately $2.5 billion in its 507-mile (811 km) Metrolink commuter rail system during the 1990s. In 2002 the commuter rail system was carrying 34,000 passengers per day. Atlanta's vaunted Marta rail system, with 219,000 daily passenger-trips in 2000, could not prevent the region from going into air-quality nonconformance in 1998. The Atlanta region has sustained substantial increases in traffic congestion and controversy about its urban sprawl in spite of its rail system. For comparison, Houston's freeway transitway system was carrying 121,086 daily passenger-trips in December 2002, with 40,185 on buses and the rest in carpools and vanpools. The capital investment for this level of transit patronage was far lower than for rail systems.[124]

The lessons from large, low-density cities that are investing in rail indicate that rail systems do not lessen the demand or need for freeways. Even though the transportation benefits of rail systems in low-density cities are marginal in spite of their huge price tags, proponents have been very successful in getting them built all over the United States. In Houston, light rail proponents now cite economic development as a key reason for building light rail, in addition to potential transportation benefits. As Houston continues its strong population growth, densification of the inner loop area will occur, possibly even fostering an urban environment that is less freeway-dependent. Light rail in Houston may very well carve out a niche serving a future dense inner-loop population.

Future Trends: the Tollwayization of Houston

Whatever happens in the future of Houston's freeways, many present trends are likely to continue. Funding will continue to fall short of needs. Available funding will tend to go where it is wanted the most—where strong political constituencies want the money and actively seek it. For the most part, this will be in suburban areas outside Beltway 8. Recent and planned construction on the Katy Freeway, North Freeway near the Woodlands, and Southwest Freeway at Sugar Land exemplify this trend. Freeway improvements inside Loop 610 will become increasingly rare.

Tollways will become the salvation of Houston's transportation system. The strong patronage of the toll road system will provide a revenue stream critical for Houston's transportation future, allowing the construction of new tollways and the expansion of existing tollways. The Harris County Toll Road Authority will become a partner in regionally important mobility projects. HCTRA's financial contribution to the Katy Freeway expansion exemplifies this trend.

Transportation facilities will continue to evolve and exhibit a wider variety of designs. Tollways will typically be small, without frontage roads when built on new alignments. Many new and upgraded freeways will adhere to the standard Houston freeway template with transitways and frontage roads. Major corridors will become more diversified with the inclusion of toll lanes and transit in addition to regular freeway lanes. There will be some movement away from frontage roads, particularly on new tollways and the Grand Parkway, but the frontage road will still remain the defining and distinguishing characteristic of Houston's freeway system.

* New urbanism promotes denser neighborhoods with narrower streets, houses on smaller lots, and inclusion of jobs and essential services within walking distance of homes so that car ownership is not a necessity. Many new urbanism communities resemble neighborhoods that were constructed before the era of widespread automobile ownership. A few neighborhoods with new urbanism characteristics have been constructed in Houston.

FREED

Freeway Metropolis

Is Houston the world's most freeway-focused city—the ultimate freeway metropolis? There are many potential competitors for such a distinction, and typically Los Angeles is automatically awarded that title. But Houston's freeways have a certain design feature that elevates the freeway to a higher level of importance in the functioning of the city than nearly all other cities. It's a feature that exists extensively only in Texas, and its prevalence in Houston is unequalled by any other city.

The freeway design feature is, of course, the freeway frontage road. Because of the frontage road, Houston's freeways are more than just transportation corridors. They form the city's principal commercial strips and business centers. Whether you are going to work, going shopping, going out to eat, staying at a hotel, seeking medical care, or buying a car in Houston, there's a good chance your destination will have a freeway address. The frontage road heavily influences development patterns, and the resulting land-use form is perhaps the most freeway-focused in the world.

The frontage road has also influenced many other distinguishing design features of Houston's freeways, including the extensive use of high mast illumination and the five-level freeway-to-freeway interchange. The frontage road increases the physical presence of the Houston freeway, adding at least 120 feet (37 m) to the freeway right-of-way footprint. Better yet, Texans and Houstonians like their frontage roads. The frontage road is almost as much a part of Texas as barbecue, cowboy boots, and oil wells, as the Texas Transportation Commission found out in 2001 when it tried to curtail the use of frontage roads on future freeways.

(Opposite page) **A Houston landscape:** The painting *Urban Landscape* (1970) by Houston artist Frank Freed depicts a web of freeways jam-packed with automobiles. Freed (1906-1975) enrolled in a basic painting class at the age of 42, intending to learn painting skills as a hobby. Over the next 27 years he produced a substantial body of work, focusing mainly on "folk art" depicting people and cultural themes. He produced very few paintings of Houston's urban landmarks. Fittingly for Houston, many of his few urban landmark paintings focus on freeways, often showing cars and freeways dominating the landscape.[1] (Source: Collection, The Israel Museum, Jerusalem)

The Parkway Ideal—Lost in the Beginning of Texas Freeway Time

As other cities developed their freeway systems starting in the 1920s, the freeways were often built around the parkway concept, with landscaping and foliage adjacent to the traffic lanes. New York City led the way with the first limited-access highways, opening the Bronx River Parkway in 1923. Through the 1930s Robert Moses built an extensive system of limited-access parkways around New York City. In the late 1930s plans for the Los Angeles freeway system began to develop. Early concepts for Los Angeles freeways showed wide park-like buffer zones separating the freeway lanes from adjacent areas. L.A.'s first freeway was a parkway, the Arroyo Seco Parkway, which opened in December 1940 and was later renamed the Pasadena Freeway.[2]

Houston, however, would march to its own drummer. The freeway designs of other cities had little, if any, influence on the design for Houston's first freeway, the Gulf Freeway, or any of its subsequent freeways. In 1942 the City of Houston Planning Department released the *Major Street Plan for Houston and Vicinity*, which discussed the need for freeways and parkways. Although it did not go into any specific design details, the document differentiated freeways and parkways. A parkway route was contemplated along Buffalo Bayou west of downtown since most of the corridor was already a park. Other discussed routes, including the Gulf Freeway, were envisioned as freeways.

The origins of the Gulf Freeway's frontage roads can be traced back to at least 1940. On April 12, 1940, Houston Mayor Oscar Holcombe announced the abandonment of the four remaining street car lines in Houston and the transfer of the title of the right-of-way of the Galveston-Houston Electric Railway to the city of Houston. In the announcement Mayor Holcombe disclosed his plans for a new superhighway on the former electric railway right-of-way. The plans included six main traffic lanes for high-speed traffic and two 2-lane sections for local traffic, which would be separated from the high-speed traffic with esplanades. It is likely that the head of the TxDOT Houston office, W. J. Van London, influenced the preliminary design of the freeway. The first detailed realization of the frontage road concept for the Gulf Freeway was reported to have been made by Van London on March 19, 1941, in a perspective drawing of the proposed Galveston-Houston Super Highway. At the time frontage roads were called service roads. Freeways having "pseudo" frontage roads

Frontage road vision, 1941: This view of the proposed Gulf Freeway was reported to be the first perspective drawing of the freeway. It was drawn by W. J. Van London, head of the TxDOT Houston office, in March 1941. Even on this semi-rural section of highway, frontage roads were in the contemplated design. In this rendering, the frontage roads are shown as two-way roadways. Plans for frontage roads on the Gulf Freeway can be traced back to at least April 1940, when Houston Mayor Oscar Holcombe announced the purchase of the Galveston-Houston Electric Railway right-of-way for use in a new superhighway. (Source: *Houston Chronicle*, August 1, 1952)

In the beginning, there were frontage roads: This circa 1950 view of the Houston's first freeway, the Gulf Freeway, shows the new freeway and its frontage roads. In the first generation of Houston's freeways, frontage roads were intended to provide local access and were not consistently continuous along freeways. In this view looking northbound toward downtown, the southbound frontage road stopped at the railroad crossing. (Photo: TxDOT)

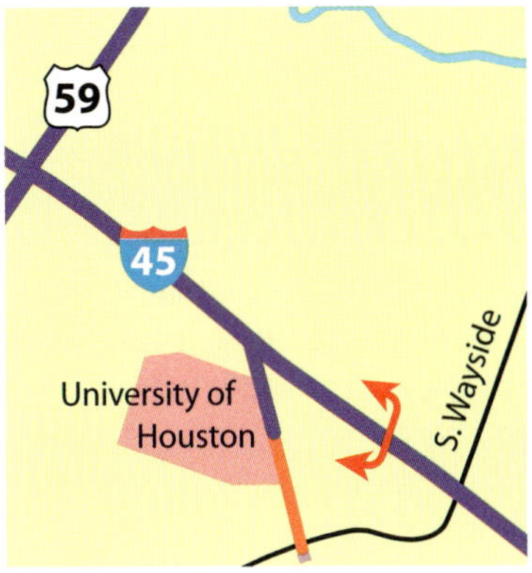

had already been built in other cities in the 1930s, particularly New York City, where several elevated freeway structures were served by ground level roads, many of which existed before construction of the elevated freeways. In Texas, and Houston in particular, frontage roads would become a standard feature of freeway design and construction.[3]

Only in Texas

Looking at the widespread usage of frontage roads on urban freeways in Texas, a question naturally arises: Why is Texas the only state in the United States to make extensive use of frontage roads? For that matter, why is Texas the only place in the world where frontage roads are used extensively? Frontage roads are used on a limited and irregular basis in many cities in the United States, but in all cases outside of Texas the frontage road mileage is a small percentage of the total freeway mileage. Only three cities with major freeway systems use frontage roads on substantially more than half of their urban freeways: Houston, Dallas-Fort Worth, and San Antonio. All are in Texas. Of the three, Houston and San Antonio have the highest percentage of frontage road freeways. Approximately 81% of Houston's freeways and tollways have frontage roads.* In Los Angeles, the freeway capital of the world, frontage roads are practically nonexistent.

The widespread use of frontage roads in Texas is the result of several factors contributing to an unwritten policy that influenced freeway design starting with the first Texas freeway, the Gulf Freeway, until the present day. Local desire for frontage roads, the financial strength to build frontage roads, access rights of adjacent landowners, traffic operation benefits, and the support of influential TxDOT head Dewitt Greer all contributed to the extensive use of frontage roads in Texas.

Until 1956 local entities in Texas were responsible for 100% of the right-of-way acquisition costs for urban freeways. Right-of-way costs were often as large as, or even larger than, actual construction costs for freeways being built through heavily urbanized areas. TxDOT generally tried to accommodate the wishes of the local governments that had made substantial financial contributions to freeways. Local interests almost always wanted frontage roads, and TxDOT generally accommodated their desires. After all, a tract of land on a freeway frontage road was often a very valuable piece of property, enriching landowners and generating more property taxes for local governments. In addition, frontage roads lessened the need for high-quality arterial streets near the freeway to serve local traffic, relieving local governments of that responsibility and cost. The Federal-Aid Highway Act of 1956 transferred right-of-way costs for the Interstate Highway

* The 81% figure is for freeways and tollways in a region extending approximately five miles outside Beltway 8. Westpark Drive is not considered a frontage road of the Westpark Tollway, but East and West Hardy are considered frontage roads of the Hardy Toll Road.

Factors that have influenced the widespread use of frontage roads in Houston and Texas

- Local governments and businesses wanted them, and TxDOT generally accommodated local desires.
- Frontage roads provide the legally required access to property adjacent to freeways.
- The father of the modern Texas highway system, Dewitt Greer, promoted highway access and the use of frontage roads.
- It was usually feasible to acquire the extra needed right-of-way in Texas' low-density cities.
- Texas has had the financial strength to shoulder the added cost of frontage roads.
- The option for staged construction: building frontage roads first and adding freeway main lanes as traffic volumes warrant.
- Traffic operation benefits.
- Texans' fondness for frontage roads and the convenience they provide.

System to federal and state agencies, and a 1957 state law made TxDOT responsible for 50% of the costs of right-of-way on non-interstate freeways. In spite of the reduced local contribution, TxDOT continued to meet the local desire for frontage roads. Paul R. Tutt, a TxDOT researcher who coauthored a comprehensive report on frontage roads in 1963, observed, "Once local people see frontage roads, they don't want freeways without them."[4]

Texas has generally had a well-funded highway program with the financial strength to shoulder the additional cost of frontage roads. This added cost applies to both right-of-way acquisition and actual construction. A report published in 2001 by the Center for Transportation Research in Austin stated, "The financial costs associated with frontage road facilities were found to be considerably higher than those associated with non-frontage road facilities." Frontage roads increase the freeway right-of-way requirement by at least 100 feet (30 m) and significantly more if modern design practices are used. Due to the low density of Texas cities, acquisition of the extra right-of-way needed for frontage roads was usually possible in the first wave of urban freeway construction from 1945 to 1980. Prior to 1956, when localities were responsible for all right-of-way costs, TxDOT could focus all of its resources on construction, making more funds available for the increased construction cost of frontage roads. When the federal government launched large-scale construction of the Interstate Highway System in 1956, TxDOT needed to ensure that federal funds would cover the cost of frontage roads. To sell the idea of widespread frontage roads along interstate highways to a somewhat skeptical Federal Bureau of Roads, TxDOT published a comprehensive research report in 1963 detailing the benefits of frontage roads. In the second wave of freeway construction (post 1980), when new freeways are built mostly through undeveloped land and property displacements typically occur along expansions of existing freeways, relatively low property values keep right-of-way costs affordable.[5]

The issue of access rights for property owners adjacent to controlled-access freeways is something that affects all states. Access-control law is a broad subject that has been influenced by legislation and many legal decisions over the years. In general, property that has access before the construction of the freeway must have its access preserved, and property that becomes "landlocked," such as when a large tract is severed into two sections, must have some kind of access provided. National-level guidelines for access control were developed after the Federal-Aid Highway Act of 1956, and each state developed its own policies and practices for handling access rights issues. Solutions to the access-rights problem included the outright purchase of affected properties, the purchase of access rights, connecting landlocked tracts to a nearby tract with access, and frontage roads. In Texas the frontage road solution was used far more frequently than in any other state. This can be attributed to strong property owner rights in Texas and the preference of officials to promote development along freeways. It is also widely believed that the construction of frontage roads has been more economical than the purchase of access rights.[6]

In Houston the influence of landowners takes on an added dimension. Houston's business, real estate, and land development interests are politically powerful and are often the driving force in getting freeways and tollways built and expanded, either directly with land donations or indirectly using their political influence. The business interests supporting freeways want frontage roads along their freeways. After all, there is probably no better way to increase the value of a property than to build a freeway with frontage roads through it.

Frontage roads also provide important traffic engineering and operational benefits to the freeway. Local traffic is kept on the frontage roads rather than entering the main lanes for short distances. In the event of an emergency freeway shutdown, traffic can be diverted to the frontage roads. Frontage roads provide the option of staged construction, where the frontage roads are built before the main lanes and the main lanes are added when traffic volumes justify their construction. The operational benefits of frontage roads were called into question in 2001, prompting an effort to revise frontage road policies.

The aesthetics of the frontage road freeway are not a problem in Texas. In the early years of freeway construction, promoting development along freeways was seen as a way to help propel a predominantly rural and underdeveloped Texas into the modern era. The idea of commercial development and commercial clutter along freeways is accepted in Texas, especially in Houston. Other states adhere to the ideal of parkway or greenbelt freeways and would never tolerate the commercialism associated with

The modern Texas frontage road: This view of the Eastex Freeway in north Houston shows a modern Houston freeway and its frontage roads. Frontage roads generally have three continuous lanes in each direction with additional lanes at merging points and intersections. This section of freeway is unusually free of commercial development. (Photo: September 2002)

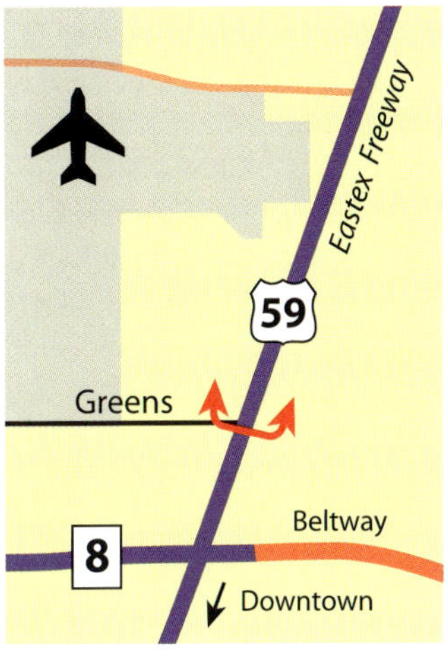

The U-turn: The U-turn is a standard part of interchange design in Texas and is a characteristic of Texas freeways that is associated with frontage roads. U-turns allow a motorist on a frontage road to turn around and proceed in the opposite direction without having to go through the intersection traffic light. This U-turn is on the North Freeway at FM 1960. (Photo: November 2002)

"Don't you mean *feeder* roads?"

Frontage roads may be the official term, but in Houston they're most commonly called feeder roads, or simply feeders, by the general population. There are numerous other terms for frontage roads, and every region seems to have its favorite. The Dallas-Fort Worth region prefers to call them service roads, a term which was also widely used in Houston in the 1960s. San Antonio prefers to call them access roads. In El Paso they're sometimes called gateways. In other parts of the United States, the terms outer roads, utility roads, and marginal roads are used. Houston's preferred name has the advantage of being more concise and can be used without the term "roads," for example, "Take the exit and stay on the feeder."

frontage road freeways. Texans like the convenience provided by the frontage road and U-turn. Motorists can easily find businesses along the freeway without having to stray onto side streets. Frontage road travel is such an entrenched part of daily driving in Texas that a state representative remarked in 2001, "I don't think any of us would know how to get around without frontage roads."[7]

Support for frontage roads started at the highest level of TxDOT in the office of Dewitt Greer. Greer presided over TxDOT from 1940 to 1968 and had a tremendous influence on the development of the Texas highway system. He is generally regarded as the father of the modern Texas highway system. Greer was a strong promoter of frontage roads and his influence helped spread them across the Texas landscape. In a March 1954 article in *Texas Highways* magazine, Greer wrote that "frontage roads are built adjacent to the right-of-way line where there is sufficient development of adjacent properties to create the need for such frontage roads to carry the local traffic to the selected points of entry into the express lanes." In Houston, frontage roads were built not only in the situation described by Greer, but also in just about every instance where they

Almost everywhere: The map below shows Houston's freeways and tollways that have frontage roads. Approximately 81% of complete limited-access facilities in a region extending about 5 miles (8 km) outside Beltway 8 have frontage roads.

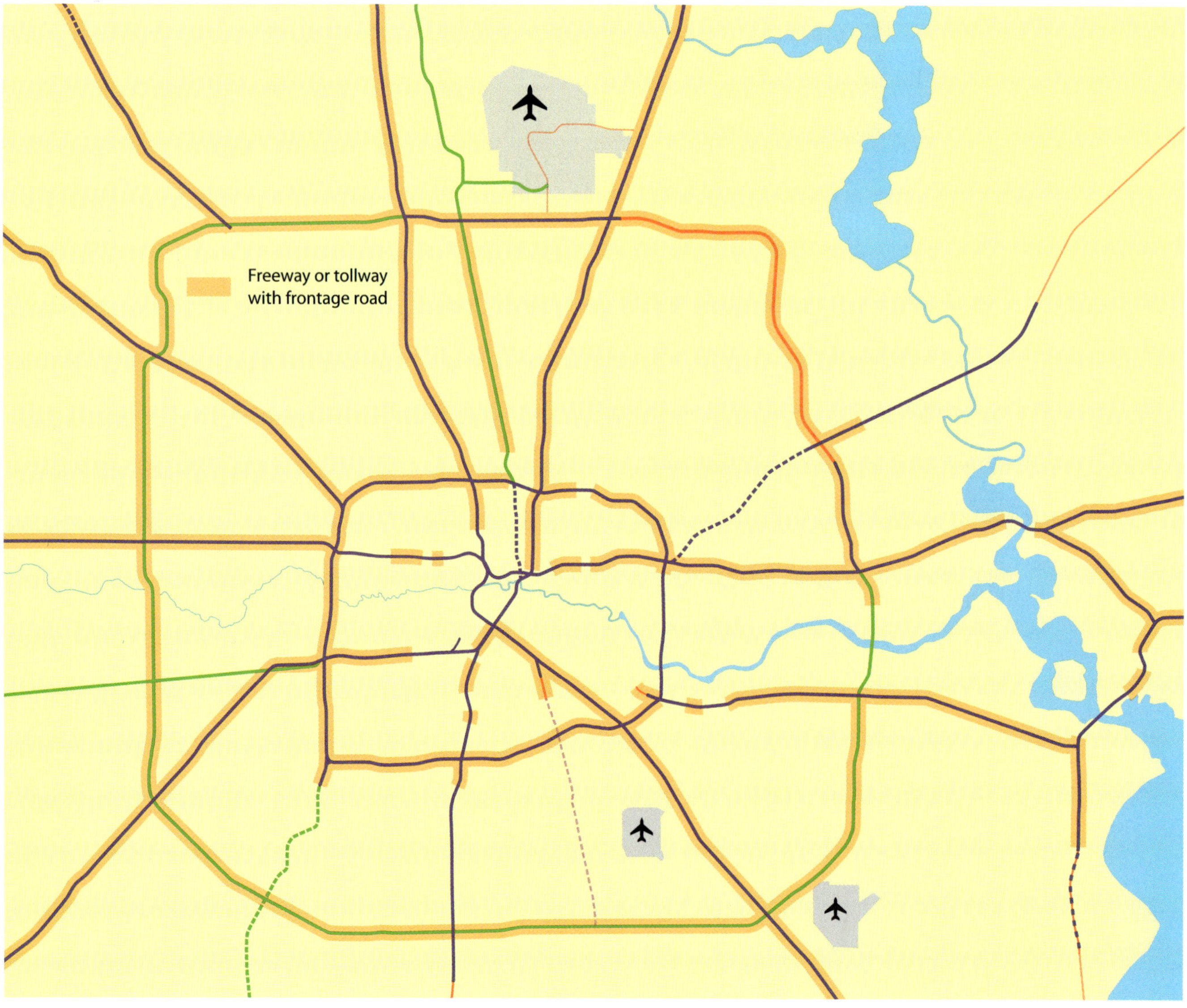

The commercial strip: Because of frontage roads, Houston's freeways serve the dual purpose of transportation corridors and commercial business strips. This 1974 photo along the Katy Freeway near Gessner shows the typical commercial development that occurs along Houston's freeways. Also in this photo, notice the military aircraft, a Lockheed F-104 Starfighter, on the Memorial City parking lot. It was part of a mall promotion. The first F-104 flew in 1954, and an F-104 set world records in speed (1,404 miles per hour, 2,246 km/h) and altitude (103,395 feet, 31,513 m) in 1958 and 1959. The aircraft was widely used by the United States and its allies in the 1960s. (Photo: Chuck Fuhs)

could practically be constructed. All the managers of the Houston offices of TxDOT, from the end of World War II to the present day, have been supportive of frontage roads and included them in freeway designs.[8]

Finally, a philosophy for Texas-sized construction—bigger and better—has surely had some influence. The frontage road freeway is the ultimate in freeway design, with the biggest footprint and the most lanes.

> "Highway frontage roads bristling with strip malls are as Texan as pickup trucks with gun racks."
>
> *Dallas Morning News*, September 6, 2001

The Frontage Road Matures and Survives Challenges

In the first generation of Houston freeways, frontage roads were intended mainly for local access. They did not continue through freeway-to-freeway interchanges, usually did not cross railroad tracks, and did not continue through areas where right-of-way acquisition was difficult. With the second wave of Houston freeways, starting around 1980, frontage roads were elevated to a higher level of service. They continued past railroads with grade separations and continued through major freeway interchanges. Major freeway expansions included the addition of missing links of frontage roads. In the most dramatic instance of adding a missing link of frontage road, the West Loop frontage roads were trenched underneath the four-level Southwest Freeway interchange in a project that began in 2001. The only all-new freeway constructed in the 1990s, the SH 249 Tomball Parkway, included full frontage roads.

Still, the era of near-universal frontage roads in planned freeways and upgrades in Houston started to come to an end in the 1980s. Houston's first freeway to be constructed mostly without frontage roads, the South Freeway, was completed in 1984. The Harris County Toll Road Authority (HCTRA) was created in 1983, and HCTRA builds tollway main lanes only. The Sam Houston Tollway has frontage roads built by TxDOT, but for the most part new facilities built by HCTRA do not have frontage roads. The Grand Parkway came back to life in the 1980s and the lack

Billboard alley: The North Freeway has distinguished itself as having one of Houston's highest concentrations of billboards and commercial clutter. This has always been a rather sore point with local economic development officials, since the North Freeway is one of the main routes leading to Bush Intercontinental Airport and provides the first impression many visitors have of Houston. On July 20, 2003, the *Houston Chronicle* reported on Houston's preparations for the 2004 Super Bowl: "Some civic leaders consider the North Freeway, with its heavy traffic and long stretches of garish signs, strip malls, car lots, topless clubs and modeling studios, such an eyesore that limo drivers may be asked to bring VIPs in on the Hardy [Toll Road]." (Photo: September 2002)

of frontage roads in its planned design was touted as one of its selling points. Without frontage roads, the Grand Parkway would not become another long commercial strip and would truly have some parkway qualities.

A much larger challenge to frontage roads emerged at the Texas Transportation Commission in June 2001. The commission approved a resolution to drastically curtail the construction of frontage roads on future limited-access highways in Texas. In the testimony at the commission meeting, it was stated that frontage roads reduce the operational efficiency of the freeway main lanes by introducing "points of conflict" at the frequent entrance ramps, exit ramps, and frontage road intersections with streets. The cost of frontage roads was not cited as a reason, but an associated resolution passed at the same meeting was intended to shift the responsibility and cost for local access to local governments. "TxDOT will work with local officials to design and construct feasible alternatives that may be less costly and provide equal or better overall operation to the transportation system," a TxDOT official stated. To most observers, this earth-shaking change in Texas policy came suddenly and without warning, but high-level TxDOT officials and the commission recognized the significance of their new policy. "It's huge; it's probably one of the biggest decisions that we've made in a long time," stated Wes Heald, administrative director of the Texas Transportation Commission. Commissioner Ric Williamson knew there would be some resistance. "We can expect to be deluged over the next year about this because the [commission] views frontage roads and access roads as part of the political milieu of the state, and we'll just all have to [be] prepared to be firm and just say we can't do this anymore."[9]

Word of the new policy started to trickle out. Retail industries were concerned that coveted freeway locations would no longer be available. Local governments were concerned about the added costs of beefing up their street networks and the need to substantially revise their street planning processes. Particularly in the Dallas-Fort Worth region, where many freeways were constructed without frontage roads, local political officials had worked long and hard to get funding for frontage roads on their freeways. With the new policy, they would never get their frontage roads. A series of public meetings was held, and public comment was overwhelmingly negative towards the new policy. In May 2002 the Texas Transportation Commission withdrew the proposed new policies on frontage roads without discussion. Instead, an access management policy would be developed to help manage traffic flow on freeway main lanes. The Texas frontage road would live, and Texas would remain unique in the world with its dedication to frontage roads.[10]

Houston's first stack interchange: The West Loop-Southwest Freeway interchange was Houston's first four-level stack when it opened in 1962. Serving the heavily congested Southwest and West Loop Freeways, it has been Houston's busiest interchange since its earliest days. But finally some relief is on the way for the beleaguered interchange. In work that began in 2001, the interchange is being upgraded to a five-level stack by trenching the West Loop frontage roads underneath the interchange (visible on the right side of the interchange in this photo). New high-level ramps will be added, and the Westpark Tollway will terminate at the west (left) side of the interchange. (Photo: November 2002)

The Commercial Strip and the Scenic Houston Freeway

Every city has its commercial strips, full of retail establishments and associated signs and clutter. Because of the extensive use of frontage roads in Houston, most Houston freeways serve the dual purpose of transportation corridors and commercial strips. Adding to the visual impact is the widespread prevalence of billboards along many Houston freeways. The resulting visual experience can almost qualify Houston's freeways as anti-parkways. If you want nature and landscaping, you shouldn't look for them along Houston's freeways. In contrast, freeways in other cities such as Los Angeles or Atlanta create an entirely different atmosphere—a noncommercial atmosphere.

Even in the 1960s, many in Houston were attempting to alter the visual atmosphere of Houston's freeways. The year 1968 seems to be the beginning of freeway beautification efforts. In June 1968, the Houston Municipal Art Commission published a small report called *Beautification Study: Freeways* in response to "the growing concern in Houston for the preservation or restoration of beauty within the city." The report was certainly practical and modest, simply recommending the addition of landscaping along freeways where it was feasible. Also in 1968, the head of the TxDOT Houston Urban Project Office, Albert Kyser, was urging the city of Houston to ban billboards on the soon-to-be-opened section of the Katy Freeway inside Loop 610, stating, "We have gone to extra expense to acquire extra right-of-way to make this freeway more beautiful, and the city should follow through and pass an ordinance prohibiting billboards along this freeway." The Houston billboard industry was very powerful, so no anti-billboard ordinance would be passed. Any beautification efforts undertaken in that era were minimal and did not last.[11]

Local officials interested in beautification focused their efforts on billboards in the 1970s. By the late 1970s, Houston had about 10,000 billboards that were not associated with the business on the property ("off-premise" billboards), prompting some people to call Houston the "billboard capital of the world." Houston's first anti-billboard ordinance was passed in 1980, banning the construction of any new off-premise billboards in Houston and placing restrictions on existing billboards. In 1985, in response to a state law, the ordinance was amended to prohibit the replacement of damaged billboards and the replacement

The drive-thru construction zone: Interchange construction is one of the most impressive events in all of construction. Unlike all other major construction events, the public is able to get right in the middle of the project on a daily basis. This is a view of the Katy Freeway-Beltway 8 interchange construction zone in 1988. (Photo: Allen Moradi)

of wooden billboards with steel billboards. In 1992 the ordinance was again amended to ban nearly all off-premise billboards. An amortization period was established—17 years for wooden billboards and 21 years for steel billboards—after which off-premise billboards would need to be removed. In 12 years, Houston had gone from being one of the most lenient cities in terms of billboards to being one of the most restrictive.[12]

Another powerful tool against billboards was the establishment of scenic districts. Beautification advocates successfully had the entire Beltway 8-Sam Houston Parkway designated as a scenic district in December 1986, preventing the placement of any billboards and strictly regulating on-premise signs. Various other freeway sections have become part of scenic districts in Houston. "Scenic" is a relative term when applied to Houston, of course. Houston can never have a freeway as scenic as the Serra Junipero Freeway (Interstate 280) in Silicon Valley, California, or Interstate 5 north of San Diego, California. Houston freeways would need to be scenic in their own way, showcasing the best that Houston has to offer.[13]

However, there was a big loophole in the anti-billboard laws—the freeway loophole. Federal law requires compensation for lost income for billboards removed along federally funded highway facilities, and it was not financially viable for government agencies to pay the estimated $300,000 to $500,000 cost for each billboard along major freeways. To further complicate the issue, Houston's freeway expansion program was taking wide swaths of right-of-way along existing freeways, more than 100 feet (30 m) wide in many cases, and large numbers of billboards were located in the right-of-way needed for freeway expansions. TxDOT was not willing the pay the huge cost for billboard removal, so a compromise was reached in 1987. When a freeway expansion necessitated the removal of a billboard, the owners were allowed to move the billboard to another location along a freeway. Billboard owners could move their billboards to any location within Houston and its extraterritorial jurisdiction where regulations allowed them. As of 1998, 202 signs had been authorized for relocation. Naturally, billboard owners tried to move their billboards to other freeway locations with maximum traffic volumes. The result: a concentration of billboards along certain sections of freeways in Houston. Houston may rid itself of most of its off-premise billboards sometime around 2012, but the freeway billboards will likely exist for a long time.[14]

While the long battle over billboards played out, other initiatives were underway to help beautify Houston's freeways. Landscaping was implemented along some

The first five-level stack: The interchange at Beltway 8 and the Katy Freeway launched the second generation of Houston stacks when it was completed in 1989. Unlike the first generation of stacks built around Loop 610, the second generation included continuous frontage roads through the interchanges, requiring an additional level. This stack will also be Houston's shortest-lived, with demolition and reconstruction scheduled for 2004-2008. (Photo: James Lyle, TTI, June 2001)

freeways on a limited basis, but more significant was the trend toward attractive freeway design features that began in the early 1990s and started to bear fruit by the late 1990s. As of the late 1980s, traditional design practices with a disregard for aesthetics were still in use. Examples include the early five-level stack interchanges along western Beltway 8, particularly at the Katy Freeway (completed in 1989) and the Northwest Freeway (completed in 1990). Initial efforts were modest and involved the use of more visually pleasing construction techniques, such as trapezoidal prestressed concrete beams and tapered bridge piers that provided a sleeker, more California-style look. These techniques appeared on the reconstruction of the Eastex Freeway, which began in the early 1990s. New stack interchanges, particularly the Beltway 8-Gulf Freeway interchange (1997) and the Beltway 8-North Freeway interchange (1997/2003), were designed with aesthetics in mind. The reconstruction of the Southwest Freeway near downtown, a project whose first phase was completed in 2002, provided Houston with the first taste of a truly aesthetic freeway with its long-span arched bridges, decorative lighting, and architectural enhancements. Plans were in development to transform the Katy Freeway into a distinctive freeway during its comprehensive reconstruction, scheduled for 2003-2008. In 2002, in conjunction with planning for the Katy Freeway project, TxDOT launched the Green Ribbon project, an effort to define beautification standards for 80 miles (128 km) of Houston's freeways and promote public-private partnerships to implement the improvements.

Stack City

Built on the coastal plain of the Gulf of Mexico, Houston is nearly a perfectly flat city. But Houston's motorists can escape the confines of flatness on the ramps of Houston's freeway interchanges. Sometimes the freeway interchanges are visible from miles away, starting as a small projection on the horizon and slowly taking form as monumental structures as the driver approaches. Other times they appear suddenly in front of the driver, almost as if to startle. From below, their gracefully curving, high-flying ramps define a unique geometry that seems to dwarf even the biggest freeway. On the connection ramps, the stack offers the thrill of an out-of-the-ordinary driving experience and a chance for vistas not normally available to the landbound driver.

The multilevel freeway-to-freeway interchange, commonly known as a stack interchange, is one of the signature structures of Houston's freeway system. With the original construction of its first wave of freeways, Houston earned strong interchange credentials with the four-level interchanges around Loop 610 and within the downtown interchange complex. But the second wave of freeway interchange construction, beginning in the mid-1980s, brought a series of new five-level stacks that

Concrete and steel tour de force: The Northwest Freeway-Beltway 8 interchange has the highest and structurally most impressive ramps in Houston. The interchange was completed in 1990. (Photo: May 2002)

(Opposite page upper) The Gulf Freeway-Beltway 8 interchange features high-flying ramps, symmetric design, a sprawling right-of-way, and architectural enhancements. The interchange was completed in 1997. (Photo: November 2002)

(Opposite page lower) This view shows a Beltway 8 frontage road at the Northwest Freeway-Beltway 8 interchange. (Photo: December 2002)

propelled Houston into the elite of stack cities.* Houston and Los Angeles are in a class of their own when it comes to the size and number of multi-level stack interchanges. Los Angeles built the original stack in 1953 and still reigns supreme with its sleek cast-in-place concrete structures. The world's most impressive group of stacks is probably the four stacks along the Century Freeway (Interstate 105) in Los Angeles. But Houston has the larger Beltway 8 collection, which is still a work in progress. If Los Angeles is the king, Houston is the queen. They are the royalty. Other contenders and up-and-comers can only reach the level of high nobility.

Houston's commitment to the multilevel stack interchange is unequalled in the United States. The cloverleaf interchange was banned from consideration in the mid-1950s and the complete stack became the only way to build freeway-to-freeway interchanges. In nearly all other freeway cities, including Los Angeles, the use of multilevel stacks is not exclusive at freeway-to-freeway interchanges. Many Los Angeles interchanges include cloverleaf connections for low-volume traffic movements or use interchange designs that minimize the number of levels or the height of structures. During the entire history of Houston's freeway system, there has been only one cloverleaf connection in all the region's freeway-to-freeway interchanges. That connection, at the Eastex-East Freeway interchange downtown, was removed in 2000 and replaced with a direct connector ramp. A few cloverleaf connections exist at freeway-to-highway intersections,

* In determining the number of levels, the ground level is counted as a level. A four-level interchange consists of a freeway at ground level and three levels of overpasses.

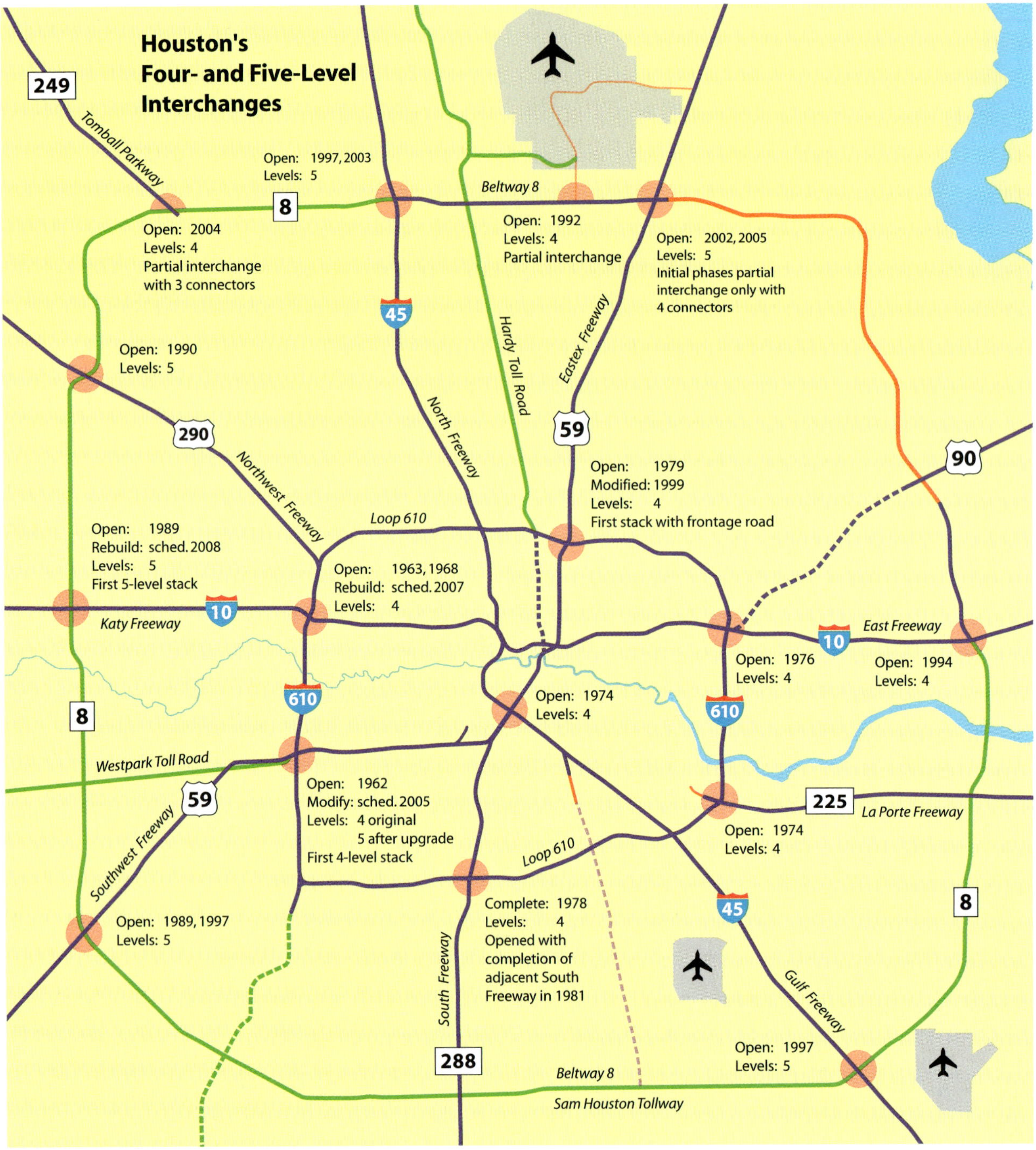

mainly along older sections of freeway. However, two new cloverleaf connections will appear at the intersection of the Sam Houston Tollway and the Westpark Tollway, scheduled for completion in 2004.

Houston's first two fully-directional freeway-to-freeway interchanges opened in July 1962 at the Loop 610-North Freeway intersection and the Loop 610-Southwest Freeway intersection. The Loop 610-North Freeway interchange was only a three-level structure and included left-lane exit ramps, a substandard design practice which was being phased out. The four-level Loop 610-Southwest Freeway interchange was Houston's first stack built to modern standards. It was the second four-level interchange in Texas, coming after the Fort Worth Mixmaster, which opened in 1958. The Houston interchange was far more modern in design than the Mixmaster, which was demolished and replaced with a new structure between 1993 and 2001.[15]

Over the next 17 years stack interchanges were built around Loop 610. By 1979, one three-level and six four-level fully-directional freeway-to-freeway interchanges were complete around Loop 610. Four partial interchanges were also complete by 1988. After the second interchange was completed at the intersection of the West Loop and the Katy Freeway in 1968, a design philosophy shift occurred. The first two four-level interchanges placed freeway main lanes at the top level of the interchange. All future stacks in Houston would have connector ramps at the upper levels. There were two exceptions to this design philosophy. The irregular Gulf Freeway-South Loop and

Gulf Freeway-US 59 interchanges placed freeway main lanes at the top level.

The frontage road has played a key role in defining the design of the second generation of stack interchanges in Houston. In the first generation of four-level freeway stacks, frontage roads did not continue through the interchanges, instead skirting around the edges or stopping short of the interchanges. In the 1980s frontage road service levels were elevated to higher standards. In the second generation of interchanges built around Beltway 8, freeway frontage roads were designed to pass underneath the interchanges. This necessitated adding an extra level, resulting in the five-level stack interchange. All the Beltway 8 interchanges are designed with the frontage roads on the first (ground) level, the intersecting radial freeway main lanes on the second level, the Beltway 8 main lanes on the third level, and connector ramps on the fourth and fifth levels. It should be pointed out, however, that the connection ramps typically do not intersect over a single point. This means that the crossing of a level four and level five ramp may occur at a point where the level four ramp is below its peak. Nevertheless, Beltway 8 still has some of the highest and longest connection ramps in the United States.

Houston's first five-level interchange was completed in 1989 at the intersection of Beltway 8 and the Katy Freeway. The interchange was particularly challenging to design due to limited available right-of-way. It features connection ramps positioned directly over the Katy Freeway frontage roads. The 1990s saw a succession of major stack interchanges being completed around the Beltway, all of them five-level structures except for the four-level interchange at the East Freeway. The interchange at the intersection of the Gulf Freeway and Beltway 8, completed in 1997, probably ranks as Houston's greatest achievement in stack interchange design. It features a sprawling right-of-way, symmetric design, and long, high-flying ramps. Could it be the perfect stack? Maybe. The Beltway 8 stack collection continues to be a work in progress in 2003. The newest addition is the interchange at the North Freeway, completed in January 2003. Work is in progress at the Eastex Freeway interchange to build three connector ramps. Three connector ramps are underway at the SH 249 Tomball Parkway interchange, which will be a partial stack only. In the long term, possibilities exist for new stack interchanges at the South, La Porte, and Crosby Freeways. However, long-range plans in 2003 do not call for construction.

Houston's ongoing program of renewal and modernization will reach a new and unprecedented level with the Katy Freeway expansion program, scheduled for 2003–2008. The expansion program will require the dismantlement and reconstruction of two major stack interchanges: the four-level stack at Loop 610 and the five-level stack at Beltway 8. Houston's first five-level stack interchange, completed in 1989, will have a life span of about 17 years—an unprecedented short life for such a major structure. In another unusual development, Houston's first four-level interchange, the 1962 Loop 610-Southwest Freeway interchange, is being modernized and converted to a five-level structure with the construction of the Loop 610 frontage roads underneath the interchange. Work is scheduled to be complete by 2005.

The future holds the tantalizing possibility of a third generation of stack interchanges along the Grand Parkway. Preliminary plans show five-level stack interchanges similar to the structures along Beltway 8. However, any stack interchanges along the Grand Parkway are in the distant future.

Urban Sprawl

Scatteration. Dispersion. Urban sprawl. These were terms the public was beginning to hear from public officials starting in the mid-1950s. Mass-produced and affordable automobiles had been available since Henry Ford launched the automobile culture in the United States, but their real impact on cities became evident in the years after World War II when rapidly rising incomes and advancing technology made widespread suburbanization possible. Around the same time that planners in Houston were formulating Houston's freeway master plan in the early 1950s, urban sprawl was becoming the new face of American cities, and Houston was no exception.

Urban planners in Houston definitely took note of the movement toward low-density development. The era just after World War II was perhaps the most remarkable time in history to be an urban planner. It was an era that provided the opportunity to define the future of cities through the development of freeway networks. In Houston the freeway network would be particularly important since there were no physical barriers and no limits to growth. Leading Houston's effort to deal with scatteration, as he more commonly called it at the time, was Houston planning director Ralph Ellifrit. Ellifrit had graduated from the University of Illinois in 1932 with a degree in landscape architecture, a degree which, at that time, included urban planning. He won the Edward L. Ryerson European Traveling Fellowship after graduation and spent nine months in England, France, Italy, Austria, and Greece studying architecture and urban planning. But the city of the future did not resemble the dense European cities he studied in any shape or form. After becoming director of planning for the city of Houston in 1940, he would lead Houston into the era of sprawl and freeways.

Ellifrit was well aware of the challenges of transportation planning in the low-density city that was taking shape. In a 1957 discussion he stated, "Scattering people so thinly poses tremendous—almost insurmountable—problems in transportation. It means that mass transit—already having rough rolling even in the most densely populated cities—will become impossible under present trends, except in the relatively small and clearly defined central city where population density offers enough passengers to support mass transit. A transit system isn't feasible because of pyramiding costs per passenger mile of pushing service further out into less and less densely

Freshly minted suburban tract homes, 1952: These new homes in the Oak Forest subdivision in northwest Houston were part of the movement toward single-family homes and low-density automobile suburbia that was rapidly transforming cities across the United States. By the mid-1950s the issue of urban sprawl was first beginning to receive attention from government officials and civic leaders in Houston. Houston's planners viewed low-density suburbanization as an unstoppable trend and the best strategy was to accommodate it—with freeways. (Photo: HMRC)

populated areas." The solution, of course, was the freeway and thoroughfare system, but Ellifrit knew that it would be difficult to keep up with demand. "Traffic is increasing far faster than facilities are being provided," he cautioned. Ellifrit had a vision of the future, and it was already starting to happen in Los Angeles. He cited the Hollywood Freeway, which had sustained a traffic volume increase from 90,000 vehicles per day in 1951 to 160,000 vehicles per day in 1957.[16]

President of the Houston Chamber of Commerce Ben C. Belt was even more dramatic in his assessment. To him, the "explosion of urban population over the countryside surrounding our major cities" was "this century's paramount problem." He continued, "This 'urban sprawl' produces a maze of metropolitan problems increasing in number and complexity as the metropolitan population increases rapidly." In the end, it seemed that only one thing could be done. "We will have to learn to live with it," said Marvin Hurley, executive vice president of the Chamber of Commerce in 1957. To planners in Houston, low-density suburbanization was an unstoppable trend and the only reasonable response was to accommodate it as best as possible.[17]

Ralph Ellifrit and all involved in Houston's planning process coped quite well, developing a comprehensive freeway plan and making great strides in its implementation through the 1960s. Houston reached its peak mobility in 1970 when the first phase of the freeway system was substantially completed. But the most important tool in accommodating urban sprawl, the freeway, would come under attack by 1970. Freeway construction was substantially curtailed in Houston and in every city in the United States starting in the 1970s in response to opposition and the political leadership's decision to allow transportation investment to decline. But urban sprawl never slowed down. In fact, it continued faster than ever, even without adequate transportation infrastructure. The opposition forces that stopped many freeways could not stop sprawl, and the net result is perhaps the worst of all worlds: sprawl with inadequate freeway infrastructure.[18]

Houston has been better equipped to deal with sprawl than most other cities, largely because of the early planning efforts of Ralph Ellifrit and others to develop Houston's freeway system. Most importantly, Houston was able to implement most of its pre-1970 planned freeways. The inclusion of Beltway 8 and the Grand Parkway were perhaps the most forward-looking features of early planning efforts, envisioning the sprawl and suburb-to-suburb commuting patterns that developed in practically all large American cities. Houston's political leadership

"This century's paramount problem."

Houston Chamber of Commerce President Ben C. Belt on the subject of urban sprawl in 1956

"The trend is toward increasing individual mobility—people with their own transportation."

City of Houston Planning Director Ralph Ellifrit, 1957

has recognized that freeways play an important role in transportation solutions for the sprawling city as well as cost-effective transit solutions. However, Ralph Ellifrit's 1957 observation that it may never be possible to keep up with the public's insatiable demand for automobile transportation infrastructure has certainly proven to be true. With all the demands placed on governments, the political issues involved in highway construction, and the limited resources available to communities, it will probably never be possible to provide the transportation infrastructure that sprawl demands.

There is perhaps another reason why Houston has been able to continue its freeway expansion program to meet its transportation needs. In its 1998 sprawl report, the Sierra Club listed the cities that it designated as the most "sprawl-threatened," including 20 cities in the category of large cities with a population more than one million. Houston did not appear on the list. Houston did not even appear in the "dishonorable mentions" list. This was certainly a surprising result since Houston is sprawling just as quickly as most other large American cities. If the Sierra Club did not find sufficient fault with Houston to place it on the top 20 list, it must be because of minimal environmental impact of sprawl in the Houston region. Certainly there are areas around Houston that have been identified as environmentally sensitive, such as the Katy Prairie. But in comparison to most major cities, Houston has fewer environmental issues. Could it be that the geographic location of Houston is near-ideal for sprawl and freeways? Maybe. And that certainly works in favor of the freeway.[19]

Flooded Metropolis

In Houston, flooding is a way of life. On a regular basis motorists have to contend with accumulations of water on Houston's streets. Usually it's just a nuisance. Every few years a major flood event occurs, inundating parts of the city, flooding houses, and submerging vehicles. And then there was Tropical Storm Allison in June 2001.

Tropical Storm Allison first visited Houston on June 5, delivering heavy rain and moderate flooding. Allison then drifted north to near Lufkin, Texas. But Allison wasn't through with Houston. The storm changed directions and began meandering southward, returning to Houston with a vengeance the night of Friday June 8. The storm dropped huge volumes of rain and unleashed Houston's worst modern flood event. Twenty-two people died. More than 70,000 homes and 95,000 vehicles flooded. Property damage topped $5 billion. Areas that had never previously flooded were submerged. One location in northeast Houston received a storm total of 36 inches (91 cm) of rain and 26.5 inches (67 cm) in the devastating 24 hours preceding Saturday afternoon. Much of Houston received between 10 and 25 inches (25 to 64 cm) of rain during the storm, and 10 to 15 inches (25 to 38 cm) during the key 24-hour period.[20]

Anything near a stream or bayou was at risk for flooding, and structures below ground level were particularly susceptible to the onslaught of water. In downtown Houston and the Texas Medical Center, basements and tunnels were submerged, causing extensive damage. Many of Houston's freeways are built below ground level, and Tropical Storm Allison unleashed the greatest freeway flood event in Houston's history. Who knows, maybe it was the greatest freeway flood event the world has ever seen. Freeways built below ground level were transformed into a chain of lakes. Abandoned cars and trucks drifted in the water, sometimes washing up on the shorelines of the freeways. Freeways provided some of the most dramatic views of the flood, but the freeways themselves were largely undamaged. After the water receded and wreckers cleaned up the carcasses of vehicles and trucks, the freeways were put back into service and were fully functional by Monday morning.[21]

A swimming pool for 18-wheelers: This view looks east along the Katy Freeway (IH 10) at T. C. Jester on June 9, 2001, during the flood caused by Tropical Storm Allison. (Photo: copyright Jim Olive/Stockyard.com)

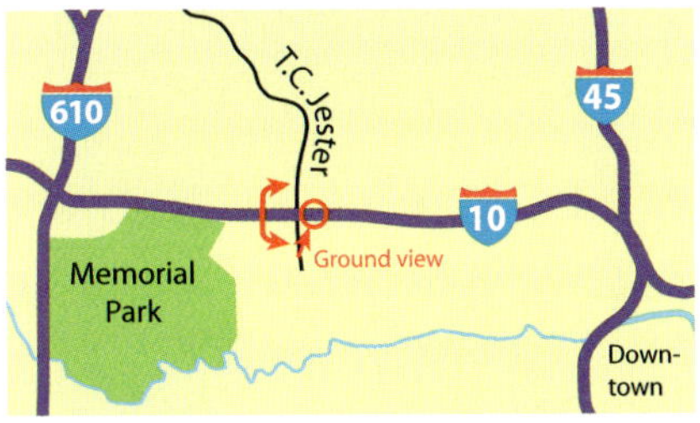

Receding water: This ground-level view shows the Katy Freeway as water receded after Tropical Storm Allison. In addition to being caught in the flood, the cab of the 18-wheeler in the foreground also caught fire and burned. (Photo: Loretta Gutierrez)

1992: Freeway flooding has been a somewhat regular event on certain sections of freeway around Houston. Depressed sections of the Katy, North, and South Freeways have a history of flooding. This view shows the Katy Freeway at T. C. Jester during a flood on March 4, 1992. The head of a swimmer can be seen in the foreground, and a hapless motorist sits on his vehicle underneath the overpass. (Photo: copyright Jim Olive/Stockyard.com)

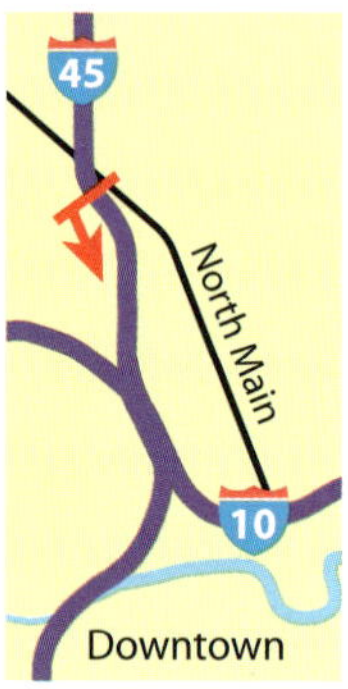

Beached SUVs: These sport utility vehicles washed up along the shoreline of the North Freeway just north of downtown during Tropical Storm Allison in June 2001. Like the Katy Freeway, this depressed section of the North Freeway has a history of flooding in major rainfall events. (Photo: F. Carter Smith/Polaris Images)

(Opposite page) **Interstate 10 disappears into the water:** This view looks west along IH 10 just west of downtown at the IH 45 interchange. IH 10 drops into the water just ahead of the red 18-wheeler and is located below the elevated structure that continues to the upper part of the photograph. White Oak Bayou overflowed its banks and inundated this section of the freeway. (Photo: copyright Jim Olive/Stockyard.com)

Multimodalism: Although the Southwest Freeway was useless for vehicles after Tropical Storm Allison, it was able to serve some waterborne traffic. (Photo: ©www.malisphoto.com, all rights reserved)

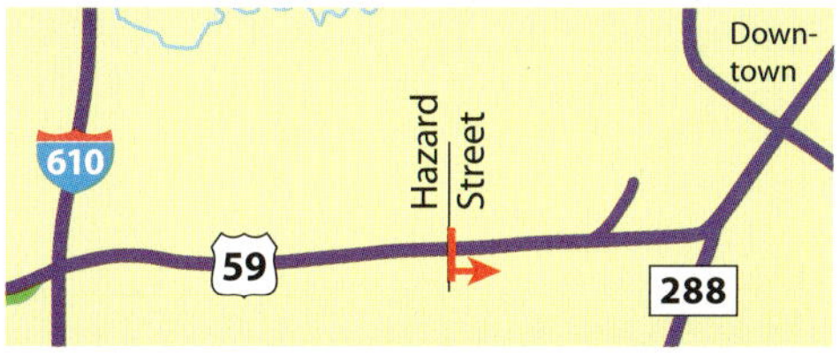

Flowing water uphill: Because Houston is nearly perfectly flat, water accumulating in freeways depressed below ground level must be pumped out. The above pump station drains the Beltway 8 trench at the present-day Westpark Tollway, which was constructed on the location of the bridge in the background. In June 2003, TxDOT has 35 operational pump stations in Houston and an additional 5 under construction. (Photo: November 2000)

Inside the pump building, there is typically a wide circular shaft that extends downward. The section of the shaft below the road level is called the well. When rainfall occurs, the depressed freeway drains water into the well, activating a floating switch which turns on one or more pumps to bring the water to the surface. New and recent pump stations use submerged pumps situated in the well area. The pumps are electric-powered, and most pump stations have backup generators to allow continued pumping if a power failure occurs.

So why do depressed freeways flood when heavy rainfall occurs? Each location has it own characteristics for susceptibility to flooding, but generally one or more of the following conditions occur: extremely high water inflow rate into the trench, flooding at the pump discharge, or water backing up through the storm drainage system. Pump stations are generally equipped with ample pumping capacity, but the pumps may not be able to keep up if a bayou is overflowing into the freeway trench. If the discharge area of the pump station is flooded, there is nowhere to pump the water. The water can be pumped up, but it goes back into the freeway trench due to localized flooding around the trench. Backup through the storm drainage system can occur if reverse flow is possible. The South Freeway (SH 288) was originally susceptible to this mode of flooding before its drainage system was modified.

Bayou breach: The South Freeway (SH 288), constructed almost entirely below grade between downtown and Loop 610, has been susceptible to flooding. In 1989 the drainage system was modified and pump stations were constructed to prevent flooding. Still, the freeway trench flooded during Tropical Storm Allison in June 2001. The South Freeway has an unusually low crossing over Brays Bayou, actually cutting in the bayou stream bank, as shown in the photo at left. The two low bridges are the South Freeway main lanes; the higher bridge in the background is the northbound frontage road. These low bridges act like a dam when the water is high, contributing to upstream flooding. After Tropical Storm Allison, plans were formulated to raise the bridge 11 feet (3.4 m). The project is expected to begin in 2004. (Photo: June 2003)

Freeway or waterway? The wide depressed freeway at the US 59-SH 288 interchange downtown filled with water. (Photo: *Houston Chronicle*)

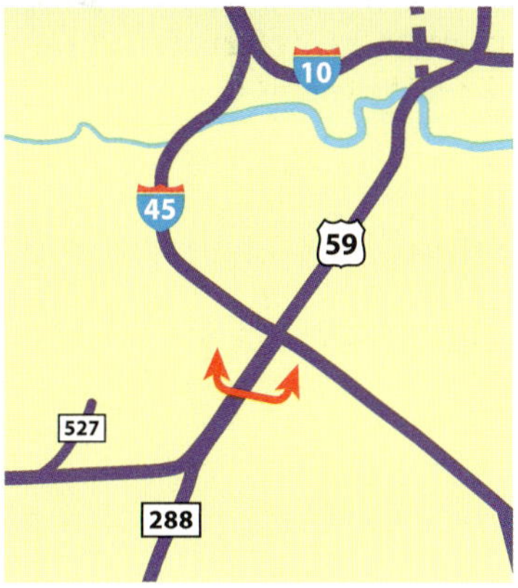

This ground-level view shows an exit along southbound US 59 downtown. (Photo: Jonathan Miller)

Downtown Freeways

Building freeways through central business districts was a costly and difficult task. Numerous freeways converged in a small area, requiring a large number of freeway-to-freeway interchanges. Extensive connections into downtown streets were necessary. The street networks around downtown would need to remain intact where they crossed freeways. Traffic volumes were high, and right-of-way was often limited. Solving these problems necessitated the most complex designs in urban freeway systems.

The designers of Houston's downtown interchange complex faced these challenges, but they had certain advantages working in their favor. Houston's downtown environment was more conducive to freeway construction than most other cities. Outside the relatively compact core of downtown, there were practically no barriers to the construction of the freeway system. Better yet, there were low-density zones around the periphery of downtown, "slum" areas that invited freeways, a natural freeway corridor on the northwest side of downtown, and a city street network capable of handling large traffic volumes coming off the freeways. The result: Houston boasts one of the nation's most extensive and impressive downtown interchange complexes.

Downtown Houston and the extensive freeway system turned out to be a match made in concrete and steel heaven. As of the late 1960s, downtown Houston was a rather unimpressive collection of mid-rise office structures, many quite old. Around the same time, the freeway system into downtown was substantially completed, providing superb accessibility. By the early 1980s downtown Houston had one of the nation's more impressive skylines. The elevated freeways skirting the edges of downtown showcase the skyline to passing motorists, providing a unique driving experience.

(Opposite page) **Mega-merge:** The US 59-SH 288 interchange south of downtown is the widest section of downtown freeway. (Photo: May 2003)

Origins

As Houston's freeway system was taking shape in the 1940s, the focus point was downtown. The Texas Transportation Commission approved the Gulf Freeway in October 1943, with the original planned alignment stopping just short of downtown. In May 1945, the commission authorized alignment studies for US 59 from Buffalo Bayou northward and US 75 (now IH 45) along the west and south sides of downtown. In October 1946, the commission approved the addition of the US 90 (now IH 10) freeway across Houston, which was planned to pass just north of the central business district. At this point, Houston's freeway system consisted of five spokes converging on downtown, and the miniloop around the central business district had three of its four sides in the development stage.

On September 30, 1948, the Gulf Freeway was opened just southeast of downtown Houston. The freeway's terminus into downtown was one of its more important design features. The freeway was directly connected to four high-capacity downtown streets—two inbound one-way streets and two outbound one-way streets. The head of the TxDOT Houston Urban Project Office, W. J. Van London, called the transition zone to the freeway the "loading platform." In November 1948, *American City* magazine stated that the freeway terminus "resembles the mouth of a stream with its delta widening out to receive the slower-moving waters." The Gulf Freeway terminus set the standard for all other freeways entering downtown, with each freeway distributing traffic onto several downtown streets.

In 1950 Van London and city officials were working to define the details of the downtown interchange complex, including exact alignments. At the time, 10 freeway spokes were envisioned for downtown Houston. The early vision included the 7 existing spokes in 2003 and Memorial Parkway, the unbuilt Harrisburg Freeway, and a route between the South and Gulf Freeways that later became the Alvin Freeway. Double-deck structures were studied, particularly for IH 10 north of downtown and some of the distribution structures into downtown. However, no double-deck structures were ultimately used.[1]

Formulation of the downtown freeway plan was part of a comprehensive local effort to define Houston's overall freeway system. In April 1953, the Texas Transportation Commission formally approved the US 59 freeway on the east side of downtown, closing the freeway loop around downtown. Houston's first comprehensive freeway plan was ready for presentation to the Texas Transportation Commission in July 1953. The initial version of the plan

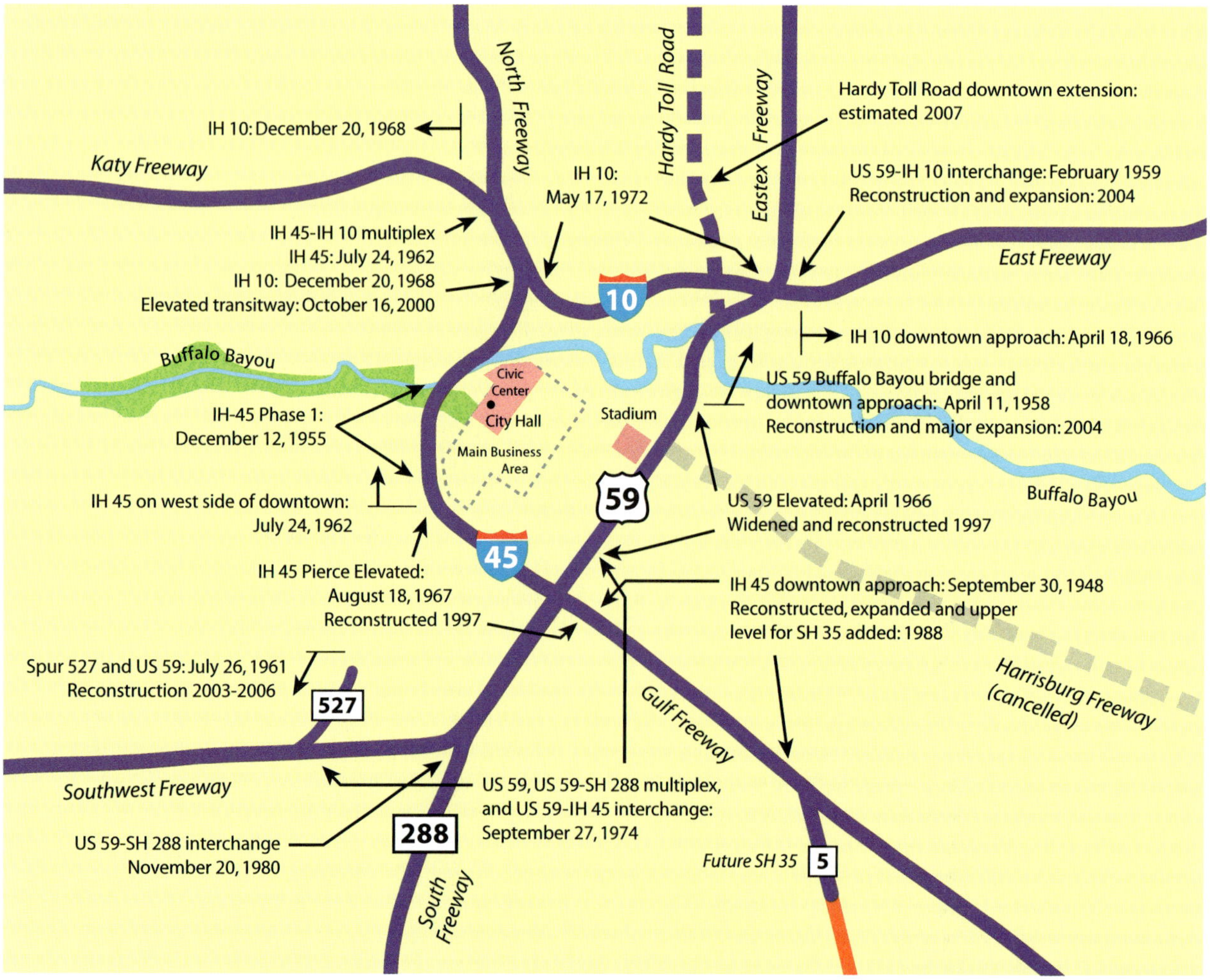

The double deck that wasn't: Early concepts for downtown Houston freeways included possible double-deck freeways, such as this one that was contemplated for IH 10 on the north side of downtown. Although downtown Houston has many elevated freeways, no double-deck structures were built. IH 10 north of downtown, opened on May 17, 1972, was the next-to-last segment of originally-planned downtown freeway main lanes to open. (Photo: TxDOT)

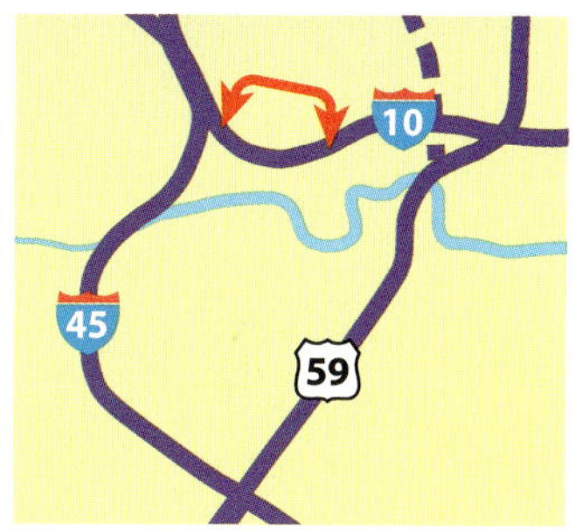

for the downtown interchange complex was slightly different from the final plan. In the 1953 plan, the Southwest Freeway followed the present-day route of Spur 527 all the way to IH 45 on the southwest side of downtown (see plan map, page 13).[2]

In 1954 the freeway master plan was adjusted, bringing the downtown interchange complex to its near final general configuration. In the final plan, Spur 527 was strictly a part of the Southwest Freeway distribution system into downtown streets. With the final and official plan in place, it was clear that Houston's downtown freeway complex would be extensive. It featured two major multiplex freeway interchanges, where two freeways would run parallel to each other—one on the north side of downtown with IH 45 and IH 10, and one on the south side of downtown with US 59 and SH 288. There would be two stack-type interchanges, one at US 59 and IH 45, and one at US 59 and IH 10. IH 45 on the west side of downtown would feature an elevated complex of ramps connecting to downtown and the parkways along Buffalo Bayou. All downtown freeways would be elevated above grade or depressed below grade.

There were still many details to be worked out. In May 1957, city of Houston planning director Ralph Ellifrit and TxDOT Houston Urban Project Office head A. C. Kyser met in Kyser's office to plan details of the downtown freeways. Two days later Ellifrit summarized their plans in a letter titled "Central Interchange Study." The letter detailed how the existing downtown street system would be fully integrated into the freeway system by assigning at least four feeder streets to each radial freeway. To increase the merging distance between the US 59-SH 288 interchange and nearby US 59-IH 45 interchange, Ellifrit recommended a southward shift of the US 59-SH 288 interchange. Most of the ideas contemplated in the May 1957 meeting, including the southward shift of the US 59-SH 288 interchange, would become part of the final plan. Houston's downtown freeways were designated as the Houston Central Interchange.[3]

In August 1960, the Texas Transportation Commission closed the last remaining gap in the downtown freeway system by formally approving US 59 between Spur 527 and IH 45. Part of this newly approved section would become the widest and most impressive freeway segment in the downtown area.[4]

In 1963 the overall plan for the downtown Houston freeway system was finalized. There would subsequently be only three significant changes to this plan. First, the interchange at the intersection of the planned Harrisburg Freeway and US 59 would never be built, since the Harris-

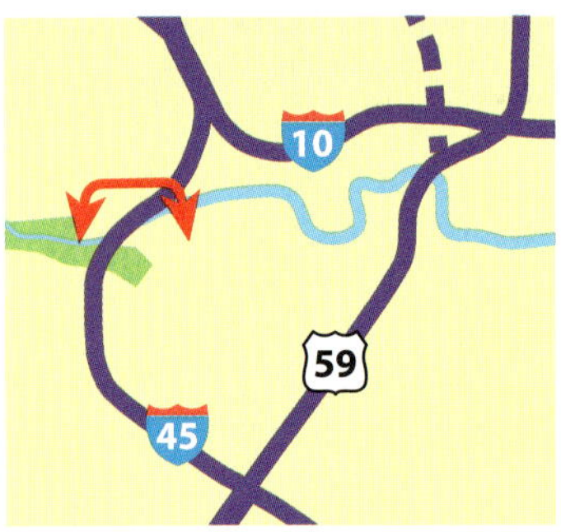

Clearing the path: Right-of-way clearance for present-day IH 45 on the west side of downtown was complete when this photo was taken on August 3, 1954. This section of freeway was the first to be constructed in downtown Houston. The parallel rows of buildings in the upper right of the photo are part of the Allen Parkway Village housing project. (Photo: Squire Haskins Photography, Dallas)

burg Freeway was later cancelled. Second, a third freeway multiplex would be added southeast of downtown along IH 45 where SH 35, the planned Alvin Freeway, merged with IH 45. And third, the Eastex Freeway interchange complex on the northeast side of downtown, including the interchange with IH 10, would be substantially expanded in work that began in the late 1990s.[5]

Clearing the Way

Most of the right-of-way clearance for Houston's downtown freeways consisted of residential structures, with some areas of low grade commercial structures and light industry also affected. The older, decayed neighborhoods around downtown Houston worked in favor of the freeway builders. Although slum clearance was much less pervasive in Houston than it was in older eastern cities, two of Houston's best examples of slum clearance occurred immediately adjacent to downtown freeways. The Allen Parkway Village project, originally known as San Felipe Courts, was opened in 1942 and IH 45 was constructed along its eastern edge in 1955.* The Clayton Homes housing project was built just northeast of

*Allen Parkway Village became the focus of a long-running, highly contentious controversy when the city of Houston began efforts to demolish it in the mid-1970s. In early 1996 a compromise plan was approved, and 677 of the 963 units were demolished in 1996 and 1997. The controversy was not related to the adjacent IH 45 freeway.

Construction begins: This 1954 photo shows the first construction on the downtown Houston freeway system—the first phase of IH 45 on the west side of downtown. This view looks east along Allen Parkway. (Photo: Greater Houston Partnership)

The first downtown freeway: This short section of IH 45 on the west side of downtown opened on December 12, 1955. This photo was taken within two years of the opening and looks southeast, similar to the perspective of the photo on the opposite page. (Photo: TxDOT)

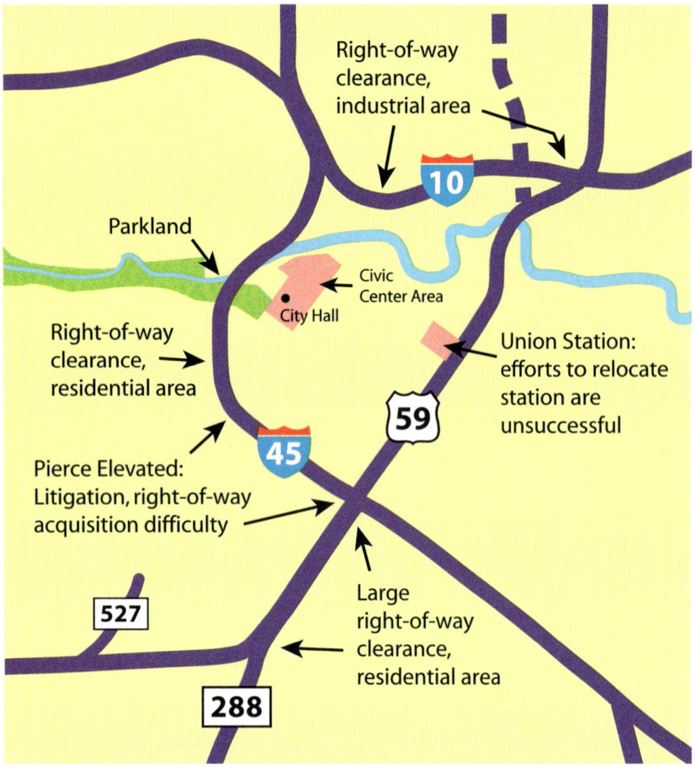

Few obstacles to freeways: Downtown Houston presented few obstacles to freeway construction. The map above shows the locations of the freeway sections that presented some difficulty.

downtown in 1952, and in 1958 US 59 was constructed alongside the property. The largest right-of-way clearance for the downtown freeway system occurred in the late 1960s on the south side of downtown, where a very wide corridor was required for the US 59-SH 288 interchange. There were generally no objections to the displacement of residences in the neighborhoods around downtown.

Another possible roadblock to freeway construction was the parkland along Buffalo Bayou just west of downtown, adjacent to the Houston Civic Center. A web of elevated structures and connection ramps was envisioned for IH 45 through that section of downtown. In a 1979 interview, Ralph Ellifrit remarked that "there was considerable discussion" about building a freeway at that location, but "there was just no way" to build the downtown freeway system without routing the freeway there. Time would prove that the freeway would not have a negative impact on the park or downtown civic center.[6]

On the east side of downtown, US 59 needed to pass over a rail yard which included Union Station, one of the city's railway stations for passenger service. In 1953 the city of Houston began efforts to relocate Union Station, which was controlled by the Houston Belt and Terminal Railroad, to Southern Pacific Railroad's Grand Central Station. City planners envisioned connecting the Eastex Freeway (US 59 north of downtown) into the street immediately adjacent to the elevated structure, Chartres, and the three parallel streets to the west, using a distribution system similar to the "loading platform" where the Gulf Freeway distributed traffic into downtown. However, Union Station blocked three of the four streets which would connect to the freeway, making the distribution system impractical. Efforts to relocate the station continued through the mid-1950s, with negotiations breaking down by 1958. In 1960 the matter needed to be resolved so construction could move forward. On November 2, 1960, officials from the city of Houston, TxDOT, and the railroad reached an agreement after a one-and-a-half hour session. The railroad agreed to relocate some tracks to provide space for the columns of the US 59 elevated structure, but the Union Station itself would stay. As a result, the Eastex Freeway would lack a full distribution system into downtown in its original construction, since the railroad station blocked the needed feeder streets. The Eastex Freeway's poor downtown connections would not be improved until the major reconstruction of the US 59-IH 10 interchange began in the late 1990s. Around the same time, construction of a new baseball stadium began at the location of the former Union Station. The stadium, opened in March 2000, blocked the originally planned street access to the new interchange from the east side of downtown.[7]

The most controversial section of downtown freeway impacted commercial property, not residential areas or parks. IH 45 on the south side of downtown, now known as the Pierce Elevated, required acquisition of property slated for commercial development by its owners. Two bank structures were planned for the corner of Main and Pierce, right in the path of the freeway. In 1956 the city of Houston refused to issue building permits to the landowners in order to keep the property clear for the freeway. However, freeway construction was estimated to be 8 to 10 years in the future, and TxDOT did not have funds available to acquire the right-of-way. The landowners couldn't improve their property, but there was no money available to buy it. In 1957 two landowners filed identical lawsuits to overturn the denial of building permits. The city was forced to rescind the "building lines"[*] it had es-

(Opposite page) **The spaghetti bowl:** The upper photo, taken in May 1961, shows construction in progress to complete IH 45 on the west side of downtown. The project included the addition of collector-distributor structures and numerous connection ramps. Engineers dubbed the new interchange the "spaghetti bowl" because of its long, stringlike roadways. The term "spaghetti bowl" had fallen into disuse by the 1970s. The lower image shows a modern view of IH 45 on the west side of downtown. The configuration of the freeway remains unchanged since its completion in 1962. (Photos: upper, TxDOT; lower, September 2002)

[*] "Building lines" were set by the city of Houston and Harris County to define the needed right-of-way for freeways. In general, building permits were not issued for improvements to property within the building lines.

Aiming for the downtown crossing: This view shows the terminus of the East Freeway (IH 10) at US 59 in May 1961. Right-of-way clearance for IH 10 on the north side of downtown had not yet begun. A substantial number of structures were in the path of the freeway in the mostly industrial area. The westward extension of IH 10 was completed on May 17, 1972. (Photo: TxDOT)

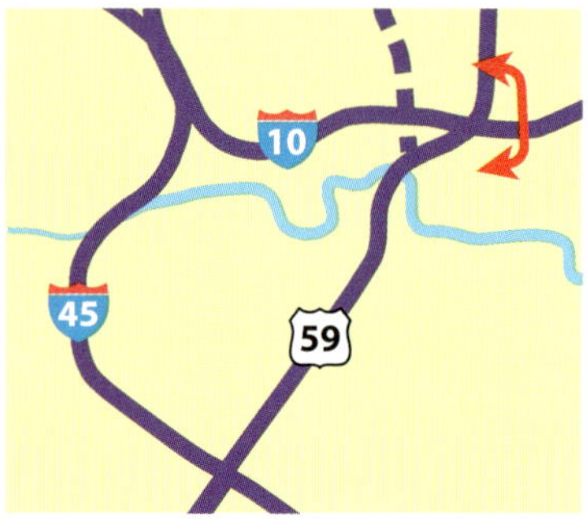

tablished, leaving the corridor at risk for development.

By October 1958, the building lines were reestablished and TxDOT began protective right-of-way acquisition. But lack of funds limited the land acquisition to only a half-block wide strip along Pierce Avenue. The half-block strip was estimated to cost $8 million and had a final cost of $11 million. A full block strip was estimated at $20 million. After adjusting for inflation in the overall U.S. economy to equivalent values in 2003 dollars, $11 million translates to about $56 million and $20 million translates to about $102 million—a large but not daunting cost. However, as a percentage of budgets at the time, $20 million for a full block along Pierce Avenue would have been a huge expenditure. The total TxDOT budget for all of Texas in 1957 totaled $242 million. In relative terms, the cost of a full block along Pierce in 1957 corresponds to about $430 million in TxDOT's 2003 budget of $5.2 billion. There was discussion of acquiring the full block, but limited financial resources ruled it out; getting the money to buy the half-block strip would be difficult enough. But the half-block strip was wide enough for a freeway with only six lanes. The consequences of the 1956 decision to acquire only half a block were being felt in 2003, when TxDOT officials were studying ways to provide relief to the now critically congested Pierce Elevated.[8]

Controversy over the Pierce Elevated persisted until 1959, when a landowner was still fighting to have the building lines revoked. By that time, local authorities were not willing to entertain any ideas that would change the alignment of the freeway. TxDOT Houston Urban Project Office head A. C. Kyser stated that changing the route was out of the question, and the Houston Chamber of Commerce rallied its support behind the Pierce Elevated. By July 1960 all challenges to the Pierce Elevated had been dismissed, and approximately half the needed right-of-way had been acquired through protective right-of-way purchases.[9]

Building the Freeways

The first construction on downtown Houston freeways was underway by late 1953, consisting of the first phase of present-day IH 45 on the western edge of downtown. The first freeway section, the main freeway lanes of IH 45, opened December 12, 1955. For the next 25 years there would be a steady progression of work on the Houston Central Interchange (see map on page 118 with completion dates). On April 11, 1958, the Eastex Freeway was extended into downtown with the opening of the Buffalo Bayou bridge. The Southwest Freeway and its distribution system into downtown via Spur 527 were completed on July 26, 1961. On July 24, 1962, IH 45 was completed from the north side of Houston into downtown, forming the first phase of the large IH 45-IH 10 multiplex. The IH 45 interchange structure on the west side of downtown was also finished in 1962 with the addition of collector lanes and connector ramps. In April 1966 the US 59 elevated structure on the east side of downtown was completed. Construction of the Pierce Elevated structure was complete in spring 1965, and it was opened to traffic on August 18, 1967, connecting the North Freeway to the Gulf Freeway. On December 20, 1968, the Katy Freeway (IH 10 West) was brought into downtown Houston, and the multiplex interchange with IH 45 was substantially complete. On May 17, 1972, IH 10 on the north side of downtown between IH 45 and US 59 was opened, completing IH 10 through Houston and the inner loop of freeways around downtown. The last major link in the downtown freeway system was finished on September 27, 1974, with the opening of US 59 between Spur 527 and IH 45. The US 59-SH 288 interchange was completed on November 20, 1980. With that project, the Houston Central Interchange was complete. Although major reconstruction of the Gulf Freeway just southeast of downtown would begin in the early 1980s, construction in the immediate downtown area would go into a state of dormancy for about the next 15 years.

Key dates in Houston's downtown freeways

1945	Official TxDOT authorization to begin study of downtown freeways.
1948	Gulf Freeway opens southeast of downtown with an extensive distribution system into downtown streets.
1954	The general plan for the downtown Houston freeway system is finalized as part of an overall freeway plan for the Houston region.
1955	First downtown freeway segment opens, a short section of IH 45 on the west side of downtown.
1963	The detailed plan for the downtown Houston freeway system is finalized.
1967	The Pierce Elevated section of IH 45 opens.
1972	IH 10 north of downtown opens, completing the loop around downtown.
1974	US 59 south of downtown opens, completing all downtown main lanes in the original plan except the cancelled Harrisburg Freeway.
1980	The US 59-SH 288 interchange opens, completing the originally planned downtown freeway system.
1997	The second wave of downtown construction begins with rehabilitation work and a large new interchange at US 59 and IH 10.
2003	Work begins on the complicated reconstruction of the Southwest Freeway (US 59) downtown approach.
2004	Completion of the US 59-IH 10 interchange.
2007	Scheduled completion of the Hardy Toll Road freeway extension into downtown, almost certainly the last new freeway for downtown.
Future	Potential improvement and expansion of IH 45 and US 59.

First Came the Freeways, then the Skyscrapers

In the mid-1960s, downtown Houston consisted of a modest collection of mid-rise office structures. By the late 1960s, most of the freeways into downtown Houston were complete and there was greatly improved access to downtown from all parts of the city. In 1970 the regional transportation planning group, the Houston-Harris County Transportation Study, published a report on the central business district which stated, "There is less traffic congestion travelling to or from the Houston central business district than in any city of comparable size. Comparative travel time studies conducted in 1960, 1965, and 1969 have shown that during the evening peak hours the travel time has been vastly improved in all directions except for the Gulf Freeway, although this facility is showing improvements compared with 1965." The reason for these improvements was, of course, the new freeway system. In terms of motorists' ability to efficiently travel during rush hour, 1970 would later be recognized as the year of peak mobility in Houston.[10]

It didn't take long before downtown Houston was transformed into one of the more impressive skylines in the United States. In 1960, the central business district had 10 million square feet of office space, increasing to about 16 million square feet in 1970. Downtown Houston was on the threshold of a boom in 1970 with 8.7 million square feet of office space planned or under construction and huge projects being launched by real estate developers.* The surge of skyscrapers mirrored the skyscraper booms in other freeway cities, especially Los Angeles and Dallas. Houston's downtown construction was further fueled by the 1970s energy industry boom.

The first major skyscraper to be constructed in Houston was the 50-floor, 714-foot-tall (218 m) One Shell Plaza in 1971. A succession of skyscrapers were built through the 1970s, culminating with Houston's tallest skyscraper, the 75-floor, 1,002-foot-tall (305 m) JPMorgan Chase Tower (formerly the Texas Commerce Tower), completed in 1982. In 2002 it was the ninth tallest building in the

* The largest proposed development was the 32-block Houston Center. Only a small part of the original proposal was ultimately constructed. Other large projects included Cullen Center, Allen Center, and towers for Shell Oil.

Pre-freeway downtown: This 1967 photograph shows the modest collection of mid-rise office buildings that comprised downtown before the completion of the downtown freeway network. By 1968 most of the spokes leading into downtown were complete, and in 1974 the final section of main lanes in the originally planned downtown freeway network was opened to traffic. Houston's skyline was transformed in the 16-year period from 1971 to 1987. (Photo: TxDOT)

Downtown Houston's tallest skyscrapers

			Height			
	Name (2003)	**Address**	**Feet**	**Meters**	**Floors**	**Completed**
1.	JPMorgan Chase Tower	600 Travis	1,002	305	75	1982
2.	Wells Fargo Plaza	1000 Louisiana	972	296	71	1983
3.	Bank of America Center	700 Louisiana	780	238	56	1983
4.	Texaco Heritage Plaza	1111 Bagby	762	232	53	1987
5.	1100 Louisiana	1100 Louisiana	748	228	55	1980
6.	Reliant Energy Plaza	1111 Louisiana	741	226	47	1974
7.	Continental Center	1600 Smith	732	223	53	1984
8.	Chevron Tower	1301 McKinney	725	221	52	1982
9.	One Shell Plaza	910 Louisiana	714	218	50	1971
10.	1400 Smith Street	1400 Smith	691	211	50	1983

Data: Skyscrapers.com

The Houston skyline, September 2002: The downtown Houston building boom came to an end in 1986 with the collapse of Houston's energy economy. Since 1987, all newly constructed office towers have been no more than 40 floors tall. The popularity of suburban, campus-style office space seems to suggest that Houston's skyline will not receive any major new high-rise additions for the indefinite future. Still, downtown Houston remained vibrant in the 1990s and 2000s with the construction of new housing, sports stadiums, and the rebuilding of basic infrastructure.

United States and the 23rd tallest skyscraper in the world. In 1983 the 71-floor, 970-foot-tall (296 m) Wells Fargo Plaza was completed. Skyscraper construction in downtown Houston came to an end in the mid-1980s with the collapse of Houston's energy industry and the resulting severe economic recession. When the 53-floor Texaco Heritage Plaza was completed in 1987, it appeared that no more skyscrapers would be constructed for a long, long time. In 2000, however, the Houston-based Enron Corporation began construction of a 40-floor skyscraper which was nearing completion in late 2001, just as the company collapsed in one of the most dramatic corporate failures in the history of the United States. Other smaller office structures were built in the 2000-2003 period. In December 2001, downtown Houston had approximately 40 million square feet of office space, including 28 million square feet of class A office space.[11]

In spite of the impressive skyline and freeway network, urban advocates in the 1980s complained that downtown Houston was hollow, becoming nearly deserted at the end of the business day. Houston's underground tunnel system kept office workers below ground, inhibiting street activity. Downtown Houston would achieve revitalization in the 1990s, a renaissance engineered by Mayor Bob Lanier, Houston's great road building advocate of the post-1980 era. After concentrating his efforts on transportation in the 1980s, Lanier was elected mayor in 1991 and focused on making the city safe and rebuilding basic infrastructure, including roads and sidewalks. When Lanier exited the mayor's office at the end of 1997 due to term limits, the revitalization of downtown was in full progress. Numerous apartment and loft-style residential structures were built, bringing long-sought urban dwellers into the core of downtown. Lanier's successor, Lee Brown, sought to capitalize on the downtown momentum by constructing a light rail line on Main Street. Construction on the rail line began in 2001 and is scheduled for opening in 2004.

The Second Wave

In the United States, construction and improvement of downtown freeways was largely dormant by the 1970s. Freeways in downtown districts had fallen into disfavor with planners who became more interested in promoting transit and pedestrian activity. Some viewed freeways as

Houston's freeways turn into lakes: Tropical Storm Allison in June 2001 unleashed a major flood event, and Houston's freeways sustained some of the most dramatic flooding. Virtually all depressed freeways around Houston filled with water. Flooding along White Oak Bayou submerged much of the at-grade interchange of IH 45 and IH 10 north of downtown, shown above in a photo taken June 9, 2001. (Photo: *Houston Chronicle*)

barriers, cutting off parts of the city from each other. Los Angeles, always the bellwether, typified this trend. Much of its downtown interchange complex dates from the early 1950s and was complete by 1962. It has seen practically no improvement since its original completion, and no major improvement is planned in the future. The Los Angeles downtown freeway complex is certainly showing its age. Phoenix, Arizona, in a project completed in 1990, placed a downtown section of IH 10 in a tunnel to minimize impacts. Boston, Massachusetts, spent a staggering $15 billion to rid its downtown of freeways by burying them underground.

But Houston would retain and improve its downtown freeway system starting in the mid-1990s. The Pierce Elevated, IH 45 on the south side of downtown, was reconstructed in 1997. US 59 on the east side of downtown was resurfaced and widened to eight continuous lanes in 1997. In October 2000, a long, elevated connector ramp for buses and high occupancy vehicles was completed over the IH 10-IH 45 interchange on the northwest side of downtown.

In 1998, the first major construction since the completion of the downtown interchange complex in 1980 was underway. A major new interchange took shape on the northeast side of downtown at the intersection of Eastex Freeway (US 59) and the East Freeway (IH 10). Included in the interchange is the downtown distribution system originally envisioned by planners in the 1950s. The operational efficiency of the distribution system may be compromised when it opens in 2004 because a new baseball stadium was constructed on the location of the former Union Station rail yard. The stadium ensures a permanent barrier to north-south traffic flow where the connection ramps reach downtown streets. The new interchange will also accommodate a planned new radial spoke for the downtown freeway system, the extension of the Hardy Toll Road into downtown Houston as a toll-free facility. Construction of the Hardy Toll Road extension is expected to begin in 2005, with completion around 2007.

In 2003 a comprehensive study is in progress to determine future improvements to IH 45 and the section of US 59 south of IH 45. The six-lane Pierce Elevated section of IH 45 has become a critical bottleneck in Houston's freeway system. The study will identify needed improvements and determine what is feasible for the Pierce Elevated, which has high-rise structures immediately adjacent to it. At the very minimum, the oldest sections of Houston's downtown interchange complex should be brought up to modern standards, and recommendations could include major expansions or additions.

Pierce, before the elevated: This view looks west along Pierce Avenue at Jackson, prior to the construction of the Pierce Elevated section of IH 45. The Pierce Elevated was constructed immediately adjacent to the south (left) side of Pierce. Although the corridor for the elevated freeway consisted mostly of vacant lots and lower-tier commercial establishments, it was the most controversial section of freeway in the downtown interchange complex. Commercial interests owning property in the path of the freeway had plans for developing their property. Lawsuits were filed, but by 1960 all challenges were dismissed and the freeway moved forward on its originally planned alignment along Pierce, opening in August 1967. (Photo: TxDOT)

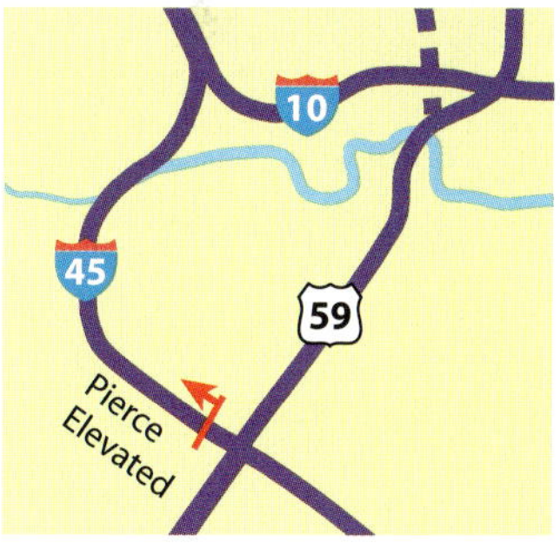

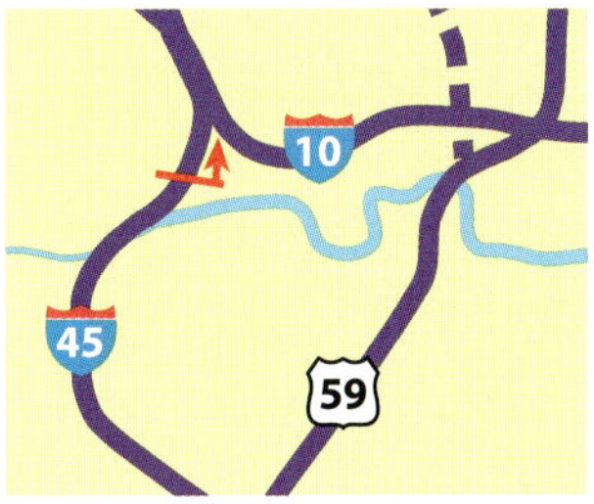

The orphan ramp, 1962-1972: This connector ramp overpass at the south end of the IH 45-IH 10 interchange was completed in 1962 with the construction of IH 45. However, the section of IH 10 which would use this ramp was not completed until 1972. The idle ramp prompted speculation from many motorists, including one person who wrote to the *Houston Chronicle* in 1967 believing that "current plans don't call for its use and it will just stand there forever."[12] (Photo: Greater Houston Partnership)

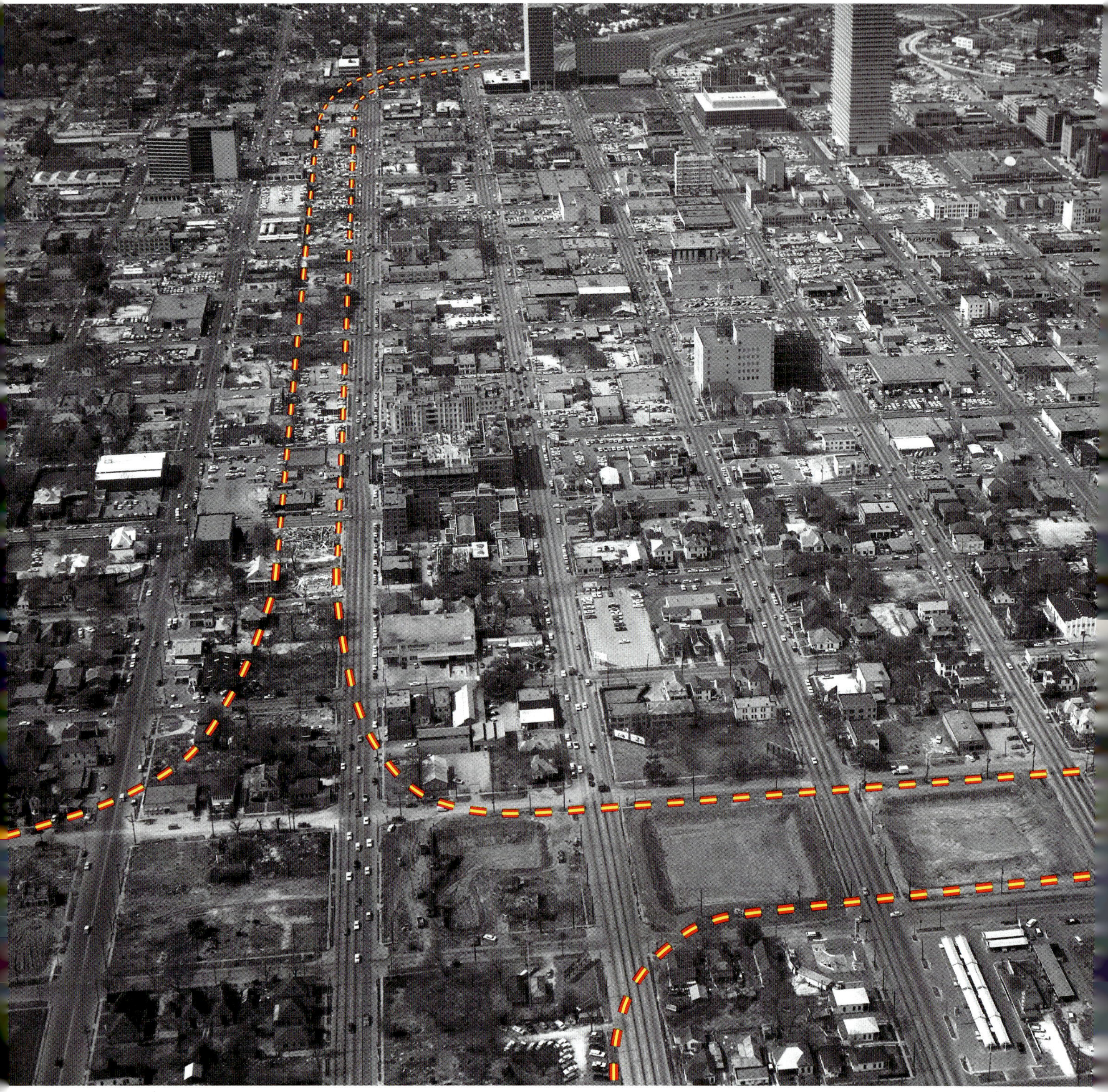

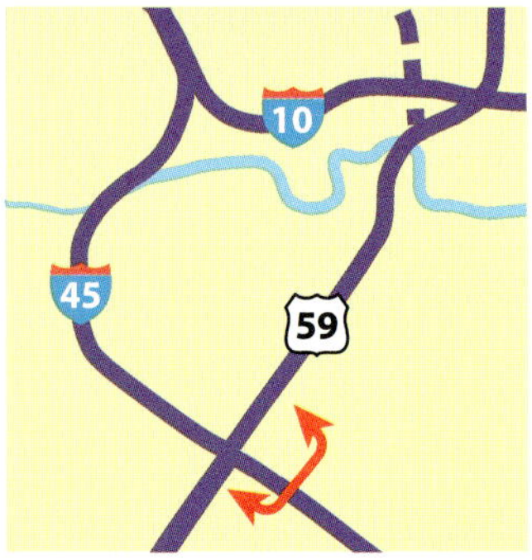

On the eve of freeway construction: This 1963 aerial view of the southeast side of downtown shows the area just before freeway construction projects would begin. The routes of IH 45 (top to bottom) and US 59 (left to right) are indicated by the dashed lines. Excavation work had already begun for US 59 in the lower section of the photo. Right-of-way clearance is in progress along Pierce Avenue. (Photo: The Positive Image)

Critical bottleneck: The six-lane Pierce Elevated, named after the street immediately adjacent to it, has become the most critical bottleneck in the downtown interchange complex. Designed to carry 75,000 vehicles per day, the freeway was serving approximately 200,000 vehicles per day in 2003. As this photo shows, large structures alongside the freeway make a major expansion very difficult. In the foreground is the interchange with US 59. The US 59 main lanes are depressed below grade and are not visible in this view. (Photo: May 2002)

IH 10 north of downtown: This section of freeway, opened on May 17, 1972, is simple and basic in comparison to other freeways in the downtown interchange complex. The dedication ceremony coincided with the retirement of Albert C. Kyser, head of the TxDOT Houston Urban Project Office from 1955 to 1972. Kyser and City of Houston Planning Director Ralph Ellifrit were the most influential persons in the design of the downtown interchange complex. (Photo: July 2002)

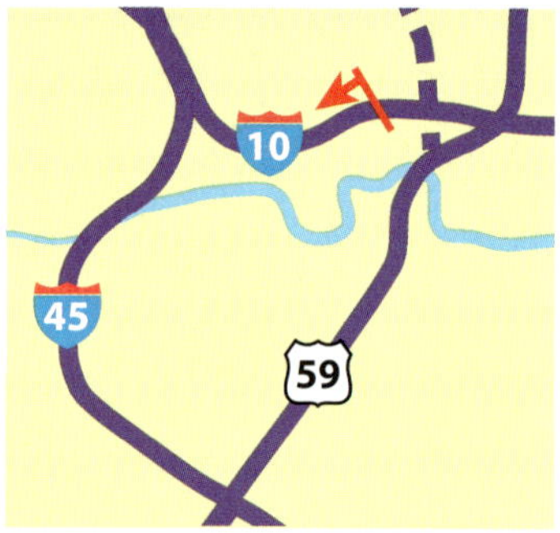

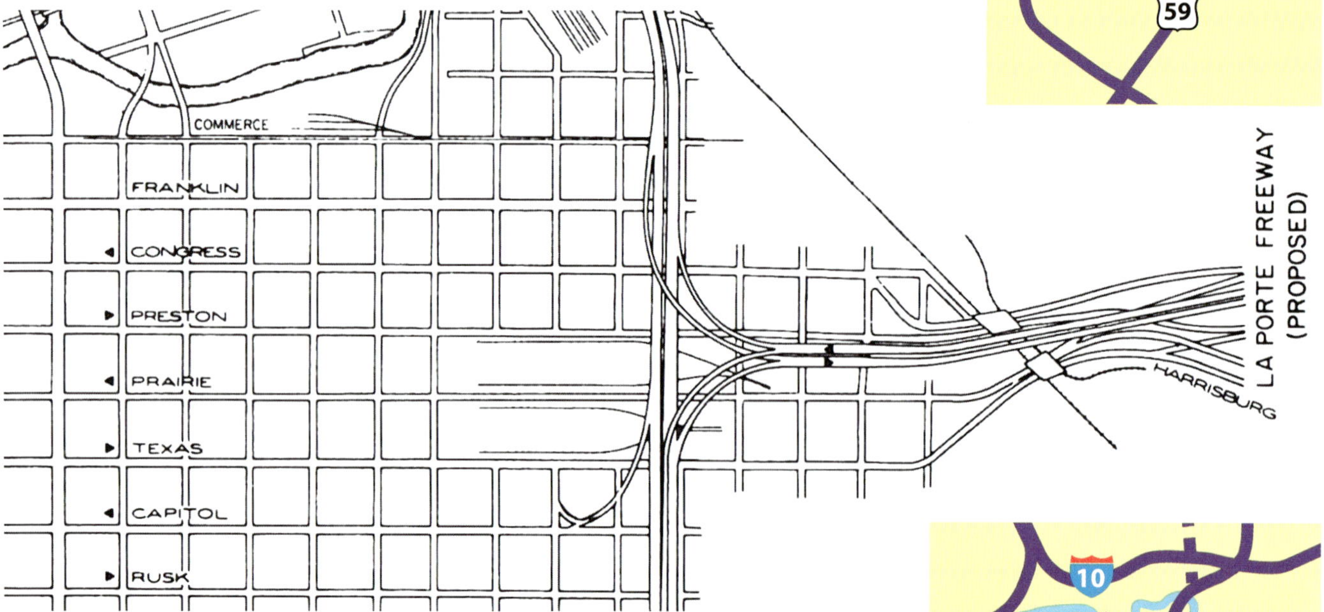

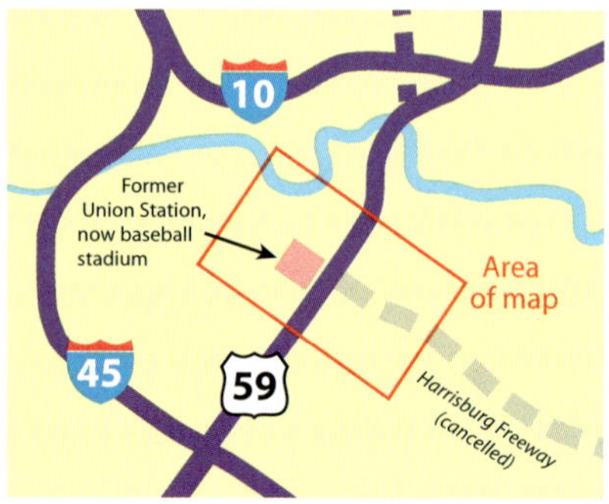

The unbuilt: This view shows a section of the final plan for the downtown interchange complex, which was approved in 1963. The La Porte Freeway, also known as the Harrisburg Freeway for the unbuilt section inside Loop 610, was planned to terminate at US 59 near what is now the baseball stadium for the Houston Astros. The La Porte Freeway was cancelled in the mid-1970s due to community opposition and funding shortages. It is the only part of the originally planned downtown interchange complex that was not constructed. (Map: HMRC)

IH 45 construction: This view shows the construction of IH 45 on the north side of downtown in May 1961 at the location of the multiplex interchange with IH 10. Initially only IH 45 was constructed through the interchange. This section of freeway parallels White Oak Bayou and followed an ideal freeway corridor due to the limited development. Houston's 1942 *Major Street Plan* identified the White Oak Bayou corridor as one of the three best potential alignments for new freeways. (Photo: TxDOT)

One freeway complete, one to come: These views show the IH 45-IH 10 multiplex interchange prior to the construction of IH 10. The wide median and adjacent cleared right-of-way would later become IH 10 and its connection ramps. The upper photo was taken just after the freeway opening in July 1962. The lower photo was taken in December 1964. (Photos: upper, HMRC MSS 157-907; lower, TxDOT)

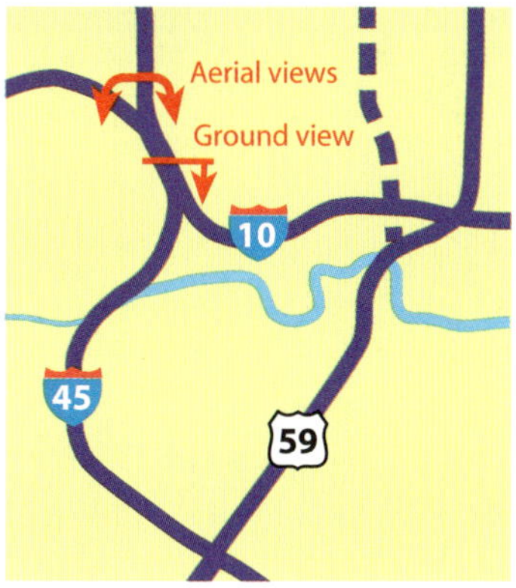

Transitways finish the interchange: The interchange remains in its original 1968 configuration, except for the addition of transitway lanes. The elevated transitway ramp for high-occupancy vehicles was opened October 16, 2000. (Photo: September 2002)

US 59 construction: This view shows construction of the interchange at US 59 and IH 45 in April 1972. The IH 45 main lanes had previously opened in 1967. (Photo: TxDOT)

(Opposite page) **Scorched earth turns into freeway:** These photos show before and after views of the US 59-SH 288 multiplex interchange on the south side of downtown. The construction of this wide freeway required the largest right-of-way clearance in the construction of the downtown interchange complex. In the upper photo, taken in January 1969, right-of-way clearance is just about complete. The freeway was opened to traffic on September 27, 1974. The entire interchange was finished in 1980 with the completion of connections to the SH 288 South Freeway, shown in the far distance in the lower photo. (Photos: upper, *Houston Chronicle*; lower, May 2002)

Lake Jackson
Freeport
Victoria
Victoria
Lake Jackson
Freeport

Construction of the US 59 Buffalo Bayou bridge, November 1957: The completion of this link on April 11, 1958, brought the Eastex Freeway into downtown Houston. The row of buildings just to the left of the freeway is the Clayton Homes housing project, which opened March 22, 1952. It replaced Schrimpf Alley, which was reportedly the worst slum district in Houston. Land acquisition costs for the Clayton Homes project were paid by Susan V. Clayton, wife of Houston cotton magnate and statesman William L. Clayton.[13] (Photo: TxDOT)

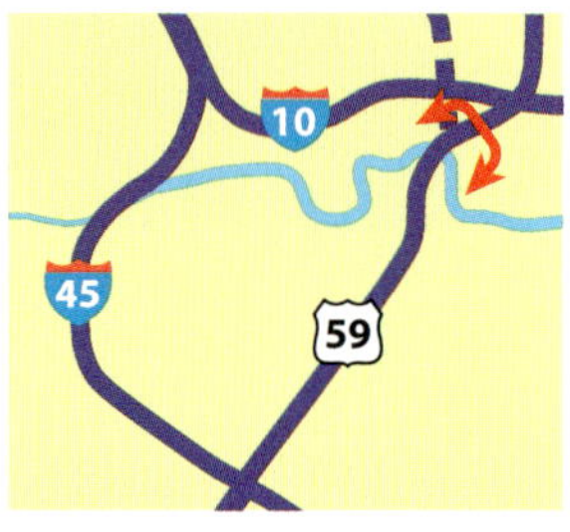

US 59 gets its downtown connections: These photos show the new interchange complex being constructed on the Eastex Freeway (US 59 North) approach to downtown. Original plans for the downtown Houston freeway system included a complete distribution system for each freeway coming into downtown, but plans for the Eastex Freeway were dropped because the Union Station rail yard blocked the needed feeder streets. In the late 1990s work began to rebuild the US 59-IH 10 interchange and add a distribution system for the Eastex Freeway.

In the upper photo, the original US 59 structure is the second from the right and was in the process of being demolished when the photo was taken in May 2002. In the photo at left, taken in November 2002, the original bridge structure has been almost completely demolished—all that remains are a few piers near the bottom of the photo. The new US 59 southbound main lanes were constructed on the location of the demolished original bridge. The project is scheduled for completion in 2004.

The US 59 downtown distribution system and accompanying reconstructed interchange at IH 10 is the first new-facility construction in downtown Houston since the original completion of the interchange complex in 1980.

Freeway with a view: The elevated structures on IH 45 skirting the west and south edges of downtown provide motorists with outstanding views. When the Pierce Elevated section of IH 45 opened in 1967, the *Houston Chronicle* took note of the impressive driving experience provided by the elevated freeways: "Not only does it speed the traffic flow, it also affords motorists a magnificent view of the business district. It's almost like going up in a cable car." [14] (Photo: May 2003)

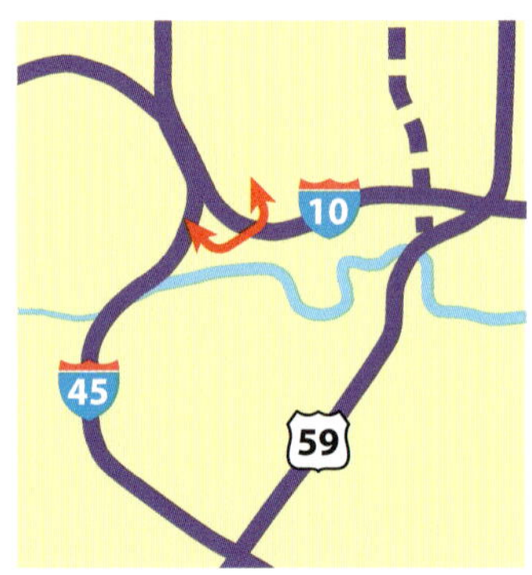

(Opposite page) **Where the freeways come together:** Looking north along the IH 45-IH 10 interchange north of downtown. (Photo: November 2002)

WATERSIDE ESTATES

The busiest and the newest: This view looking west near Chimney Rock shows two of Houston's spoke freeways: the Southwest Freeway, Houston's busiest freeway, and the Westpark Toll Road, the newest and smallest addition to Houston's freeway system. (Photo: November 2002)

The Spokes

In the world of the loop and radial freeway system, downtown is the center of the universe. Call it the hub. Houston's eight spokes converging on downtown are just about the limit of what a downtown freeway interchange system can handle. But as space opens away from downtown, the next tier of six spokes takes root, mostly along Loop 610. Even the Beltway gives birth to a spoke. Houston's system of spokes is one of the most extensive and well-balanced among the major cities in the United States.

In fact, Houston's freeway system does look a lot like a compass. This similarity didn't escape the freeway system's designers, who gave many of Houston's radial freeways compass-point names. The North, Northwest, Southwest, South, East, and Northeast-Crosby freeways are all key increments in Houston's freeway compass.

Houston's spoke freeways exhibit their own form of magnetism by being powerful attractors of development. Pervasive frontage roads along the spokes have brought commercial development right up to the edges of the freeways. The spokes are the scene of some of Houston's most fascinating freeway history, as well as some of the biggest plans for the future.

Gulf Freeway, Interstate 45 South

The Gulf Freeway launched Houston into the freeway era on September 30, 1948. As the history of the Gulf Freeway unfolded, it became a classic story of the rise of the American urban freeway and the myriad of issues that would accompany it. Intertwined in the story of the Gulf Freeway is the demise of the urban electric railway, the unprecedented demand for new freeways in the postwar era, huge suburban development, malls, traffic jams, the development of better freeways, urban protest, and the never-ending battle to catch up to demand. The newly dedicated Gulf Freeway was ahead of its time, yet it couldn't keep up with the times. It would be brought into the modern era, not just once, but twice.

While the Gulf Freeway was at the forefront of many national freeway issues, it also became the prototypical Houston freeway. From the very beginning it was a freeway with continuous frontage roads, spawning the development and commercial clutter that would become characteristic of Houston's freeways. The ongoing—some would say never-ending—improvement of the freeway would become a way of life for Houston's freeway system. Perhaps as a fitting reward for its first 50 years of struggle, in 1997 it received the most impressive instance of Houston's newest freeway trademark: the five-level stack interchange. The Gulf Freeway story, while presently in a quiet mode, is certainly not over. A wave of expansion will push southward from Beltway 8 in the future, giving the freeway its widest and most modern sections yet.

Gulf Freeway	
Previous designation	US 75
Designated as freeway	1943
First freeway section open	September 30, 1948
Freeway complete	1970 (to NASA 1) 1976 (to Galveston)
Reconstruction	1959–1997
Max traffic volume, 2001	266,000 vehicles per day
Future construction	Expand freeway from Beltway 8 to Galveston

Freeway champion: Houston Mayor Oscar Holcombe was the strongest promoter of the Gulf Freeway in the late 1930s and early 1940s. In 1940, he negotiated the city of Houston's acquisition of the right-of-way of the electric railway operating on the corridor. (Photo: HMRC RGD5-7481)

Origins

The beginnings of the Gulf Freeway corridor can be traced to the formation of the Galveston-Houston Electric Railway (GHE) in 1905. Stone and Webster, which at that time actively purchased and consolidated electric railroads and power utilities, took control of GHE in 1906 and immediately changed the originally proposed route of the railway. Instead of following the shore of Galveston Bay, the GHE would have a direct route between Houston and Galveston. The route selected in 1906 was 50.5 miles (81 km) long and featured a 34-mile (54 km) perfectly straight section. In 1908 GHE entered into agreements to participate in the new concrete Galveston causeway. Passenger service on the GHE began December 5, 1911. The GHE distinguished itself as one of the fastest interurban railways in the nation, earning the title of "Fastest Interurban in the United States" in 1925 and 1926 in a nationwide speed contest conducted by *Electric Traction* magazine. The last day of service on the GHE was October 31, 1936. The track inside the Houston city limits from downtown to Park Place was taken over by Houston's transit operator, the Houston Electric Company, and continued to be used for streetcar service. The Gulf Freeway would ultimately follow the alignment of the GHE from downtown Houston to near the present-day Bellfort intersection, just south of Loop 610. Between Bellfort and Galveston, the Gulf Freeway closely parallels the GHE alignment.[1]

Even before the demise of the Galveston-Houston Electric Railway there was talk of a "super-highway" between Houston and Galveston. Ross Sterling, Houstonian and chairman of the Texas Transportation Commission, proposed the highway in 1930. The alignment contemplated at the time would have begun near the present-day Houston Hobby Airport and generally followed the Gulf Freeway corridor to the intersection of SH 6 and SH 146—the Texas City Y. However, the Depression dashed

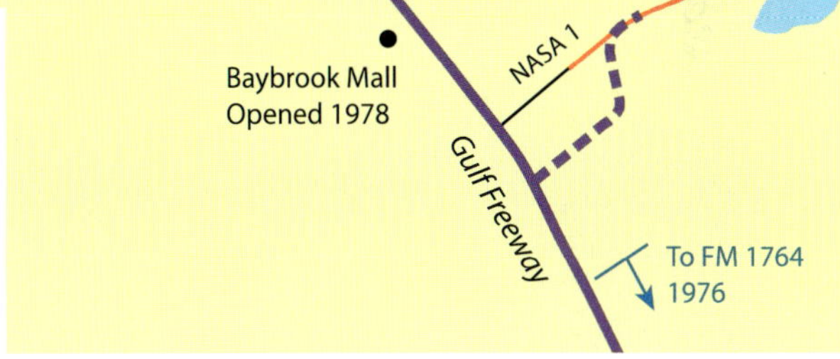

any hopes of constructing the highway. Ross Sterling became governor of Texas in 1931 but saw his proposal for road construction bonds die at the end of the 1931 Texas legislative session.[2]

By the late 1930s a new champion for the Gulf Freeway emerged: Houston Mayor Oscar Holcombe. Oscar Holcombe served as mayor for a total of 22 years between 1921 and 1958. During his terms from 1939 to 1944, he set the wheels in motion for construction of the Gulf Freeway. On April 12, 1940, Holcombe announced that the city of Houston had reached an agreement with the Houston Electric Company for the termination of electric streetcar service in Houston. Houston Electric would be allowed to abandon its four remaining streetcar routes, including the Park Place line on the former GHE right-of-way, and the city of Houston would be responsible for the rail removal. In return, the city would receive a $50,000 payment from Houston Electric and ownership of the former GHE right-of-way. The GHE railway right-of-way varied in width from 60 to 100 feet (18 to 30 m), and a 230-foot-wide (70 m) corridor would be needed for the six-lane freeway with frontage roads. Because of the increased tax burden due to World War II, it would not be possible to immediately raise the funds needed to purchase the additional right-of-way along the corridor. But planning for the freeway continued, and in October 1943 the Texas Transportation Commission entered into an agreement with the city of Houston and Harris County to develop the facility, referring to it as the relocation of US 75. The Federal-Aid Highway Act of 1944 revived the project by providing funding for new highway construction. Funds were released in October 1945 after the conclusion of World War II.[3]

By early 1946 plans had been formulated for the freeway, and on January 31 TxDOT released drawings of the planned design. The next day the *Houston Post* featured a large report on the project to build the "highway of the future," showing artists' depictions of the freeway, its overpasses, and the frontage roads. The inclusion of continuous frontage roads as a matter of design policy was a characteristic that would set the Gulf Freeway apart from most emerging freeways in the nation and would also set

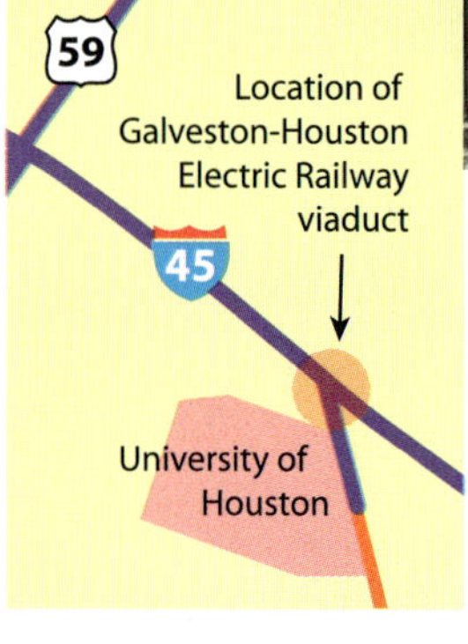

Before the freeway: These photos show the Galveston-Houston Electric Railway, which occupied the Gulf Freeway corridor in Houston prior to the construction of the freeway. The interurban service operated from December 5, 1911, to October 31, 1936. The tracks in Houston continued to be used by Houston Electric, Houston's streetcar operator, until the shutdown of streetcar operations on June 9, 1940. In the upper photo, passengers board the Galveston Flyer in downtown Houston circa 1929. The lower photo shows the electric railway viaduct at the railroad crossing near the present-day University of Houston. The viaduct was demolished after abandonment of the railway, and construction of the freeway began in 1946. (Photos: upper, HMRC MSS 200-375; lower, HMRC MSS 145-4)

The first freeway generation sees the future take shape: Schoolchildren cross the Gulf Freeway construction zone at Telephone Road in October 1948. At the time of this photo, the first section of the freeway north of Telephone Road had just opened. (Photo: *Houston Chronicle*)

the standard for future Houston freeways. Although previous freeways outside of Houston had been constructed with frontage road or "quasi" frontage road sections, particularly in New York City and Detroit, those frontage road sections were more on a case-by-case basis and were not part of a policy for comprehensive use of frontage roads. The terminus of the freeway at downtown Houston would distribute the traffic among four city streets in an arrangement that the project engineer called the loading and unloading platforms.[4]

The first construction contracts were awarded in September 1946. On September 30, 1948, the first section of the Gulf Freeway from downtown to Telephone Road was ready for service. This was not only the first freeway in Houston, but also the first in Texas. A crowd of about 500 gathered on the Calhoun Road overpass for a 7:00 P.M. ceremony that the state of Texas had never seen before: a new freeway dedication. In the spirit of the occasion, speakers at the ceremony praised the freeway as the "best that money can buy." The president of the Chamber of Commerce stated, "There is nothing finer in the construction line in the United States than this highway." After speakers made their statements, Mayor Holcombe threw a circuit breaker to turn on the freeway lights. "The expressway will be in use from now on," he said as a phonograph played "Happy Days are Here Again," his political theme song. In a few moments, the trailer on which the speaker's stand was located was removed. Automobiles which had backed up for about one-half mile in each direction were now free to drive on the city's first freeway.[5]

With the freeway now in service, it quickly became apparent that something was missing: the freeway had no name. Press reports called it the Interurban Expressway, referring to the Galveston-Houston interurban railway that previously occupied the corridor. Mayor Holcombe

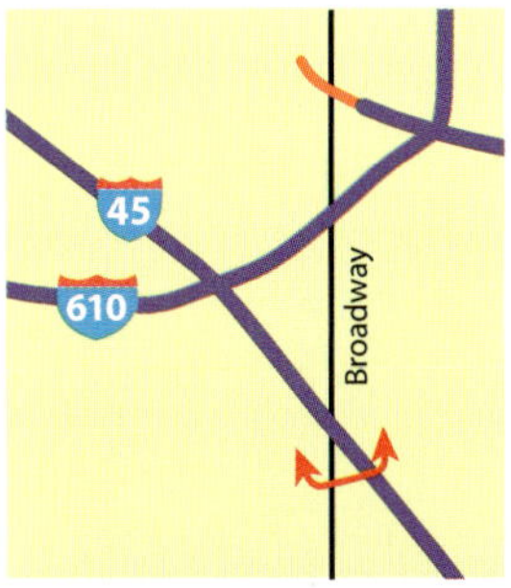

A Houston rarity: This view looks north at the Broadway intersection in early 1952. The intersection design included a traffic circle, and the same basic design remains in service in 2003. It is the only remaining traffic circle on Houston's freeway system. (Photo: Greater Houston Partnership)

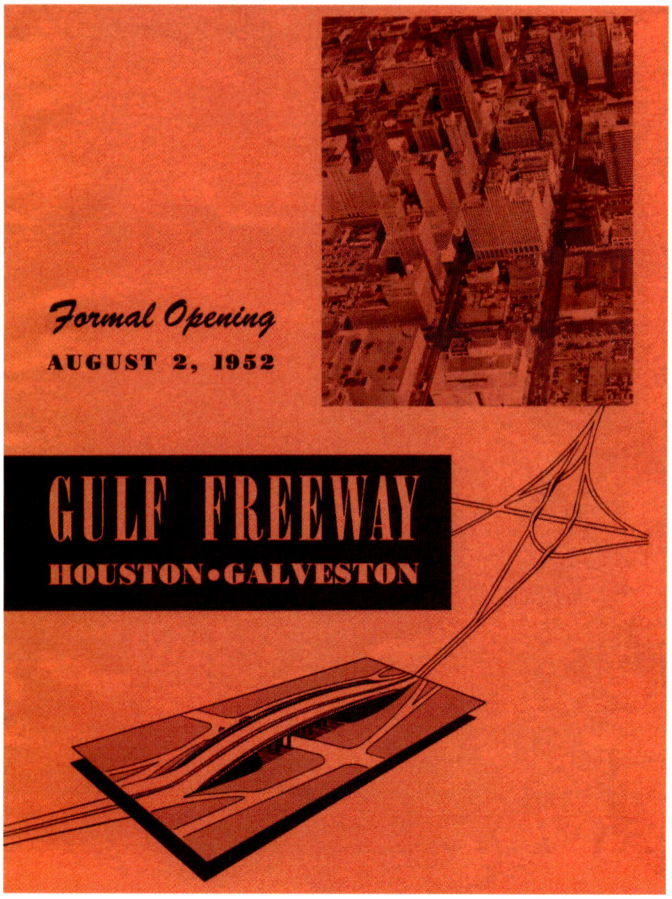

Official dedication: The official opening of the Gulf Freeway between Houston and Galveston took place on August 2, 1952. Motorcades from both Houston and Galveston met at the approximate midpoint near Dickinson for the ceremony, which was held on the FM 517 overpass. A full-sized 12-page brochure with an orange cover (shown at left) was distributed. The designation as a freeway was somewhat of a misnomer, however. Only an 8.5-mile (14 km) section in Houston was actually a freeway, and there were 32 at-grade crossings between Houston and Galveston. But it just wouldn't have sounded right to call it the Gulf Highway. (Source, photo and brochure: *Galveston Daily News)*

Not really a freeway: This view, near the present-day El Dorado Boulevard, shows a typical section of the Gulf Freeway in July 1956. After its official dedication, most of the Gulf Freeway was in fact a highway without access control. TxDOT did not purchase access rights along the highway, so frontage roads would need to be added for the entire length to Galveston to make the Gulf Freeway a true freeway. (Photo: TxDOT)

quickly took action to solve the problem and launched a contest to name the freeway. There was no shortage of suggestions. The six judges sifted through 86 pages with the 13,000 entries, and on December 17, 1948, the winner was announced. The freeway would be named the Gulf Freeway. The exact name "Gulf Freeway" was suggested by only one entry, that of Miss Sara Yancy of the Heights. She received a $100 prize and, perhaps more importantly, placed her mark on Houston's freeway system. Only one other Houston freeway, the Eastex Freeway, would be named by a contest.[6]

Completing the Highway to Galveston

After the opening, work progressed quickly on the rest of the corridor to Galveston, which was a new 300-foot-wide (90 m) alignment bypassing the cities between Houston and Galveston. The new highway took on a mild dog-leg form on its route, in contrast to the existing highway which was almost perfectly straight. At the time, highway experts believed that arrow-straight highways over a long stretch created driver monotony and laxness, and gentle curves in the highway alignment were desirable. In some cases, a dog-leg was needed to steer the roadway to easily-acquired right-of-way or provide a better alignment for a stream crossing.[7]

A large dedication ceremony was held on August 2, 1952, for the completion of the highway between Houston and Galveston. Convoys departed from both Houston and Galveston, meeting at the approximate midpoint at FM 517 near Dickinson where the official dedication was held. Even though the facility was widely called the Gulf Freeway, most of the facility was not a freeway but a four-lane divided highway. The Gulf Freeway was a true limited-access facility for only 8.5 miles (14 km) from downtown Houston to the Houston city limits near the present-day Airport Boulevard. Although there were no traffic signals between Houston and Galveston, there were 32 at-grade crossings, a fact emphasized by officials who worried that motorists might not exercise appropriate caution at the intersections along the route. Those fears were realized when 15 fatalities occurred in the first six months after the official opening. In 1959, TxDOT officials launched a program to eliminate the 30 at-grade crossings that still existed along the route. Complicating the upgrade program was the fact that TxDOT had not purchased access rights along the Gulf Freeway. In order to provide landowners with the right of access and make the freeway limited-access, the Gulf Freeway would need frontage roads for its entire length to Galveston. The final at-grade crossing between Houston and Galveston would not be eliminated until 1976. But in 1952, no one was complaining about the at-grade crossings. The "dream highway" was enough to make everyone happy.[8]

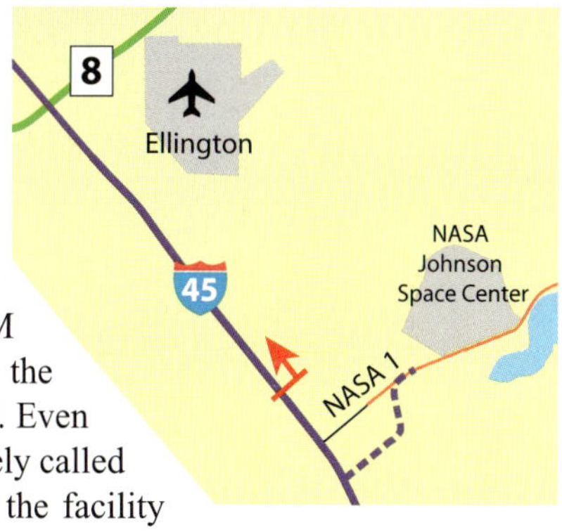

The prototypical Houston freeway: This 1950s aerial view looking toward downtown illustrates one of the key features that would define urban freeways in Texas, and especially in Houston: the frontage road. Although freeways with frontage roads had previously been built in other cities, those instances were typically special cases, often the result of previously existing roads. On the Gulf Freeway, frontage roads were part of a policy and design philosophy that frontage roads should be built whenever possible. This view shows that the original frontage roads did not extend over the railroad tracks, consistent with the policy of serving local traffic, not through traffic, with frontage roads. *(Photo: TxDOT)*

The loading and unloading platform: This view shows the terminus of the Gulf Freeway just southeast of downtown Houston on August 3, 1954. The chief engineer for the Gulf Freeway called the connection into downtown the loading and unloading platforms, since four city streets were integrated into the freeway.[9] (Photo: Squire Haskins Photography, Dallas)

Increasing Traffic Load

The Gulf Freeway corridor served southeast Houston, the city's most active region in the 1940s and 1950s. The corridor included the Houston International Airport (now Hobby Airport), Ellington Field, and the burgeoning petrochemical complexes along the Houston Ship Channel. Later, in the early 1960s, NASA's Manned Spacecraft Center (now the Johnson Space Center) would be constructed in the Gulf Freeway corridor.

It didn't take long for Houston motorists to develop a love for their new freeway. On October 2, 1948, the second day of operation, 28,800 vehicles used the Gulf Freeway at its busiest point. A traffic count meter resembling a big thermometer was erected alongside the freeway to keep motorists up-to-date on the increasing traffic volume. Traffic increased rapidly, and by September 1950 the traffic meter had reached 66,300 vehicles per day. Local officials celebrated the 100 millionth vehicle-mile of travel in September 1950 with a ceremony next to the freeway and a chart that showed the growth in cumulative traffic volume.

Officials had not yet realized that rapidly increasing

The who's who of Houston celebrate freeway traffic: In August 1950, 21 months after the freeway opening, local officials celebrated the 100 millionth vehicle-mile of travel on the Gulf Freeway. Unexpectedly heavy demand for the freeway was cited as evidence of the need for more new freeways. Officials would soon realize, however, that rapidly increasing traffic volumes were not necessarily something to celebrate. By 1954 the freeway was operating at full capacity during rush hour, and the era of the Houston freeway traffic jam had begun. The following are included in the photo: William P. Hobby, former Texas governor and owner of the *Houston Post* (second from left in the white suit); W. J. Van London, head of the Houston TxDOT office and chief designer of the freeway (immediately to the left of the sign); Houston Mayor Oscar Holcombe, one of the freeway's strongest backers (immediately to the right of the sign); and Jesse Jones (second to the right of the sign). Jones was often called "Mr. Houston" in recognition of his contribution to the development of Houston in the first half of the 20th century. He was instrumental in the construction of the Houston Ship Channel, a leading real estate developer in the downtown area, and owner of the *Houston Chronicle*. His political connections brought the 1928 Democratic National Convention to Houston, and he served in powerful, high-level federal positions under President Franklin Roosevelt. In 1999 the *Houston Chronicle* named Jesse Jones the Houstonian of the Century.[10] (Photo: Center for American History, University of Texas at Austin, CN 11063)

traffic volume was not necessarily something to celebrate. Six years after the opening, in November 1954, 90,000 vehicles per day were using the freeway at its busiest point. A report issued by W. J. Van London, head of the TxDOT Houston Urban Project Office, stated that the "freeway is carrying about full capacity for 30 minutes between 7:00 and 7:30 in the morning on the inbound lanes and for about 20 minutes on the outbound lanes in the evening from about 5:00 to 5:20." The era of the freeway traffic jam was about to arrive. And so began the never-ending battle against freeway congestion in Houston.[11]

By the early 1960s, the Gulf Freeway became heavily congested and officials were recommending the construction of parallel relief freeways. The proposed extension of the La Porte Freeway (SH 225) into downtown had its origins in a 1961 traffic study. In 1965 the Alvin Freeway (SH 35) was recommended. Neither freeway would be constructed in that period, and the La Porte Freeway extension would ultimately be cancelled.[12]

By the mid-1960s, average rush-hour speeds on the Gulf Freeway would drop to about 22 miles per hour (35 km/h). In spite of traffic management efforts by TxDOT, traversing the five-mile (8 km) section of freeway from just south of Loop 610 to downtown could take 30 to 60 minutes in 1969. The freeway recorded 150,000 vehicles per day just south of downtown in 1969, far busier than Houston's second busiest freeway, the West Loop, which had a volume of 113,000 vehicles per day just south of the Katy Freeway. After 1969, a shift in Houston's freeway travel patterns occurred. Without a precipitating event, Gulf Freeway traffic volume just south of downtown dropped to 142,000 vehicles per day in 1971. In the same period from 1969 to 1971, traffic on the West Loop increased to 146,000 vehicles per day. Something larger was happening in Houston. The center of Houston's population growth and development had shifted away from southeast Houston and became focused on the west and southwest sides of the city, with huge commercial real estate developments underway at Greenway Plaza on the Southwest Freeway and in the Galleria area on the West Loop. Traffic

Keeping count: A traffic count indicator was placed alongside the freeway to keep motorists up-to-date on freeway usage. At the time of this photo, circa 1951, traffic volume had reached 75,000 vehicles per day. Unfortunately for motorists, the traffic count would go off the scale by 1954 and then continue to increase, soon causing traffic congestion. (Photo: TxDOT)

Houston's first congested freeway: By the early 1960s rush-hour traffic jams were part of everyday life on the Gulf Freeway. This 1965 photo shows the evening rush-hour backup at the beginning of the freeway in downtown. Researchers at the Texas Transportation Institute used the Gulf Freeway to investigate experimental traffic management techniques between 1963 and 1969. (Photo: HMRC MSS 157-622)

Crossing over to safety: The most serious design flaw of the original Gulf Freeway was the lack of a center guardrail. The upper left photo shows a typical section of the original freeway with its curbed median. Prior to the installation of the barrier, there were 60 crossover accidents, resulting in 9 fatalities and 70 injuries. The upper right photo shows the installation of the median guardrail in 1956. The lower photo shows the freeway in the late 1950s. In the three years after the guardrail installation, there were no crossover accidents. (Photos: TxDOT)

Ramping up to danger: This 1963 photo illustrates design and safety hazards in the original Gulf Freeway. The freeway on-ramp has practically no merging space and was the most accident-prone spot in Houston in 1963. Also shown is a row of light fixtures positioned within about 12 feet (3.7 m) of the main traffic lanes. Breakaway fixtures had not yet been developed in 1963, and these obstructions were a serious hazard to motorists.[13] (Photo: *Houston Chronicle*)

demand on the Gulf Freeway would resume its upward trend in 1972, but the Gulf Freeway would never again be Houston's traffic volume leader. From the traffic congestion perspective, this was good news. However, major relief from traffic congestion would not take place until the freeway expansion was completed in the 1980s.[14]

First Generation Freeway Design

Even though the Gulf Freeway was touted as the "best that money can buy" in 1948, its shortcomings quickly became apparent. First and foremost was the absence of a median barrier. In the 1940s and early 1950s, it was common to construct freeways with only a raised median with curbs, not only in Texas but also in other states, including California. By 1956, there had been 60 accidents caused by vehicles crossing the median into oncoming traffic, resulting in nine fatalities and more than 70 injuries. A 1954 report from the Houston office of TxDOT reported that six of the seven median crossings in 1953 occurred on overpasses, and it was believed that drivers were using the elevated position to observe the cityscape rather than pay attention to the road. In 1956, a two-foot-high (61 cm) metal barrier rail was installed in the median. Three years later there were no reports of vehicles jumping over or running through the barrier.[15]

There were numerous other design deficiencies. The merging distances at the freeway on-ramps were too short. The freeway, which had been called a roller-coaster, had poor sight lines, especially at overpasses. Bridges, and in many cases ground-level lanes, lacked shoulders. The Gulf Freeway would later be designated as a "first generation freeway" by TxDOT.* The lessons learned on the Gulf Freeway would be applied to the next generation of freeways in Houston, which would start to take shape in the late 1950s. However, many of the first-generation sections of the Gulf Freeway would remain in service until the mid-1980s.

A Match Made in Heaven

As the widespread ownership of private automobiles and development of freeways began to gain momentum across the United States after World War II, new concepts

* For a complete history of the first generation freeway and subsequent generations of freeways, see "Building Better Freeways," starting on page 71.

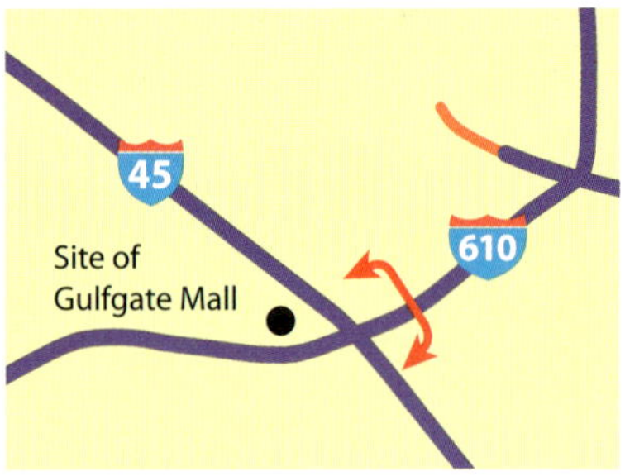

The mall arrives in Houston: Houston's first shopping mall, Gulfgate Shopping City, opened on September 20, 1956. A huge crowd estimated at 150,000 to 200,000 visited the mall on opening day. This photo shows the original, open-air mall in the late 1950s. The Gulf Freeway crosses from left to right in the lower half of the photo, and the South Loop would later be built in the corridor just to the left of the mall. Gulfgate Mall was highly successful, but went into decline starting around 1982. In May 2001 the mall was demolished and redeveloped with retail stores in a strip-center configuration. (Photo: TxDOT)

in suburban land development began to appear. Levittown, New York, launched in 1947, typified the trend toward low-density, mass-produced, single-family housing. Another new product of the era was the auto-oriented shopping center. Crenshaw Center in Los Angeles, also opened in 1947, was one of the first auto-oriented shopping centers of the new automobile suburbia. Northgate on Interstate 5 in Seattle, Washington, which opened on April 21, 1950, was the next step in the evolution of the regional shopping center with its arrangement of stores along a central mall. The first regional shopping mall is generally recognized to be Northland Center, opened in 1954 in Southfield Township, Michigan, near Detroit. The open-air mall was designed by architect Victor Gruen. The first fully enclosed, climate-controlled regional shopping mall is generally recognized to be Southdale Center in Edina, Minnesota, just southwest of Minneapolis. Southdale, also designed by Gruen, opened in 1956.

Freeways and shopping malls were a match made in heaven, so it was only a matter of time before the Gulf Freeway would get its first mall. In March 1954 ground was broken for Gulfgate Shopping City, an open-air mall located at the intersection of the Gulf Freeway and the planned South Loop. The grand opening of Gulfgate Shopping City and its 62 stores took place on September 20, 1956. A huge crowd converged on the shopping center opening day. The parking lot was filled to capacity for the 10 A.M. opening, and by 11 A.M. automobiles were spilling over onto nearby streets and fields. An estimated 50,000 people attended the dedication, and an estimated 150,000 to 200,000 people visited the mall on opening day. Houstonians quickly adopted the shopping mall lifestyle, and Gulfgate Shopping City was highly successful. Houstonians would get their first fully enclosed air-conditioned mall, Sharpstown Mall on the Southwest Freeway, on September 14, 1961.[16]

The Gulf Freeway as a Guinea Pig

One morning in May 1956, motorists on the Gulf Freeway were greeted by something very strange. A 50-foot (15 m) tower had been erected immediately adjacent to a freeway overpass, and on top of the tower were a large clock and two movie cameras. What was this all about? It was just one of the many studies and experiments conducted on the Gulf Freeway over the years, distinguishing the Gulf Freeway as the guinea pig of Houston's freeways. The 1956 study filmed vehicle movement to help researchers get a better understanding of vehicle behavior at freeway entrance and exit ramps. In addition,

Freeway laboratory: Since the Gulf Freeway was the first freeway in Texas and the first to experience many freeway-related problems, it became a laboratory for experimentation and research. In 1956, researchers erected this 50-foot-tall (15 m) tower at the Cullen overpass to film traffic movements before and after the installation of a center guardrail. Extending from the tower platform is a clock, which was included in the camera view. Plans for the tower were publicized in advance to minimize the disruption caused by curious motorists. (Photo: TxDOT)

the study examined the effects of the addition of a central median barrier which would soon be installed. The clock was mounted so that it was visible in all camera views of the traffic flow, and white lines painted on the roadway allowed researchers to estimate vehicle speed.[17]

Intensive study of the Gulf Freeway actually began soon after its completion. In July 1949 the city of Houston released a report titled *Economic Evaluation of the Gulf Freeway*, documenting the time savings provided to motorists. The first study to determine the effect of the freeway on property values took place in 1950. The study found that freeway property values in the "zone of influence" increased by 103% between 1945 and 1950, while land values outside the freeway zone of influence increased by only 50%. Accident rates were also carefully studied, and the freeway sections of the Gulf Freeway were found to have substantially fewer fatal accidents than the national average. The benefits found by these studies were used to justify the cost of the freeway and to help secure funding for future freeways.[18]

The traffic congestion that developed on the Gulf Freeway by the early 1960s made it an ideal laboratory for studying new traffic management techniques. The period from 1963 to 1969 was the most active for freeway research. The first ramp control experiments were conducted in August 1964. Selected ramps were closed during peak hours and police officers regulated the flow of vehicles entering the freeway at ramps which remained open. Results indicated that ramp control was effective in improving freeway operation. In September 1965, ramp metering signals began operation at freeway on-ramps, with the timing of the signals still controlled manually by operators with remote control radio. After full implementation, researchers reported a 30% increase in average speed and a 12% increase in roadway capacity. In March 1966, automatic ramp metering was tested by using computer-controlled signal actuation, metering vehicles into the freeway based on available capacity and gaps in the traffic. The 1960s-era computer occupied two six-foot-tall (1.83 m) racks. Ultimately, fully automatic operation of

Freeway and train tracks part ways: This aerial view from the mid-1950s looks south at the Bellfort crossing. Just south of Bellfort, the Gulf Freeway veers away from the right-of-way of the Galveston-Houston Electric Railway. At the top center of the photo, evidence of the railway right-of-way can still be seen. (Photo: TxDOT)

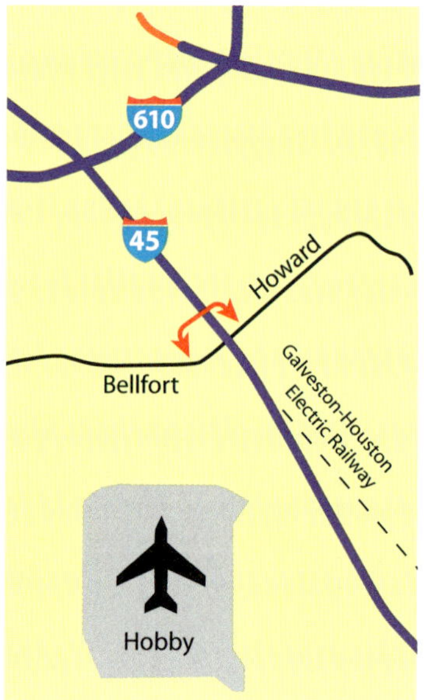

the ramp meters was achieved using fixed timing.

In conjunction with the ramp metering project, a closed-circuit television system with 14 cameras was put into operation in December 1966, providing a continuous video view of a 6.5-mile (10 km) stretch of freeway. Near the end of the study in November 1968, researchers reported that the new traffic management techniques had increased the average speed during the morning rush hour from 22 to 34 miles per hour (35 to 54 km/h).[19]

Space Freeway

On May 25, 1961, President John F. Kennedy summoned both houses of Congress for a Special Message on Urgent National Needs. His speech focused on the Cold War, the threat it imposed, and steps that were needed to ensure national security. But near the end he delivered a statement that would distinguish this speech as perhaps one of the most significant in the nation's history: "I believe that this nation should commit itself to achieving the goal, before this decade is out, of landing a man on the moon and returning him safely to the earth. No single space project in this period will be more impressive to mankind, or more important for the long-range exploration of space; and none will be so difficult or expensive to accomplish." With that statement, President Kennedy upped the stakes in the space race between the United States and the Soviet Union—a contest in which the Soviet Union had established its lead with the October 4, 1957, launch of Sputnik, the world's first orbiting spacecraft.

Though few would know it at the time, Houston would soon become a key player in the nation's space effort. Influential Texans in Washington, D.C., including Vice President Lyndon Johnson and Houston Representative Albert Thomas, worked to designate Houston as the location for the new Manned Spacecraft Center that would lead the nation's effort to reach the moon. On September 19, 1961, the announcement was made in Washington, D.C. that a site southeast of Houston, just off the Gulf Freeway, would house the new complex. Throughout the 1950s, the Houston Chamber of Commerce had promoted Houston as "America's Industrial Frontier," printing the statement on its publications. In an instant, Houston became Space City, USA.

The original seven Mercury Program astronauts, who

The original seven: America's first seven astronauts were introduced to the public in Washington, D.C. on April 9, 1959. In 1961, Houston was designated as the location of the Manned Spacecraft Center (now the Johnson Space Center), and on July 4, 1962, the astronauts were welcomed to Houston with a parade. Five of the astronauts were auto enthusiasts, and two in particular, Gordon Cooper and Gus Grissom, were racing fanatics and were known to race frequently on the roads around the space center, including the Gulf Freeway. Front row, left to right: Walter Shirra, Deke Slayton, John Glenn, and Scott Carpenter. Back row, left to right: Alan Shepherd, Gus Grissom, Gordon Cooper. (Photo: NASA)

had been selected in 1959, were introduced to Houston on July 4, 1962, with a large downtown parade and a barbecue at the Sam Houston Coliseum.* While the space program was pushing science and technology to new levels, five of the astronauts had a more down-to-earth interest: fast cars. Gordon Cooper, Gus Grissom, Walter Shirra, Alan Shepherd, and Scott Carpenter were all automobile enthusiasts. Two in particular, Cooper and Grissom, were especially fanatical about auto racing and were partners in an Indy car racing team that participated in major racing events. The racing enthusiasts didn't limit their speed contests to the race track. The roads around NASA, especially the Gulf Freeway and State Highway 3, were the scene of frequent high-speed races between the astronauts.[20]

The Manned Spacecraft Center, which was renamed the Johnson Space Center in 1973, also served as the launch pad for Houston's first large-scale, master-planned real estate development. In late 1962, Del E. Web Corporation and Friendswood Development announced the 15,000-acre Clear Lake City community, dwarfing the 6,500-acre Sharpstown development which had begun along the Southwest Freeway in 1955.[21]

Entering the Anti-Freeway Era

Among its many firsts, the Gulf Freeway was also Houston's first freeway to generate a substantial amount of community opposition over expansion plans. The year was 1972, and TxDOT was formulating plans to expand the antiquated section of freeway just south of downtown, which had opened in 1948, and run the northern terminus of the planned Alvin Freeway on the same right-of-way as the Gulf Freeway. Approximately 100 feet of right-of-way would be taken on the south side of the Gulf Freeway, displacing about 170 homes with 500 to 800 residents and 22 businesses. The residents to be displaced were almost entirely black and low-income, and many were elderly.

Leading the black community's efforts against the project was the director of the Martin Luther King Community Center, Madgelean Bush, a feisty woman who fought for improvement of the black inner city and told people she had a degree in "ghettology." After the first public hearing

Eight days in a space capsule? No problem. The Gulf Freeway? Too dangerous! This cartoon appeared in the *Houston Chronicle* on August 10, 1969, just after Apollo 11's moon landing with its crew of Neil Armstrong, Edwin "Buzz" Aldrin, and Mike Collins.

---

* The Sam Houston Coliseum was located downtown at the present-day location of the Hobby Center for the Performing Arts, 800 Bagby. It was constructed in 1937 and demolished in 1998.

The Gulf Freeway transitway: In the late 1960s planners at TxDOT were already envisioning a comprehensive busway system for the Gulf Freeway, fully integrated with park-and-ride lots and connecting ramps along its length. Twenty years later in May 1988, the original concept of the fully integrated bus lane was realized with the opening of the transitway from downtown to Loop 610. The transitway was extended southward to south of Beltway 8 during the 1990s. This morning rush-hour view looks north along the freeway at the connection ramp to the Monroe Transit Center, which is approximately halfway between Loop 610 and Beltway 8. This view shows the mix of buses and high-occupancy automobiles that use the transitway. (Photo: Metropolitan Transit Authority)

in February 1973, two organizations were formed to oppose the freeway: the Anti-Freeway Coalition and the Third Ward Preservation Association. A third group known as the Urban Bunch, consisting primarily of architects, also opposed the freeway expansion. The opposition groups enlisted the help of recently elected minority representatives Barbara Jordan of the United States House and State Representative Mickey Leland. Additional public hearings were held, often with heated dialog and racial overtones. Fueling the opposition was a belief that freeways benefited affluent white suburban communities and contractors, while disrupting low-income inner-city neighborhoods. At one meeting an irate black woman exclaimed: "This will be the last time a white man's bulldozer moves through the black man's bedroom!"[22]

The opposition was not able to stop the Gulf Freeway expansion, however. The 1973 environmental impact statement for the project stated, "An analysis … would indicate that there is no widespread opposition to the project. It is also concluded that those opposing the project have received little or no widespread support even though some of those opposing the project have received wide publicity through the news media." The expansion project to build a dual freeway with the Gulf Freeway at ground level and the lanes for the future Alvin Freeway on elevated structures would proceed, although construction would not begin until the early 1980s.

"The Gulf Freeway, billed as the Highway of the Future at its inception in 1945, is still the highway of the future, for it has never been completed."

Houston Chronicle, August 10, 1969

50 Years of Construction

When the Gulf Freeway was officially dedicated on August 2, 1952, construction on the Gulf Freeway was far from complete. In fact, work had really just begun. The first dedication ceremony in 1948 marked the beginning of 50 years during which there was almost always some construction in progress. Finally, in 1997, ongoing construction came to a temporary halt with the completion of the magnificent stack interchange at Beltway 8. But in those 50 years, the Gulf Freeway became known as the freeway that was never finished, prompting one of Houston's longest-running one-liners: "I just hope to live long enough to see the Gulf Freeway finished."[23]

As sections of the Gulf Freeway were opened starting in 1948, many were obsolete as soon as they opened or shortly thereafter. First generation freeway design standards and rapidly increasing traffic volume rendered the freeway inadequate. The Gulf Freeway "was the first freeway in the state, and we didn't have freeway experi-

The Gulf Freeway wishbone: This transitway access structure was completed in 1997 just south of Beltway 8. (Photo: James Lyle, TTI, June 2001)

ence," explained A. C. Kyser, manager of TxDOT's Houston Urban Project Office, in 1969. A few years later in 1973, another engineer observed, "Very little was known about freeways when we started. And who knew about the Space Center location or the rate of growth Houston would have?" In 1976 a construction worker on a project joked, "I think we'll get through with this in about the next 20 years." His joking prediction turned out to be surprisingly accurate.[24]

Soon after the 1952 dedication, work began to provide grade separations at many of the intersections on the highway. Starting in the late 1950s, TxDOT began to reconstruct the freeway one small section at a time to bring it up to standards and convert it from a highway to a true freeway. In 1960 the first widening was completed when a short section of freeway was widened to six lanes from the present-day Loop 610 southward to Sims Bayou. From the late 1960s to the late 1970s, traffic was shifted to the frontage roads as long sections of main lanes were reconstructed all the way to Galveston. The process began all over again in the early 1980s, with work beginning downtown and proceeding southwards to rebuild and expand the freeway to eight main lanes with a central, barrier-separated transitway lane. The slow progression finally reached Beltway 8 in the mid-1990s with the completion of the huge stack interchange in 1997. Finally, the Gulf Freeway was construction-free. But it was not finished. In fact, the Gulf Freeway may never be truly "finished."

Future Plans

A comprehensive study completed in 1999 set the wheels in motion for the next round of major reconstruction. From Beltway 8 to Galveston, the freeway will be modernized and widened. The section from Beltway 8 to

> "Like the pyramids of Egypt, the freeway has consumed much of the working careers of many of the workers involved in its construction."
>
> *Houston Post*, March 25, 1976

Key dates in the history of the Gulf Freeway

1930	The Houston-Galveston super highway receives its first serious consideration.
1940	The city of Houston purchases the right-of-way of the former Galveston-Houston Electric Railway.
1943	TxDOT officially designates the route as a freeway.
1948	The first freeway section opens on September 30.
1952	The highway to Galveston is dedicated.
1959	The first major reconstruction and expansion begins.
1961	NASA announces the Manned Spacecraft Center near the Gulf Freeway.
1976	The entire distance to Galveston meets freeway standards.
1988	Transitway operation begins.
1997	Major freeway construction temporarily ends with the completion of the interchange at Beltway 8.
Future	Expansion between Beltway 8 and Galveston.

FM 518 in League City, currently a 6-lane freeway with 2-lane frontage roads in each direction, will have 10 general purpose main lanes, 2 diamond lanes for high occupancy vehicles, and 3 frontage road lanes in each direction. From FM 518 in League City to Galveston, the freeway will be expanded from its present 6 main lanes to 8 main lanes, and the Galveston Causeway will be reconstructed to an 8-lane facility. Work on the $136 million causeway reconstruction project began in the summer of 2003, and work on the new NASA 1 bypass freeway, which will involve reconstruction of the Gulf Freeway main lanes, could begin in 2004. Those projects will kick off another 20 years of work on the freeway.[25]

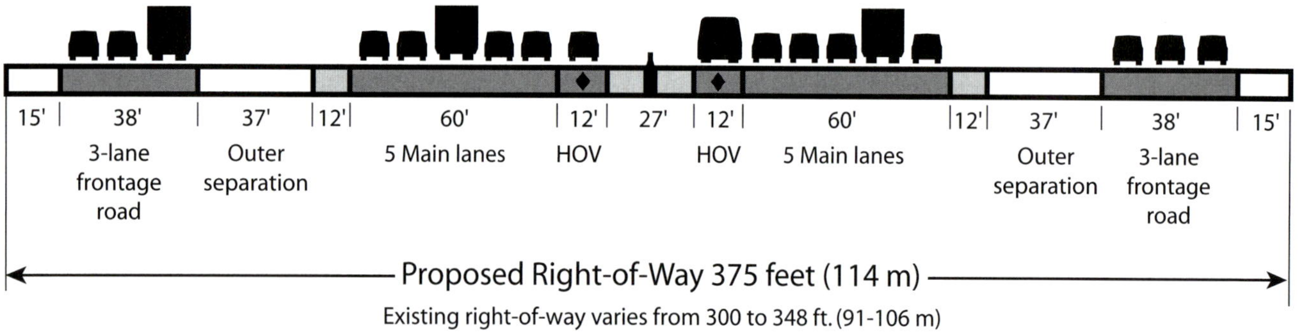

Future construction: In 2003 the Gulf Freeway is free of major construction, but plans are being developed for the next wave of construction which will expand the freeway from Beltway 8 to Galveston. The 9-mile (14 km) section immediately south of Beltway 8 will be expanded to the configuration shown above, with 10 general purpose main lanes, 2 non-barrier-separated HOV lanes, and 3-lane frontage roads in each direction. Construction will likely take place between 2010 and 2015. (Source: adapted from *I-45 South Corridor Major Investment Study Executive Summary*)

The downtown approach: The downtown approach of the Gulf Freeway is actually a multiplex with the planned Alvin Freeway. The elevated structures between the Gulf Freeway main lanes and the frontage roads serve as the terminus of the Alvin Freeway and the downtown exit for both the Gulf and Alvin Freeways. The early 1980s image at right shows this section of freeway before reconstruction began. The freeway was still generally in its original 1948 configuration. The main lanes were expanded and the elevated structures were added. This project generated the first major community opposition over a planned freeway expansion in Houston when public hearings were held in 1973. (Photos: above, May 2002; right, Chuck Fuhs)

The ultimate stack interchange? The impressive interchange at the Gulf Freeway and Beltway 8 features high-flying ramps, symmetric design, and a sprawling right-of-way. This view looks northwest along the freeway. The completion of the interchange in 1997 capped 50 years of nearly nonstop construction on the Gulf Freeway. (Photo: May 2002)

Overall view: This wide-angle view shows the overall configuration of the interchange. (Photo: May 2002)

Ground view: This view looks west along the Beltway 8 main lanes, with the frontage road at ground level at left. (Photo: November 2002)

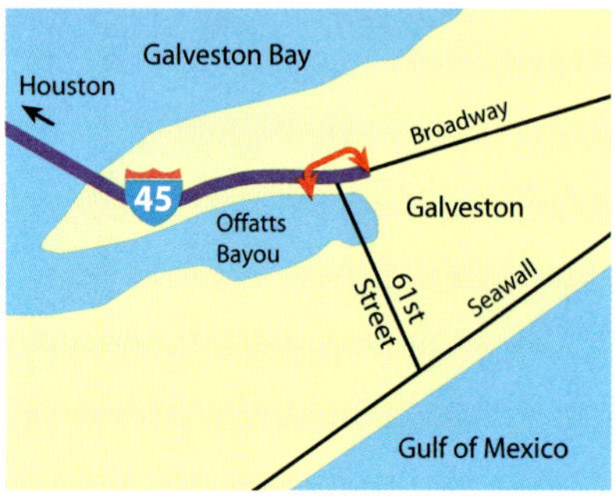

Gulf Freeway namesake: This view of the Gulf Freeway in Galveston shows the freeway's namesake, the Gulf of Mexico. The body of water in the center of the photo is Offatts Bayou, which is really more of an inlet from Galveston Bay than a bayou. The water in the upper part of the photo is the Gulf of Mexico. The road running vertically is 61st Street. The Gulf Freeway ends just to the left of this photo and does not actually reach the Gulf of Mexico. (Photo: James Lyle, TTI, June 2001)

Southwest Freeway, US 59 South

The relationship between land developers and freeways has always been a cozy one. The freeway gets built, the land becomes prime for real estate development, and developers make millions. But what if the freeway doesn't come, or what if it is indefinitely delayed? If you are the real estate developer, you do whatever it takes to get the freeway built. That includes offering to donate the land for the freeway and aggressively lobbying political officials to build it.

In 1957, the Southwest Freeway became the recipient of Houston's first big freeway land giveaway when real estate developer Frank Sharp pulled together a large coalition of landowners to donate a 10.5-mile (17 km) right-of-way strip for the freeway. Sharp needed the freeway to get shoppers to his planned mall and suburbanites to their new houses in his huge Sharpstown community. Sharp probably never envisioned that the Southwest Freeway would someday become Houston's freeway-era main street and one of the busiest freeways in the United States.

Southwest Freeway	
Designated as freeway	1953
First freeway section open	1961
Freeway complete	1975 (to Sugar Land)
Reconstruction	1989–ongoing
Max traffic volume, 2001	337,000 vehicles per day
Future construction	Depress elevated freeway into trench (in progress in 2003); expand freeway in Fort Bend County.

Origins

The need for a freeway serving the southwest part of Houston was first identified in a 1953 origin and destination study that served as the basis for the formulation of Houston's first freeway master plan. Local authorities presented the master plan to the Texas Transportation Commission in July 1953. In that plan, two separate freeways were envisioned for southwest Houston: one followed the present-day Southwest Freeway and Westpark Tollway to far west Houston and the second freeway would be located on the US 90A ("Alt-90")-South Main corridor (see map on page 13). In September 1953, the Texas Transportation Commission authorized route determination studies for US 59 from the Fort Bend County line to US 75 (now IH 45) in downtown Houston.[26]

In late 1953 the plans for southwest Houston were being reviewed. In a December 29, 1953, letter to the Houston office of TxDOT, City of Houston Planning Director Ralph Ellifrit reported that his department had devised a new alignment for the Southwest Freeway which curved southward from the Westpark corridor and connected with the present-day US 90A near Sugar Land. "We have made some very rough studies of such a location and would like

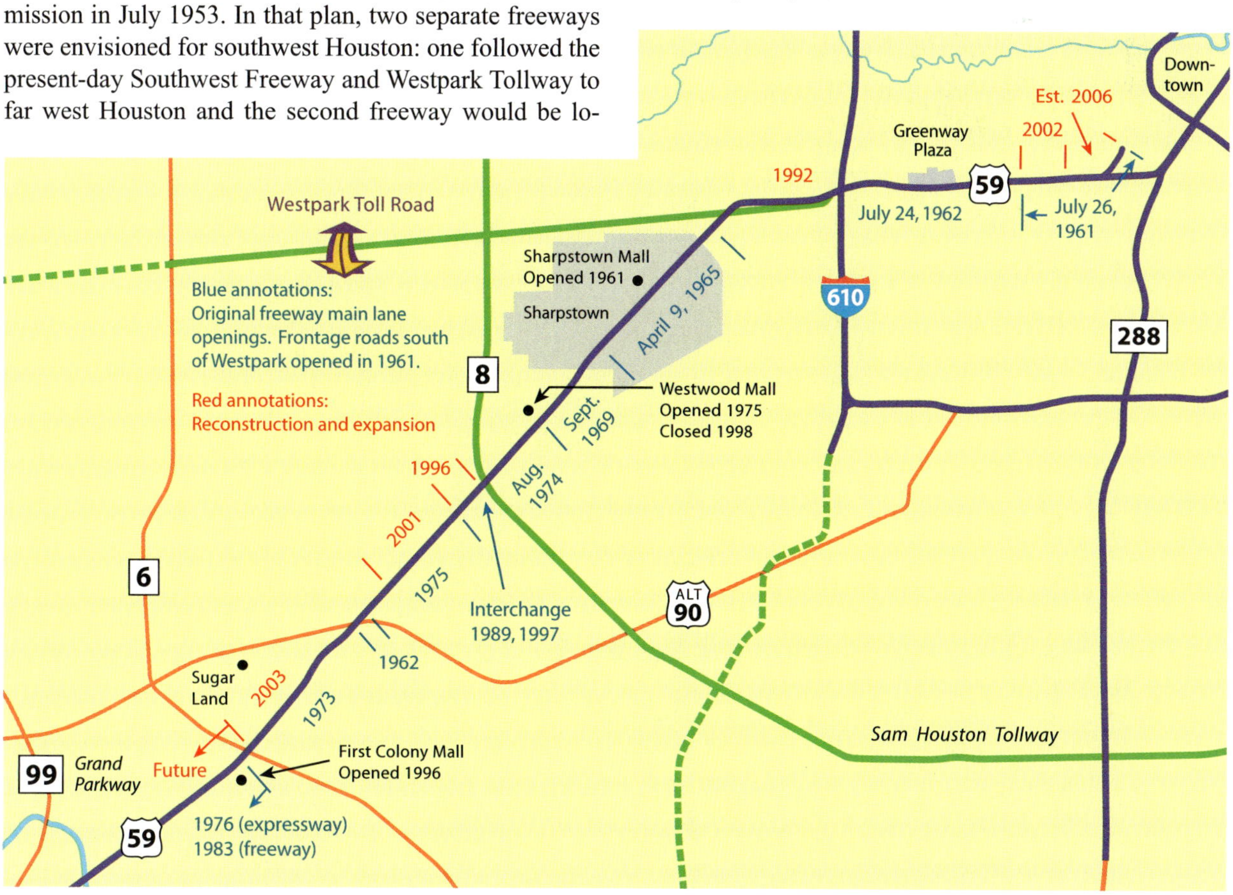

Pre-freeway US 59: Before the freeway era, major highways were routed through the city on ordinary city streets. This view, from around 1950, shows US 59 where it followed today's South Main. (Photo: HMRC MSS 227-519)

to show it to you after the first of the year to see what you think about it," Ellifrit wrote.[27]

The new alignment of the Southwest Freeway as proposed by Ellifrit closely resembled the final alignment and would be adopted in the March 1954 revision of the master plan for Houston's freeways. The Westpark Freeway and US 90A Freeway were eliminated in the 1954 plan. The alignment appearing on the city of Houston's 1955 *Major Thoroughfare and Freeway Plan*, the first edition of the plan to include the full freeway system, was still not final, however. Plans for the freeway alignment in 1955 showed the freeway veering southward in the vicinity of Weslayan and proceeding westward along Glenmont Street, which is approximately one-half mile (0.8 km) south of the actual constructed alignment. In January 1956 the final alignment was adopted. In August 1960, the Texas Transportation Commission formally approved the closure of the downtown gap in US 59, connecting the Southwest Freeway to the Eastex Freeway.[28]

Freeway Suburbia

Just southwest of the urbanized area of Houston was a huge tract of land whose destiny would be shaped by the imminent age of the freeway and freeway-driven suburban development. The newly designated Southwest Freeway was aligned to cut a diagonal path right through the property. For land developer Frank Sharp, the owner of the tract, the freeway would be the backbone of his new vision of urban America—a vision he modestly named Sharpstown.

Sharp and his partners were reported to have performed the first studies for Sharpstown in early 1954, about the same time that Houston's original 1953 freeway master plan was modified to place the Southwest Freeway

Texas-sized suburbia: This is an excerpt from the full-page announcement for the grand opening of Sharpstown that appeared in the Houston newspapers on March 13, 1955. The advertisement touted Sharpstown as the world's largest residential development, a "city within a city" with 25,000 homes planned. The opening ceremony featured the sealing of a time capsule to be opened on March 13, 2000. When the time capsule, a metal enclosure, was opened, moisture infiltration had damaged or destroyed all the contents.

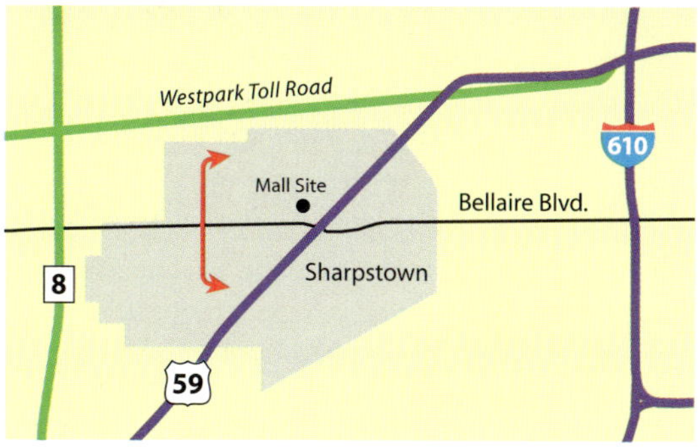

Opening a path to suburbia: This 1959 view looks east over Sharpstown, with the Southwest Freeway frontage roads crossing diagonally across the photograph. As part of the deal in which landowners donated 10.5 miles (17 km) of right-of-way for the Southwest Freeway in 1957, TxDOT agreed to begin construction of the frontage roads within one year. In this view, the first section of Sharpstown is at right center, abutting the frontage roads. Within five years of this photograph, nearly all the vacant land in this view had been developed with housing, apartments, and shopping centers. (Photo: The Positive Image)

on its alignment through Sharpstown. Speculation that the Southwest Freeway was realigned for Sharp has been a freeway planning legend in Houston. Available documents suggest that the Sharpstown alignment was a technically superior alignment, and Sharp's cooperation in preserving the freeway corridor worked to the benefit of both Sharp and Houston's freeway program. For local authorities who faced the difficult and expensive job of right-of-way acquisition for freeways, it was logical to align the freeway where there was a single, cooperative landowner.[29]

Sharp and his partners formally acquired a 4,000-acre tract known as Westmoreland Farms in June 1954 and added another 2,500 acres nine months later. Plans for the new project were first announced on July 10, 1954. Whereas many freeways in the 1950s cut paths through established neighborhoods, the Southwest Freeway would be the lifeblood of Sharpstown, not a destructive force. The entire community would be designed around the freeway.

The sheer size of what was touted as the "world's largest residential development" was impressive, with 25,000 homes planned for the 6,500 acres.* The planned community would be substantially larger than the record holder

* Only about 6,800 single-family homes were ultimately constructed in Sharpstown. Land was converted to commercial or educational use, some of the original acreage was sold, and a planned development of high-rise housing north of Sharpstown Mall was not built. There probably was also some marketing hype in the original announcements.

The birth of Houston's busiest freeway: This December 1959 view looks east over the beginnings of the Southwest Freeway-West Loop interchange. This interchange would be the first four-level interchange in Texas with modern design characteristics. The four-level Fort Worth Mixmaster at IH 30 and IH 35W was completed in 1958, but it was built to first-generation freeway standards with left exits and poor geometrics. The section of the Southwest Freeway to be built in the immediate foreground would go on to become Houston's busiest freeway starting in 1993 and continuing in 2003, with 337,000 vehicles per day in 2001. (Photo: TxDOT)

"Sharpstown [is] completely planned and geared for the motor … ."

Houston Post, March 14, 1955

at the time, William Levitt's 16,000-home Levittown, Pennsylvania. But more significant was the integration of all urban amenities into Sharpstown, including a shopping mall, schools, churches, hospitals, a country club, and of course, the freeway. A representative from the Federal Housing Administration remarked, "The primary consideration is not size, but the satisfaction of developing the kind of city which will stand as a model in years to come." It was, he said, "a new experiment in our way of life." The large-scale development of virgin land also provided savings that impressed out-of-town reporters. The real estate writer for the *Cleveland Press* observed that "developed lots which cost $2,000 here would start at $3,500 to $4,000 around Cleveland." Also noteworthy to reporters was Sharpstown's country club, which would bring country club living to all residents of Sharpstown for an affordable price.[30]

But perhaps the most significant aspect of Sharpstown was succinctly stated by the *Houston Post* on March 14, 1955: "Sharpstown [is] completely planned and geared for the motor … ." This was automobile suburbia taken to a new level of scale. The architect for the planned community shopping center stated, "We have designed it to maintain four or five square feet of parking space to every square foot of shopping space." Main Street was gone, and residents would soon be greeted by a small ocean of a parking lot at the neighborhood shopping mall.[31]

On March 13, 1955, Frank Sharp launched his Sharpstown community in a large, locally televised ceremony attended by thousands. Two-bedroom and three-bedroom homes were priced from $12,500 to $16,800. A month later 54 homes were complete and 420 homes were under construction. A year later about 1,000 homes were complete. Sharp celebrated the first anniversary of his community with a big birthday party featuring a 20-foot-wide by 15-foot-tall birthday cake and appearances by film star Walter Pidgeon and actress Jarma Lewis.[32]

Getting the Machinery in Gear

With the formulation of Houston's freeway plan in 1954, there was a huge job ahead for the agencies responsible for getting the freeways built. Getting the freeway construction machinery in motion and up to full speed proved to be problematic in the mid-1950s, slowing down many projects, including the Southwest Freeway. The alignment of the Southwest Freeway from Weslayan to Hillcroft was in a state of flux until January 1956, preventing the protection and purchase of right-of-way. TxDOT officials were contemplating ideas for the freeway design, including a nine-lane facility with three reversible lanes inside Loop 610. Agencies responsible for acquiring right-of-way were understaffed and overwhelmed with the workload. In the meantime, development in southwest Houston continued at a furious pace. On September 19, 1955, Ralph Ellifrit, director of the city of Houston's Planning Department, sent a letter to TxDOT identifying the Southwest Freeway inside the loop as the "most critical point" in terms of potential loss of right-of-way to development, saying, "There has been too much uncertainty as to design and exact location."[33]

In early 1957, Harris County took steps to expedite right-of-way acquisition by authorizing the hiring of 20 additional appraisers. A total of 547 parcels needed to be acquired for the freeway, and efforts to that point had acquired only a small fraction of the total needed. In July 1957, the right-of-way acquisition program was still struggling and Eugene Maier, director of the city of Houston's Traffic and Transportation Department, sent a letter to Mayor Oscar Holcombe recommending that every effort be made to develop a "crash" program to acquire necessary right-of-way for the freeway. At a freeway planning meeting in July 1957, local officials designated the inner loop section of the freeway as the top priority for construction. The process began to pick up momentum soon afterward.[34]

Let's Make a Deal

Sharpstown was a huge success and developed rapidly, but there was one big problem. Plans for the Southwest Freeway outside Loop 610 were stuck in neutral as all efforts focused on the section inside the loop. Frank Sharp had to put his plans for a new air-conditioned shopping center on hold since the freeway would be critical in providing customers to the mall. To make matters worse, Fort Bend County southwest of Houston announced in early August 1957 that it could not provide funds for right-of-way for the section of the Southwest Freeway within its boundaries.[35]

By late summer 1957, real estate interests in southwest Houston realized that they would need to take matters into their own hands to get the freeway built. While Frank Sharp had offered to donate a 300-foot-wide (91 m) right-of-way strip through 3.2 miles (5.1 km) of his community, there was still a large section of the 11.1-mile (18 km) freeway segment for which no right-of-way was in hand. On September 6, 1957, a group of landowners on the originally planned freeway alignment submitted an offer to donate a total of 10.5 miles (17 km) of right-of-way for the freeway—Sharp's original 3.2 miles plus 7.4 miles of new donations. Only one-half mile (0.8 km) of right-of-way would need to be acquired by authorities. The offer hinged on the stipulation that construction of the freeway

Frank Sharp finally gets his freeway: On April 9, 1965, the Southwest Freeway main lanes through Sharpstown were dedicated. The freeway frontage roads, completed in 1961, had kept development in Sharpstown running at full speed. By the time the main lanes were dedicated, about two-thirds of the land in Sharpstown was already developed. Visible in the photo from left to right: Louie Welch, mayor of Houston; Gail Whitcomb, president of the Houston Chamber of Commerce; Herbert C. Petry, chairman of the State Highway Commission; Frank Sharp; J. H. Kultgen, member of the State Highway Commission. (Photo: TxDOT)

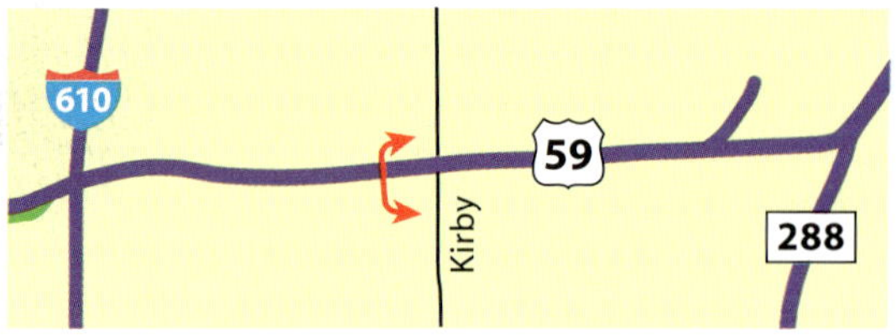

Moving into the major league of freeways: This May 1961 photograph looking east shows construction in progress at the Kirby Road intersection. This section of freeway was the first in Houston to be constructed to modern design standards with high traffic-carrying capacity. The depressed section of freeway in the upper part of the photo, with 10 freeway lanes, is nearly complete and was opened to traffic in July 1961. The section of freeway under construction in the foreground was opened in July 1962. (Photo: TxDOT)

frontage roads would begin within one year. On September 26, 1957, the landowners and local authorities presented their offer to the Texas Transportation Commission in Austin. The offer was accepted. In March 1958, TxDOT made the Southwest Freeway its top priority among non-interstate highways, allocating $6.2 million of the $15.6 million in available Houston-area funds to the freeway, the largest chunk going to any single freeway. Construction contracts for the frontage roads through Sharpstown were awarded in the next few months.[36]

It was nothing new for real estate interests to be supportive of freeways. What was new was the donation of such a large amount of right-of-way to expedite the freeway construction. This arrangement would be prophetic of the future of freeway construction in Houston, as real estate interests would step forward with land donations to get freeways built when government resources for new freeways dwindled starting in the 1970s.

Progress on the inner loop section of the Southwest Freeway also began to gain momentum in late 1957. By August 1958, 188 of the 547 right-of-way parcels had been acquired. In September 1959 the last parcel was acquired. The path was clear for construction. With the freeway right-of-way in hand and construction underway, the suburbanization of southwest Houston had the green light to proceed at full speed.[37]

Building the Freeway

The freeway section inside Loop 610 would be the largest and most modern freeway segment in Houston at the time. The first full freeway segment from downtown to Kirby, a three-mile (4.8 km), 10-lane facility with both elevated and below-grade sections, was dedicated on July 26, 1961. On December 22, 1961, the freeway frontage roads from Sharpstown southward to US 90A in Sugar Land were dedicated, although sections had opened in advance of the official dedication. On July 24, 1962, a dedication ceremony underneath the West Loop overpass marked the completion of the freeway from downtown to just south of Westpark near Sharpstown, where traffic exited to the frontage roads. Motorists now had an uninterrupted roadway from downtown to far southwest Houston, and Frank Sharp had his coveted freeway access to Sharpstown. The interchange at Loop 610 was only the second four-level interchange in Texas and was far more modern in design than the Mixmaster in downtown Fort Worth, which had opened in 1958. On April 9, 1965, the freeway main lanes through Sharpstown, from south of Westpark to Beechnut, were opened. Frank Sharp finally had his freeway. By then, residential development in Sharpstown was about two-thirds complete. The main lanes pushed southward section by section in the following years and were complete to Sugar Land in 1975. The expressway south of Sugar Land opened in 1976.[38]

Big traffic jam, tiny car: Amidst all the hulking American cars in this 1962 photograph, the driver of the BMW 600 must have felt a little overwhelmed. The BMW 600 was an enlarged version of the Motocoupe, a popular minicar of the 1950s that originated with the Italian company Iso in 1953 and was licensed to numerous automobile manufacturers, including BMW in Germany. Approximately 8,500 Motocoupes were exported to the United States by BMW starting in 1957. What was believed to be the first Motocoupe in the United States arrived at the Port of Houston in late 1956 and was put on display in the lobby of the Houston Bank and Trust Company. The 772-pound (350 kg) Motocoupe had an air-cooled one-cylinder motorcycle engine capable of speeds up to 60 miles per hour (96 km/h). The BMW 600 was slightly larger than the Motocoupe, weighing about 1140 pounds (515 kg) and powered by a two-cylinder engine providing a top speed around 63 miles per hour (100 km/h). Notice that the Motocoupe and BMW 600 do not have driver-side doors. The front panel of the vehicles including the steering wheel hinged open to let the driver in. The BMW 600 has a passenger-side door for its rear seat (not visible in the photo), while the Motocoupe has only the front door. As for the traffic jam, it was probably due to an accident or the abrupt end of the freeway just ahead.[39] (Photo: upper, *Houston Chronicle*; below, Greater Houston Partnership)

A Neighborhood in the Wrong Place

In the 1960s, a story of a freeway, a neighborhood, and big-time real estate development began to unfold on the Southwest Freeway inside Loop 610. The neighborhood was Lamar Weslayan, a community of modest post-World War II tract houses constructed around 1950. In 1962 the Southwest Freeway cut a path through the south end of the Lamar Weslayan neighborhood, displacing 52 homes, but the neighborhood remained intact with houses on both sides of the freeway. In 1963 a 41-acre tract at the Southwest Freeway and Buffalo Speedway, just east of Lamar Weslayan, was purchased for development. Real estate developer Kenneth Schnitzer and his Century Properties development firm began development of office towers

First the freeway, then the real estate developers: The Lamar Weslayan subdivision had the unfortunate luck of being in the path of the Southwest Freeway and the westward expansion of the Greenway Plaza office complex. The above 1960 view looking east shows right-of-way clearance through the subdivision. Most of the subdivision remained intact, for the moment. The Greenway Plaza real estate development was launched in the late 1960s on the vacant tract of land in the upper left of the above photo. The developer, Kenneth Schnitzer, needed more land, so he bought all the homes in the Lamar Weslayan subdivision to make way for expansion of the office complex. In the 1973 photo at the top of the opposite page, the first displacements had occurred for a new sports arena, which was just beginning construction. By the late 1970s all the homes in the neighborhood had been removed. The lower photo on the opposite page shows Greenway Plaza in September 2002. (Photos: above, HMRC RGD6-892; opposite upper, HMRC; opposite lower, September 2002)

SHELL

No relief in sight: This section of the Southwest Freeway opened in 1962, and by 1968 severe traffic congestion had already developed around Loop 610 at rush hour. A 1968 study recommended an aggressive program of improvements to avoid "complete paralysis of traffic on the Southwest Freeway." This photograph, looking east towards the Loop 610 interchange, was taken in 1972 when TxDOT was discussing plans for widening the freeway to 10 main lanes from Hillcroft all the way into downtown. Then came the 1970s highway funding crisis and population boom in Houston. The Southwest Freeway became a commuter's nightmare. In 1992, 20 years after TxDOT first started planning improvements, the expansion of this section was finally completed.[40] (Photo: National Archives NWDNS-412-DA-10534).

on the tract in the late 1960s. By 1968 Schnitzer had big plans for his development known as Greenway Plaza, proclaiming, "There is no office development in the country that compares in size, scope, and accessibility, with the exception of Century City in Los Angeles." But there was one problem. Forty-one acres would not be enough for his big plans.[41]

So Schnitzer looked west—to the Lamar Weslayan neighborhood. During the 1960s he began to buy up the 350 houses in Lamar Weslayan, often personally closing deals with homeowners. By the early 1970s houses were being cleared for development, including the land for the Summit sports arena, which opened in 1975 and was renamed Compaq Center in 1997. However, Kenneth Schnitzer was unable to use his charm to close the deal on one house. The homeowners, Jim and Dorothy Lee, were demanding $500,000 for their home, which would otherwise have been worth about $26,000. The asking price translates to about 1.51 million in 2003 dollars. Finally in 1974, Schnitzer gave in to the homeowner's demand so development could move forward.[42]

By the late 1970s the Lamar Weslayan neighborhood had been wiped off the map. While the Southwest Freeway was directly responsible for only a small number of displacements, its presence provided the impetus for real estate development that displaced the entire neighborhood. It was perhaps one of the most interesting twists in neighborhood displacement resulting from freeway construction.

Too Successful for Its Own Good

The Southwest Freeway quickly became Houston's most important freeway corridor in terms of commercial real estate development, activity centers, and population. But with success comes traffic, and lots of it. In 1968, just six years after the opening of the freeway, a study by a consulting firm reported that peak-hour speeds averaged less than 10 miles per hour and an aggressive program of improvements was necessary to "avoid complete paralysis of traffic on the Southwest Freeway between Buffalo Speedway and the West Loop." In 1972, just 10 years after its opening, TxDOT officials were discussing plans to

Key dates in the history of the Southwest Freeway

1953	A freeway for southwest Houston is proposed as part of the freeway master plan. The freeway is approved by TxDOT in September 1953.
1954	The general alignment of the freeway is defined. The alignment is finalized in 1956.
1957	A group of landowners led by Frank Sharp donates right-of-way for 10.5 miles (17 km) of freeway.
1961	The first freeway section opens.
1962	The West Loop interchange, the first four-level stack in Houston, is completed.
1975	The freeway main lanes are completed to Sugar Land.
1989	Reconstruction and expansion begins. A long segment from Shepherd to Beltway 8 is completed in 1992.
1993	The Southwest Freeway becomes Houston's busiest.
1997	The interchange at Beltway 8 is completed.
2002	Work begins to place the elevated section of freeway at Montrose into a trench.
Future	Expansion in Fort Bend County.

expand the 8-lane sections of the freeway to 10 lanes and revise ramp configurations. Officials stated that the expansion might begin in four to six years. The highway funding crisis of the 1970s caused an indefinite postponement of plans for improvements. In the meantime, Houston's 1970s energy boom unleashed a massive amount of development along the Southwest Freeway corridor and in the nearby Galleria area, feeding more and more traffic onto the already overloaded freeway. The 6-lane freeway outside Loop 610 became a critical choke point in Houston's freeway system.[43]

In 1981 the Southwest Freeway broke the 200,000-vehicle-per-day barrier at a point just outside Loop 610. Traffic continued to build during the 1980s with no funding available to make any improvements to the freeway. In 1989 work finally began on a major expansion project to make the full length of the freeway inside Loop 610 ten lanes wide, the section outside the loop to the Westpark curve twelve lanes wide, and the remaining section to Beltway 8 eight lanes wide. A barrier-separated transitway in the center of the freeway was also added. With the completion of the project in 1992, the Southwest Freeway was positioned to become Houston's traffic volume leader, overtaking the West Loop in 1993 and breaking the 300,000-vehicle-per-day barrier in 1996. Traffic volume was 337,000 vehicles per day in 2001 just outside Loop 610.

Expansion continued south from Beltway 8 in the mid-1990s. Plans were also being formulated to reconstruct the freeway near the downtown exit and add a transitway. Original plans called for the construction of

Braided ramps: This photograph just outside Beltway 8 of a section of freeway reconstructed in 2001 shows a frequently used design in Houston's current generation of freeways: the braided ramp. Braided ramps allow both entrance and exit ramps to be built at the same location and are particularly suited where there is limited distance between intersections. (Photo: November 2002)

an elevated structure above the existing elevated freeway lanes at Montrose Boulevard, prompting substantial protest from the surrounding neighborhoods. The final adopted plan called for the demolition of the 10-lane elevated freeway, with the expanded freeway including transitway lanes depressed into a new trench. Elevated freeways have been demolished in several places in the United States, including Fort Worth (Texas), San Francisco, and Boston. However, this appears to be the first case where an existing elevated freeway is removed and replaced with a trench at the same location. The new trench will feature the distinctive long-span arched bridges that were added during the

Excavation: This view dating from around 1960 shows the excavation for the depressed section of freeway near downtown. At the time, this excavation for the 10-lane freeway was the largest and most impressive construction project to occur on Houston's freeway system. (Photo: HMRC MSS 334-1153)

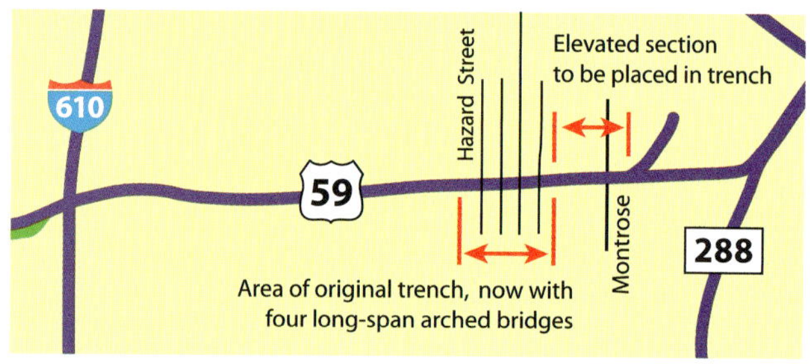

Houston's most appropriately named street: For 40 years Hazard Street was, in fact, one of Houston's most hazardous streets. The Hazard Street bridge over the Southwest Freeway had a clearance of only 13 feet, 11 inches (4.24 m) and was frequently the scene of collisions with oversize vehicles. The damage to the bridge structure (underneath the height signs) visible in the photo was not part of the demolition process—it was the result of collisions during the previous 40 years. The replacement span is an arched structure with improved vertical clearance. On August 12, 2000, the original bridge was demolished. (Photo: Chuck Fuhs)

Tropical Storm Allison:
When the storm hit in June 2001, it unleashed a major freeway flood event in Houston and filled the Southwest Freeway trench nearly to ground level. At that time, construction was in progress to widen the trench. Most motorists were able keep their vehicles out of the water, but the construction contractor, Williams Brothers Construction, wasn't so lucky. It lost 42 pieces of equipment, including 22 pieces of machinery (including large cranes, such as the one shown in the lower photo), 17 trucks, and 3 message boards. (Photos: Robert Cowart)

The Montrose elevated, 1961-2004:
The elevated structure in the photo will be demolished and replaced with a trenched freeway. The project, which began in late 2002, is perhaps the first instance in the United States of an elevated freeway being removed and replaced with a trenched facility at the same location. Elevated freeways in other cities have been relocated to new locations or placed in tunnels. The reconstructed freeway will have the same design as the adjacent section of freeway and will feature two more long-span arched bridges. (Photo: July 2002)

Houston's busiest freeway: The Southwest Freeway just outside Loop 610 is Houston's busiest freeway, with 337,000 vehicles per day in 2001. The reconstruction to its present configuration with 12 main freeway lanes was completed in 1992. At the right side of this photo, construction on the Westpark Tollway, an elevated structure at this location, is in progress. (Photo: September 2002)

reconstruction of the adjacent depressed section, completed in December 2002.

Efforts to designate a new Interstate 69 corridor from Indianapolis, Indiana, to the Texas-Mexico border also gained momentum in the 1990s and a study completed in 2000 identified options for the routing of Interstate 69 through Houston. Most of US 59 in Texas will ultimately be absorbed into Interstate 69, but it remains to be seen how US 59 in Houston will be affected.

Sharpstowns of the Future

Freeway history has a way of repeating itself. People still want suburban housing, and developers still want to make money. Further south along the Southwest Freeway in Sugar Land, the next big real estate development, First Colony, began to take shape in the late 1970s. Houston's second loop, Beltway 8, connected to the Southwest Freeway in 1988. Intensive development in and around Sugar Land in the 1980s and 1990s overwhelmed the freeway, and by the late 1990s the city of Sugar Land was doing everything possible to get the freeway widened. Whereas Frank Sharp brought land to the table, Sugar Land brought money. Major expansion in Sugar Land began in 2000 and was completed in May 2003. Further south along the freeway is the intersection with Houston's third loop, the Grand Parkway. The Sharpstowns of the future are a glimmer in the eye of land developers.

Sugar Land: The photos on the opposite page show the Southwest Freeway in Sugar Land, the suburban city named after the sugar mill that operated there from 1843 to 2002. The upper photo shows construction of the Southwest Freeway in 1970. The view shows the first major real estate development in Sugar Land, Venetian Estates, which included the finger lakes visible in the photo. The freeway opened in 1973. With the opening of the freeway, other large real estate developments followed. The lower photo, taken in 2000, shows the development which has occurred, including the Sugar Creek golf course community, which was developed shortly after the freeway opening. The large First Colony real estate development was launched in 1977 and other developments followed in the 1980s and 1990s. The extensive development necessitated the freeway expansion, which began in 2000 and was completed in May 2003. (Photos: The Positive Image)

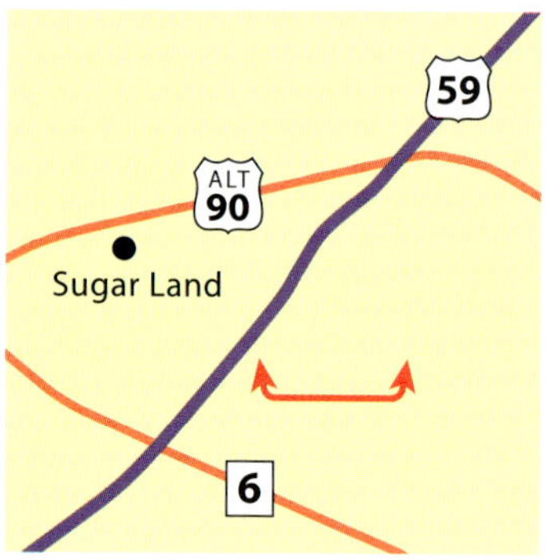

Beltway interchange: This view looks southwest along the Southwest Freeway at the Beltway 8 interchange. This interchange was built in two phases, with the first completed in 1989 and the second completed in 1997. (Photo: May 2003)

Overarching style: This view looks west along the freeway with the Mandell overpass at the bottom. (Photo: May 2003)

South Freeway, SH 288

South Freeway	
Designated as freeway	1953
First freeway section open	1980
Freeway complete	1984 (expressway south of Loop 610)
Reconstruction	none
Max traffic volume, 2001	161,000 vehicles per day
Future construction	Freeway upgrades south of Beltway 8

As Houston's freeway construction machine was running at full speed in the early 1960s, the highway engineers at the Houston TxDOT office may have had a song chiming in the back of their heads. As Frank Sinatra sang it: "The best is yet to come."

In the 1960s engineers watched as many of their new freeways became clogged with traffic only a few years after opening. These 6- and 8-lane freeways were hopelessly inadequate, and due to the limited right-of-way width of the freeway corridors, it would be difficult to expand these freeways to more than 10 main lanes. By the mid-1960s, TxDOT had a vision of the freeway of the future, and it could be best described by a saying that rings true especially in Texas: "Bigger is better."

Houston's highway engineers had studied freeway designs around the country. The Dan Ryan Expressway in Chicago, opened in 1962, was influential in the design of large freeways, and Houston's engineers paid particular attention to California. Houston had equaled California standards with the inner loop section of the Katy Freeway, which began construction in 1965. But the South Freeway would exceed anything California could offer and would propel Houston to the top in terms of freeway design. The South Freeway would be not just one freeway, but two freeways in one—a "dual freeway" with a regular freeway on the outside and a wide median for a future express freeway on the inside. The initial phase of the freeway would include only the outer section of the dual freeway. Nevertheless, this would be a highly impressive facility. The wide freeway right-of-way corridor required for the facility would make the South Freeway one of the most distinctive in the nation.[44]

But just as the South Freeway represented the vision of the freeway of the future, it would also be intertwined with perhaps the most dramatic story in the history of race relations in Houston. It's a story of integration, white flight, and urban transformation. And it's a story where the South Freeway would have the final word, at least in terms of ground zero for the events that would unfold over a 15-year period.

Neighborhoods, Integration, and the Freeway

The Riverside Terrace neighborhood was initially developed in 1924 and included a selection of housing for the mid-range and affluent markets. Just to the south of Riverside Terrace on the south side of Brays Bayou was Riverside, an exclusive community including many estate homes that rivaled Houston's elite River Oaks neighborhood. Jews had been denied housing in River Oaks in that era, and many of Houston's wealthy and prominent Jewish families resided in Riverside. Riverside Terrace and Riverside were the heart of the Jewish community in Houston.

The year was 1952. Successful black cattleman Jack

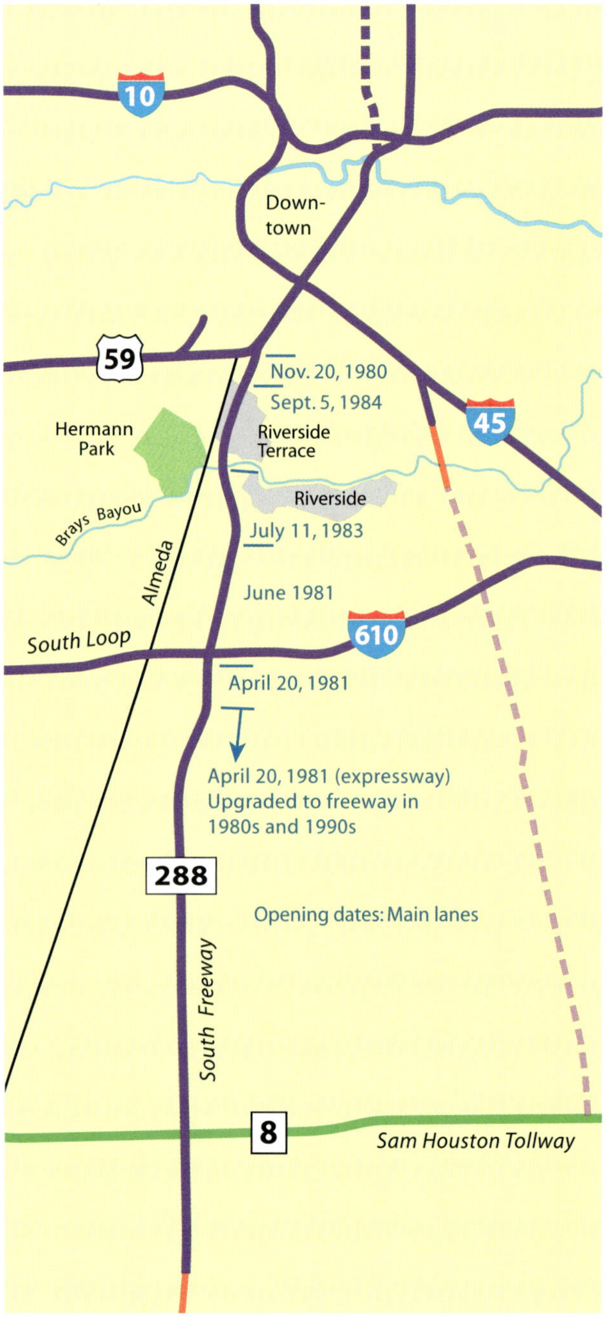

Caesar was looking for a nice home to match his success, but found that his options were limited because blacks were restricted to black neighborhoods. He set his sights on a nice brick home on the south side of Houston in the Riverside Terrace neighborhood, but there was one problem—it was in a white area. Caesar arranged to have his white male secretary purchase the home. The home title was then transferred to Caesar. He and his family moved into the home in the middle of the night in 1952.

After Caesar moved into Riverside Terrace, there was unease among the white residents. However, Caesar and his family maintained a low profile, so the neighborhood residents generally adopted a "wait and see" attitude. That all changed in the early morning hours of April 17, 1953. A bomb consisting of four sticks of dynamite was exploded on Caesar's porch. The bomb blew out the windows and destroyed the porch area, but there were no injuries. The bomb seemed to be intended to frighten rather than to kill.

If the bomber had intended to frighten Caesar, it had exactly the opposite effect. Caesar would stay, and nervous whites began selling their homes. There was talk among white residents of formulating an agreement not to sell to blacks, but neighborhood transformation had begun and "white flight" was underway. As whites moved out, many of the homes on larger lots were torn down and replaced with low-quality apartments, many without air-conditioning. The predominantly black, low-income residents of the apartments were forced onto the street at all hours of the day and night in Houston's warm, humid climate. Homes along the main streets in the neighborhood were converted to businesses. The character of the neighborhood changed, accelerating white flight and sending the neighborhood into a downward spiral.

While this process was in progress, local planners were reexamining the alignment of the South Freeway. The original alignment along Almeda Road was shifted eastward in late 1959, sending it through the eastern end of Riverside Terrace. Jack Caesar's house would be in the path of the freeway.

Riverside Terrace would become nearly entirely black by the early 1960s. Residents of the exclusive Riverside neighborhood adjacent to Riverside Terrace felt that the bayou separating the two neighborhoods would be a barrier to further southward black migration, but by the mid-1960s white flight was in full progress in Riverside. Unscrupulous real estate agents fueled the fire by playing on the fears of white residents and actively soliciting sellers. Many of the white residents banded together to state their commitment to the community and their openness to an integrated neighborhood. These residents posted signs in their yards stating, "This is our home, it is not for sale," openly defying the real estate agents promoting white flight. It was a nice morale booster, but the transformation of the neighborhood continued. Black professionals, many of them professors at nearby Texas Southern University, took advantage of the collapsing real estate prices to move into the beautiful neighborhood. By the early 1970s the posh old homes along Brays Bayou were occupied almost entirely by blacks, with only "one or two" whites remaining, according to a real estate agent interviewed for a 1971 *Houston Chronicle* report. Riverside became to the black community what it formerly was to the Jewish community: the center and focus of the community's leaders and prominent citizens.[45]

Moved by the freeway: When Jack Caesar first moved into Riverside Terrace, he did it quietly in the middle of the night. His arrival launched the transformation of the neighborhood. But in the end, the South Freeway would have the final word. The home Caesar moved into, as well as many others, was displaced to make way for the freeway. (Photo: *Houston Chronicle*)

In the late 1960s right-of-clearance for the wide South Freeway corridor was underway. Jack Caesar's house was relocated to south Houston and occupied by another owner. For Houston it was perhaps a fitting end for the home that unleashed the storm—the freeway had the final word.

Origins

The South Freeway was part of Houston's original freeway plan formulated in the early 1950s and was included in the plan presented to the Texas Transportation Commission in July 1953. On September 23, 1953, TxDOT entered into an agreement with the city of Houston and Harris County to further the development of three proposed freeways, including SH 288 from downtown Houston southward to the Fort Bend County line—the South Freeway. In the agreement, TxDOT would define the freeway routes and pay for construction, while the city and the county were responsible for right-of-way

(Opposite page) **The freeway and ground zero:** This view shows the South Freeway corridor in June 1972, looking north. Right-of-way clearance had begun in the late 1960s and was nearly complete at the time of this photo. The location of Jack Caesar's house at the southeast corner of Wichita and Hutchins is indicated in red, and is now approximately in the center of the freeway. Caesar's house became a flash point for racial integration in Houston in 1952 when Caesar and his family became the first blacks to move into the all-white, heavily Jewish Riverside Terrace neighborhood. In April 1953, a bomb exploded on Caesar's front porch. The bomb caused only minor property damage and no injuries, but it launched full-scale white flight from the Riverside Terrace neighborhood. (Photo: The Positive Image)

acquisition. The South Freeway first appeared on the city of Houston's *Major Thoroughfare and Freeway Plan* in 1955. The original alignment showed the freeway following Almeda Road southward out of downtown, veering west of Almeda Road just north of Loop 610, and then rejoining Almeda Road near Beltway 8.[46]

In the 1950s planning efforts were focused on Houston's more rapidly growing areas and the freeways serving those areas, especially the Southwest Freeway and the West Loop. The South Freeway corridor was largely rural and devoid of any significant population, so it took lower priority. Little—if any—work had been done on the South Freeway when in March 1959 City Planning Director Ralph Ellifrit proposed an eastward realignment of the freeway. The main reason for the re-alignment was to shift the freeway closer to the Gulf Freeway to relieve some of its traffic. The new alignment would also avoid an oil field, prevent the need to displace either a hospital or a biscuit factory on Almeda Road, traverse less-developed land, and be situated closer to development in southeast Houston. Ellifrit's alignment still used Almeda Road for most of the route inside Loop 610. Near Old Spanish Trail it veered east from Almeda Road rather than west.[47]

On May 22, 1959, TxDOT authorized a full restudy of the South Freeway alignment. In late 1959, the South Freeway relocation study was complete and recommended a new alignment entirely to the east of Almeda Road. This would become the final freeway alignment with only slight modifications. On August 14, 1963, TxDOT held the first public meeting for the freeway alignment. The *Houston Post* reported that "protests were negligible," with most comments coming from landowners in the south part of the corridor. The newspaper report mentioned only one homeowner who spoke up at the meeting, and apparently the subject of impacts on the Riverside Terrace neighborhood never arose. The public also got its first glimpse of the big plans for the freeway at the 1963 meeting. The schematics on display showed the eight-lane freeway and its wide center median. A second meeting was held on July 21, 1965.[48]

The full details of the proposed freeway design were presented at a public hearing on January 27, 1966. The head of the TxDOT Houston Urban Project Office, A. C. Kyser, provided a summary of the design from downtown to south of Loop 610. The freeway would be a dual freeway, which meant that TxDOT would be putting "two freeways on one right-of-way." On the outside would be a regular 8-lane freeway with entrance and exit ramps. A

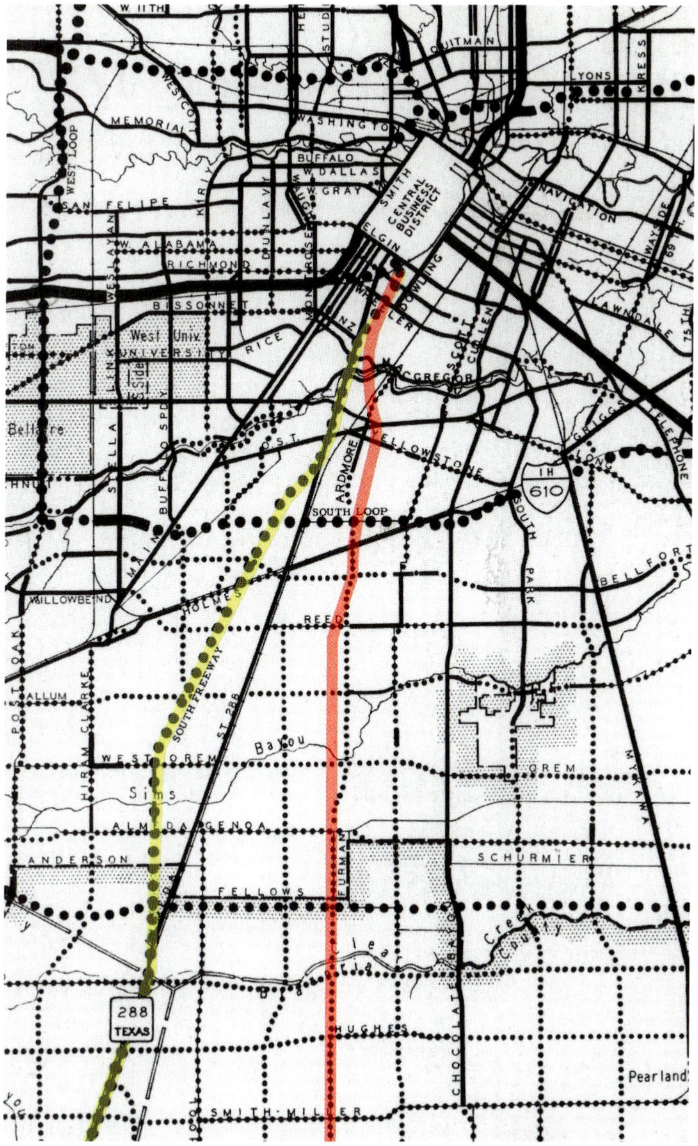

Alignment shift: This view from the city of Houston's 1959 *Major Thoroughfare and Freeway Plan* shows the original alignment of the South Freeway, highlighted in yellow. The original alignment was along and west of Almeda Road. The revised alignment, shown in red, was defined in a special study completed in late 1959. The South Freeway is Houston's only freeway to sustain a major realignment. (Map: HMRC)

Freeway construction in slow motion: This September 1976 photo shows the cleared right-of-way corridor north of Brays Bayou. Right-of-way clearance began in the late 1960s and was complete by 1972, but the corridor sat vacant for more than five years due to funding shortfalls at TxDOT. The existence of such a wide corridor of clear land so close to the central business district for most of the 1970s was a remarkable novelty and is something that will probably never again be seen in any city in the United States in the context of freeway construction. (Photo: Chuck Fuhs)

new feature of this freeway would be the numerous "criss-cross" entry and entrance ramps, where one ramp would be constructed over the other, a design which is now called a braided ramp. On the inside would be a 6-lane express freeway with no entrance or exit ramps between Loop 610 and downtown. The entire freeway, inside and outside, would be depressed below grade for nearly its entire 4.7-mile (7.5 km) length inside Loop 610. The initial phase would construct the 8-lane outer freeway, leaving a wide median for the future inner freeway. But with the space reserved for the inner freeway, the design would accommodate a freeway with 14 main lanes. "This freeway is being designed, as far as ultimate capacity, [to] probably exceed the capacity of any freeway I know of, anywhere," said Kyser.[49]

> "This freeway is being designed, as far as ultimate capacity, [to] probably exceed the capacity of any freeway I know of, anywhere."
>
> TxDOT Houston Urban Project Office head A. C. Kyser, at the public meeting for the design of the South Freeway, January 27, 1966

Building the Freeway

If there was one thing an observer would notice about the plans for this freeway, it was its width. The right-of-way corridor would be approximately 500 feet (152 m) wide. At the January 1966 public meeting Kyser stated that the right-of-way cost for the freeway was estimated at $12 million, and the cost of actual construction for the large freeway was estimated at $10 million. Purchasing and clearing the right-of-way for the freeway was expected to cost more than the freeway construction itself. Kyser reported that right-of-way had already been acquired for the southern section. The real challenge in terms of right-of-way would be the northern half, which traversed through the Riverside Terrace neighborhood.

A key event in the construction of the freeway was the passage of a Harris County bond referendum on January 29, 1966. The bond issue provided funds for the county's 50% portion of the right-of-way acquisition cost. Without passage of the proposition, the project would not have been able to move forward, and any delay may have had unknown consequences for the future of the freeway. The window of opportunity for building large freeways through inner-city neighborhoods was rapidly closing, not just in Houston but nationwide. New policies would soon take effect and the political environment would soon

Construction finally gets underway: After a delay that lasted for much of the 1970s, construction of the South Freeway was finally making progress by the late 1970s, as shown by this May 1978 photo. (Photo: TxDOT)

change to be much less favorable to this type of freeway. A two-thirds majority was needed to approve the bonds, and voters passed the proposition with 72% of the vote. The South Freeway project would proceed.

The right-of-way clearance for the north section of the South Freeway and adjacent section of US 59 was one of the largest right-of-way clearance events in the history of Houston's freeways. It took place over approximately five years from the late 1960s to the early 1970s. The final environmental impact statement (EIS) for the South Freeway, completed in 1974, reported that 295 families were displaced for the South Freeway. But the EIS also reported that "a large majority [of those displaced] agreed that they now have equal or better facilities than they had before displacement." The transition from neighborhood to freeway is not always an orderly process, especially when the process plays out over many years, as was the case for the South Freeway corridor. The scene can take on a surreal look, with some houses still standing among vacant lots, houses in various states of demolition, and piles of rubble scattered about. The South Freeway right-of-way clearance spawned small industries in salvage, demolition, and relocation as it reached its peak around 1970. Scavengers sifted through the remains of buildings looking for valuable materials or collectibles. Transients occupied abandoned buildings. The scene generated a unique ecology as the forces of displacement played out.[50]

By 1972 the right-of-way for the South Freeway was clear from downtown to Brays Bayou. In the past, right-of-way clearance was promptly followed by freeway construction, but in the budget-constrained 1970s work would proceed much more slowly. For several years the wide right-of-way stood vacant, a remarkable novelty in a close-in urban area and a scene that will likely never again be seen in a major urban area in the context of freeway construction. A well-known example of a similar situation took place on the corridor for the Century Freeway (IH 105) in Los Angeles, which is about 7 miles (11 km) south of downtown Los Angeles. The 18-mile (29 km) Century Freeway corridor stood vacant for years as the California

Complete: The final section of the South Freeway was formally opened on September 5, 1984, with a ribbon cutting ceremony. Holding the ceremonial scissors is Bob Lanier, who was chairman of the Texas Transportation Commission at the time and would later serve as mayor of Houston from 1992 to 1998. Also participating in the ribbon cutting are representatives from the Houston Citizens Chamber of Commerce. From left to right are Al Calloway, president; Lanier; Lorene Lancelin; and Ernest Clauser. Clauser is holding a certificate conferring the "Road Hand Award" to him. The Road Hand Award recognizes individuals whose efforts promote Houston-area road projects. (Photo: HMRC RGD6).

Department of Transportation dealt with financial problems and litigation. In 1975 the *Los Angeles Times* described the right-of-way strip as "largely a wasteland." The Century Freeway would not be completed until 1993. Right-of-way corridors cleared for freeways that were never built still exist in a few places around the United States, but in terms of its sheer width and closeness to the urban center, the South Freeway corridor was perhaps the most distinctive.[51]

Finally, on September 22, 1975, a ground breaking ceremony was held for the first actual construction on the South Freeway—the frontage roads at Brays Bayou. The first section of main lanes, a short segment just south of the US 59 interchange, opened in November 1980. The four-lane divided highway south of Loop 610 opened on April 20, 1981. Inside Loop 610 work proceeded slowly on small segments. A section of freeway immediately north of Loop 610 was completed in June 1981, and the previously completed interchange at Loop 610 was fully opened to traffic. On July 11, 1983, a section of freeway from Brays Bayou to Yellowstone, just south of Old Spanish Trail, was opened, providing a continuous facility from downtown southward. However, traffic still had to use the

Key dates in the history of the South Freeway

1953	The South Freeway is proposed as part of Houston's freeway master plan.
1959	The South Freeway is realigned.
1963	Plans are unveiled for the dual freeway design.
1968-1972	Right-of-way is cleared through Riverside Terrace.
1975	Construction begins.
1984	The freeway is completed, but is only an expressway south of Loop 610.
1980s, 1990s	Expressway sections south of Loop 610 are upgraded to full freeway status.

Room for the interior freeway: This view shows the South Freeway as it passes through the Riverside Terrace area. The South Freeway is depressed below grade for nearly its entire length inside Loop 610. The wide median is reserved for future express freeway lanes which are part of the "dual freeway" design developed for the freeway in the early 1960s. It remains to be seen if the interior freeway lanes will ever be constructed. As of 2003, there appeared to be no need for the interior freeway based on long-range demand estimates. The downtown freeway interchange complex may not be capable of handling the additional vehicles that the interior freeway would deliver. (Photo: James Lyle, TTI, June 2001)

frontage roads for a one-mile (1.6 km) section north of Brays Bayou. The closure of this last gap took place on September 5, 1984, and the freeway was complete. The expressway south of Loop 610 was upgraded to freeway standards in the 1980s and 1990s.

On its completion, the South Freeway appeared to be the last, all-new inner-city freeway to be constructed in Houston. It was an impressive facility. Even though only the eight-lane outer freeway was constructed, the wide center median reserved for the future inner freeway and high design standards distinguished the South Freeway as one of the best in the nation. In 2000, plans were formulated to extend the Hardy Toll Road inside Loop 610 as a toll-free facility with an estimated completion date of 2007, adding one final new freeway inside Loop 610.

Expansions of the North Freeway and the Eastex Freeway in the 1990s increased the corridor widths of those freeways to the range of 350–400 feet (107–122 m). The planned expansion of the Katy Freeway will have a corridor width averaging 475 feet (145 m) between Loop 610 and Beltway 8. With its right-of-way typically 500 feet (152 m) wide, the South Freeway will still reign as the widest freeway corridor in Houston and will probably never be dethroned.

Perhaps the most interesting question to ponder is the future of the wide freeway median. Will the inner freeway ever be built? As of 2003, TxDOT had no plans—not even long term—to construct the inner freeway. Although the South Freeway corridor has been slow to urbanize, development has accelerated in recent years, prompting TxDOT to initiate a corridor feasibility study in 2003 to determine future improvements to meet the transportation needs of the corridor. The South Freeway inner freeway is also included on a list of potential toll corridors to be considered by the Harris County Toll Road Authority. Only time will tell if the freeway demand will support the extra capacity of an inner freeway or tollway. If and when the need ever arises, Kyser's dual freeway design will have the potential to "exceed the capacity of any freeway, anywhere."

No frontage roads: The South Freeway is the only freeway in Houston constructed without frontage roads for most of its length. (Photo: September 2002)

Lake 288: This view shows the South Freeway after Tropical Storm Allison in June 2001. In its early years the South Freeway was susceptible to flooding caused by water from Brays Bayou backing up into the freeway drainage system. Modifications to the drainage system and more powerful pumps alleviated most flooding but were not enough to prevent flooding during Allison. (Photo: Jonathan Miller)

The freeway beginning: This view shows the start of the South Freeway at its interchange with US 59 just south of downtown. The South Freeway branches to the left, and the Southwest Freeway branches to the right. This interchange normally has a backup of vehicles trying to get on the Southwest Freeway, as this photo shows, but motorists typically face no delays as they enter the South Freeway. This interchange was completed on November 20, 1980. The Southwest Freeway through this location opened in 1974. (Photo: May 2003)

La Porte Freeway, SH 225

The La Porte Freeway has earned itself a unique distinction among Houston's freeways. It is the only freeway in the region that has sustained a permanent cancellation within the city of Houston. The most notable feature of the La Porte Freeway is its abrupt end just inside Loop 610, where the freeway was truncated. The dramatic freeway end serves as a fitting monument to the cancelled freeway. It is also a monument to one man who relentlessly fought the freeway and ultimately succeeded in doing what no one else had done before in Houston—and no one else has done since.

The story of the La Porte Freeway has deep roots, with its origins going back to the 19th century. In 2000, more than 100 years after the La Porte Freeway corridor was first used for transportation, construction on the freeway was finally completed. Unlike other Houston freeways, the La Porte Freeway is largely devoid of strip shopping centers or any kind of commercial development. Instead, the freeway is lined with petrochemical plants and refineries, providing an impressive industrial landscape of catalytic crackers, distillation towers, and tank farms.

La Porte Freeway	
Designated as freeway	1953
First freeway section open	1966
Freeway complete	2000
Reconstruction	none
Max traffic volume, 2001	141,000 vehicles per day
Future construction	None planned

The La Porte Freeway

The development of the La Porte Freeway has been a slow evolution over a century. The route started out as a dirt road that generally followed the railroad corridor from La Porte to Houston, which opened in 1894. By 1907 the route was paved with crushed rock from Houston to approximately today's Beltway 8. Sylvan Beach Park on Galveston Bay in La Porte was a popular destination for Houstonians, and there was increasing interest in extending the paved section all the way to La Porte to provide reliable automobile access to the park. In 1910 paving to La Porte was completed with a shell surface. The road followed the alignment of the present-day freeway to about one mile (1.6 km) west of today's SH 146, where it veered south.[52]

On November 5, 1927, a concrete road on much of today's freeway alignment was dedicated and designated as the La Porte-Houston Highway. The new concrete road was extended eastward where the previous shell road veered south, so the 1927 roadway defined today's freeway alignment for the full length of the corridor all the way to SH 146. It was probably no coincidence that Ross Sterling's mansion was completed the same year at Morgan's Point, which is just east of the eastern terminus of the highway. Ross Sterling was one of Houston's wealthiest and most politically influential citizens. In 1909 he founded Humble Oil, the predecessor to ExxonMobil, the world's largest corporation in 2001 and second largest in 2002. He sold his interests in Humble Oil and purchased the *Houston Post* newspaper in 1925. In 1926 he purchased the *Houston Dispatch* newspaper and then merged it with the *Houston Post*. Sterling was appointed chairman of the Texas Transportation Commission in February 1927 and served in that position until he became

The first three-level interchange in Texas: The interchange at the eastern terminus of the La Porte Freeway at SH 146 was the first three-level traffic separation on the Texas state highway system when it was completed in 1953. TxDOT officials in Austin were initially hesitant to approve the three-level design because they felt that traffic volumes did not justify the "elaborate installation" and that it was being urged for its novelty. But it was approved, and within a few years three- and four-level interchanges would become common. Texas still had some catching up to do in 1953: that same year, California completed the four-level stack interchange at the intersection of the Harbor-Pasadena and Hollywood Freeways.[53] (Photo: TxDOT)

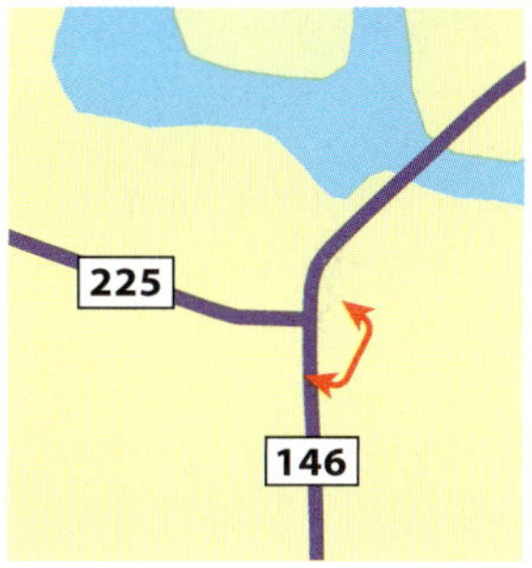

governor of Texas in 1930. For many years the highway was known as the Ross Sterling Highway.[54]

The freeway era for the La Porte Highway began in May 1945 when the first section of freeway immediately east of the Gulf Freeway was authorized by the Texas Transportation Commission. The section of freeway was called the La Porte cutoff and connected the existing La Porte Highway to the Gulf Freeway. It opened in 1952 and would later be absorbed into the south Loop 610. Today's La Porte Freeway from the east Loop 610 to SH 146 was authorized by the Texas Transportation Commission on September 18, 1953, as part of the proposed freeway system that Houston officials had presented to the commission in July 1953.[55]

The first major construction on the present-day La Porte Freeway occurred in conjunction with the Baytown Tunnel on SH 146, which opened on September 22, 1953. SH 225 intersected SH 146 just south of the tunnel, and an interchange was completed in late 1953 to provide access to the tunnel. The interchange featured the first three-level traffic separation on the state highway system in Texas. Texas still had some catching up to do in terms of multi-level interchange design, since Los Angeles completed its first four-level stack interchange at the intersection of the Harbor-Pasadena and Hollywood Freeways in the same year.[56]

As authorities worked to define the exact alignment of the freeway in the late 1950s, they ran into a problem. At the time, the biggest risk to the alignment of freeways was parkland. In particular, many of Houston's large parks, including Memorial, Hermann, and Macgregor, had been donated to the city of Houston, and the terms of donation often included "reverter clauses," which stated that the land would revert to the original owner if it was not used as parkland. The La Porte Freeway would need to cut through the 74-acre Milby Park, which had been donated to the city of Houston by Charles W. Milby in 1937. Included in the terms of the parkland donation was a stipulation that no facilities for anything except specific park purposes could ever be constructed or used on the park site, or the park site itself would revert to the donor

An impressive industrial landscape: The La Porte Freeway passes through one of the densest concentrations of refineries and petrochemical plants in the United States. Approximately 13.2% of the nation's refining capacity and nearly 50% of the nation's base petrochemical manufacturing capacity is located in the Houston area. This view shows the La Porte Freeway crossing through a tank farm just east of Loop 610. Completed in 1966, this was the first section of the freeway opened to traffic. (Photo: September 2002)

(Opposite page) **Pre-freeway and modern SH 225:** The upper photograph from 1951 shows SH 225 at the Shell petrochemical plant, which is just east of Beltway 8. The freeway would later be built on the strip of land between the industrial plant and the two-lane highway, as shown in the lower photo. The freeway main lanes were completed in 1992. Work on the Shell refinery in Deer Park began in 1928 when the first team of engineers arrived from Illinois and set up offices in an abandoned schoolhouse on SH 225. In August 1929 the Shell refinery was operational. The above photo shows the Shell chemical plant, which produced its first product in September 1941, just in time to fill World War II demand for high-octane aviation fuel, toluene, and synthetic rubber. (Photos: upper, Shell Oil; lower, September 2002)

or his heirs.[57]

Original plans for the La Porte Freeway showed its alignment severing the park, using 6.5 acres for the freeway and isolating another 10 acres north of the freeway. In 1958, the city of Houston reviewed the legal implications of right-of-way acquisition through the park and determined that it could not sell or transfer the right-of-way. In order to minimize the impact on the park, the alignment of the La Porte Freeway was shifted northward along the edge of the park, reducing the impact to only 8 acres. Because the city would not sell the property, TxDOT obtained it through condemnation proceedings in 1960, clearing the way for the freeway. The daughter of the park donor then sued the city of Houston, arguing that the parkland should be returned to the estate of the donor since the terms of the original agreement had been violated. In August 1961, a court ruled that the Milby estate was not entitled to reclaim the remainder of the park.[58]

Another difficulty in the construction of the La Porte Freeway through Pasadena was the relocation of the railroad that was situated along the western half of the corridor. Harris County and the Port of Houston entered into an agreement to relocate the tracks in October 1958, and the Texas Transportation Commission authorized TxDOT financial participation in the relocation in November 1959. From just east of Loop 610 to just east of Beltway 8, a 7.5-mile (12 km) section of railroad was relocated about 0.5 mile (0.8 km) north. The actual track relocation

Finishing the freeway: The final push to complete the La Porte Freeway occurred in the 1990s. This view looks west from the Battleground Road overpass. This section was completed in 1992. The full La Porte Freeway was completed in 2000. (Photo: July 2002)

proceeded with no difficulty, but litigation ensued and was not cleared up until 1967.[59]

By 1964 construction on the freeway could begin. In November 1966 the first 2.5-mile (4 km) freeway section opened from Loop 610 eastward. Other sections subsequently opened, pushing the freeway eastward through the central part of Pasadena to east of Red Bluff Road by July 1971. Then the freeway went into hibernation due to the 1970s highway funding crisis. Work resumed in the early 1980s, pushing the freeway east of Beltway 8 by 1984. The freeway then went into hibernation once again. In the early 1990s the final push began to complete the freeway. Three sections opened during the 1990s, culminating with a new interchange at SH 146 in 2000 that completed the freeway. Forty-seven years after the official freeway designation in 1953, the La Porte Freeway was finished. The next phase of improvements for the La Porte Freeway will be defined in a corridor study scheduled to begin in 2003.

The La Porte Freeway traverses through an impressive industrial landscape of petrochemical plants and refineries. Call it ugly or call it beautiful, it is certainly a distinctive stretch of freeway. The La Porte Freeway is generally free of traffic congestion, a rare distinction in Houston. But the freeway has a more dubious distinction—the part that wasn't built.

Key dates in the history of the La Porte Freeway

1894	Railroad tracks are constructed on the La Porte Freeway corridor.
1927	The La Porte-Houston highway, most of which became the La Porte Freeway, was paved with concrete.
1952	The La Porte cutoff on the Gulf Freeway is completed. It later becomes part of Loop 610.
1953	The La Porte Freeway is proposed as part of Houston's freeway master plan and officially designated as a freeway by TxDOT.
1966	The first section of freeway opens.
2000	The freeway is completed.

Harrisburg Freeway

The Harrisburg Freeway had its origins in the 1960 traffic and transportation study for the Houston area. The study recommended the addition of two new freeways, the extension of the La Porte Freeway from Loop 610 to downtown and the southward extension of the West Loop. In August 1961, the Houston City Planning Commission officially delivered its recommendation for the new freeways in a report called *Freeway Phase*. The recommendation explained that the traffic study found "a demand on the Gulf Freeway far beyond its reasonable capacity

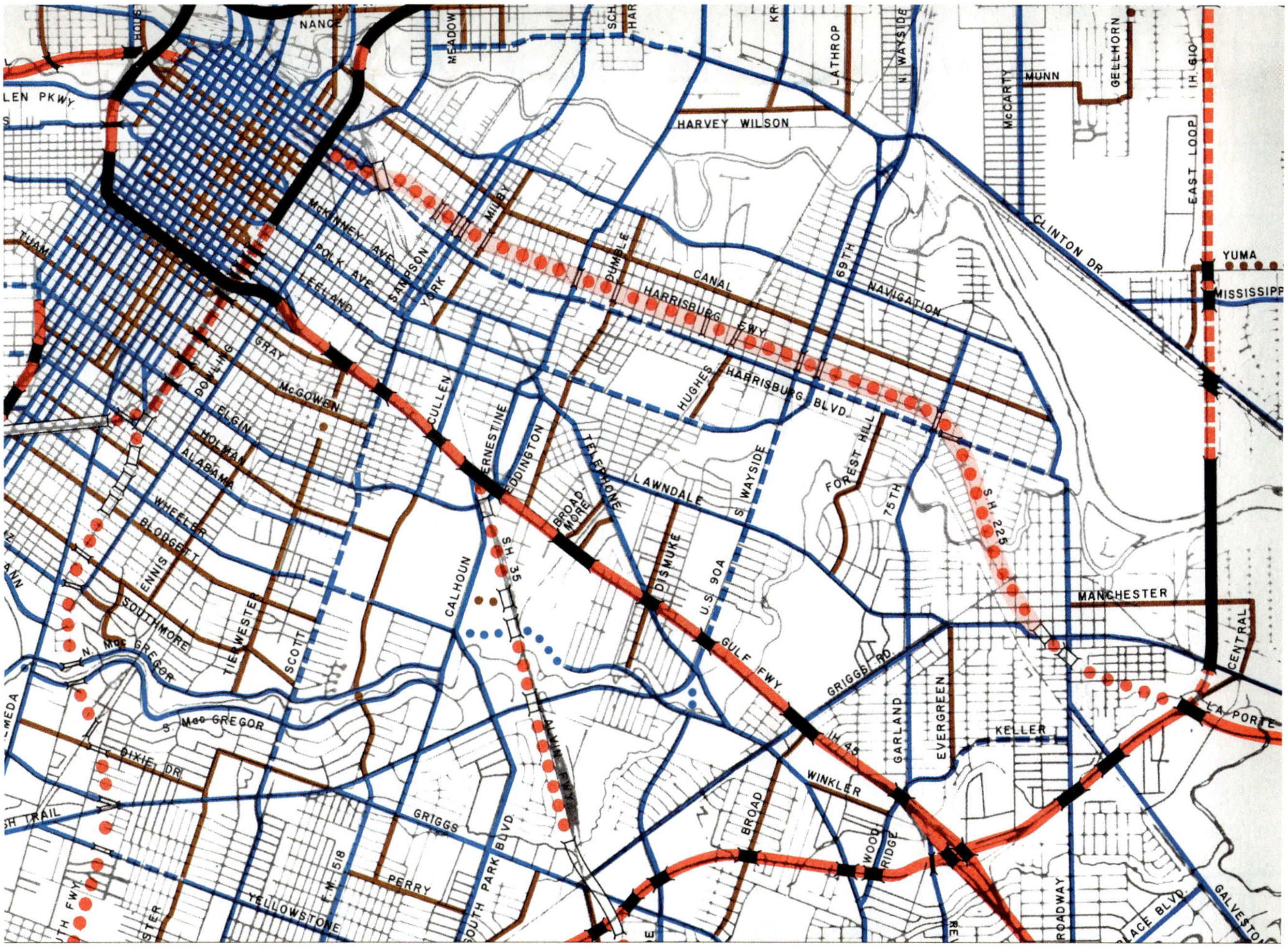

The planned alignment: This November 1971 map from the *Houston-Harris County Transportation Plan, Volume 3* (1971) shows the planned alignment of the Harrisburg Freeway. The unbuilt section is indicated with the pink highlight. This alignment is identical to the originally proposed 1964 alignment and the final alignment recommended in January 1973.

due to the large area it must serve and the tributary nature of the La Porte Freeway." In a December 1963 letter to TxDOT, city of Houston department heads wrote, "It is our opinion that the most effective measure for the immediate relief of the traffic overload in the southeast corridor would require the extension of the La Porte Freeway westerly, in the vicinity of Harrisburg Boulevard, to the Central Business District Loop (US 59)." The La Porte Freeway extension would become known as the Harrisburg Freeway since it generally followed the alignment of Harrisburg Boulevard.[60]

The local officials then needed to persuade TxDOT to add the new freeway section to the state highway system. In January 1962 TxDOT agreed to provide interchanges at the intersections of the Harrisburg Freeway with US 59 downtown and Loop 610. On March 31, 1964, Houston Mayor Louie Welch led a Houston delegation that made a presentation to the Texas Transportation Commission in Austin to request the inclusion of the freeway segment into the state highway system. The Texas Transportation Commission did not grant Houston's request. In November 1964, A. C. Kyser, head of the TxDOT Houston Urban Project Office, reported that the Harrisburg Freeway was "just a line on a map and a report" and said the Texas Transportation Commission had not yet reviewed the project. An effort had been made to include the route in the federal Interstate Highway System, but that request had been rejected by the U.S. Bureau of Public Roads. The project was at least six to eight years in the future because TxDOT was overcommitted for work on the state highway network.[61]

Saying No to the Freeway

Richard Holgin was born just north of downtown Houston in the area then known as the Fifth Ward. He moved to the Magnolia neighborhood east of downtown near Harrisburg Road as a child and attended the local public schools. Nearly all of his extended family, which included about 100 cousins, aunts, and uncles, lived in the east side neighborhoods along the proposed freeway path. He purchased his home in 1958 on Rusk Street, two blocks south of Harrisburg Boulevard, and started a family.

One day in the early 1960s, Holgin first heard rumors that a new freeway was planned for the Harrisburg corridor. Holgin immediately became concerned. Just around that time, the right-of-way clearance for IH 10 was cutting

Ramp to a dead end: This view along Loop 610 at the SH 225 interchange shows the exit and high-level connection ramp to SH 225 west, the planned Harrisburg Freeway. The freeway comes to a dead end just after the ramp returns back to ground level. (Photo: May 2002)

a path through the predominantly Hispanic neighborhoods just north of downtown, near the neighborhood where he was born. Would the east end neighborhoods, a developing Hispanic area, be destined for the same fate? Holgin set out to find some answers. He started by contacting the city of Houston, but soon found that the city was just one player in Houston's freeway construction machine, and the city couldn't or wouldn't answer his questions. So he tried contacting TxDOT, and then Harris County. Still, he couldn't get the information he wanted. The freeway was a "done deal," he was told, and it was best that he get out of the way. It seemed there was no place to focus a coordinated effort to stop the freeway. Holgin and the neighborhood were forced into an apprehensive wait-and-see situation for the rest of the 1960s.[62]

But luck would be on Holgin's side. In spite of the efforts of the city of Houston and Harris County, the Harrisburg Freeway had not been adopted into the state highway system by TxDOT. The city of Houston, Harris County, and TxDOT were all overwhelmed by the huge task of acquiring right-of-way and building the approved freeways in the 1960s, so the Harrisburg Freeway was put on the back burner and remained just a line on the map.

By 1969, when the 1960s freeway construction binge was winding down, TxDOT was in a better position to add new freeways to the state highway system. The Texas Transportation Commission formally accepted the Harrisburg Freeway on April 2, 1969, just in time, it seemed, for the harsh new freeway climate that was developing. On January 1, 1970, President Nixon signed the National Environmental Policy Act of 1969 (NEPA) into law. NEPA and provisions of the 1968 and 1970 Federal-Aid Highway Acts had dramatically changed the way freeways were planned. TxDOT would now have to prepare an environmental impact statement and provide a comprehensive public hearing process. The Harrisburg Freeway would be the first in Houston to go through the new planning process.[63]

The freeway was coming back to life, and Richard Holgin was ready for it. He was determined to do everything he could to stop the freeway, and he was now empowered by NEPA and other new regulations that gave communities a large voice in the process. With his bachelor's degree in business administration from the University of Houston, Holgin was one of the few college-educated Hispanics in the mostly blue-collar area. He was the ideal person to bridge the divide between the Spanish-speaking community and the freeway construction establishment. Leading the opposition effort provided Holgin with an opportunity to work at the grassroots level, improve his visibility, and build name recognition in the community. It was a perfect opportunity for a political launch pad. His opposition to the freeway and political aspirations turned out to be a perfect match. Whatever his greater motivation was, he brought dedication and passion to the anti-freeway effort.

Holgin knew very little about freeway construction and the newly adopted freeway planning processes required by NEPA. He found an ally and consultant in Alfred Davey, a private planning consultant who knew the ins

A monument to the cancelled freeway: This view shows the dramatic end of the La Porte Freeway just inside Loop 610. The main lanes come to an abrupt halt, and the frontage roads continue a little further until they dead-end at Lawndale Avenue. If the La Porte Freeway extension had been constructed, it would have continued to downtown Houston in the distance. The project was in serious trouble by 1973, and by 1976 the highway funding crisis effectively put the nails in the coffin of the freeway. (Photo: May 2002)

and outs of the planning process. With advice and counsel from Davey, Holgin gained the knowledge to work in the system and deal with city and state officials. Holgin and the five or six community members who actively opposed the freeway designated themselves as the La Porte Freeway Extension Information Committee.

The first public meeting for the Harrisburg Freeway was held in March 1970. There was considerable support for the freeway, especially among the local business establishment. Most residents were more interested in knowing if and when they needed to move, rather than opposing the freeway. But Holgin felt certain that there was opposition to the freeway, especially in the Hispanic community. He needed to inform the neighborhood that the world had changed. It was now possible to stop a freeway.[64]

After that first public meeting, Holgin and Davey met one day with their cameras in hand to take a tour of the neighborhood. They followed the proposed freeway corridor and took photographs of homes and businesses that would be displaced. Holgin then arranged a community meeting for May 1970, which was held at a neighborhood community center called the Ripley House. The TxDOT representatives used their allotted 15 minutes to give a brief overview of the project. Then, for the next hour

Anti-freeway crusader: Richard Holgin stands along Harrisburg Boulevard in May 2002. If the project had moved forward, the 400-foot-wide (122 m) freeway corridor would have been constructed in the area behind him. Today, the neighborhoods along the corridor are mostly low-income. The neighborhood has not been revitalized like many close-in urban areas in Houston, but retains its modest houses and scattering of businesses that serve a predominantly Hispanic and immigrant population.

and 45 minutes, Holgin's supporters presented the anti-freeway case. To start their program, Holgin and Davey presented the audience with a slide show of the properties along the freeway path. Barbara Streisand's "People" ("people who need people") was used as background music. As one slide after another was shown, the impact of the freeway registered with the audience. It became more poignantly clear that their homes, friends, and neighbors would be displaced. It was an emotional appeal, and it worked. Holgin was unlocking the latent opposition that he knew existed. Also at the meeting, Davey presented a plan he had developed for an alternative freeway alignment along the Houston Ship Channel which did not impact the neighborhood. If TxDOT wasn't yet aware that this freeway controversy would be different from ones in the past, then Holgin's first meeting—which became known as the "Ripley House meeting"—was certainly an eye-opener.[65]

In July 1970 TxDOT launched its official study with a multidisciplinary team. This was something entirely new for TxDOT. Previously, plans and alignments for Houston's freeways had been determined by engineers and local political authorities who generally took the most direct or least expensive routes. But now, with the National Environmental Policy Act in force, selecting an alignment for the freeway would be a much more complex process. The multidisciplinary team was appointed in July 1970 to study the route and recommend an alignment, taking into consideration the social, economic, environmental, and physical implications of the project. The team included an urban sociologist, an economist, a landscape architect, an urban planner, a public information specialist, and the usual staff of designers and engineers. A team member noted that they had "never been involved in anything like this" and the length of the study was uncertain because there was "no history of this type of undertaking."[66]

Included in the multidisciplinary team was TxDOT project engineer Dexter Jones. Like Holgin, Jones had deep roots in the Harrisburg community, having been born there and having several immediate family members, including his grandmother, still living in the area. More than anyone else on the study team, Jones had a personal passion about getting the freeway built. Jones believed that the only way to reverse the decline that was afflicting the Harrisburg area was to build the freeway. The narrow streets and general inaccessibility of the area had choked off investment. The freeway, he believed, wasn't just about moving cars. It was about saving the community he knew so well by making it more attractive to businesses and homeowners. Jones and Holgin both shared a common goal of revitalizing the Harrisburg area, but were at opposite ends of the spectrum about how to achieve the goal. The stage was set for a battle that would play out over the next three years.[67]

Six months later in January 1971, everything seemed to be going well for TxDOT. The study's progress report stated, "With the exception of the La Porte Freeway Extension Information Committee, all of the civic groups and virtually all of the private citizens who have contacted the team have expressed a pressing need for a freeway to serve the Harrisburg area." During this period Holgin contacted the Federal Highway Administration in Washington, D.C., and discovered that the person monitoring the Harrisburg Freeway study held the belief that there was no opposition to the freeway. Holgin still had work to do.[68]

In early 1971 events began to work in Holgin's favor.

In February Houston Mayor Louie Welch proposed to make the freeway a tollway in order to get it constructed more quickly. The proposal was dismissed almost immediately by everyone involved in the process, and it invigorated opposition to the freeway. Many in the blue-collar neighborhoods along the freeway were already ambivalent about having a freeway in their backyards, but they became incensed about the possibility of being forced to pay to use it. At a February 1971 public information meeting, Holgin found that his organization was no longer alone in its battle against the freeway. An organization called Response in City Hall, or RICH, turned out a large contingent of people to oppose the alignment through the neighborhood and support an alignment along the Houston Ship Channel and Buffalo Bayou. RICH had collected thousands of signatures on a petition opposing the toll road proposal. The president of RICH, Jerry McGee, was skeptical of TxDOT's study and stated that TxDOT was "coming into my community not to consult, but to brainwash." Regarding TxDOT's statement that the bayou route was being considered, Holgin shared McGee's skepticism, stating, "I think they're saying that to pacify people." Another organization, the Urban Bunch, a group of architects and planners, was expressing concerns about bringing more traffic into the center of the city.[69]

Dexter Jones was still confident of the ultimate success of the freeway project. At the meeting he stated, "The freeway is necessary and it will be built." He then paused while the statement was translated into Spanish. "We don't know how, where, or when, but it will be built."[70]

Holgin had renamed his organization the East End Preservation and Development Association, a name which better conveyed Holgin's broader goals for the area. In April 1971 he organized a meeting featuring United States Congress Representative Bob Eckhardt, a Democrat who favored environmental and neighborhood causes. Holgin and his group vented their frustrations to Eckhardt, stating their belief that they were not being heard by the TxDOT study team. Following the meeting, Eckhardt stated, "I feel the freeway should go [along the ship channel], and not through a residential area."[71]

In 1972 Holgin made a run for political office. His effort to seek the office of state representative fell short, as he lost to Ben Reyes, who would go on to larger prominence in Houston politics. Reyes was convicted of bribery, conspiracy and mail fraud in 1998 and was sentenced to nine years in prison. Holgin's political run was unsuccessful, but the freeway fight would go on.

By the summer of 1972 the list of potential alignments had been reduced to two options: the Harrisburg Boulevard route and the ship channel route. In January 1973 the recommended alignment was announced. The freeway would take the Harrisburg route. And in the end, the study team recommended what it knew best—a state-of-the-art freeway facility with a 400-foot-wide (122 m) right-of-way and main lanes depressed below grade to minimize impacts to the surrounding area.[72]

Key dates in the history of the Harrisburg Freeway

1961	The Harrisburg Freeway is recommended by a regional transportation study.
1962-1968	Local officials attempt to gain TxDOT approval of the freeway.
1969	TxDOT accepts the route into the state highway system and official designates it as a freeway.
1970	The Harrisburg Freeway becomes the first Houston freeway planned under new federal regulations and the National Environmental Policy Act. Opposition to the freeway develops.
1973	The Harrisburg Boulevard alignment is recommended for the freeway after three years of study. The EPA rejects the draft environmental impact statement for the freeway.
1974	A full interchange is completed at Loop 610.
1974-1976	The highway funding crisis forces many freeway cancellations and postponements. By 1976 the Harrisburg Freeway is dead due to public opposition and lack of funding.
1992	The Harrisburg Freeway is deleted from Houston's *Major Thoroughfare and Freeway Plan.*

The environmental impact statement (EIS) reported that 1,244 residential units, including 617 single family residences, would be displaced, affecting approximately 4,000 people. In addition, 47 industrial buildings, 40 commercial buildings, 2 churches, 2 Masonic lodges and 2 fire stations were in the path of the freeway. The EIS reported, "The project as recommended will avoid any disruption or division of established communities." The team's rationale for making that statement was that the freeway followed a corridor that "constituted a long-established boundary between two communities that have distinctly different characteristics." This boundary had been recognized by other agencies, including the city of Houston and its Model Cities Department. To Holgin, that statement was nonsense. He felt the freeway would destroy the neighborhood, and he continued his efforts to stop or delay the freeway. He appeared at the meeting of the Houston-Galveston Area Council, the regional planning organization that approved all transportation plans, and appealed to the council to delay approval, saying that the freeway would "destroy a developing Mexican-American community." Looking back on that meeting, Holgin knew that the outcome of the council vote was never in doubt. "It was a matter of rubber stamping the process. It was a good old buddy system." The council approved the freeway alignment.[73]

Holgin had, however, received some encouraging news in this period. His sources at TxDOT were telling him about TxDOT's developing financial crisis, which became substantially worse with the loss of revenue resulting from the 1973 Arab oil embargo. Even if the freeway's EIS received approval, there was a good chance there would be no money to build it. Then on September 1, 1973, word came of a huge victory for Holgin. The Environmental Protection Agency had rejected the Environmental Impact Statement for the Harrisburg Freeway and labeled it inadequate. The EPA said the EIS did not contain sufficient information to adequately assess two aspects of the impact of the freeway. The first item was the impact of the displacement of persons and buildings, not on the Harrisburg corridor itself, but on the areas where displaced residents would move to. The second item was the impact on current and future air, noise, and water pollution in the corridor. Finally a governmental agency had put itself on Holgin's side. The *Houston Chronicle* quoted Holgin, "The EPA ruling on the draft report is a victory for the people in the Harrisburg community. It shows that somebody is looking out for us." The head of Houston's urban highway program, Bill Ward, was confident the problem could be overcome, saying, "We will do everything we can to answer the EPA's objections."[74]

The EPA ruling turned the tide in favor of the opposition, but it was not a final decision and both Holgin and TxDOT continued their efforts. Holgin appeared before Houston City Council on May 22, 1974, to ask the council what action it planned to take on the freeway. The city of Houston reiterated its support for the freeway.[75]

The Death of a Freeway

But the freeway was doomed. TxDOT's budget shortfall was developing into a full-blown crisis, and TxDOT was forced to put the Harrisburg Freeway and other projects on hold. In September 1976 Bill Ward stated in a project status letter, "It is assumed that further development of the Harrisburg Freeway is suspended until the funding situation is cleared up." Looking back on the period, Ward remembered, "There was no point to pursue a project that had opposition since there were no funds to build the freeway." With political support for the highway program at an all-time low, it didn't make sense to continue to generate negative publicity.[76]

The freeway still had widespread public support in April 1975, however. A study by the Texas Transportation Institute found that 70% of a sample of residents in the freeway corridor supported the freeway. Sixty-six percent of Hispanic residents, who comprised 61% of the survey participants, favored the freeway. The Harrisburg Freeway Study Team issued a final route report in August 1976, stating its final case for the freeway.[77]

Holgin closely monitored any new developments regarding the freeway, but there would be none. The freeway was dead. The freeway didn't burn out, but it faded away, a victim of declining financial support for the highway system and public opposition. TxDOT moved on. Dexter Jones moved on, focusing on highway safety research. Holgin moved out of the area in 1986.

In the final analysis, it can be concluded that insufficient highway construction funding was the principal cause of the demise of the Harrisburg Freeway. But when the freeway's future became tenuous due to the funding situation, Richard Holgin's opposition probably was the decisive factor in the ultimate decision to abandon the freeway. Had there been no opposition or if there had been visible community support, the Harrisburg Freeway probably would have moved forward, slowly but surely.

The Harrisburg Freeway remained on official planning maps until 1992, when it was finally deleted from the city of Houston's *Major Thoroughfare and Freeway Plan*. Left behind at the eastern end of the proposed Harrisburg Freeway is a dramatic reminder of the cancelled freeway. Just to the west of a four-level interchange completed in 1974, the freeway grinds to an abrupt halt and traffic exits to the frontage roads. The frontage roads proceed for another half-mile (0.8 km), with the land once designated for the freeway main lanes now vacant.

Is Houston a better place without the Harrisburg Freeway? Is the Harrisburg community a better place because the freeway was not built? To Richard Holgin, the answer is yes.

To others, the answer is no. Frank Mancuso, Houston City Council member during the period, looked back in 1999 and stated about the freeway, "I still think there's a need, more than ever." While other neighborhoods close to downtown experienced a renaissance during the 1990s, particularly the Heights area near the Katy Freeway, the Harrisburg corridor was largely left behind. Perhaps the area remained just as Richard Holgin wanted it, serving its low-income and immigrant residents with modest homes and a scattering of small businesses. The decline and stagnation of the Harrisburg area was a particularly bitter pill for Dexter Jones to swallow. In a career that included many successful safety improvements to the highway system, including Houston's leading highway lighting program, the demise of the Harrisburg Freeway stood out as the greatest single disappointment. For him, it was an opportunity lost—an opportunity to revitalize a neighborhood to which he had a personal connection.[78]

Katy Freeway, Interstate 10 West

All great freeway cities need a great freeway. Chicago has the Dan Ryan Expressway. New York City has the New Jersey Turnpike. Los Angeles has the El Toro Y. Toronto has the 401 Freeway. If Houston is to join the ranks of the world's great freeway cities, it needs a big, monumental freeway. And if the Katy Freeway expansion moves forward as planned in mid-2003, Houston will get its mega-freeway to propel it into the ranks of the freeway elite.

Residents of west Houston have suffered for decades on the small, antiquated, six-lane Katy Freeway. As expansion and reconstruction proceeded on nearly all other Houston freeways, the Katy Freeway has remained in its 1960s time warp, hopelessly incapable of meeting modern-day demand. However, sometimes being last in line can have its advantages. In the case of the Katy Freeway, which will be Houston's last 1960s-era freeway to undergo comprehensive rehabilitation, the best was saved for last. The Katy Freeway reconstruction will be the largest freeway expansion to occur in Houston and will transform the corridor into one of the nation's most impressive. Houston's freeway embarrassment will become its showcase…if 2003 plans move forward to construction, that is.

The Katy Freeway in 2003 isn't all bad. The 10-lane section inside Loop 610, opened in December 1968, is still modern and impressive by today's standards. The congested section outside Beltway 8 has earned itself a more positive distinction as Houston's "Energy Corridor" due to the large number of oil, gas, and petrochemical firms with large offices along the freeway. The freeway also serves many of Houston's most affluent neighborhoods, which are located between the Katy Freeway and Westheimer, 3.5 miles (5.6 km) to the south.

Katy Freeway	
Previous designations	SH 73 US 90 West Freeway (1956–65)
Designated as freeway	1946 (inside Loop 610) 1953 (outside Loop 610)
First freeway section open	1956
Freeway complete	1968
Reconstruction	Pavement replacement only
Max traffic volume, 2001	219,000 vehicles per day
Future construction	Major expansion and tollway construction beginning in 2003

Origins

The Katy Freeway had its origins in SH 73, which was designated as the "new" San Antonio Highway in 1939 and was generally located along the route of today's IH 10. In 1940 there was discussion of a Memorial Park entrance for SH 73, but by August 1940 TxDOT had dropped the plan. The city of Houston then acquired land north of Memorial Park near Washington Street for a new interchange with US 290, but those plans were later changed because TxDOT purchased that land for a new district headquarters. In 1941 the route of SH 73 in west Houston was redesignated as US 90.[79]

The original freeway plans for SH 73 west of Post Oak Road were formulated in 1940. The plans called for a 175-foot-wide (53 m) right-of-way for the freeway with six main lanes and a frontage road on the south side of the freeway only, due to the railroad track on the north side of the corridor. The Katy Freeway's narrow right-of-way

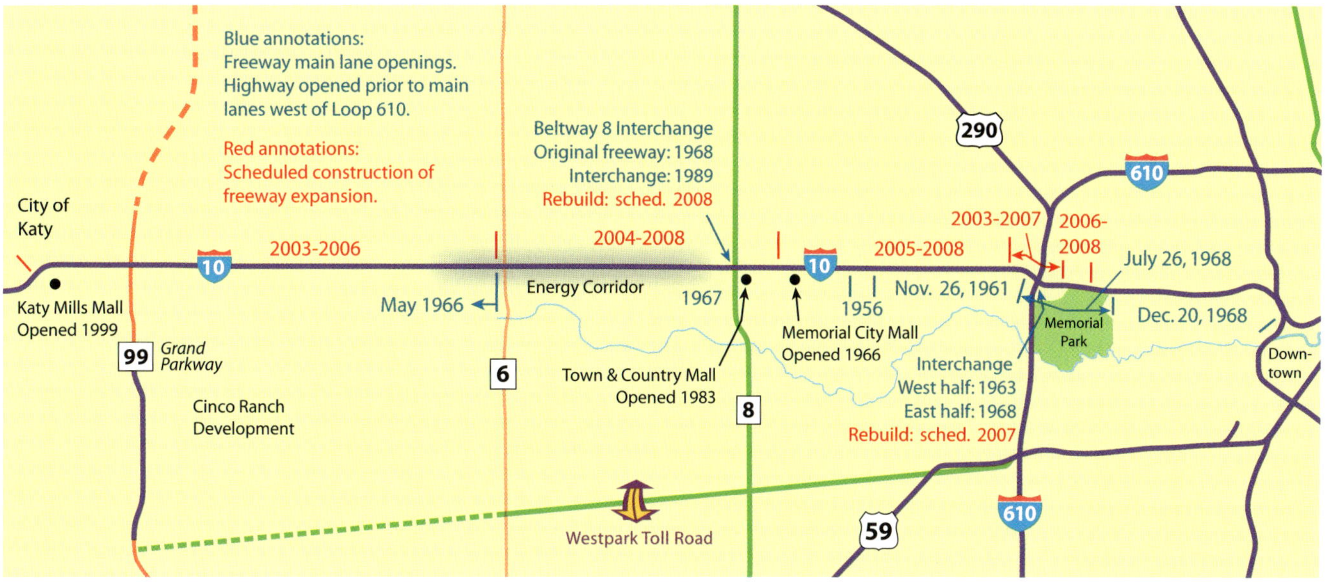

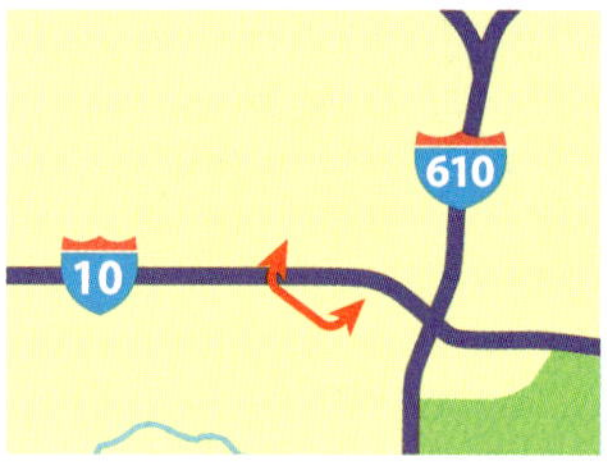

The original US 90: This 1954 view shows the original US 90 four-lane divided highway just west of today's Loop 610. Houston's freeway system carries its heaviest traffic volumes just outside Loop 610. After the completion of the IH 10 expansion, this location will likely become the busiest point in Houston's freeway system. (Photo: TxDOT)

had its roots in the early plans. The 175-foot width was widened along most of the corridor when the facility was later upgraded to a full freeway, but the extremely narrow right-of-way persisted on one key section and is one of the reasons the Katy Freeway was never expanded or improved.[80]

The need for an east-west freeway across Houston was one of the earliest freeway needs identified. In June 1942, the Texas Transportation Commission formally approved the north section of the city's planned Defense Loop, which was intended as a bypass for US 90 and designated as the "Loop on US 90." In October 1946, the Texas Highway Commission authorized US 90 as a full freeway within the city limits of Houston, which at the time extended approximately to the location of Loop 610. In June 1953, the Texas Transportation Commission authorized US 90 as a full freeway westward to the city of Katy. The route was officially designated as IH 10 in 1959.[81]

In the early days of the Katy Freeway's development outside Loop 610, there was a dispute between the two highest-ranking TxDOT managers in the Houston district about the required corridor width for the freeway. Wiley Carmichael, who managed projects outside of Loop 610, wanted to construct IH 10 within the available US 90 right-of-way, while A. C. Kyser, who managed the Houston Urban Project Office and was responsible for projects inside Loop 610, recommended a wider corridor. The project was under the jurisdiction of Carmichael and he got his way, but the use of the narrow corridor proved to be very costly to Houston in the long run, necessitating a large and costly right-of-way clearance for the planned expansion and delaying the project about 25 years after it was needed.

The first work to upgrade US 90 to a full freeway began in 1954. The Campbell and Blalock Road overpasses were completed in 1956. The section from Campbell Road to Post Oak was officially dedicated on November 13, 1962. The freeway pushed westward over the following years and, by April 1968, the freeway was completed to the city of Katy, west of Houston.

The Beltway 8 frontage roads first connected to the Katy Freeway in 1968. However, construction of modern frontage roads and the main lanes of the Sam Houston Tollway did not get underway until the late 1980s. The Beltway 8-Katy Freeway interchange was Houston's first five-level freeway-to-freeway interchange. It was partially opened in 1988 and fully opened in 1989.

The Inner Loop Section

While Wiley Carmichael was building the minimal 6-lane freeway with 2-lane frontage roads through west Houston, A. C. Kyser was developing big plans for the section inside Loop 610. More than any other freeway in Houston, it would be similar in design to a Los Angeles freeway. In fact, the plans were probably influenced by California freeway design, since Kyser had visited California to study its freeways. The freeway would be big, with a minimum of 10 general purpose lanes. A 1.7 mile (2.7 km) section of the 4.7-mile (7.5 km) length would be depressed below grade. The freeway would have superior geometrics with gentle curves and grades. The design even featured an unusual, wide deck across the freeway to allow the crossing of a railroad which intersected the freeway with only a 27-degree angle from the freeway centerline. The freeway was built underneath the railroad due to grade lines, scarcity of embankment for an overpass, and the superior characteristics of a depressed freeway in terms of entrance and exit, since downward-sloping entrance ramps help vehicles accelerate and upward-sloping exit ramps slow down vehicles .[82]

Right-of-way clearance for the inner loop section of the Katy Freeway was one of the more extensive in the history of Houston's freeways. Impacts to adjacent areas were minimized by depressing the main lanes below grade. By 1965 most of the right-of-way had been cleared and construction could begin. The freeway was dedicated to service on December 20, 1968. A 1971 study on the effects of the freeway on adjacent neighborhoods found that 75% of those surveyed thought the freeway had been properly located with respect to their neighborhood, and positive freeway effects dominated the negative freeway effects.[83]

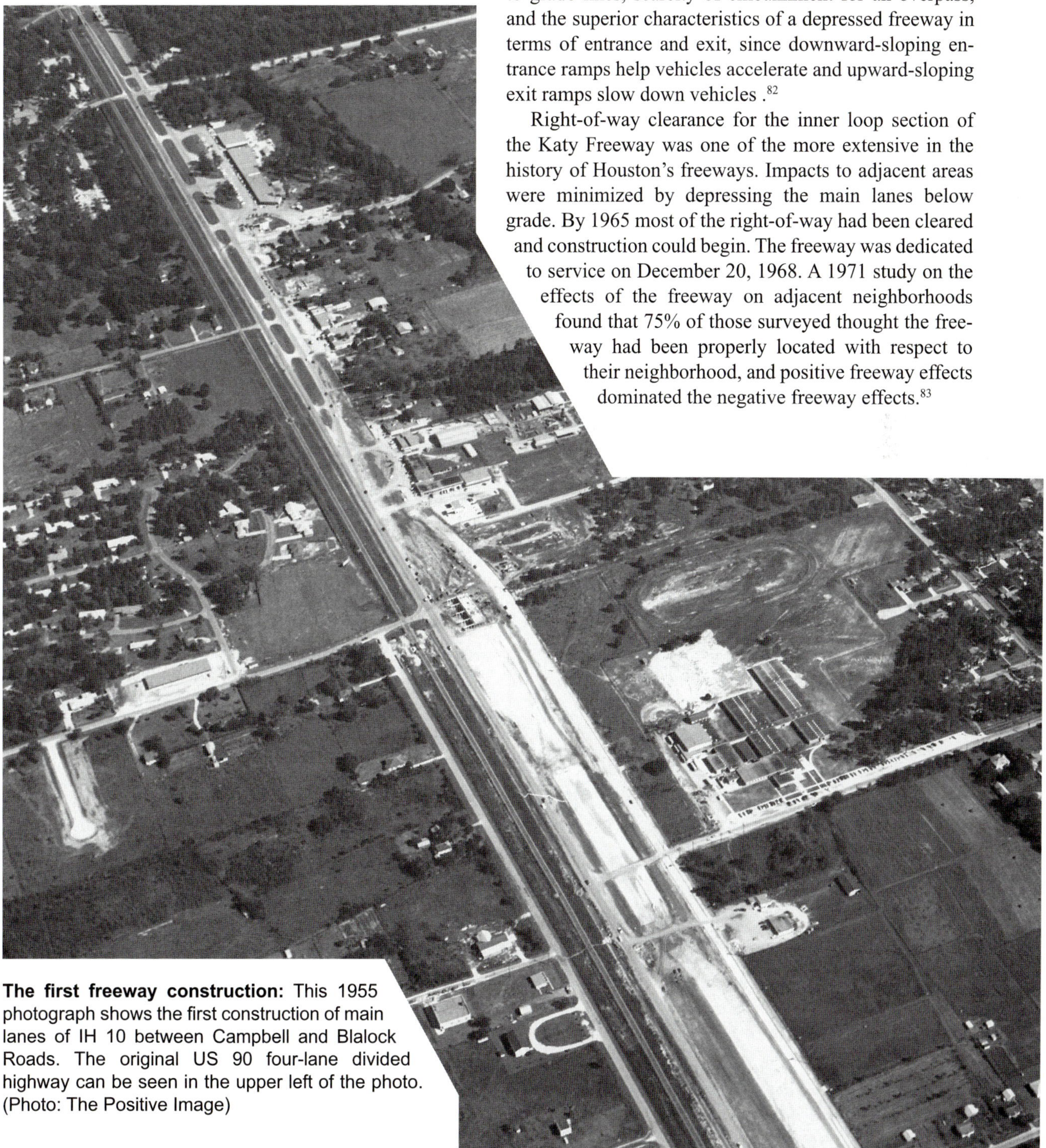

The first freeway construction: This 1955 photograph shows the first construction of main lanes of IH 10 between Campbell and Blalock Roads. The original US 90 four-lane divided highway can be seen in the upper left of the photo. (Photo: The Positive Image)

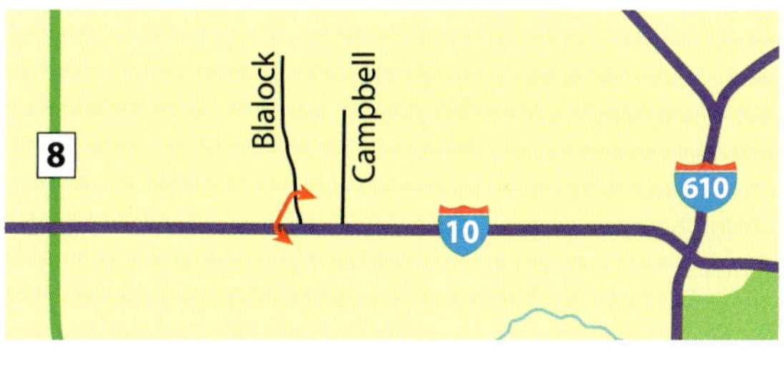

1961 construction: This view shows construction of the Katy Freeway in 1961, looking west with the Wirt Road overpass at center and Bingle/Voss near the top. (Photo: The Positive Image)

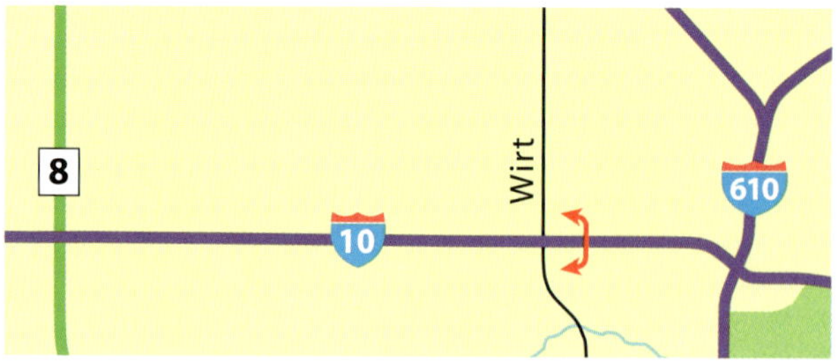

The Pain and Misery of the Katy Freeway

While A. C. Kyser's 10-lane inner loop section of the Katy Freeway was well-designed to take care of traffic needs far into the future, Wiley Carmichael's underdesigned section outside Loop 610 soon became overwhelmed with traffic and eventually degenerated into Houston's worst traffic nightmare. Houston's other congested freeways had been expanded and improved through the 1980s and 1990s, but the Katy Freeway outside Loop 610 remained in its originally constructed configuration with 6 general purpose main lanes and 2 frontage road lanes in each direction. The freeway reached a traffic volume of 179,000 vehicles per day in 1981, and in 1988 it first broke the 200,000-vehicle-per-day barrier, reaching 214,000 vehicles per day. Traffic volumes would not increase much further, however, in spite of substantial growth and development in the Katy Freeway corridor. It simply was not possible to squeeze more cars through the narrow freeway. The only relief for the Katy Freeway came in September 1984, when the interior shoulders were converted into a reversible transit lane during a pavement rehabilitation project. The Katy transitway became Houston's most heavily used transitway, serving 10,398 vehicles and 30,241 passenger trips daily in the third quarter of 2002. The only good news for area dependent on the Katy Freeway was that it remained a vibrant corridor for real estate development. Starting in the 1970s, major energy firms began locating offices along the Katy Freeway. This trend continued through the 1980s and 1990s, and the section between Beltway 8 and SH 6 became known as the "Energy Corridor." The Grand Parkway, Houston's third outer loop, connected to the Katy Freeway in August 1994. New housing development continued near the city of Katy, including the 7,000-acre master-planned Cinco Ranch community.[84]

(Opposite page) **Halfway there:** These 1964 photos show the newly completed first phase of the interchange at the Katy Freeway and West Loop. Just west of the interchange, the Katy Freeway eastbound traffic is diverted off the freeway to Old Katy Road via a temporary roadway. The full interchange and the eastward extension of the Katy Freeway were completed in 1968. (Photos: TxDOT)

610 NORTH
610 SOUTH
EXIT
40
M.P.H.

Right-of-way clearance: This 1965 view looking east toward downtown shows the right-of-way clearance required for the inner loop section of the Katy Freeway. This section of the Katy Freeway required one of the larger clearances in the history of Houston's freeway system. The low density of the area minimized the number of displacements, and the freeway was built below grade level to reduce impacts to the surrounding area. The freeway was opened to traffic on December 20, 1968. A 1971 study examining the effect of the freeway on the adjacent neighborhoods found that 75% of the area residents felt the freeway had been properly located and the majority of residents felt the neighborhood had been "made better off" by the freeway. (Photo: The Positive Image)

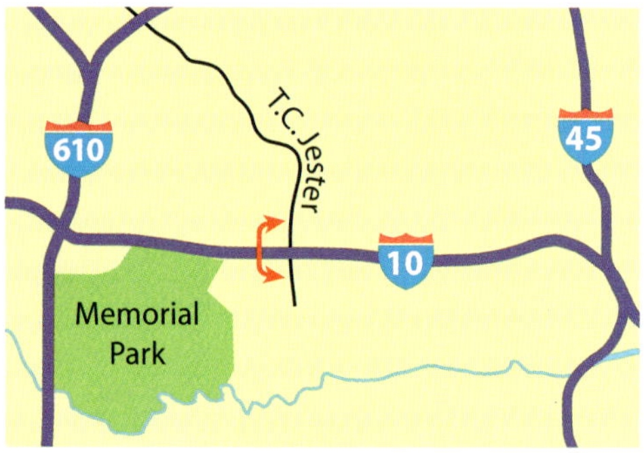

Slip-form comes to Houston: This view shows paving of IH 10 north of Memorial Park in December 1967. This paving project is believed to be the first use of slip-form paving in Houston. In slip-form paving, there are no molds to hold the concrete in place. A stiff concrete mixture is used and the paving machine forms the mold as it "slips" along the section of highway being paved. Slip-form paving became standard for concrete paving of highways. (Photo: TxDOT)

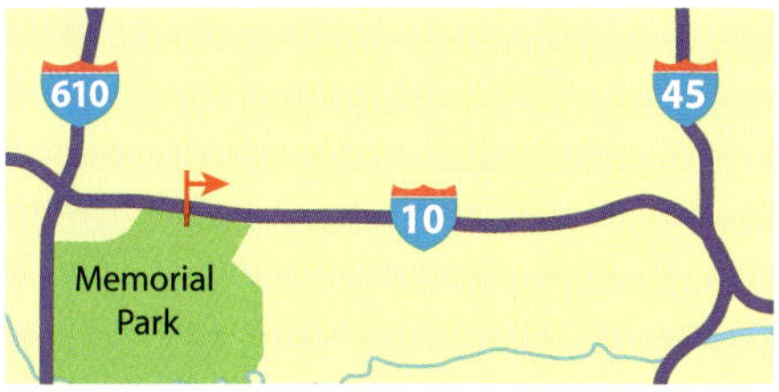

Houston's first sustained 10-lane freeway: This view looks east along the inner loop section of the Katy Freeway. When this section of freeway opened in December 1968, it was the first Houston freeway to have 10 main freeway lanes for a sustained distance. The Southwest Freeway, opened in 1961, had 10 lanes for a short distance. (Photo: May 2003)

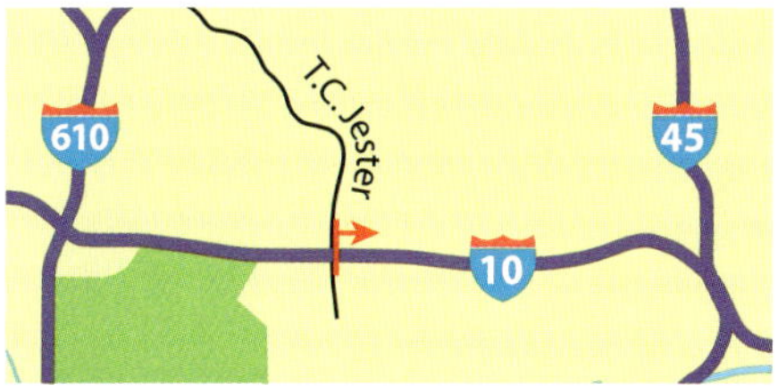

A traffic nightmare: The Katy Freeway outside Loop 610 is the most congested freeway in Houston. The freeway is still in its original minimal configuration with six general-purpose traffic lanes for most of its length. The central reversible transitway, added in 1984, has been the only improvement to the freeway since its construction in the 1950s and 1960s. Heavy commercial and residential development has occurred since the original construction, overwhelming the freeway. (Photo: September 2002)

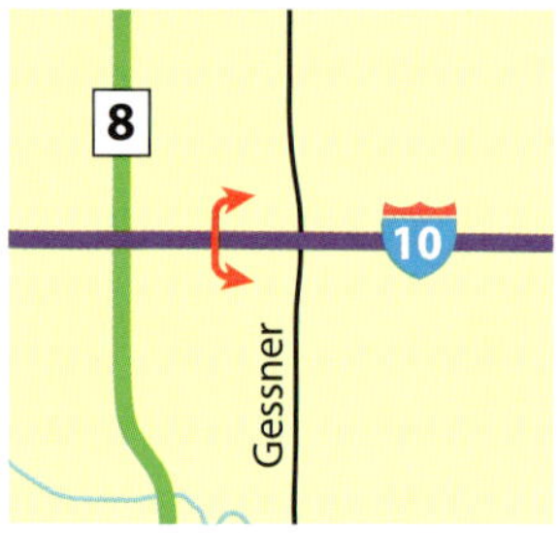

Fixing the Problem

By the mid-1980s studies were underway to determine how best to expand the freeway. The West Houston Association, a group representing real estate and business interests in west Houston, sponsored a study in 1985 which recommended the addition of elevated lanes for a 17-mile (27 km) section west of Loop 610. The idea was well received by local political leaders. Shortly thereafter in 1986, TxDOT launched its own study which proposed three options for the freeway expansion. The first option involved widening the freeway at ground level to 10 general-purpose lanes and 6 express lanes. The cost for construction (excluding right-of-way acquisition costs) was estimated at $500 million, approximately 750 million in 2003 dollars. The second and third options both added 6 express lanes on elevated structures in slightly different configurations. The cost was estimated to be $1.1 to 1.3 billion, approximately 1.65 to 1.95 billion in 2003 dollars. The high price tag and impacts of either the at-grade or elevated options made quick action unlikely. At a public meeting TxDOT Houston District head Omer Poorman said, "I'm not promising anybody anything" regarding when improvements could actually take place. It's a good thing he didn't make promises, because it would be a long road to the beginning of actual construction to improve the freeway.[85]

In 1992 a key event took place. In late December, after three years of "very hard negotiations," TxDOT and Union Pacific Railroad reached an agreement on the price for the purchase of the railroad tracks on the north side of the Katy Freeway. TxDOT paid $78 million to buy 28 miles (45 km) of track and agreed to pay an additional $25 million for "environmental remediation and substitute transportation costs." The railroad right-of-way was 100 feet (30 m) wide for most of the corridor length. The land purchase eliminated the need for elevated structures in the freeway expansion. By the end of 1992, TxDOT had formulated a preliminary conceptual plan for the freeway expansion with 10 general-purpose lanes and 6 express lanes.[86]

However, study of the Katy Freeway corridor had only just begun in 1992. In 1991 Congress passed the Intermodal Surface Transportation Efficiency Act, or ISTEA, which changed the way major new projects were developed. Now, a "major investment study" would be re-

Katy Freeway-West Loop interchange, 1968-2004: In a project that began in summer 2003, this four-level interchange completed in 1968 will be demolished and replaced with a new interchange having geometrics and ramp capacities to accommodate the expanded Katy and West Loop Freeways. This will be the second replacement of a four-level interchange in Texas, following the replacement of the downtown Mixmaster in Fort Worth, a project completed in 2001. (Photo: November 2002)

quired. A major investment study is a comprehensive corridor study that considers a range of options and all modes of transportation. TxDOT launched the study in 1995, and a series of large public meetings were held along the corridor. By late 1995 it appeared that public opinion was in favor of a major expansion of the freeway. In 1997 TxDOT announced the locally preferred option, which included a minimum of 8 general-purpose lanes and 4 special-use lanes. When auxiliary lanes were included, the proposed freeway had at least 10 continuous general purpose lanes for most of the key section between Loop 610 and SH 6. The planned freeway corridor width averaged about 475 feet (145 m) from Loop 610 to Highway 6. Even with the railroad right-of-way and the Old Katy Road right-of-way north of the railroad tracks, the expansion would require substantial right-of-way clearance, mainly affecting businesses along the north side of the corridor. Moving the freeway expansion forward to construction became the number one priority of several area politicians.[87]

But the proposed design was far from complete. The design continued to evolve as TxDOT worked to accommodate public input. Two additional non-barrier-separated lanes for high occupancy vehicles were added to address public concerns that the four express lanes would be inaccessible to most motorists. The design appeared to be complete on March 27, 2001, when TxDOT held

Key dates in the history of the Katy Freeway

1939	SH 73 along the present-day freeway alignment is constructed.
1946	Official freeway designation inside Loop 610.
1953	Official freeway designation outside Loop 610.
1956	The first freeway section opens.
1968	Freeway completed.
1984	The Katy Freeway transitway opens. It becomes Houston's busiest transitway.
1985	The first formal study for improving the Katy Freeway is conducted. TxDOT performs an initial study in 1986.
1989	Houston's first five-level interchange is completed at the Beltway 8 intersection.
1992	TxDOT purchases the railroad right-of-way north of the freeway.
1997	A major investment study of the freeway corridor recommends expansion.
2002	Plans for a tollway in the center of the freeway are finalized.
2003	The first construction contracts are awarded.
2008	Scheduled completion of the Katy Freeway expansion.

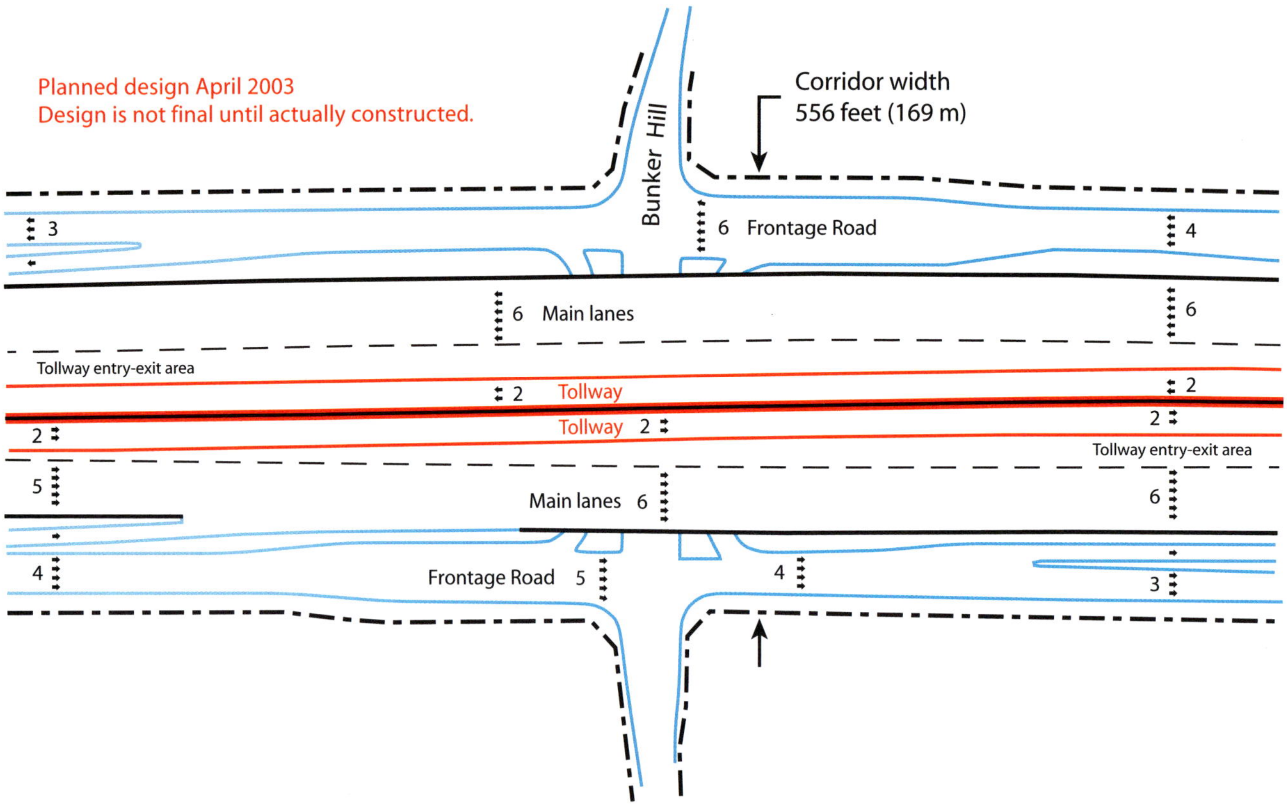

Widest right-of-way: This view shows the planned Katy Freeway at the Bunker Hill intersection about 1.5 miles (2.4 km) inside Beltway 8, where the right-of-way for the expansion reaches its widest point: 556 feet (169 m). Between Loop 610 and SH 6, the corridor is typically 475 feet (145 m) wide, with five continuous freeway lanes and two continuous toll lanes in each direction. The Bunker Hill interchange and adjacent sections of freeway are scheduled to be awarded for construction in February 2005. However, a legal challenge must be resolved before construction can begin. (Source: diagram adapted from TxDOT schematic of the planned design, dated March 10, 2003.)

Biggest Freeways in North America

Freeway	Location	Complete in present or planned configuration	Comments
Dan Ryan Expressway Interstate 90/94	Chicago, Illinois	December 15, 1962	The first super-size freeway in the United States.
New Jersey Turnpike Interstate 95	New Brunswick, New Jersey	Circa 1967 for opening of first dual-dual section; major expansions or extensions of dual-dual roadway circa 1973, 1990, and 1996.	Sustained length of widest section is approximately 12 miles (19 km); dual-dual roadway is approximately 35 miles (56 km).
401 Freeway	Toronto, Canada	Circa 1967 for opening of first dual-dual section; ongoing extension of dual-dual section with major extensions in 1985 and 1990s.	Dual-dual section is approximately 34 miles (55 km) in 2003, with further extensions planned.
El Toro Y Interstate 5/405 merge	Irvine, California (near Los Angeles)	March 1997	Freeway width is reduced quickly south of interchange.
Interstate 5/805 merge	San Diego, California	Scheduled 2007	Wide section is about 3 miles (5 km) long.
Katy Freeway Interstate 10	Houston	Scheduled 2008	Will have continuous frontage roads and a sustained distance of 12 miles (19 km).
Lyndon B. Johnson Freeway Interstate 635	Dallas, Texas	Estimated sometime between 2010 and 2020	Will include twin deep-bored tunnels.

Note: The term "dual-dual" refers to a facility where each traffic direction has separate inner and outer sections. The inner section is intended for express traffic, and the outer section is intended for local traffic.

An unprecedented short life: This view looks west along the Katy Freeway at the Beltway 8 interchange, completed in 1989. Due to minimal available right-of-way, many of the connector ramps are built over the frontage roads. The interchange will be demolished and rebuilt during the Katy Freeway expansion and reconstruction, scheduled for 2003–2008. The strip of land on the north (right) side of the freeway was formerly a railroad and was purchased by TxDOT in 1992 for the Katy Freeway expansion. As this photo shows, the geometrics and pier placement of the interchange will not accommodate the expanded freeway, which will extend across the available right-of-way. The life of this interchange will be about 17 years. This appears to be the shortest lifespan of a major interchange in the United States. (Photo: May 2002)

a public meeting to display the "final" schematic. Within days of the meeting, a proposal surfaced to convert the four central express lanes into a toll facility. The financial participation of the Harris County Toll Road Authority (HCTRA) provided the possibility of an expedited construction schedule. A favorable public response to the toll road proposal was reported.[88]

In the construction industry, certain expressions have always brought smiles to everyone. "On time" and "on budget" have always been near the top of that list. But in the modern era of the transportation industry, there is perhaps an expression that trumps all others as a cause for celebration: "record of decision." After the record of decision, everything else is usually anticlimactic. The record of decision, or the "ROD," is the final approval from the Federal Highway Administration, certifying compliance with environmental standards and authorizing local agencies to proceed with a project. For the Katy Freeway, the magic words arrived on January 15, 2002. Two months later, on March 7, 2002, the Federal Highway Administration approved the use of a value pricing pilot program on the Katy Freeway, which cleared the way for a tollway in the center of the freeway. It was reported to be the first approval of the addition of toll lanes on a free interstate highway, although similar projects had been implemented on non-interstate highways, most notably the California 91 Riverside Freeway outside Los Angeles.

A freeway project as large as the Katy Freeway expansion is certain to generate opposition. Opponents were relatively quiet during the study phase but began to organize after the record of decision to mount a last-ditch effort to stop or substantially alter the project. The opposition formed the Katy Corridor Coalition and filed a lawsuit on September 17, 2002, to halt the project and conduct fur-

Future frontage road: The westbound frontage road of the reconstructed Katy Freeway will be positioned on the location of the present Bunningham Street. This is the only area where residences will be affected by the Katy Freeway expansion. As of March 2003, most of the homes along Bunningham were cleared. (Photo: March 2003)

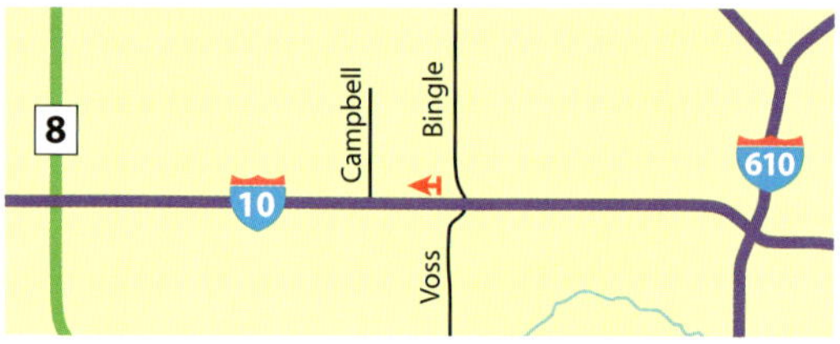

ther studies of the environmental impacts of the project. In January 2003, the coalition unveiled its plans for the corridor. The alternative plan used a narrower right-of-way for the freeway, depressed the freeway lanes below grade, included a set-aside for a light rail line, and added tree plantings between the frontage roads and main lanes. A TxDOT analysis of the proposal in April 2003 ruled it infeasible, mainly due to a cost increase estimated at $500 million and an additional three years of construction to implement the depressed freeway design.

In mid-2003 project supporters were confident that the years of study and public meetings would ensure defeat of the opposition lawsuit. Key elected officials were solidly behind the TxDOT/HCTRA plan. Providing relief from the severe traffic congestion of the Katy Freeway remained the top priority of U.S. Congressional Representative John Culberson and Harris County Judge Robert Eckels. But in the world of freeway construction, nothing is final until it is built. Whatever the resolution of the lawsuit, the project outside Beltway 8 should remain unaffected and proceed as planned. On January 30, 2003, the Texas Transportation Commission officially approved a funding agreement with HCTRA whereby HCTRA would contribute up to $500 million to the overall project cost. The first construction contract for a section between SH 6 and the Grand Parkway (SH 99), with a value of $208 million, was awarded in May 2003. An $83 million contract for work near the Grand Parkway and a $262 million contract for the reconstruction of the Loop 610 interchange were awarded in July. Also in July, the first reports of cost overruns and possible delays were reported. The project cost estimate increased from $1.47 billion to $1.71 billion, mainly due to increases in right-of-way acquisition and utility relocation costs. Difficulty acquiring the 426 right-of-way parcels threatened to delay the project completion beyond the originally scheduled date of late 2008.[89]

If built in its entirety as planned, the Katy Freeway will become one of the nation's most impressive urban freeway corridors for a sustained distance. It will rank among the world's great freeways and be a distinguishing freeway for Houston. With the Katy Freeway, Houston will undoubtedly be among the world's freeway elite. Residents of west Houston will finally be served by the state-of-the-art freeway for which they have waited so long. As the old proverb says, patience is a virtue.

North Freeway, Interstate 45 North

Most Houston freeways have developed a distinguishing characteristic over the years. The West Loop has its impressive and ritzy Uptown Houston skyline. The Katy Freeway is the Energy Corridor. The La Porte Freeway has its industrial complex. When most Houstonians think of a freeway, something comes to mind, whether it is the Astrodome on the South Loop or the perennial construction on the Gulf Freeway. So what do Houstonians think of when the North Freeway is mentioned? In 2003, that honor goes to "Gallery Furniture, 6006 I-45 North between Tidwell and Parker," a retailer with a fast-talking owner who saturates Houston's media with advertisements, always including the North Freeway location in his sales pitches. Aside from Gallery Furniture, Houstonians are apt to think of billboards, commercial clutter, and lower-tier commercial establishments. No one has ever called the North Freeway glamorous or scenic. The most notable structure along the North Freeway, a Goodyear blimp hangar, was dismantled in 1994 and replaced with "big box" retail structures. Perhaps a 1999 *Houston Chronicle* article on the North Freeway found the right word for the freeway, calling it a "workhorse."[90]

But in terms of transportation, the North Freeway has always been one of Houston's most important freeways. It is one of the main routes to Bush Intercontinental Airport and serves as the link to Houston's fast-growing northern suburbs. It connects Houston to its cross-state rival, Dallas. It served as the location for Houston's contraflow lane transit experiment which launched Houston's transitway system. A major reconstruction and expansion of the freeway began in the early 1980s, and the freeway north of Beltway 8 is among the widest and most modern in Houston. The North Freeway also has an asset that motorists on every freeway would love to have: a traffic relief valve running parallel to it. The Hardy Toll Road, completed in 1988, runs parallel to the North Freeway for 21 miles (34 km), helping keep the North Freeway free-flowing and preventing severe traffic congestion. Still, the North Freeway ranks as one of Houston's busiest, with a traffic volume of 281,000 vehicles per day just outside Loop 610 in 2001.

North Freeway	
Previous designation	US 75
Designated as freeway	1945, 1950, 1952
First freeway section open	1959
Freeway complete	1963
Reconstruction	1982–ongoing
Max traffic volume, 2001	281,000 vehicles per day
Future construction	Continued expansion north of Houston; long-term upgrade inside Beltway 8

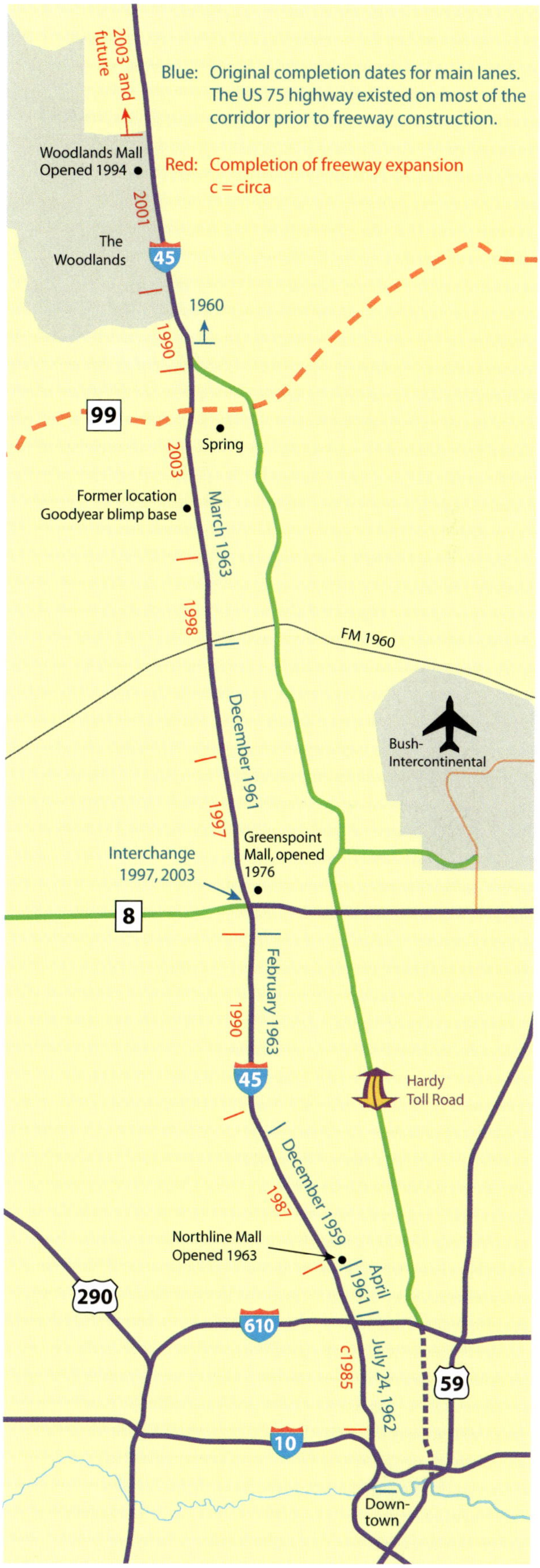

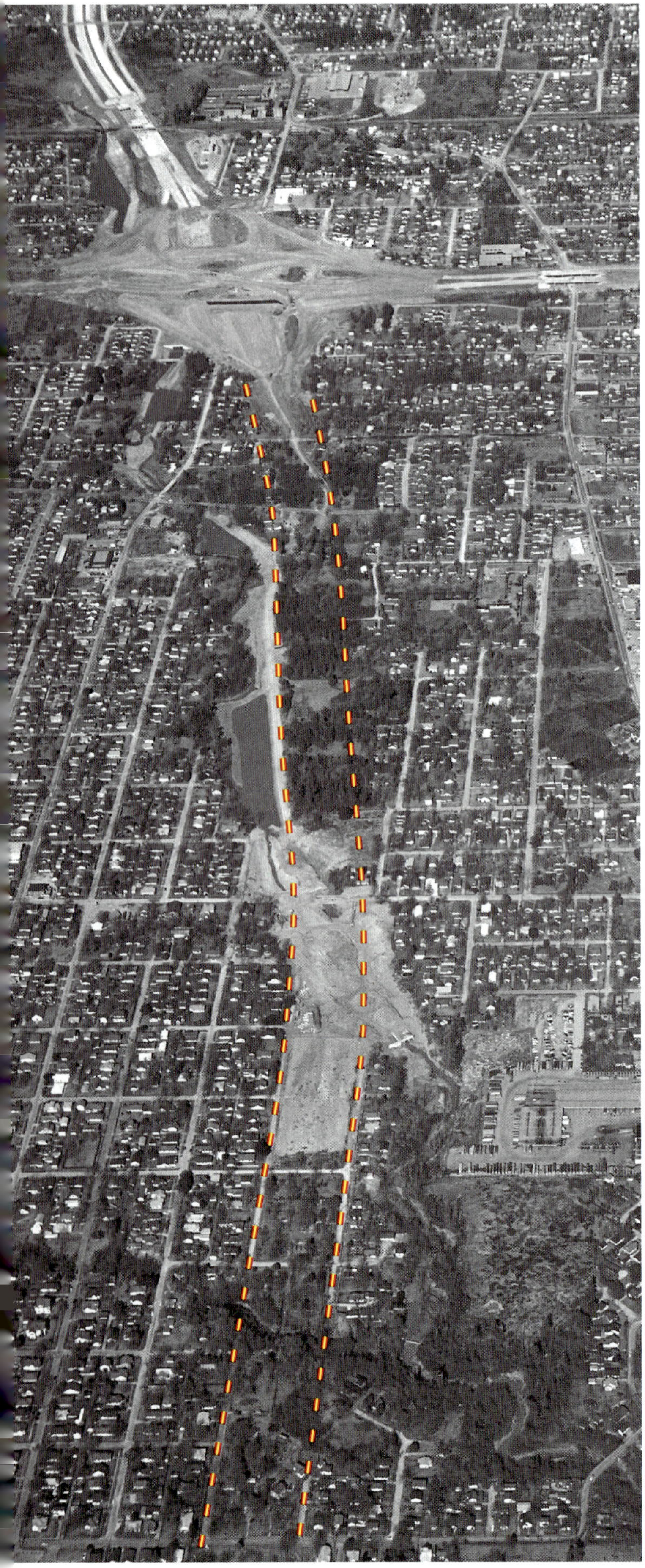

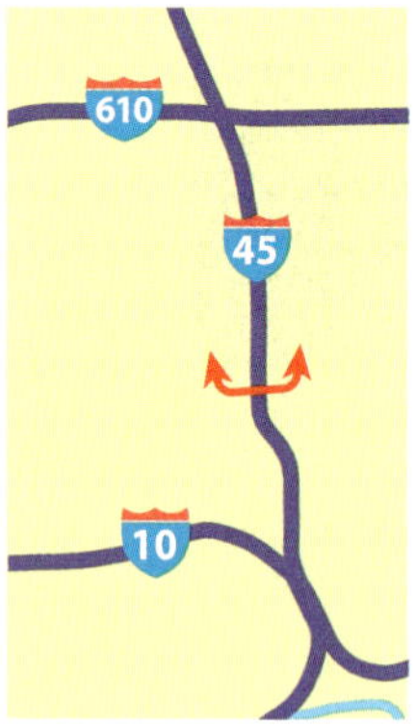

Pre-freeway: This view looks north along the North Freeway corridor in 1960 with the Loop 610 interchange construction zone in the upper part of the image. The path of the North Freeway is indicated by the dashed lines. Relatively few displacements were required for the freeway since its alignment followed Little White Oak Bayou. (Photo: The Positive Image)

Origins

A short section of the North Freeway was in the first group of Houston freeways to be approved by the Texas Transportation Commission in May 1945. The approved section of the North Freeway was a segment from downtown northward to the present-day intersection with North Main Street. At the time, the route was designated as US 75. In October 1946, the commission authorized route determination from North Main to the north city limit, which at the time was about two miles (3.2 km) north of the present-day Loop 610. That section was authorized for full freeway status in March 1950. From the north city limit to the Harris County line, route determination and right-of-way maps were authorized in February 1949. In June 1952, the commission designated all of US 75 between Houston and Dallas as a full freeway.[91]

The entire North Freeway was built in a remarkably short time—faster than any other freeway in Houston. The first 2.8-mile (4.5 km) section of freeway opened in December 1959, and the entire freeway to Spring in far north Harris County was complete by February 1963.

In its early years the North Freeway corridor remained largely devoid of any major commercial developments or landmarks. Up until the early 1970s it served a mostly lower- and middle-income area of northside Houston, and Northline Mall, opened in 1963, was the only notable commercial development. The Greenspoint office and retail complex began to take shape in the mid-1970s at the intersection with Beltway 8, but the office structures were not immediately adjacent to the North Freeway and seemed to be more closely associated with the Beltway.

The 1970s brought the oil boom to Houston, and the forests along the North Freeway became very active with residential development. As the FM 1960 corridor was urbanized and other large developments such as the Woodlands got underway, the North Freeway became overwhelmed by traffic. By the mid-1970s the North Freeway had succumbed to severe traffic congestion, setting the stage for the unique experiment that would heavily influence the future of Houston's freeways and transit.

The Contraflow Lane

As Houston descended into its traffic crisis in the 1970s, political leaders were under pressure to do something. But there was a problem. No money was available,

Loop 610 interchange construction: This view looks north at the three-level interchange with the North Loop in May 1961. The North Freeway north of Loop 610 had opened a month earlier, and the interchange was partially open to traffic. The North Freeway and the full interchange were opened in July 1962. (Photo: TxDOT)

and the prospects for increased highway funding in the future were slim. Officials knew that transit needed to play a role in any solution, but the defeat of the Houston Area Rapid Transit Authority in 1973 by Houston voters eliminated the possibility of any big programs.

Something fast and cheap was needed, and the contraflow lane met both those requirements. It took away the inside traffic lane from the non-peak traffic direction, using it for buses and vanpools travelling in the peak direction. The contraflow lane was separated from oncoming traffic only by pylons spaced at 40-foot (12 m) intervals. There was no barrier.

In November 1974 Houston Mayor Fred Hofheinz asked the Houston office of TxDOT to evaluate the feasibility of a contraflow lane on the North Freeway. The project looked feasible, and authorities worked as quickly as possible within the regulatory environment to move it forward. With approvals required from the Urban Mass Transit Administration, Federal Highway Administration, and TxDOT headquarters in Austin, pulling the project together was no small task. Finally in November 1977, after three years of effort, bids were received for the $2.1 million project. Construction began in February 1978 and the 9.6-mile (15 km) contraflow lane from downtown to Shepherd Drive was officially opened on August 28, 1979. Two park-and-ride lots were also included in the project, bringing the total project cost to $6.8 million, roughly 15.6 million in 2003 dollars.[92]

Contraflow operation was successful, and usage began to increase after a modest 1,458 passenger-trips, 27 bus trips, and 96 vanpool trips the first day. After one year, daily usage had increased to 8,724 passenger-trips, 125 bus trips, and 412 vanpool trips. The facility overcame concerns that it would be unsafe, in spite of a collision that occurred on April 25, 1980, when an automobile lost control, slid into the contraflow lane, and was broadsided by a van. The automobile driver was killed. Three years after its opening, the contraflow lane was serving 15,600

The North Freeway contraflow lane: One lane from the off-peak direction was marked off with pylons and opened to buses and vanpools travelling in the peak direction. This view looks south, just north of Loop 610. The contraflow lane operated from 1979 to 1984, when it was replaced by a barrier-separated transitway in the center of the freeway. The success of the contraflow lane accelerated the implementation of transitway lanes on most of Houston's radial freeways. (Photo: Metropolitan Transit Authority)

passenger-trips per day. The contraflow lane ended its five-year run on November 23, 1984, when traffic was shifted to a central barrier-separated lane. By that time, planning was well underway for an extensive transitway system on Houston's freeways—a system that was inspired and influenced by the success of the North Freeway contraflow lane.[93]

Rebuilding the Freeway

As one of Houston's early freeways, the North Freeway was built to 1950s standards and was soon hopelessly inadequate to meet the needs of the rapidly expanding suburbs north of Houston. Between 1970 and 1980 the population in the North Freeway corridor north of Beltway 8 and within Harris County increased by 302%. The freeway had only 6 main traffic lanes between Loop 610 and FM 1960, and only 4 main lanes in its northern section near Spring and the Woodlands. Frontage roads typically had only 2 lanes in each direction. In the early 1980s freeway reconstruction began, increasing lane capacity to 8 and 10 general purpose lanes with 1 central barrier-separated transitway lane. Whereas the original construction of the freeway occurred within a very short time period, the reconstruction would play out over more than 20 years as funding became available. Work began just north of downtown Houston in 1982 and proceeded northward, segment by segment. By 1990 the expansion had been completed to Beltway 8. The freeway expansion between downtown and Beltway 8 did not involve any substantial right-of-way acquisition, and the freeway section between Loop 610 and Shepherd Drive remained on an unusually narrow right-of-way. That section still does not meet the standards of modern 1980s freeway design due to the constraints imposed by the narrow right-of-way.[94]

In 1986 TxDOT finalized plans for the expansion north of Beltway 8. After considering elevated express lanes, an at-grade plan was recommended. The plan called for expanding the main lanes from 6 to 10 from Beltway 8 to FM 1960; from 4 to 8 from FM 1960 to the Hardy Toll Road; from 4 to 10 from the Hardy Toll Road to the Woodlands Parkway; and from 4 to 8 from the Woodlands Parkway to the south edge of Conroe. The entire length would have a central barrier-separated transitway lane and 3 continuous frontage road lanes in each direction. The

Houston still far away: This 1967 view looks south along the North Freeway at the community of Spring, north of Houston. This view shows that FM 1960, in the upper part of the photo, was still truly a farm-to-market road. Suburbanization of this part of north Harris County began in the 1970s. Expansion of the foreground section of freeway was completed in 2003. (Photo: The Positive Image)

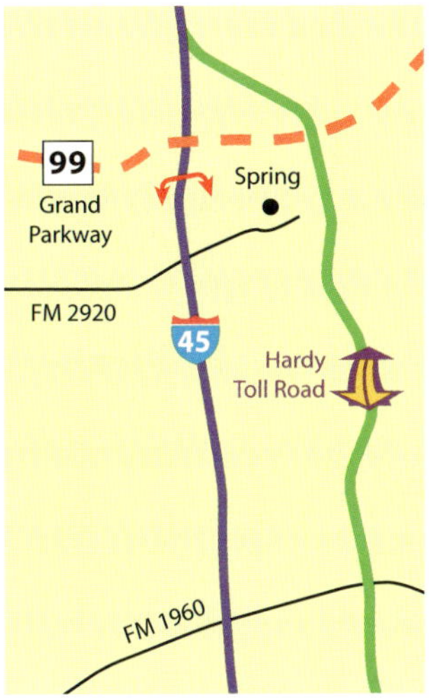

existing freeway corridor ranged from 256 to 276 feet (78-84 m) wide. The expanded freeway right-of-way would be increased to 377 feet (115 m) for the 10-lane sections and 353 feet (108 m) for the 8-lane sections. Since most of the land along the freeway was undeveloped, only 102 businesses and 9 residences were displaced.[95]

Work north of Beltway 8 began in 1993 and will continue through approximately 2005 to reach the south end of Conroe. With the completion of the freeway expansion from Beltway 8 to FM 1960 in 1998, north Houston finally had its first section of the long-awaited wide, modern freeway. The first phase of the stack interchange at Beltway 8 was finished in 1997, and the interchange was fully completed in early 2003. Initial work at the interchange, consisting of two connector ramps, was completed in 1970 to serve traffic going to Houston Intercontinental Airport, which opened in June 1969. In 2003 a study was in progress to determine future improvements from downtown to the Woodlands. Preliminary options emphasized transit and using all available capacity of the Hardy Toll Road. The study is expected to recommend freeway expansion to accommodate future demand that cannot be served by transit or the Hardy Toll Road.

The Ultimate Freeway Suburb?

In 1972, architect Robert Heineman began to sketch out a plan for a large new community directly adjacent to the North Freeway on 27,000 acres of forest in far north Houston. The community he designed would be way ahead of its time. It would incorporate design features that would later be promoted as essential to sustainable development and the goal of better urban design and "livable" communities. Harmony with the natural environment. The concept of putting shops, services, and jobs close together in one community. Shared and efficient infrastructure. In today's terminology, it would be called smart growth.[96]

Originally opened in 1974, the Woodlands received numerous national and international awards for its responsible and innovative design. Suburban home buyers have made the Woodlands Houston's perennial leader in new home starts through most of the 1980s and 1990s, and continuing through 2003. The Hardy Toll Road was

Typical view: This view looks north along the freeway at Richey Road. Expansion of this section of the freeway was completed in 1998. (Photo: November 2002)

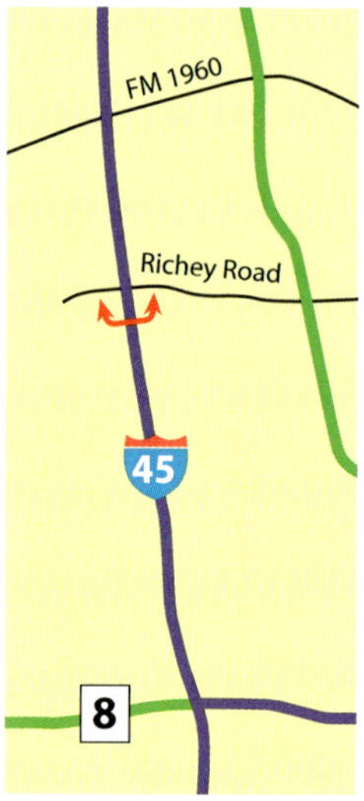

completed in 1988 and ended practically at the doorstep of the Woodlands, providing quick transportation access to Houston and further accelerating development.

When the major expansion of the North Freeway adjacent to the Woodlands was completed in 2001, the Woodlands had reached a sufficient population base to support the fully integrated community that Heineman envisioned 29 years earlier. Approximately 17,600 acres of the 27,000 acres had been developed, and 24,055 new homes had been built. The area's 1,000 businesses employed 27,000 people, and the population reached 70,000. Completion of the Woodlands is projected to occur around 2012.[97]

When the freeway era arrived in the 1950s, the concept of freeway suburbia came along with it. Developer Frank Sharp launched his Sharpstown community on the Southwest Freeway in 1955, bringing large, freeway-driven

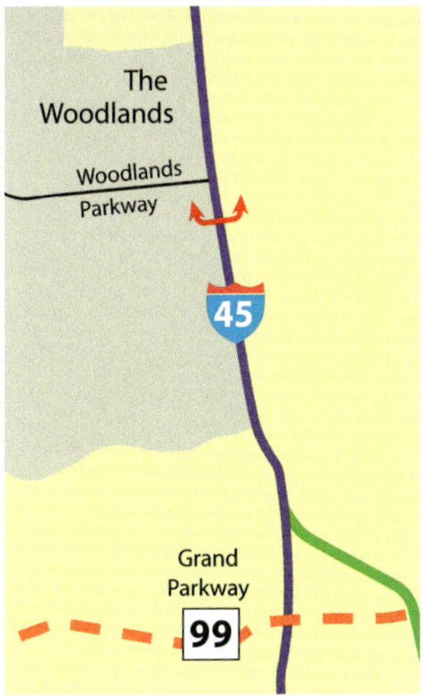

The North Freeway at the Woodlands: Expansion of this section of freeway was completed in 2001 and features a direct connector ramp to the Woodlands Parkway. The Anadarko Petroleum building, completed in 2002, towers over the Woodlands Town Center and nearby Woodlands Mall. The 27,000-acre Woodlands master-planned community was started in 1974 and was about 65% developed in 2002. (Photo: May 2002)

suburban development to Houston. A succession of bigger and better communities followed. But the Woodlands will almost surely go down in Houston's history as the largest and most successful suburb, and one of the more notable suburban developments in the United States. And it took a freeway to make it all happen.

Key dates in the history of the North Freeway	
1945	First designation as a freeway. Other sections are designated in 1950 and 1952.
1959	First freeway section opens.
1963	Freeway complete.
1974	The Woodlands real estate development is launched.
1979	Contraflow lane operation begins and continues until 1984.
1982	Reconstruction and expansion begins.
1988	The Hardy Toll Road is completed from Loop 610 to near the Woodlands.
2003	The interchange at Beltway 8 is completed.
Future	Expansion north of Houston, upgrades inside Beltway 8.

Airship base: The Goodyear Blimp hangar was a fixture on the North Freeway for 25 years from 1969 to 1994. When this photo was taken in September 1991, commercial development and strip shopping centers were encroaching on the blimp base. The base was closed in April 1992 when the airship *America* was relocated to Akron, Ohio. In March 1994 the hangar was dismantled and relocated to San Diego, California. Two large retail establishments, a Home Depot and Lowe's Home Improvement Center, were subsequently constructed on the blimp base land.[98] (Photo: The Positive Image)

No slouch for construction: Although IH 45 South, the Gulf Freeway, has the reputation for being Houston's most construction-affected freeway, the North Freeway has had its share of construction. Since 1982 there has been construction in progress nearly continuously. This view looks north from the Louetta Road overpass in March 2003, with construction on the freeway expansion nearing completion. Construction continues further north on IH 45. (Photo: March 2003)

Dallas Freeway? No way! In 1965 there was an effort to rename several Houston freeways to eliminate compass-point names. The North Freeway was proposed to be renamed the Dallas Freeway. A *Houston Chronicle* editorial came out against the change, saying, "Why advertise Dallas, for goodness sake?" The North Freeway name change was rejected, as well as all the other proposed changes except a renaming of the West Freeway to become the Katy Freeway. This view is on the northbound lanes just north of Beltway 8, with the Rankin Road overpass just ahead.[99] (Photo: November 2002)

North Belt interchange: This view looks north along the North Freeway at the North Belt interchange, which was completed in early 2003. (Photo: September 2002)

Eastex Freeway, US 59 North

Eastex Freeway	
Designated as freeway	1945, 1953
First freeway section open	1953
Freeway complete	1970 (to Bush Intercontinental Airport)
Reconstruction	1991–ongoing
Max traffic volume, 2001	192,000 vehicles per day
Future construction	Expansion north of Houston; complete interchange at Beltway 8

Clearing out the old and bringing in the new is a regular event on Houston's freeways, but sometimes the transformation is especially satisfying. In the case of the Eastex Freeway, the "old" was the original, antiquated 1950s-era design, Houston's last vestige of a first generation freeway. The "new" is a state-of-the-art freeway, among the best in Houston. Getting from old to new was a particularly long and painful process due to the seemingly interminable delays during construction, but by the late 1990s Eastex Freeway motorists were finally driving on their new freeway. After the transformation was complete, there was surely no doubt: new *is* better.

Origins

The Eastex Freeway was Houston's second freeway to be constructed, after the Gulf Freeway. The section from downtown to Loop 137, just north of present-day Loop 610, was authorized by the Texas Transportation Commission in May 1945 when the commission approved a small group of freeways that would form the basis for Houston's early freeway plan. By November 1948 city and state officials had determined that the US 59 freeway would be routed just east of Jensen Drive, leading to the informal designation of the route as the Jensen Drive Freeway. The freeway was also informally called the Humble Expressway. Early plans showed the Eastex Freeway terminating at the planned east-west freeway (now IH 10) with connections being made into downtown via ramps from the east-west freeway. In 1948 officials were already contemplating the idea of bringing the freeway into downtown on its own structure. By early 1952 an alignment along Chartres Street east of downtown was identified, and by April 1953 the continuation of US 59 into downtown had been officially approved into Houston's freeway plan. In October 1953 the Texas Transportation Commission officially authorized US 59 as a full freeway from Houston to the Liberty County line, 40 miles (64 km) north.[100]

Building the Freeway

In the early 1950s, the city of Houston was responsible for acquiring all right-of-way for new freeways within the city. Since the Eastex Freeway was being built through an urbanized area, clearing the corridor was a costly and time-consuming task. The city of Houston sold $1.3 million in bonds around 1950 to pay for the right-of-way acquisition, and a team of 10 people worked "day and night" to complete all necessary transactions to clear the corridor for freeway construction. The first construction contract was awarded in December 1951 for the frontage roads on a 1.9-mile (3 km) segment just south of the present-day Loop 610 and a short segment of main lanes at Loop 610. An official dedication was held on December 22, 1953, and Houston Mayor Roy Hofheinz announced the winning name in a contest to name the freeway. From

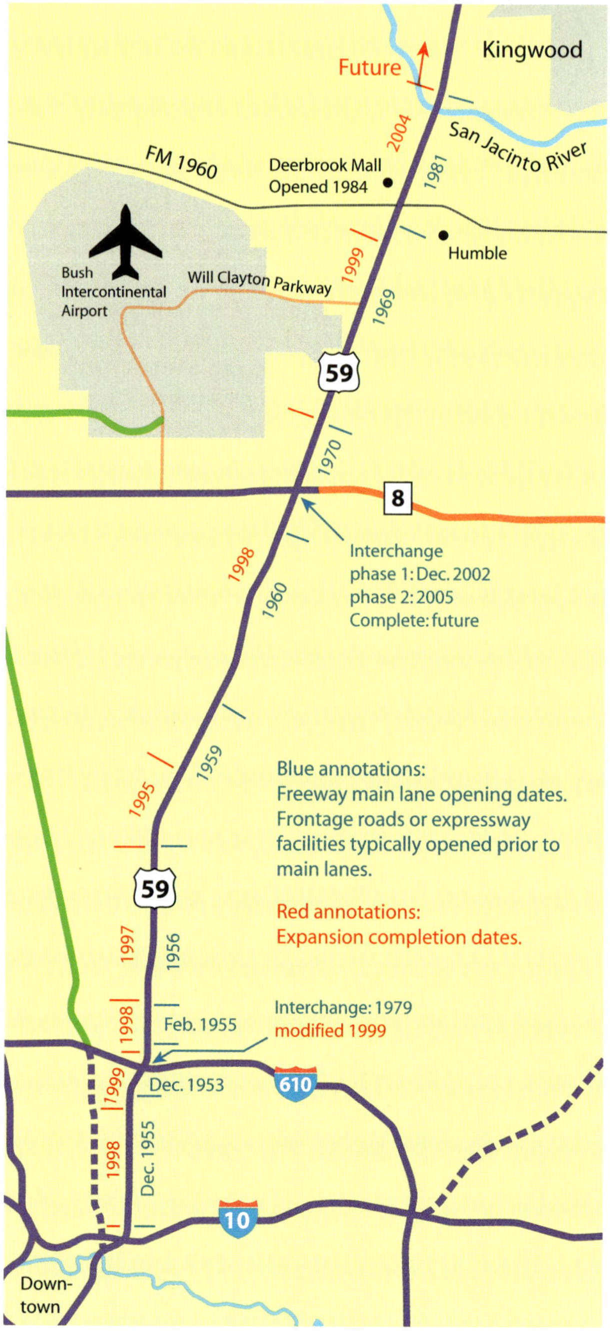

The beginnings of the Eastex Freeway: This view shows the cleared corridor for the Eastex Freeway just north of downtown in January 1954. The frontage road section in the upper part of the photograph and a short section of main lanes at the top of the photo were officially dedicated on December 22, 1953. On the opening day, the winning entry in a contest to name the freeway was announced. (Photo: *Houston Chronicle*, January 11, 1954)

CLEVELAND
59
LIBERTY
JENSEN DR.
LOOP
137
90
LOOP
137

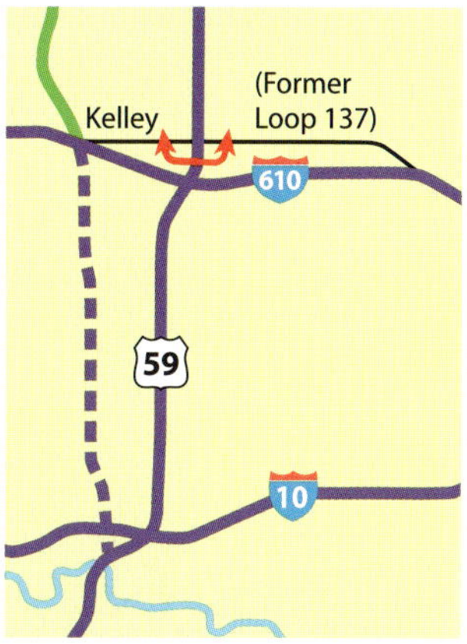

The motorized merry-go-round: These views look north along the Eastex Freeway at Kelley Road. The upper view on the opposite page shows construction on August 3, 1954. At the time Kelley Road was Loop 137, the predecessor to Loop 610. A traffic circle was used for this interchange. It was a good thing that traffic circles were rare in Houston, because Houston motorists and traffic circles don't mix well. From 1964 to 1966 the intersection was the most dangerous in Houston, with 121 accidents in 1965 and 136 accidents in 1966. In 1969 the *Houston Post* reported, "Few other intersections can begin to compete with the more than 100 annual smash-ups on this motorized merry-go-round." The signage along the Eastex Freeway shown in the lower photo attempted to help motorists navigate the intersection. The completion of the North Loop in 1975 finally diverted traffic from the traffic circle, allowing TxDOT to remove it and replace it with a standard bridge, shown in the photo above. The Eastex Freeway was built to its present configuration in 1998.[101] (Photos: opposite page upper, Squire Haskins Photography, Dallas; opposite page lower, TxDOT; above, September 2002)

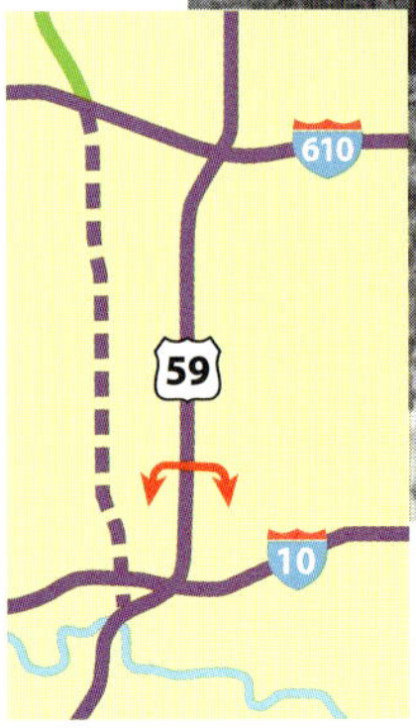

Freeway construction: This August 3, 1954, view looks south along the original construction of US 59 just north of IH 10. Expansion of this section of freeway was completed in 1998. (Photo: Squire Haskins Photography, Dallas)

that day on, the freeway would be known as the Eastex Freeway. Houston's first two freeways, the Gulf and the Eastex, would be the only Houston freeways to be named by contests.[102]

By the end of 1953 work was underway both to the north and south of the completed section. Freeway openings became a regular occurrence during the rest of the 1950s, pushing the freeway northward toward Beltway 8. North of Houston, the Eastex Freeway was initially constructed as an expressway and later upgraded to full freeway standards. By March 1970, the Eastex Freeway was a full freeway from downtown to the Will Clayton Parkway entrance of Bush Intercontinental Airport. At the time, Will Clayton Parkway was called Jetero Boulevard.

A Quiet Corridor

As one of Houston's earliest freeways, the Eastex Freeway suffered from the low standards of first generation freeway design. The freeway was narrow, overpasses lacked shoulders, and ramps had poor geometrics. There were only six freeway lanes from downtown to just north of Loop 610, and only four freeway lanes north of Loop 610. The low capacity of the freeway did not immediately become a problem because the Eastex Freeway corridor remained quiet in terms of real estate development. The Eastex Freeway became one of the two main access routes to Houston Intercontinental Airport (now Bush Intercontinental Airport) when the airport opened in June 1969, but the airport opening spawned almost no commercial development along the Eastex Freeway. In 1971 Friendswood Development (then a division of the Exxon Corporation) launched the 14,000-acre Kingwood master-planned community in far north Houston just east of the Eastex Freeway. The community became highly successful. The Eastex Freeway would remain devoid of major commercial development until the opening of Deerbrook Mall in 1984 in Humble near Kingwood.[103]

By the 1980s the Eastex Freeway frontage roads had become lined with junkyards, used car lots, and other low-grade commercial establishments. The vintage 1950s-era freeway design and unsightly clutter had become an embarrassment to Houston, especially since it was one of the main routes to Houston's principal airport. The reconstruction of the Eastex Freeway became one of the top priorities of local leaders in the early 1980s when plans for the next generation of improvements to Houston's freeways were being formulated. When Houstonian Bob Lanier became chairman of the Texas Transportation Commission in 1983, he made the reconstruction of the Eastex Freeway one of his priorities, moving the project forward in the line for state funding.

The Expansion

The Eastex Freeway expansion required the largest right-of-way clearance in Houston since the early 1970s. The freeway corridor right-of-way width, which was typically 256 to 300 feet (78 to 91 m), would need to be widened to 400 feet (122 m) for the 10-lane sections of freeway and 376 feet (115 m) for the 8-lane sections. Nine churches, 196 businesses, 194 single-family homes, and 240 units of multifamily housing were to be displaced. Even with the depressed real estate prices following Houston's oil bust in the mid-1980s and the generally low property values in that section of Houston, the cost of the right-of-way acquisition and relocation was estimated to be \$205 million in 1985—308 million in 2003 dollars. There was some protest among businesses and the usual complaints from inner-city interests that the freeway would primarily benefit suburban communities like Kingwood, but the project had strong political support and moved forward without difficulty. The expansion had the big benefit of cleaning up much of the decay and clutter alongside the freeway, drastically improving its appearance.[104]

> "The Eastex Freeway widening has been a theater of the glacial."
>
> *Houston Chronicle*, April 26, 1998

The first construction contract for the Eastex Freeway expansion was awarded in 1991. The project turned out to be a long-running, drawn-out affair, with construction between downtown and Beltway 8 taking nearly all of the 1990s. The project was divided into numerous contracts with the usual three- to four-year construction period. The contractor on most of the construction contracts, Williams Brothers Construction, generally took six to seven years to complete the work on each of its contracts. A comprehensive report by the *Houston Chronicle* in 1998 cited numerous reasons for the delay, including utility relocations, unexpected environmental cleanup, TxDOT's project sequencing, and design changes after work began. But those factors couldn't account for a three-year delay, the report stated. The report speculated that the very low penalties for late completion of the contracts provided little incentive for the contractor to complete the projects on time, and actually made it advantageous for the contractor to divert resources to other, more profitable jobs. Whatever the reason, work proceeded at a glacial pace. The Eastex Freeway reconstruction project prompted TxDOT to use higher penalties for late completion on its future contracts, and Williams Brothers would go on to demonstrate an exemplary record for timely completion of projects in the following years.[105]

Key dates in the history of the Eastex Freeway

1945	First official designation as a freeway. Full length designated as a freeway in 1953.
1953	First section opens and freeway named in a contest.
1979	Interchange at Loop 610 is completed.
1991	Reconstruction and expansion begins.
Future	Expansion north of Houston, complete interchange at Beltway 8.

Old and new: The photo at left looks north along the Eastex Freeway from the Greens Road overpass in 1978. The lower image shows the exact same location in 2002 after completion of the expansion. The Greens Road overpass was removed and the Eastex Freeway main lanes now go over Greens Road. The reconstructed Eastex Freeway has 10 general-purpose main lanes, a central transitway, and frontage roads. (Photos: left, TTI; lower, December 2002)

Before, during, and after: These views show the Eastex Freeway-North Loop interchange in 1973, September 1977, and 2002. Completed in September 1979, this interchange was the last of Houston's first generation of four-level stack interchanges. It was, however, the first Houston interchange to include a frontage road—the Eastex Freeway northbound frontage road. All subsequent interchanges in Houston would include the frontage roads of both intersecting freeways. (Photos: upper, TxDOT; middle, Chuck Fuhs; lower, November 2002)

Tropical Storm Allison: The storm unleashed a huge flood event on Houston in June 2001. The Eastex Freeway drops below ground level at only one point. As shown above, the underpass was completely filled with water. (Photo: copyright Jim Olive/Stockyard.com)

Original Eastex: This view dated August 1981 shows the original four-lane freeway just north of Aldine Mail Road. Expansion of this section was completed in 1998. (Photo: Texas Transportation Institute)

Finally in 1998 sections under contract to Williams Brothers were being completed. By 1999, work had been completed from IH 10 downtown to the Will Clayton Parkway entrance of Bush Intercontinental Airport. The end result was worth the wait. The freeway features 10 continuous general purpose lanes for most of its length to the airport, with 8 on the remainder. It includes a fully integrated transitway facility and continuous frontage roads built to high standards. What was Houston's last vestige of a first generation freeway in 1990 had been transformed into one of Houston's most impressive freeway corridors. In 2003 work continued on expansion near the Kingwood community. Houston's newest five-level stack interchange was also under construction at Beltway 8, with half of the planned eight connectors being built in the current phase. Future plans will extend the freeway expansion further northward and complete the stack interchange at Beltway 8.

The Rest of the Spokes

East Freeway, Interstate 10 East

Origins

The early planning for the East Freeway* was a combination of work done for SH 73 and US 90. In October 1946, the Texas Transportation Commission authorized US 90 as a full freeway within the city limits of Houston, launching plans for an east-west freeway across Houston. At the time the eastern Houston city limit was located approximately at the present-day east Loop 610. However, US 90 veered northeastward at McCarty Road, just inside Loop 610. SH 73 provided a route that left Houston in an eastward direction. IH 10 absorbed the US 90 alignment west of McCarty Road and the SH 73 alignment east of McCarty Road.[106]

SH 73 was originally planned as a direct route to Port Arthur and was designated as the Port Arthur Short Line, or simply the Port Arthur Freeway. The first right-of-way for the Port Arthur Freeway in Harris County was purchased in 1946, and the first construction contract was awarded in 1948. In October 1953 the Texas Transportation Commission authorized SH 73, which began just inside Loop 610, as a full freeway eastward from Houston. Also in 1953, the first freeway section opened near today's Beltway 8. Inside Beltway 8 the route was initially constructed as a four-lane divided highway, not to freeway standards. In 1956 a new freeway-quality bridge over the San Jacinto River was opened. By early 1958, the route was substantially complete as a divided highway or freeway from just inside Loop 610 to a point 38 miles (61 km) east of Houston near Anahuac. In 1961 the SH 73 designation was officially dropped on its section that was absorbed into IH 10.[107]

East Freeway	
Previous designations	US 90 (west of McCarty) SH 73 (east of McCarty)
Other names	Baytown-East Freeway
Designated as freeway	1946 (US 90) 1953 (SH 73)
First freeway section open	1953
Freeway complete	April 18, 1966
Reconstruction	Late 1970s–ongoing
Max traffic volume, 2001	206,000 vehicles per day
Future construction	Widening east of Houston, completion of interchange at IH 610 and US 90

Bringing the Freeway into Downtown Houston

Constructing SH 73 east of Loop 610 presented no particular challenges since it traversed a nonurbanized area. The SH 73 freeway itself was built to early 1950s standards with four main lanes and frontage roads. Bringing the freeway into downtown Houston would be much more challenging, however, since it cut through a heavily urbanized area. Freeways inside Loop 610 were under the jurisdiction of the Houston Urban Project Office, which was headed by A. C. Kyser. Kyser's freeways incorpo-

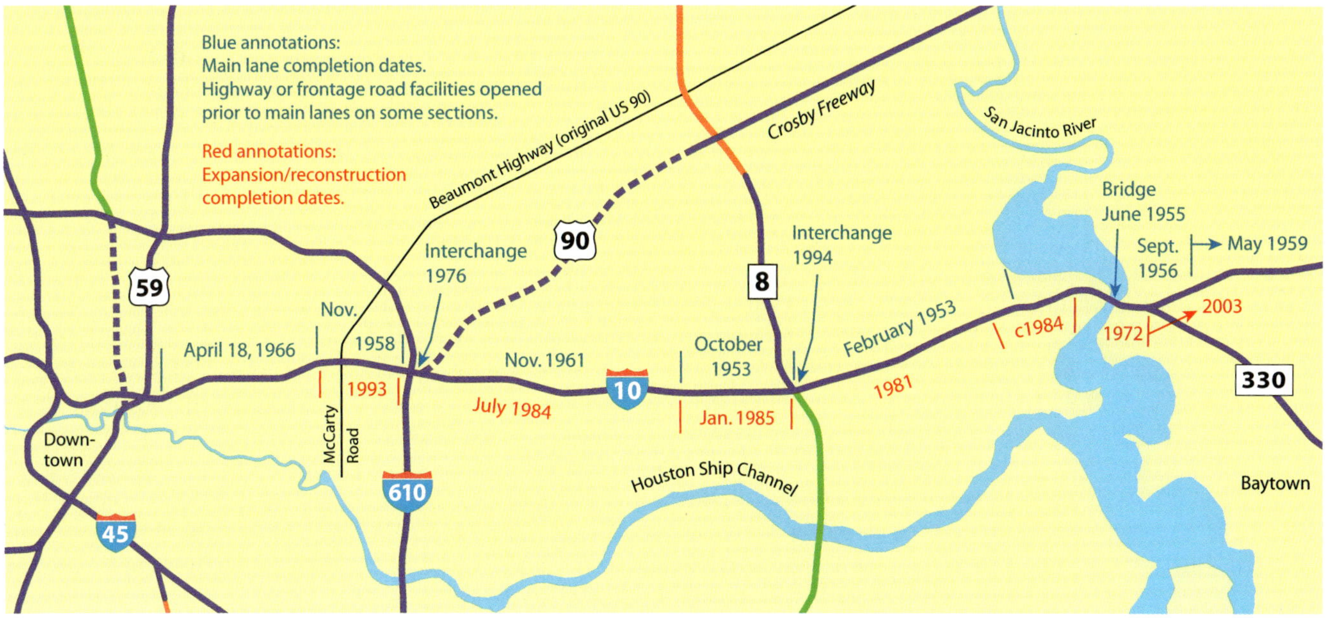

* The East Freeway is also called the Baytown-East Freeway. The designation "East Freeway" is used here.

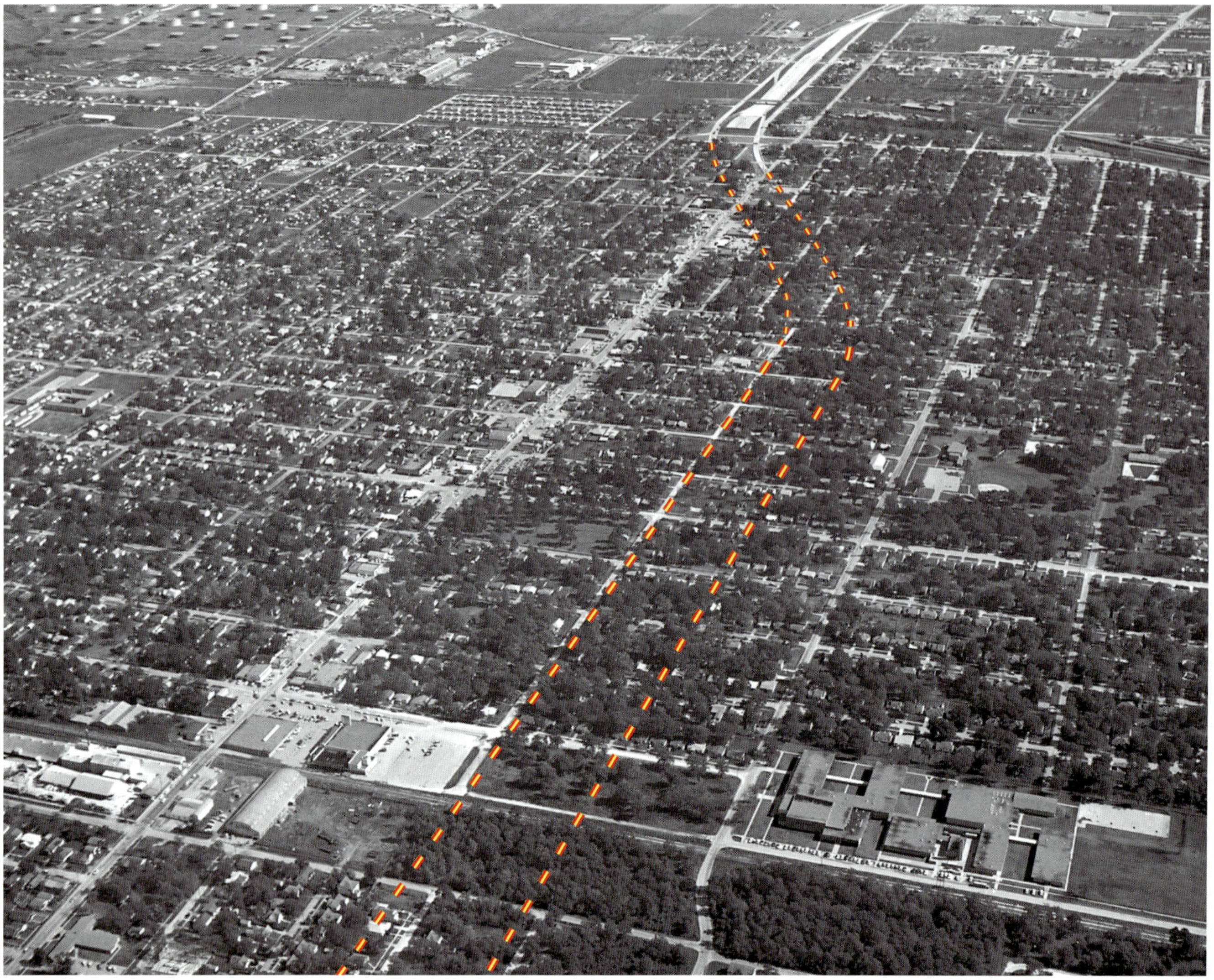

Freeway coming through: This view looks east over the Denver Harbor area in 1958, just before right-of-way clearance for the East Freeway began. In the distance a completed section of the freeway can be seen. The section indicated by the dashed lines was opened to traffic on April 18, 1966. (Photo: The Positive Image)

rated the latest design and safety features, typically equaling the quality of the best freeways in the United States, including those in Los Angeles. The inner loop section of IH 10 would be no exception. By 1963 right-of-way had been cleared and construction was underway. On April 18, 1966, a 3.5-mile (5.6 km) segment of eight-lane freeway east of downtown was opened from US 59 to the previously completed freeway at Wayside Drive. The chairman of the Texas Transportation Commission, Herbert Petry, remarked that the freeway is "of the most modern design." The freeway featured a long section depressed below grade and a long section elevated on embankment, eliminating the frequent overpass humps found on many Houston freeways. The freeway incorporated the latest safety features, including breakaway signs, reflective lane markers, raised markers on emergency shoulders to alert straying vehicles, and illumination fixtures integrated into the median barrier.[108]

Modernizing the Antiquated Freeway

With the April 1966 opening of the inner loop section, the East Freeway was complete. However, the older sections of the freeway were built to low standards and by 1970 reconstruction and expansion were needed. The 1970s transportation funding crisis and new federal regulations delayed the start of major reconstruction work until 1976. The freeway would be expanded to eight main lanes from Loop 610 to Beltway 8 and to six main lanes from Beltway 8 to the San Jacinto River Bridge. Gaps in the frontage roads would be closed. Work on the expansion proceeded very slowly, leaving the East Freeway in a seemingly perpetual state of a construction war zone. This was the first comprehensive major reconstruction of

Construction: This view looks east near Waco Street on December 3, 1964. (Photo: TxDOT)

Downtown approach: This view looks west with the Waco Street overpass just ahead. (Photo: July 2002)

Beltway 8 interchange: This view looks southeast with the East Freeway crossing from lower right to upper left. This interchange was completed in 1994. (Photo: September 2002)

a Houston freeway, and traffic management techniques for reconstruction were inadequate or nonexistent. Local officials complained about poorly marked detours, dangerous ramps, large potholes in the pavement, and 10-foot (3 m) drop-offs inches away from traffic lanes. The complaints prompted an official inquiry into the project by the Texas Legislature in 1980. Construction would drag on until 1985.[109]

In the 1990s work focused on the interchanges with Loop 610 and Beltway 8. In 1993 improvements in the vicinity of the Loop 610 interchange were completed. The interchange at Beltway 8 was completed in 1994. Unlike the other Beltway 8 stack interchanges, the stack at the East Freeway is a four-level design rather than a five-level design, but still a very impressive structure. In 1998 the East Freeway received validation as a major traffic artery by breaking the 200,000-vehicle-per-day barrier at a point just outside Loop 610. Future work on the East Freeway will complete the interchange at Loop 610 and the Crosby Freeway, adding the fifth leg (the Crosby Freeway) to the interchange. East of Houston, sections of freeway with only four main lanes will be upgraded to six main lanes. The entire corridor is slated for a comprehensive study to determine long-term improvements. The study is expected to be underway by 2004.

Northwest Freeway, US 290

Northwest Freeway	
Designated as freeway	1953
First freeway section open	1963 (Loop 610 interchange) 1975 (main lanes)
Freeway complete	Scheduled 2005
Reconstruction	none
Max traffic volume, 2001	245,000 vehicles per day
Future construction	Long term widening from Loop 610 to Grand Parkway; potential toll road along Hempstead Highway

If any Houston freeway is deserving of a big yawn, it's the Northwest Freeway. There is nothing particularly interesting about its history, its associated urban development, or the freeway itself. In spite of being relatively new, it is one of Houston's smallest freeways. The only notable feature of the Northwest Freeway corridor is the huge stack interchange at Beltway 8. In 1999 a *Houston Chronicle* report on the freeway stated, "The Northwest Freeway is, in some senses, the anonymous freeway. Unlike others in Houston, it doesn't extend all the way downtown. It isn't part of a major national corridor like an interstate. It was built and opened in small chunks in four different decades and simply carries huge numbers of commuters from the northwest suburbs to their jobs."[110]

The potential for even more commuters, however, may just propel the Northwest Freeway to become a prominent freeway someday. City of Houston planning maps dating back to the 1960s suggested that urban sprawl in Houston would spread most quickly towards the northwest along the Northwest Freeway, reaching into far northwest Harris County. Although a substantial amount of development occurred, the predicted urbanization along the distant stretches of the Northwest Freeway corridor never materialized.

With land along other freeway corridors filling up, the Northwest Freeway will probably get its long-anticipated development surge. In 2002 a major investment study for future freeway improvements planned for the eventual realization of the long-predicted urbanization along the corridor, recommending an expansion of the freeway and construction of a new, parallel tollway.

Origins

The Northwest Freeway was added to the Houston freeway system during the development of the first comprehensive freeway plan for Houston, which was presented to the Texas Transportation Commission in July 1953. The 1953 plan showed the Northwest Freeway aligned along the present-day Hempstead Highway for its entire length, penetrating into Loop 610 and terminating at IH 10 near the northeast corner of Memorial Park. In December 1953 the Texas Transportation Commission officially approved the US 290 freeway. By that time, the Houston TxDOT office already had additional studies of the exact alignment in progress. The alignment of the Northwest Freeway was shifted north of Hempstead

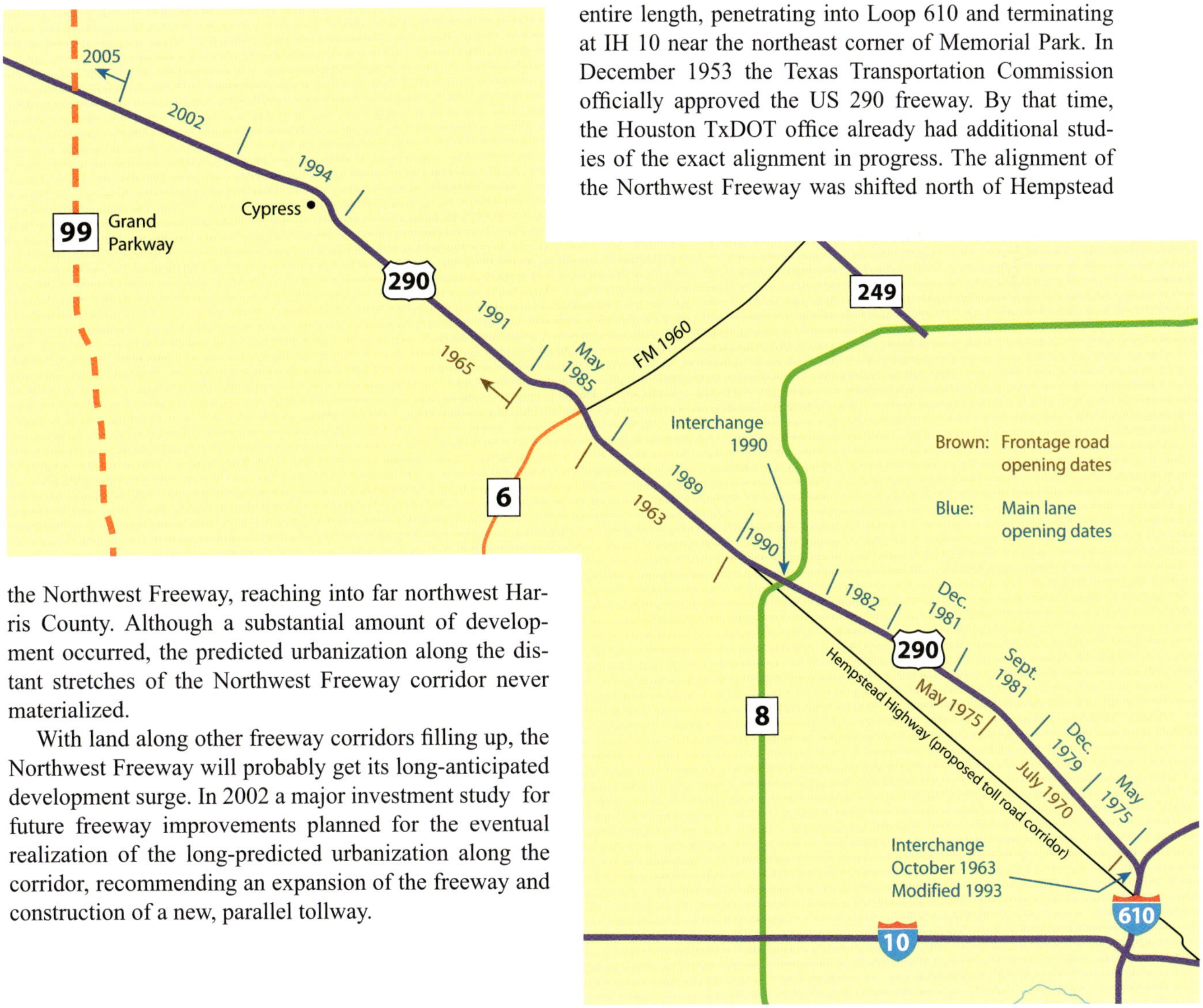

Freeway beginning: This December 1964 view looks northwest along the freeway at the Loop 610 interchange, where the freeway begins. Of all Houston's freeways, the Northwest Freeway holds the distinction of having the longest running, most drawn-out construction phase. The freeway was completed piece by piece through the 1970s, 1980s, 1990s, and 2000s. In 2003 work is in progress to build the freeway main lanes in far northwest Houston near the planned Grand Parkway. (Photo: TxDOT)

Highway between Loop 610 and Beltway 8. In addition, the revised alignment terminated the Northwest Freeway at Loop 610 without extending it inside the loop. When the near-final version of Houston's freeway plan emerged in 1954, the Northwest Freeway was shown on its present-day alignment.[111]

In March 1959 a public hearing was held for the construction of the Northwest Freeway from Loop 610 to FM 1960-SH 6, but construction was far from imminent. Building the freeway would be a slow-motion affair playing out over a long period of time, with construction of freeway main lanes not even beginning until the 1970s. A very short section of the freeway that was part of the Loop 610 interchange was opened in October 1963. After that, the Northwest Freeway was dormant for many years. The first section of frontage roads, from near Loop 610 to Pinemont, halfway to Beltway 8, was opened in 1970. In 1975 the frontage roads were extended to near Beltway 8. Also in 1975, the first section of freeway main lanes opened, a one-mile (1.6 km) section just west of Loop 610. Construction of the Northwest Freeway would proceed slowly during the 1970s, 1980s, and 1990s, with main lanes completed to Cypress in far northwest Houston in 1994. The only distinguishing feature of the Northwest Freeway, the huge interchange with Beltway 8, was completed in 1990. The Northwest Freeway transitway lane, which was built on the location of the interior shoulders, was opened in two phases in 1988 and 1990. In 2003 work continues to slowly extend the freeway westward and convert the remaining highway section in northwest Harris County to a freeway.[112]

It is somewhat surprising that most of the Northwest Freeway was constructed to relatively poor design standards with only six main freeway lanes. Houston's other freeway planned at about the same time—the South Freeway—is far more advanced in terms of design and has provisions for far more capacity. The Northwest Freeway, it seems, was the last hurrah for early 1960s freeway design and capacity.

In 2002 a major investment study of the Northwest Freeway corridor was completed. The study recommended the expansion of the freeway to 10 main lanes between Loop 610 and Beltway 8, and to 8 main lanes outside Beltway 8 to the Grand Parkway. The study also recommended the construction of a new tollway/managed lane facility along the Hempstead Highway corridor from Loop 610 to the Grand Parkway. The study indicated a need for a 4-lane tollway facility, as well as provisions for future transit facilities. If constructed, the new tollway would make the Northwest Freeway corridor similar to the North Freeway corridor with both a freeway and a separate tollway. Outside Beltway 8, where the Northwest Freeway follows the Hempstead Highway alignment, the combination of the two facilities on adjacent rights-of-way will make the Northwest Freeway a very wide and impressive freeway corridor. So maybe someday, the Northwest Freeway will qualify to be among the best of Houston's freeways.

Too small, but relief on the way (someday): In spite of being relatively new, the Northwest Freeway was constructed to very low design and capacity standards with only 6 main lanes for most of its length. This view looks east along the Northwest Freeway toward downtown and shows the regular afternoon traffic backup on the freeway. In 2002 a study recommended widening this section of the freeway to 10 main lanes and the construction of a parallel tollway. Major construction will likely occur after 2010, although the tollway could be constructed sooner. (Photo: September 2002)

Northwest Freeway-Beltway 8 interchange: Although the Northwest Freeway itself is not impressive, the huge interchange at Beltway 8 is one of Houston's most impressive stack interchanges. The interchange was completed in 1990. This view looks northwest along the Northwest Freeway. The ramp in the center of the freeway (with the van and two vehicles) provides access to the transitway and is not part of the interchange. (Photo: March 2003)

Hardy Toll Road

Hardy Toll Road	
Designated as tollway	1983 (voter approval)
First segment open	September 20, 1987
Tollway complete	Scheduled 2007 (downtown extension)
Max traffic volume, 2001	58,000 vehicles per day
Future construction	Downtown extension, reconstruction and widening north of Beltway 8, possible northward extension

If any major transportation facility in Houston can be attributed to the efforts of one individual, it is the Hardy Toll Road. Jon Lindsay made this project happen. Not only did then Harris County Judge Lindsay make the Hardy Toll Road happen, but he did it in the face of substantial opposition to the project. Building an all-new, limited-access transportation facility is a big accomplishment in the modern era. Doing it nearly single-handedly is even more impressive.

The product of Jon Lindsay's efforts is a facility that is somewhat unusual in Houston. Unlike the Sam Houston Tollway, the Hardy Toll Road was not part of Houston's original freeway plan, so it exhibits none of the characteristics associated with a Houston freeway. No conventional frontage roads. No commercialization. No billboards. An almost parkway-like feel at the northern end. And except during rush hour, no heavy traffic. For most of its existence since its opening in 1987, the Hardy Toll Road did not pull its weight financially. But by 2002 it had approached or exceeded financial self-sufficiency. The Hardy Toll Road has served the important role of providing a relief for the North Freeway—something that motorists on Houston's other crowded freeways can only dream about. It has

provided an express route from central Houston to the far northern suburbs, ending practically at the doorstep of the Woodlands, Houston's perennial leader in new housing construction.

Origins

The Hardy Toll Road was first mentioned as a possible tollway in the 1970 Texas Turnpike Authority (TTA) annual report. However, the project did not move beyond the discussion phase at the time. By the mid-1970s there was a renewed interest in tollways as Houston descended into a traffic crisis and public funding for new freeways was dramatically reduced. In the mid and late 1970s local officials asked the TTA to study several projects in the Houston area, including the Hardy Toll Road. In October 1979 Wilbur Smith and Associates completed an initial feasibility study of three potential toll road projects in Houston: the Hardy Road corridor, the Westpark Drive corridor, and the Harrisburg Boulevard corridor. The Harrisburg corridor was the extension of the SH 225 La Porte Freeway into downtown, a freeway project that had recently been cancelled by TxDOT. The study concluded that the Hardy Road project had the best potential of the three projects and "would appear to warrant further study." The Westpark Drive project was not recommended for further study at the time, and the Harrisburg route was declared to be infeasible as a tollway.[113]

The potential feasibility of the Hardy Toll Road was good news to many. The nearby North Freeway (IH 45) was severely congested, and the Metropolitan Transit Authority (Metro) had recently begun a contraflow lane project in a desperate effort to provide some relief to the corridor. Running parallel to the North Freeway about 1.5 miles (2.4 km) to the east, the Hardy Toll Road would provide an alternate route for commuters on the North Freeway. It was also very attractive to real estate interests in far north Houston.

But the proposed Hardy Toll Road was not attractive to everyone, and by early 1980 substantial opposition was developing. Further complicating the issue were plans by Metro to construct a rail system. The Hardy Road corridor was the leading candidate for a north side rail line. A tollway in the corridor had the potential to preclude rail service, throwing Metro's entire rail project into limbo.* Lining up against the project were United States Congressman Bob Eckhardt, State Representative Gene Green, and Houston City Council members Dale Gorczynski, Ernest McGowen, and Ben Reyes. The Metropolitan Organization, an organization representing inner-city interests, was strongly against the Hardy Toll Road and actively worked against the project.[114]

On July 29, 1980, the TTA entered into an agreement with a consulting firm to perform a preliminary engineering feasibility study for the corridor. After hearing the opposition from the Metropolitan Organization, Harris

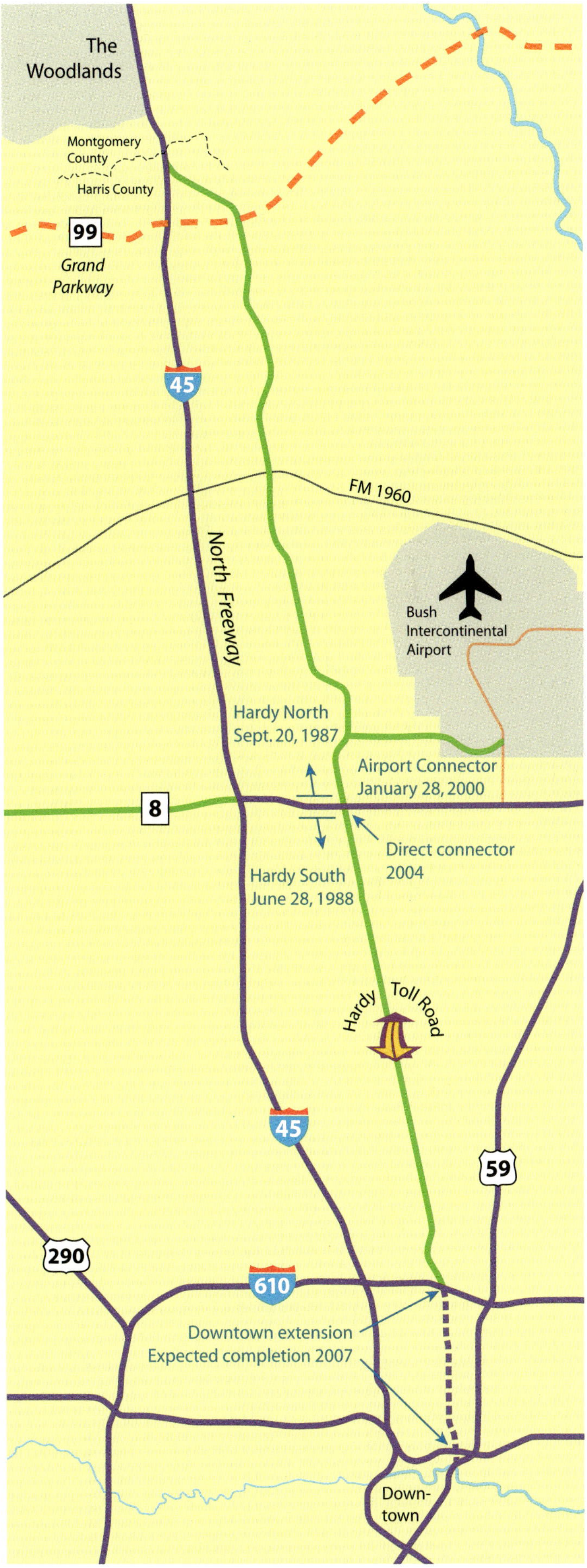

* The final design for the toll road included a right-of-way set-aside for a potential future rail line.

Alongside the railroad: The Hardy Toll Road follows the Union Pacific railroad tracks inside Beltway 8. The railroad and adjacent high-voltage right-of-way made the corridor ideal for the tollway, minimizing the displacements required for its construction. Nevertheless, there was a substantial amount of opposition to the tollway in the inner city. (Photo: September 2002)

County Commissioners Court, with Judge Jon Lindsay absent, voted 4-0 to oppose the toll road on August 7. Upon his return, Lindsay introduced a measure in support of further studies which passed on a 3-2 vote on August 25.[115]

During 1981 and 1982 positions hardened on both sides of the issue. The classic battle that had played out all across the nation during the 1970s took shape once again. Low-income inner-city interests opposed the facility and the displacements it would cause, instead favoring mass transit solutions. The inner-city interests perceived the tollway as benefiting affluent, predominantly white suburban areas. A Hispanic editorial columnist for the *Houston Post* summarized the feelings of many when he wrote, "There is perhaps nothing more galling in covering minority community affairs than to watch as county and city officials use their positions to benefit the well-to-do affluent suburbs rather than improve conditions in barrios like the northside." Lining up in favor of the tollway were politicians and civic groups representing suburban areas, the Houston Chamber of Commerce, and the project's staunchest supporter, Jon Lindsay.[116]

With all the controversy raging, Lindsay pointed out, "It's an ideal corridor. The benefits [of the toll road] are tremendous." Since it followed a railroad corridor that had an adjacent high-voltage easement for much of the route, the displacement of property along the corridor would be minimal. It truly was an ideal corridor for a new transportation facility. For its 21.6-mile (34.6 km) length, the project would displace only 70 residences, 54 businesses, and two churches. Of those 70 residences, 6 were mobile homes and 4 were abandoned houses. Inner-city neighborhoods would get crossings over the busy railroad track, better connecting the neighborhoods separated by the existing corridor.[117]

Unusual for Houston: The northern section of the Hardy Toll Road does not have frontage roads, preventing the proliferation of commercial development. Parts of the tollway are like a parkway as it passes through forests. (Photo: March 2003)

By late 1982 it appeared that the opposition would succeed in killing the project. On November 9, 1982, Houston Mayor Kathy Whitmire publicly stated her opposition to the project, saying that no city funds should be used and resources should instead be channeled toward a rail system: "I oppose building the toll road because it would compete with the rail system and in fact preclude the rail system from being built, and I want to see the rail system built." The TTA's detailed study of the corridor was now complete and concluded that the tollway was feasible, but on November 10, 1982, the *Houston Post* reported that the TTA was planning to drop the project due to lack of support from the city of Houston and Metro. About the same time, Lindsay was publicly stating that if the TTA would not build the project, Harris County would attempt to proceed on its own.[118]

It was time for Jon Lindsay to step up to the plate and get the tollway built. First, however, new legislation would be needed to enable the county to take over the project. That came with Texas Legislature Bill SB970, signed by Governor Mark White in June 1983. In the meantime, the Metropolitan Organization and its supporters continued efforts against the toll road. On April 28 an anti-tollway crowd estimated at 1,200–2,000 attended a meeting featuring Governor Mark White.[119]

Lindsay pushed the Harris County Toll Road Authority forward as fast as he could. A $900 million bond referendum for the creation of the Harris County Toll Road Authority (HCTRA) was placed on the ballot for a special election on September 13, 1983. On August 24 Houston City Council endorsed the Hardy Toll Road on a 13-2 vote. The Houston Board of Realtors sent a letter to its membership in support of the bond issue, stating, "Its passage is vital to the solution of the horrendous traffic situation in our city and county." Most Harris County voters shared the view of the realtors and were influenced by the severe traffic congestion afflicting the area. The $900 million bond issue passed with 69.7% of the 137,161 votes cast. The Hardy Toll Road and the western Beltway 8 tollway would be built.[120]

Building the Tollway

The design of the Hardy Toll Road was based on the December 1982 detailed design report which had been prepared by HNTB consultants for the TTA. The recommended feasible section was a 21.6-mile (34.6 km) segment from Loop 610 to IH 45 near the Harris-Montgomery County line. At the south end the report recommended terminating the tollway at Loop 610 due to the high number of displacements and high cost of extending the facility into downtown, which was estimated at $115 million in 1982, approximately 204 million in 2003 dollars. The report concluded that the existing freeways and streets between downtown and Loop 610 would be adequate to handle the traffic. On the north end an extension into Montgomery County had been considered and mapped out in detail but was ruled infeasible. The tollway would connect with IH 45 just south of the Harris-Montgomery County line, keeping the entire project within Harris County.

A ground breaking ceremony for the Hardy Toll Road was held in September 1984. The north section from Beltway 8 to IH 45 opened on September 20, 1987. The south section from Beltway 8 to Loop 610 opened on June 28, 1988.

Looking for Drivers

It didn't take long before the traffic pattern on the Hardy Toll Road became evident. In the morning there was a surge of traffic going into Houston. In the evening there was a surge of traffic headed for the suburbs. The rest of the time the Hardy Toll Road was nearly empty. In 1990 the head of the Houston TxDOT office observed, "The problem with the Hardy is that in the off-peak hours there is almost nothing there." During off-peak hours motorists preferred to use the toll-free North Freeway. A commuter traffic load would not be enough to pay the bills.[121]

By 1989, just a year after the completion of the Hardy Toll Road and the first segment of Beltway 8, it was unclear if revenue on the Harris County Toll Road system would be sufficient to cover bond payments. Revenue was running more than 50% lower than 1984 estimates. The need for a subsidy from the Harris County general tax fund was a distinct possibility. A financial crisis was averted by a dramatic increase in systemwide toll collections in the 1990s. However, the increase in revenue was driven by the phenomenal success of the western sections of the Sam Houston Tollway. In 1998 the Hardy Toll Road still was not paying for itself, according to a July 1998 analysis of Hardy Toll Road revenue by the *Houston Chronicle*. The report showed annual expenses (bond payments and operating costs) of \$30.1 million and toll revenue of \$20.4 million, indicating that a 48% increase in toll-paying traffic was needed to close the revenue gap. By 1998 HCTRA had achieved a very strong financial position, so the shortfall was easily covered. Patronage of the Hardy Toll Road increased 50% from 1997 to 2001, and the Hardy Toll Road was at or near financial self-sufficiency by 2002. HCTRA operates its finances on a systemwide basis and does not provide detailed financial information for each facility.[122]

In the late 1990s Harris County was able to move forward with projects to fully integrate the Hardy Toll Road into Houston's transportation system. The airport connector was studied as part of the original 1982 design study but was not included in the original construction of the toll road. For most of the 1990s the project was frequently discussed but did not move forward. Finally on August 25, 1998, Harris County awarded the first construction con-

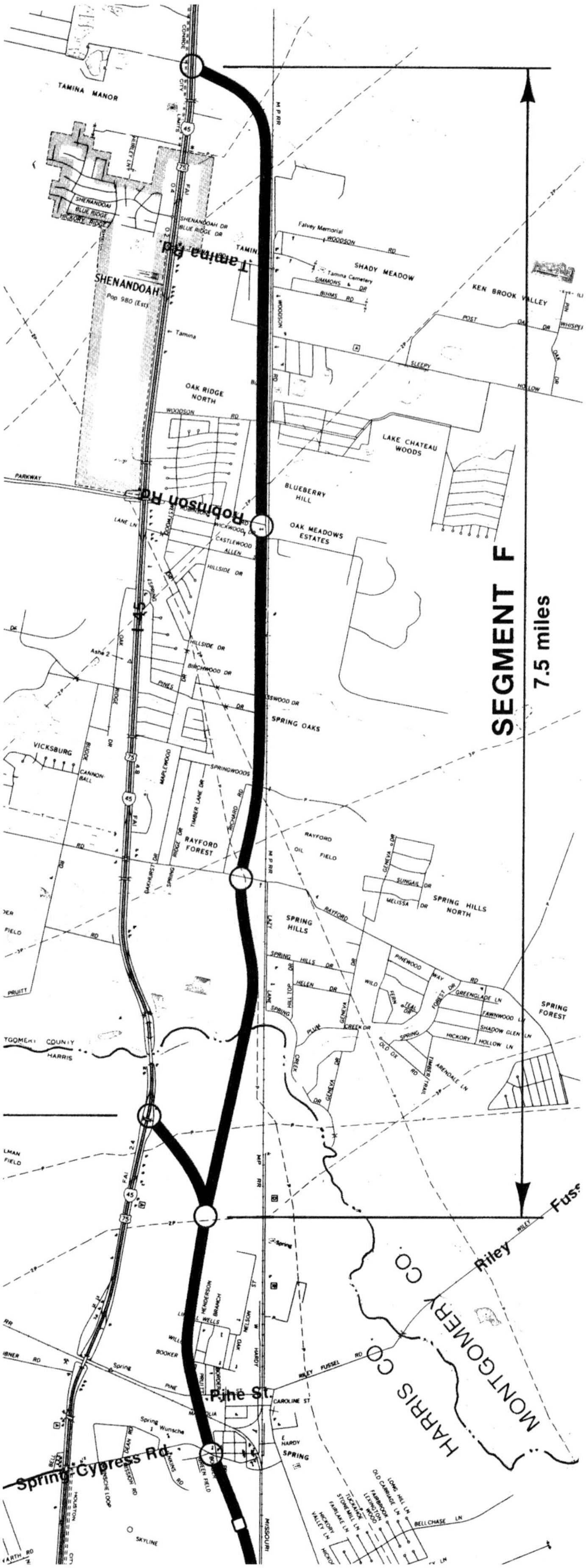

The unbuilt: The 1982 engineering study for the Hardy Toll Road considered a section of tollway north of the present terminus. The study determined that the north section was financially infeasible. The northern extension remained an item under study in 2003. If constructed, the northern extension is expected to follow a different alignment than the one proposed in the 1982 study. (Source: *Hardy Tollway Preliminary Engineering Feasibility Study.*[123])

tract for the airport connector. The $41.7 million project was opened on January 28, 2000, and features a full interchange with connectors at the intersection with the Hardy Toll Road. Also in the late 1990s work began to expand lane capacity on the Hardy Toll Road north of Beltway 8. Sections experiencing congestion were widened to three lanes in each direction. A direct connector ramp at Beltway 8 is under construction in 2003 and is scheduled for completion in 2004.

In the early 1990s there was an effort to extend the Hardy Toll Road into downtown Houston. The project was declared dead by Lindsay in September 1994 because Metro was not willing to make a substantial financial contribution to the project. The project came back to life in 1999 when HCTRA initiated new engineering studies for the corridor. A preliminary design was complete by October 2000, and in 2001 TxDOT agreed to contribute $17 million to the estimated $101 million project cost. As of 2003, plans were being developed to construct the Hardy Toll Road extension as a toll-free facility. Delays in right-of-way acquisition and railroad relocation have pushed back the expected start of construction to 2005, with completion around 2007. In May 2003 HCTRA launched a new feasibility study of the far north section of the tollway, originally considered in 1982. The potential 10-mile (16 km) extension would start at the present northern terminus and extend north to Conroe.[124]

The Hardy Toll Road has certainly required more patience and nurturing than the go-getting Sam Houston Tollway. As HCTRA better integrates it into Houston's freeway system and urbanization north of Houston continues, the Hardy Toll Road will become an increasingly important part of Houston's transportation network.

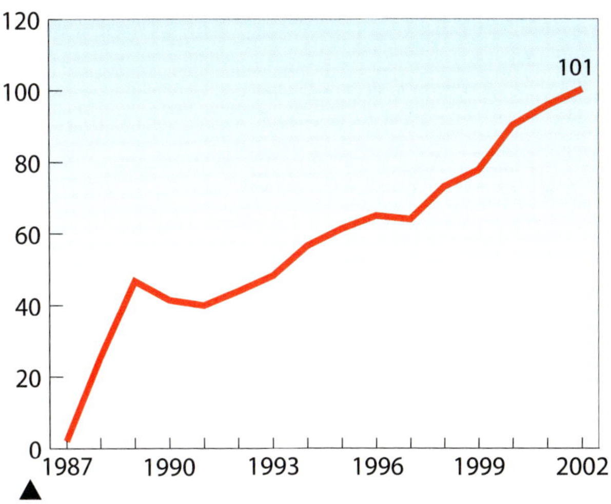

Hardy Toll Road Average Daily Vehicle Transactions, thousands
Sum of traffic for north and south main toll plazas
Data: Harris County Toll Road Authority

Revenue laggard no more: During its early years the Hardy Toll Road was known for its lack of traffic and toll revenue that didn't meet projections. Patronage steadily increased during the 1990s and by 2002 the tollway had approximately reached financial self-sufficiency.

Airport connector: The Hardy Toll Road airport connector opened in January 2000. This view looks west with Hardy Toll Road interchange in the distance. (Photo: November 2002)

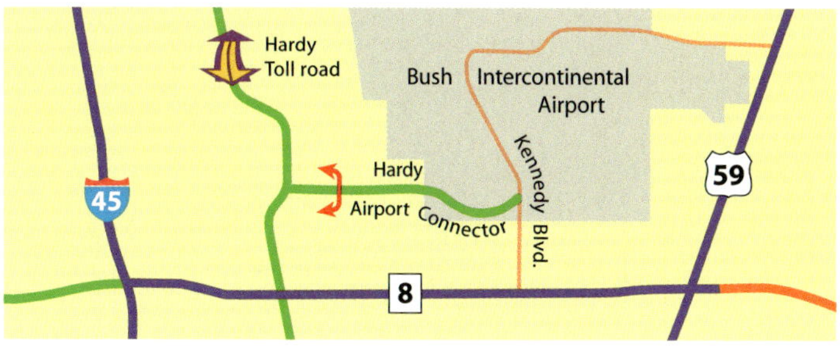

Crosby Freeway, US 90

In 1954 Houston's core freeway network was defined and construction progressed quickly on most of the freeways. But there was one freeway in the original plan that was left behind in the construction spurt of the 1960s: the Northeast Freeway. Nearly 50 years later the Northeast Freeway, which was renamed the Crosby Freeway in 1988, still remains unfinished. The Crosby Freeway may have been left behind, but it was not forgotten. The freeway outside Beltway 8 was completed in 1991, and most of the remaining section is scheduled for construction before 2010, although much of it may have frontage roads only.

The first proposal for Houston's freeway master plan, presented to the Texas Transportation Commission in July 1953, showed the Northeast Freeway following the alignment of McCarty Road and the present-day Beaumont Highway (see map on page 13). The original plan also extended the freeway inside Loop 610 to a terminus on IH 10. The final plan, which was formulated in early 1954, placed the Northeast Freeway in its general present-day alignment, beginning the freeway at the intersection of Loop 610 East and IH 10. The Northeast Freeway went into a state of dormancy after its initial inclusion in the freeway plan. There was very little development in northeast Houston, so there was no pressing need to move forward with construction. In 1961 a TxDOT study recommended the long-term upgrading of US 90 to a controlled-access freeway.[125]

Crosby Freeway	
Previous designation	Northeast Freeway (1956-1987)
Designated as freeway	1966
First segment open	February 25, 1991
Freeway complete	Estimated after 2010
Max traffic volume, 2001	26,000 vehicles per day
Future construction	Frontage roads and main lanes inside Beltway 8; connections at interchange with IH 10 and Loop 610

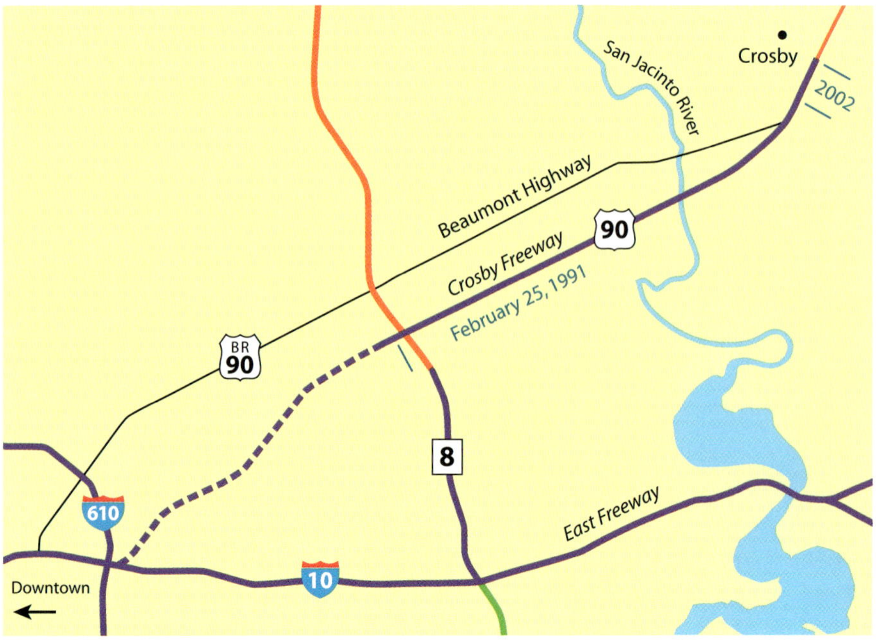

In 1966 planning began for the Northeast Freeway. The Texas Transportation Commission designated it as a controlled-access freeway in March 1966. A public hearing on the alignment of the proposed freeway was held on April 18, 1966, and based on public input, the alignment was adjusted and approved by the Transportation Commission in October 1966. TxDOT and Harris County entered into an agreement in November 1971 in which Harris County would be responsible for right-of-way acquisition and TxDOT would be responsible for construction. By 1972 local interests were pushing for the construction of the Northeast Freeway from Crosby to the present-day Beltway 8 because of the need for a modern crossing at the San Jacinto River. One of the factors that slowed progress of the freeway, lack of development in the region, turned out to be very beneficial for the design. A wide right-of-way corridor was easily acquired.[126]

In 1975 it looked like construction would begin when the Texas Transportation Commission authorized funds for construction of the frontage roads near the freeway's beginning at the interchange with Loop 610 and IH 10. But in March 1975, funds were diverted from the Northeast Freeway to the South Freeway (SH 288). TxDOT was at the peak of its financial crisis in 1975 and a reevaluation of the department's priorities placed the Northeast Freeway near the back of the line for construction funding. The Northeast Freeway once again went into dormancy.[127]

The Northeast Freeway came back to life in the late 1980s. In 1987 funding was finally in place to begin construction from Beltway 8 to the city of Crosby, about nine miles (14 km) east. In late 1987, the Crosby-Huffman Chamber of Commerce began lobbying to change the name of the freeway from the Northeast Freeway to the Crosby Freeway to increase awareness of the community of Crosby and help promote development. Harris County agreed to the request in December 1987. On February 25, 1991, the first section of the Crosby Freeway was opened. The freeway was constructed without frontage roads for most of its length due to the lack of development along the freeway corridor. In 2002 a short extension near Crosby was completed.[128]

The need for the section of freeway inside Beltway 8 was minimized by the completion of the Beltway 8 main lanes between the Crosby Freeway and IH 10

Crosby Freeway: This view looks east along the semi-rural freeway. (Photo: September 2002)

in 1994, which provided an alternate freeway route into Houston. Inside Beltway 8 only a small section of frontage roads exists along a section of the freeway that shares its route with the C. E. King Parkway. The next wave of construction on the Crosby Freeway is scheduled to begin in 2005 and will complete the freeway between Beltway 8 and Loop 610. Initially frontage roads will be constructed. Someday, probably between 2010 and 2020, the main lanes will be constructed and the Crosby Freeway will be completed.

Awaiting the final push: The Crosby Freeway grinds to a halt just inside Beltway 8. This section has stood idle since 1991, and it could be 10 years or more before the main lanes are pushed westward toward Houston. (Photo: June 2003)

Tomball Parkway, SH 249

In May 1988 a group of Houston's key transportation policymakers met secretly in a downtown conference room. One after another, the top officials committed millions of dollars to a new transportation initiative to benefit a key local interest. An hour and a half later, when the dealing was done, \$235 million had been put on the table. Included in the bag of goodies was acceleration of a brand-new, \$253 million freeway.[129]

Why the secrecy? Why the big money? To these officials, nothing less than the economic future of Houston was at stake. Houston's economy had been devastated by the collapse of oil prices in the mid-1980s, but there was one shining star on the Houston economic landscape: Compaq Computer. Since its founding in February 1982, Compaq Computer had become one of the fastest-growing startups in the history of American business, generating about 4,000 new jobs in Houston by 1988. Compaq had even bigger plans for expansion, but there was no assurance that the new production would stay in Houston. Transportation was a big issue for Compaq, which was served by inadequate country roads in far northwest Harris County. The new freeway was the crown jewel in a comprehensive incentive package designed to keep Compaq's expansion in Houston. Key local officials kept their efforts secret so as not to tip off competing cities which were also offering packages. On June 1, 1988, top local officials offered their deal to Compaq. Compaq officials were amazed. They would stay in Houston, and Houston would get an all-new freeway.

Tomball Parkway	
Previous designation	FM 149
Designated as freeway	1988
First segment open	1990
Freeway complete	Estimated after 2010
Max traffic volume, 2001	115,000 vehicles per day
Future construction	Extend freeway north, ultimately to Navasota; possible toll road extension inside BW 8

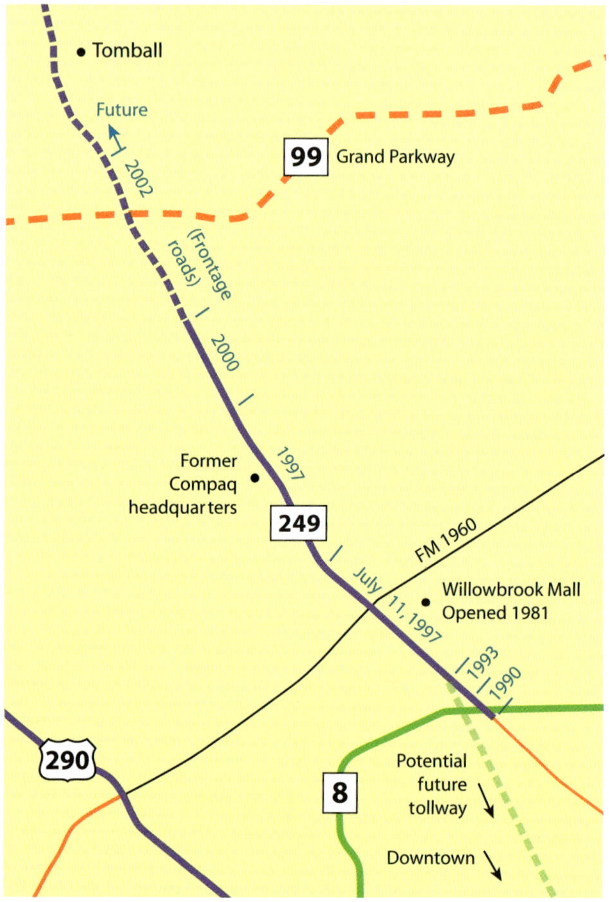

Origins

The Tomball Parkway corridor had historically been Farm-to-Market Road (FM) 149, a rural road crossing through the pine forests and pastures of northwest Harris County. Approximately halfway between the US 290 Northwest Freeway and IH 45 North Freeway, it seemed to be an ideal location for a freeway. But all the official transportation plans had passed over the FM 149 corridor, never designating it as a freeway. The only document identifying a need for a freeway in the corridor was the city of Houston's *Houston Preliminary General Study Plan for 1990*, an artist's depiction of potential future roads, parks, and land usage published in 1968. The map showed a freeway starting at Beltway 8 and proceeding northward just east of FM 149. The FM 149 freeway was gone from a similar map published in 1972.

During the 1970s, as financial woes forced the cancellation of several planned freeways, there was no chance of adding new freeway routes to the planned system. In February 1982 a key event occurred that would ultimately lead to the construction of the Tomball Parkway freeway: Compaq Computer was founded. Compaq initially focused on portable personal computers which were compatible with the IBM standard. The first bulky portable computers were shipped in January 1983. Over the next few years Compaq delivered record-setting revenue growth and in 1986 joined the Fortune 500 faster than any company in history at the time. The initial phase of the Compaq main campus on FM 149 was completed in 1986.[130]

The explosive growth of Compaq Computer and expanding development in the area were soon putting FM 149 under heavy traffic pressure. By the time officials were preparing the incentive package in 1988, it was clear that a freeway would be needed to meet traffic demand. One of the key participants in the May 1988 meeting where the transportation incentive plan was developed was Houstonian and Texas Transportation Commissioner John Butler. Butler quickly secured official state approval of the 19-mile (30 km), \$253 million freeway on June 28,

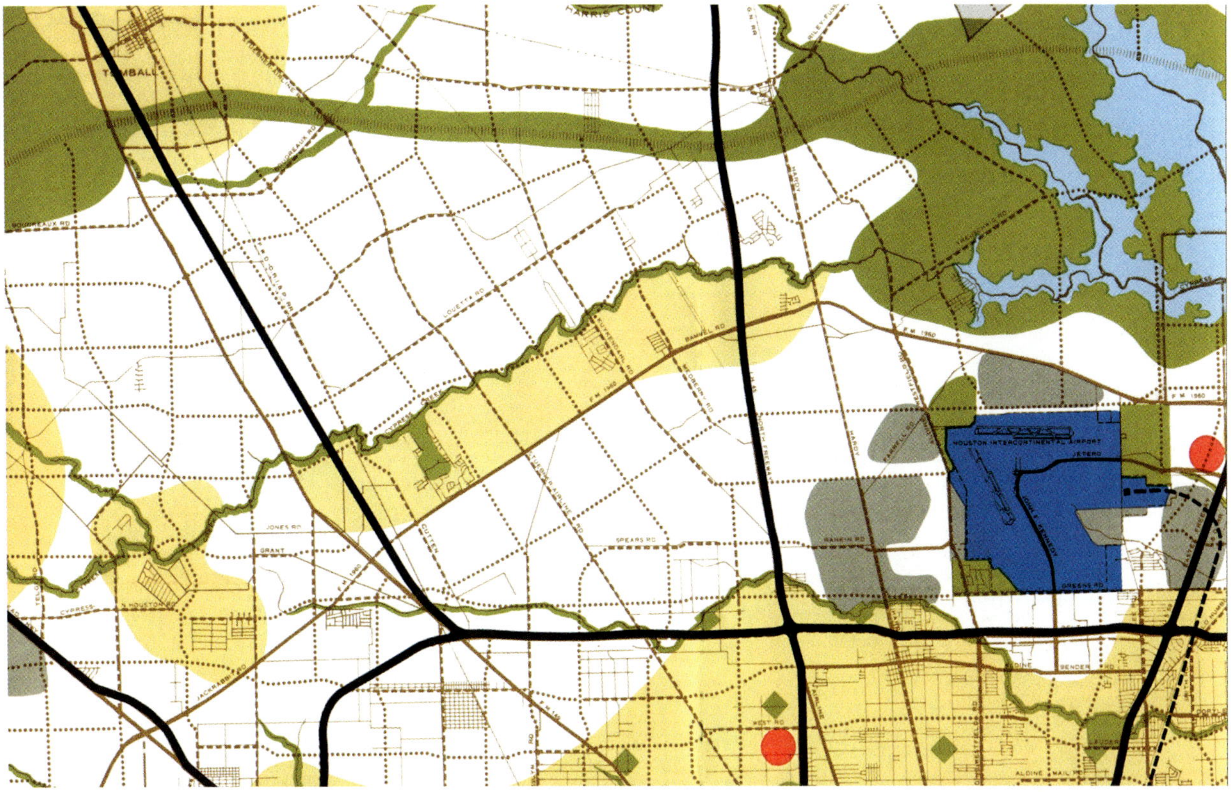

First signs of the Tomball Parkway: This 1968 map published by the city of Houston titled *Houston Preliminary General Study Plan for 1990* appears to be the first recognition that a freeway would be needed in the corridor. The 1968 map was not an official transportation planning document, but an artist's depiction showing potential future roads, parks, and land use. The freeway in the SH 249 corridor was gone from a similar 1972 map.

1988. The freeway was envisioned as part of a freeway corridor that would ultimately extend all the way to College Station, 100 miles (160 km) northwest of downtown Houston. In conjunction with the official freeway designation, the route number for the freeway was changed to State Highway (SH) 249. The corridor was already known as the Tomball Parkway, and this designation was retained even though the SH 249 freeway would have continuous frontage roads and few, if any, parkway-like qualities.[131]

Construction could not begin immediately, however. As with all projects, it would need to wait in line for available funds. Also, a substantial amount of right-of-way acquisition would be necessary to clear the corridor for the freeway, which generally required a 350-foot-wide (107 m) right-of-way strip. In the vicinity of the FM 1960 intersection, many businesses and strip shopping centers would need to be purchased and cleared.

Before the freeway could be built, there was considerable pain for local motorists. Traffic on the Tomball Parkway continued to increase rapidly and by the early 1990s the intersection of the Tomball Parkway and FM 1960 had become the most severely congested in Houston. Freeway construction began in conjunction with the Sam Houston Tollway main lanes in the late 1980s. The main lanes at the Sam Houston Tollway intersection were completed in 1990. A short section of the freeway, an overpass just north of Beltway 8, opened in 1993. Then after four agonizing years of construction, a long section of freeway including the FM 1960 intersection opened on July 11, 1997. The notorious traffic monster at FM 1960 and the Tomball Parkway had been tamed. In its place was a wide, modern freeway and a three-level interchange with FM 1960 passing below ground level.[132]

In the following years the freeway pushed northward one section at a time. As of 2003, about eight miles (13 km) of freeway main lanes and additional two miles (3.2 km) of frontage roads were open to traffic. A study to determine the alignment of the freeway north of Tomball was completed in 2002. Previously in 1999 the alignment of the future freeway to Navasota, approximately 80 miles (128 km) northwest of downtown Houston, had been identified.

In May 2002 the freeway's reason for being—Compaq Computer—ceased to exist as an independent company when it was purchased by Hewlett Packard in a highly controversial corporate takeover. As of 2003, it appeared that the former Compaq operations would remain an important part of Hewlett Packard. Whatever the long-term viability of the former Compaq operations may be, the Tomball Parkway is now the backbone of northwest Harris County and will help the region absorb any corporate downsizing. Construction of three direct connectors at the Beltway 8 interchange began in late 2002. Work on a bypass around Tomball is expected to begin in 2004, with the initial phase consisting of frontage roads only. Long-term plans of the Harris County Toll Road Authority show a potential toll road corridor aligned along a railroad track which would extend the Tomball Parkway inside Beltway 8 to Loop 610.

Reason for being: Compaq's phenomenal success and explosive growth after its founding in 1982 prompted the construction of a large corporate campus in 1986 along FM 149 (campus shown above). The rural FM 149 was soon overwhelmed, and plans for a freeway were approved in 1988. This section of freeway opened in 1997. Compaq computer was purchased by Hewlett Packard in 2002 in a highly controversial corporate takeover, but the Houston campus appeared to remain an important part of Hewlett Packard's operations. (Photo: May 2003)

(Opposite page) **Country crossroads no more:** The upper photo shows the intersection of FM 149, the predecessor to today's SH 249 Tomball Parkway freeway, and FM 1960 in October 1980. The view looks east with FM 149 crossing from lower left to upper right. FM 149 was still a two-lane rural roadway with some minor improvements in place at the FM 1960 intersection. But big changes were on the way. The construction of Willowbrook Mall was in progress, and Compaq Computer was founded in 1982. The success of Compaq Computer fueled a boom in the corridor and by the early 1990s the intersection of FM 149 and FM 1960 was the most congested in Houston. The SH 249 freeway at this location, shown in the lower photo, opened on July 11, 1997. (Photos: upper, Texas State Library and Archives Commission; lower, September 2002)

(Right) This view looks north along SH 249 with the FM 1960 intersection in the foreground. There are two pairs of braided ramps just north of FM 1960. (Photo: May 2003)

Westpark Tollway

Near the beginning of Houston freeway time, when the freeway system was being created, planners drew a line on the master plan map all the way along the Westpark corridor to far west Houston. Soon afterwards most of that line was erased when the Southwest Freeway alignment was finalized. Many years later a vision of the Houston of the future showed a freeway in far west Houston, roughly along the Westpark corridor. That line was also soon forgotten. Years after that, the idea came back to life again and was officially studied, but then quickly faded away. But the dream of that west Houston freeway never died. Fifty-one years after that first line was drawn on the master plan map, motorists will take their first drive on the Westpark Tollway.

From the early 1950s to the mid-1990s, the idea of a major transportation facility along the Westpark corridor seemed to be a dream that simply couldn't get enough momentum to move beyond the concept stage. Even mass transit options for the corridor failed, including a heavy rail line defeated by voters in 1983 and a proposed monorail line in 1991. But finally in the 1990s, a series of events opened a window of opportunity for building the tollway, and barely—just barely—the tollway managed to seize the opportunity.

The Westpark Tollway is a very un-Houston-like facility. It is very narrow, with only four lanes for its entire length, and it is squeezed onto the tightest possible right-of-way. It seems more like something that would be built in a high-density urban environment such as New York City. But the Westpark Tollway probably represents the wave of the future for all-new transportation facilities in Houston. Small tollways with no more than six lanes will be squeezed into narrow corridors where there are few, if any, environmental issues. Funding will come from many sources, but toll revenue will be the key source. Size will be kept small to keep costs manageable. But is something small better than nothing? For west Houston interests which fought long and hard to make the tollway a reality, the answer is surely yes.

Origins

When Houston's freeway master plan was being formulated in 1953, the original plan presented to the Texas Transportation Commission that July showed a freeway along today's Westpark corridor all the way to far west Houston, following the San Antonio and Aransas Pass (S.A. & A.P.) Railroad and adjacent Alief Road. By late 1953, however, plans for the Southwest Freeway were being formulated, and by early 1954 the preliminary Southwest Freeway alignment was defined, roughly following the railroad to just west of Loop 610 then veering southward into Sharpstown, a huge real estate development that was about to be launched. The Alief Road freeway to far west Houston was not part of any major intercity highway system, and perhaps more importantly, it did not serve any powerful real estate constituencies that could promote and support it. So the freeway went southwestward to serve the Sharpstown community, and west Houston was left without a freeway.[133]

The next sign of life for a west Houston freeway came in 1968 when the city of Houston published a map titled *Houston Preliminary General Study Plan for 1990*, an artistic depiction of the existing and proposed roads, freeways, transit, parks, lakes, and development around Houston. The map showed the west Houston freeway starting at Beltway 8 just north of the railroad corridor, then proceeding west and southwestward to the Grand Parkway south of the railroad corridor. This freeway appears to have been just a fleeting line on a map, however. The next revision of the map, published in 1972, did not

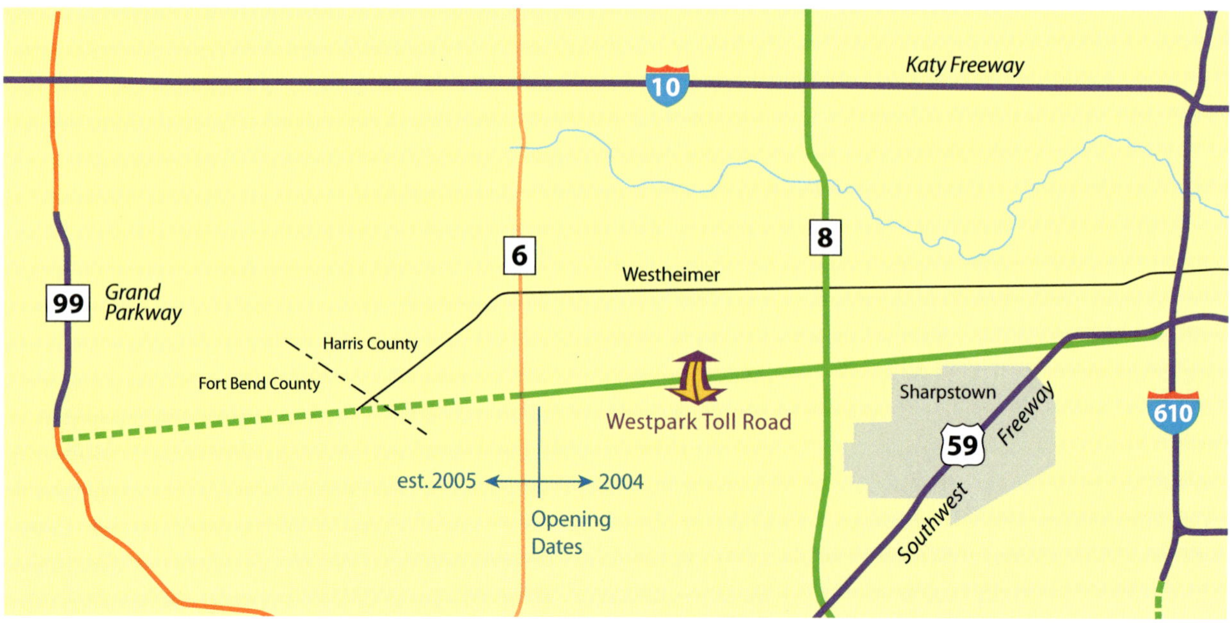

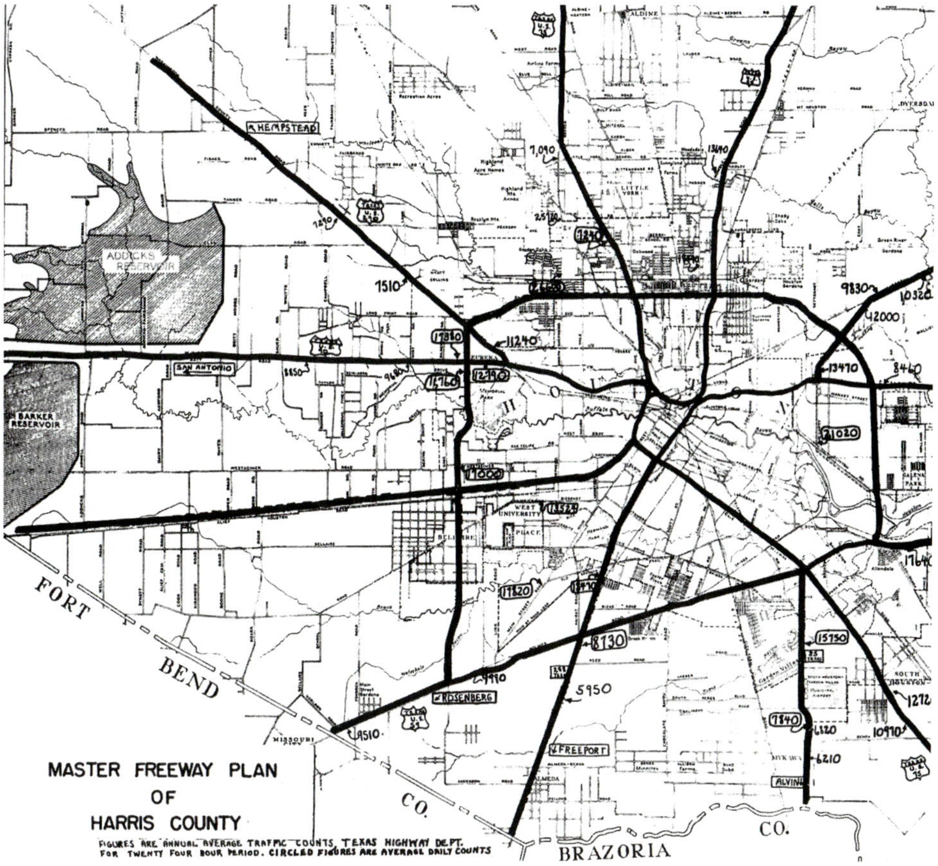

An early casualty: This 1953 map of the proposed Houston freeway system shows a freeway along the full length of the Westpark corridor. By early 1954, the Westpark Freeway in west Houston had been deleted and replaced with the Southwest Freeway, which veered away from the Westpark corridor just west of Loop 610 and bisected the large Sharpstown real estate development.[134] (Source: Houston Chamber of Commerce/Greater Houston Partnership)

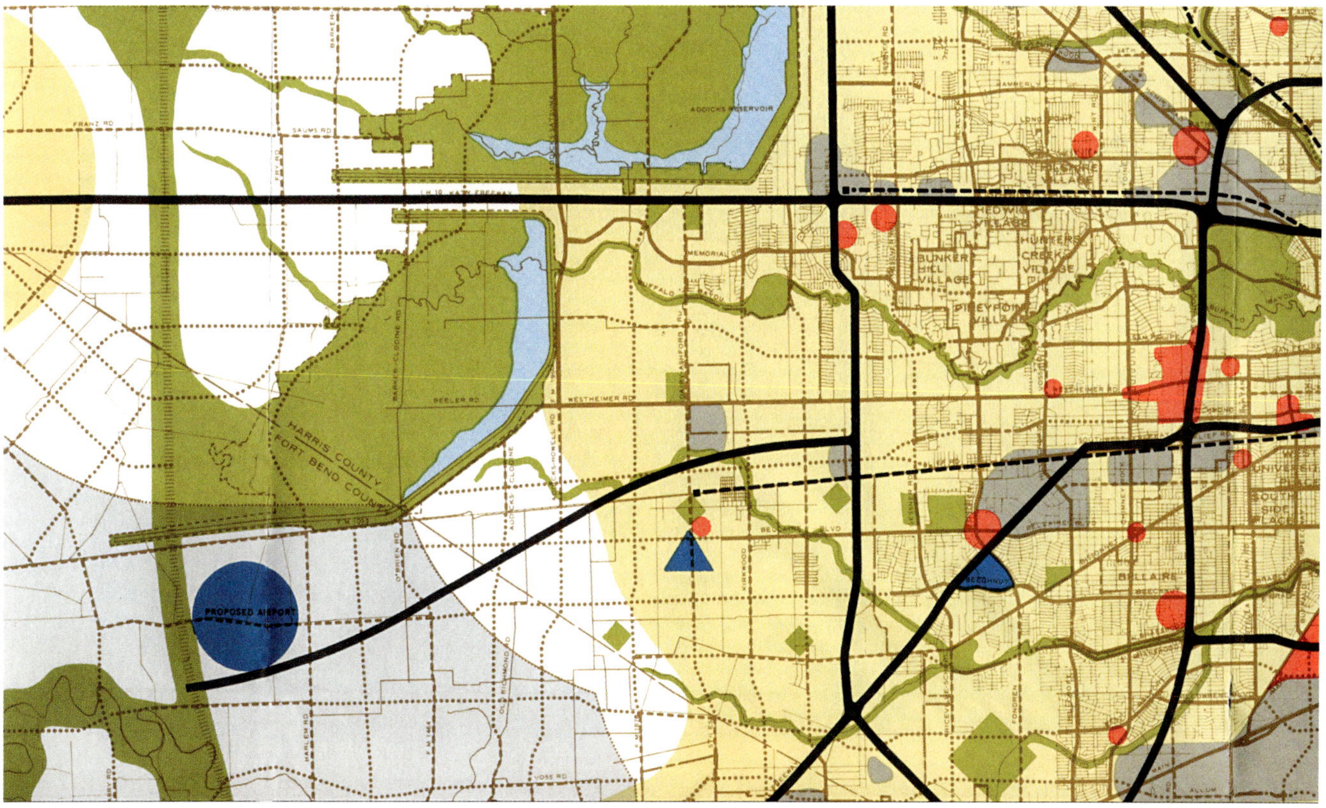

1968 freeway vision: A map published by the City of Houston Planning Department in 1968 titled *Houston Preliminary General Study Plan for 1990* showed a freeway in west Houston extending from Beltway 8 to the Grand Parkway, which is depicted by a linear park strip. The proposed alignment did not follow the present-day Westpark corridor but served the same general area. The blue circle near the western end of the contemplated route is identified as "proposed airport." The west Houston freeway was gone from a similar map published by the city of Houston in 1972.

Pre-tollway Westpark: The above view looks west along the Westpark Drive corridor from the Southwest Freeway overpass in July 2001, just before tollway construction began. The Harris County Toll Road Authority acquired 50 feet (15 m) of the 100-foot (30 m) railroad corridor, roughly the section between the railroad bed and Westpark Drive on the right. The left views show the corridor before the railroad tracks were removed, with the upper view looking east at Beltway 8 and the lower view looking east at Old Westheimer Road. (Photos: upper, July 2001; left, November 2000)

show the freeway, and no other agency took an interest in the proposed freeway.[135]

The planners in 1953 and 1968 were right about one thing. They anticipated the need for a major transportation facility in west Houston, and their predictions proved to be correct. Heavy residential and commercial development occurred in west Houston starting in the 1970s, and the arterial street network became incapable of handling the traffic load. The need for a freeway-type facility continued to grow, but constrained transportation resources prevented any meaningful plans from being formulated.

In 1979 the idea of a Westpark Tollway was given its first serious consideration. The Texas Turnpike Authority authorized a preliminary study of three corridors in Houston: the Hardy corridor, the Westpark corridor, and the Harrisburg corridor, location of the cancelled Harrisburg Freeway. The study found that the Hardy corridor warranted further study, the Harrisburg corridor was infeasible as a tollway, and the Westpark corridor was a marginal project. For the Westpark corridor, the study envisioned a

The golden shovels go to work: Officials gather on June 21, 2001, for the official ground breaking for the Westpark Tollway. Second from the right is Wesley Freise, who at the time was executive director of the Harris County Toll Road Authority. More than anyone else, Freise was responsible for moving the project forward. Third from the right is Harris County Judge Robert Eckels, who also played a key role in getting the tollway built. (Photo: Harris County Toll Road Authority)

six-lane toll road on a 100-foot-wide (30 m) right-of-way, extending from the US 59 downtown split at Spur 527 to SH 6, a distance of 15.5 miles (25 km). Construction cost was 158 million and total cost was 244 million in 1979 dollars, approximately 335 million and 517 million in 2003 dollars. The study concluded that the Westpark corridor estimated revenues were "below the levels generally considered adequate for successful financing," but the project would be able to meet its financial obligations over its lifetime and there was potential to reduce the cost substantially by using the railroad right-of-way and parts of the Westpark Road right-of-way. The study concluded that the project would warrant further study if the cost could be reduced or if financial assistance could be provided by other agencies. Around the same time, proposals for improving Westpark Road and adding bus lanes were being considered, still keeping it as an arterial street.[136]

The Westpark Tollway went into dormancy during the 1980s, and the focus turned to mass transit for the corridor. In June 1983 Houston's transit agency, the Metropolitan Transit Authority (Metro), placed a $2.35 billion bond issue on the ballot, with the bulk of the money earmarked for an 18.5-mile (30 km) heavy rail line with elevated structures along Westpark and a downtown subway. Voters defeated the bonds with 61% opposing. In the late 1980s Metro began new efforts to build a rail line. This time the agency was promoting a monorail line that would generally follow the Westpark corridor. In March 1991 the Metro board approved the monorail project and began negotiations with engineering and construction firms. The monorail became highly controversial and was a major issue in the 1991 mayoral election. Mayoral candidate Bob Lanier, former chairman of the Texas Transportation Commission and strong highway advocate, actively campaigned against the project. When Lanier won the election, the monorail was dead. Lanier then set the wheels in motion for the events that would ultimately lead to the construction of the Westpark Tollway.[137]

Setting the Wheels in Motion

When he took office on January 2, 1992, Mayor Bob Lanier promptly replaced three of the five city of Houston representatives on the Metro board of directors, breaking the previous pro-rail majority on the board. Metro had accumulated about $650 million in a reserve fund which it intended to use for the $1.2 billion monorail line. Lanier had plans to spend nearly all of the reserve funds on projects that he felt would have more benefit for Houston, including improved police protection and increased spending on roads, seemingly opening the door for a Metro

Beam placement: This view shows construction of the connector ramp from the Westpark Tollway to the Southwest Freeway, July 23, 2002.

contribution to the tollway.[138]

By February 1992 Metro had entered into negotiations with four railroads in the Houston area to buy or lease rights-of-way for a potential commuter rail system. Included in the negotiations was Southern Pacific, owner of the railroad corridor along Westpark. The idea of a Westpark Toll Road once again came to life. The chairman of the Metro board of directors, Billy Burge, was touting the corridor's potential as a toll road. At the end of December 1992, Metro and Southern Pacific approved a deal in which Metro would purchase the railroad corridors along Westpark Road and Hempstead Road in northwest Houston and obtain operating rights on other tracks for a price that could reach $113 million, depending on final details. Metro subsequently purchased the 58 miles (93 km) of track on the Westpark corridor for $45 million but declined to complete the other purchases in the 1992 agreement, leading to lawsuits from both sides in March 1994. The final price for the Westpark corridor land was $72 million when the litigation was finally settled in 1997.[139]

In the meantime, west Houston business interests had proposed a transit-friendly toll road for the corridor in 1993 in an effort to persuade Metro to donate the land for toll road use. Harris County Judge John Lindsay was of the opinion that the Harris County Toll Road Authority (HCTRA) should be the lead agency for any toll road. However, HCTRA was not yet financially strong and was focusing its attention on the purchase of the Beltway 8 ship channel bridge and the completion of the southern section of the Sam Houston Tollway. Metro was developing plans to build a one-lane reversible transitway for buses and high occupancy vehicles on the land and by 1995 was in the process of finalizing the plans. Metro authorized a detailed study of the feasibility of a toll road in 1996. The preliminary results of the study concluded that the toll revenue alone would not be sufficient to fund the project and large contributions from other agencies, most likely Metro, would be needed to make the project work. The study also indicated that the most financially viable option was a conventional toll road, not a toll-transit hybrid that some envisioned. Also complicating the situation were funding agreements with the Federal Transit Administration for the planned dedicated bus transitway. Metro began to solicit bids for the one-lane reversible transitway in late 1997 for a projected contract award in February 1998. Term-limited Mayor Bob Lanier would have to leave office at the end of 1997 and his successor, Lee Brown, was pro-rail. Brown immediately changed the composition of the Metro board of directors when he entered office in 1998, making it much more rail-oriented than it was in the Lanier era. Contributions to the tollway from Metro now seemed unlikely. Although the planned transitway was designed so that it would not preclude the addition of toll lanes in the future, it appeared that the Westpark Tollway was perhaps finally dead.[140]

West Houston business interests mounted a last-ditch, intense effort to stop the transitway construction. They succeeded in obtaining a nine-month delay in early 1998. The make-or-break period for the Westpark Tollway was just ahead. Harris County Judge Robert Eckels provided key support for the project during this critical period, and HCTRA executive director Wesley Freise led negotiations

Westpark Tollway construction: This view looks east toward the Southwest Freeway interchange in March 2003.

between HCTRA and Metro. Eckels was a strong advocate of road and tollway construction, and as chairman of the area's Regional Transportation Council he was the most powerful official in the Houston area on transportation issues. In spite of the modest projections for usage of the tollway, an estimated 26,000 vehicles per day for a projected opening in 2001, Eckels felt that a tollway would not incur any additional cost to Metro but would provide greater benefit to both west Houston and Metro. By 1998 HCTRA had achieved a very strong financial position and was capable of taking on financially marginal projects.[141]

Negotiations between HCTRA and Metro continued into 1999. In September 1999 a deal was reached. Metro would sell a 50-foot (15 m) strip of the Westpark corridor, half of the 100-foot (30 m) corridor width, to HCTRA. HCTRA would pay $14.3 million for 13 miles (21 km) of the 50-foot strip. Metro would retain 50 feet for future transit use, which was envisioned as a light rail line. On November 18, 1999, the Metro board of directors formally approved the land sale. The Westpark Tollway would be built.

Building the Tollway

It was a near-miracle that the Westpark Tollway would be built. Only one other Houston transportation facility, the Fort Bend Parkway Tollway, has had a longer and more tortuous route to construction. In November 2000 voters in Fort Bend County southwest of Houston approved bonds to extend the Westpark Tollway all the way to the Grand Parkway. The 1953 concept for a freeway along the entire Westpark corridor and the 1968 vision for a limited-access facility to the Grand Parkway would be realized.

A ground breaking ceremony for the tollway was held on June 21, 2001. The Westpark Tollway from Loop 610 to the Harris County line had an approximate construction cost of $265 million. By Houston standards, the Westpark Tollway has a very unusual design. It is very small, having no more than four continuous main lanes. It is shoehorned in a narrow strip of land between the existing Westpark Road and the 50-foot (15 m) strip set-aside for future transit, in some places taking right-of-way from Westpark Road in order to fit. Most unusual is the interchange at the Beltway 8-Sam Houston Tollway. Whereas the Sam Houston Tollway features numerous five-level stack interchanges with high-flying ramps, the connector ramps between the Westpark Tollway and the Sam Houston Tollway are all below grade, with one connection even two levels below grade. It is more like a reverse stack, going down instead of up.

But perhaps a little variety in the design of Houston freeways and tollways is not such a bad thing. The Westpark Tollway will be a minimalist facility. But just five miles north of the Westpark Tollway is the Katy Freeway, which is planned to be transformed into one of the nation's widest freeways for a sustained distance—a fully paved corridor averaging 475 feet (145 m) in width.

Southwest Freeway interchange: This view looks southbound along the Southwest Freeway (US 59), showing the connector ramps to the Westpark Tollway. The Westpark Tollway is in a trench at this point and is not visible in this view. (Photo: ©Scott Teven/www.HoustonPhoto.com, May 2003)

Fort Bend Parkway Tollway, SH 122

Persistence. Determination. Perseverance. Multiple lives. If a freeway could have personality traits, these would be the traits of the Fort Bend Parkway. No other freeway in Houston has overcome as much adversity as the Fort Bend Parkway in its long road to construction. It has been rejected by TxDOT, left for dead by the city of Houston, brought back to life by Fort Bend County, and then put into limbo by increasingly complex environmental regulations and funding shortages. But in November 2000, the time for the Fort Bend Parkway finally arrived with voter approval of bonds to construct the parkway. After 42 years of struggle it would prevail, although it would be a tollway rather than a freeway.

Of course, the personality traits that ultimately pushed the parkway to construction were those of the backers of the project. The relentless efforts of Fort Bend County officials over 15 years and Houston's growing acceptance of tollways finally made the Fort Bend Parkway a reality. The story of the Fort Bend Parkway also illustrates the huge difficulty in getting a new limited-access facility constructed in today's regulatory environment. Even in Houston, it was a long, costly struggle.

Origins

The Fort Bend Parkway had its origins in the 1960 traffic and transportation study for the city of Houston. The 1960 study was the first in Houston to make use of a computer to compile data and test different scenarios for modifications to the freeway system. By June 1961, the data had been analyzed and a deficiency in freeway service in south Houston in the area between the Gulf Freeway (IH 45) and the Southwest Freeway (US 59) was identified. This wide arc, 13 miles (21 km) along Loop 610 and 22 miles (35 km) along Beltway 8, would be served by only one freeway, the South Freeway (SH 288). The addition of a new freeway extending south from the southwest corner of Loop 610 would provide the recommended density of freeways, which was one every four to six miles (6.4 to 9.6 km). The freeway was designated as the West Loop Extension. The recommendation for the freeway was included in an August 1961 report called *Freeway Phase*.[142]

In 1963, the city of Houston added the West Loop Extension to its *Major Thoroughfare and Freeway Plan*, showing it extending from the West Loop to the South Belt. The city established the corridor right-of-way for the proposed freeway and began the process of acquiring property. In 1964, the city of Houston and Harris County began efforts to get the route adopted into the state highway system and on March 31 a presentation was made to the Texas Transportation Commission. The commission did not grant Houston's request to adopt the facility into the state highway system. On April 5, 1966, the City Planning Commission officially adopted a revision of the *Major Thoroughfare and Freeway Plan*, extending the proposed alignment of the West Loop Extension southwest from the South Belt into Fort Bend County and across the Brazos River toward Bay City. Around this time the freeway began to be called the Bay City Freeway.[143]

The first signs of trouble for the West Loop Extension appeared in 1968. On March 4, city of Houston department heads recommended to the mayor and city council that the right-of-way plans established on the West Loop Extension be repealed. New design standards for freeway construction required a wider, more costly corridor on the urbanized section near Loop 610, and the city of Houston could no longer bear the financial burden of protecting the right-of-way for the proposed freeway. Houston City Council took no action on the recommendation. On June 3, 1969, city of Houston department heads again recommended the repeal of the original right-of-way plan

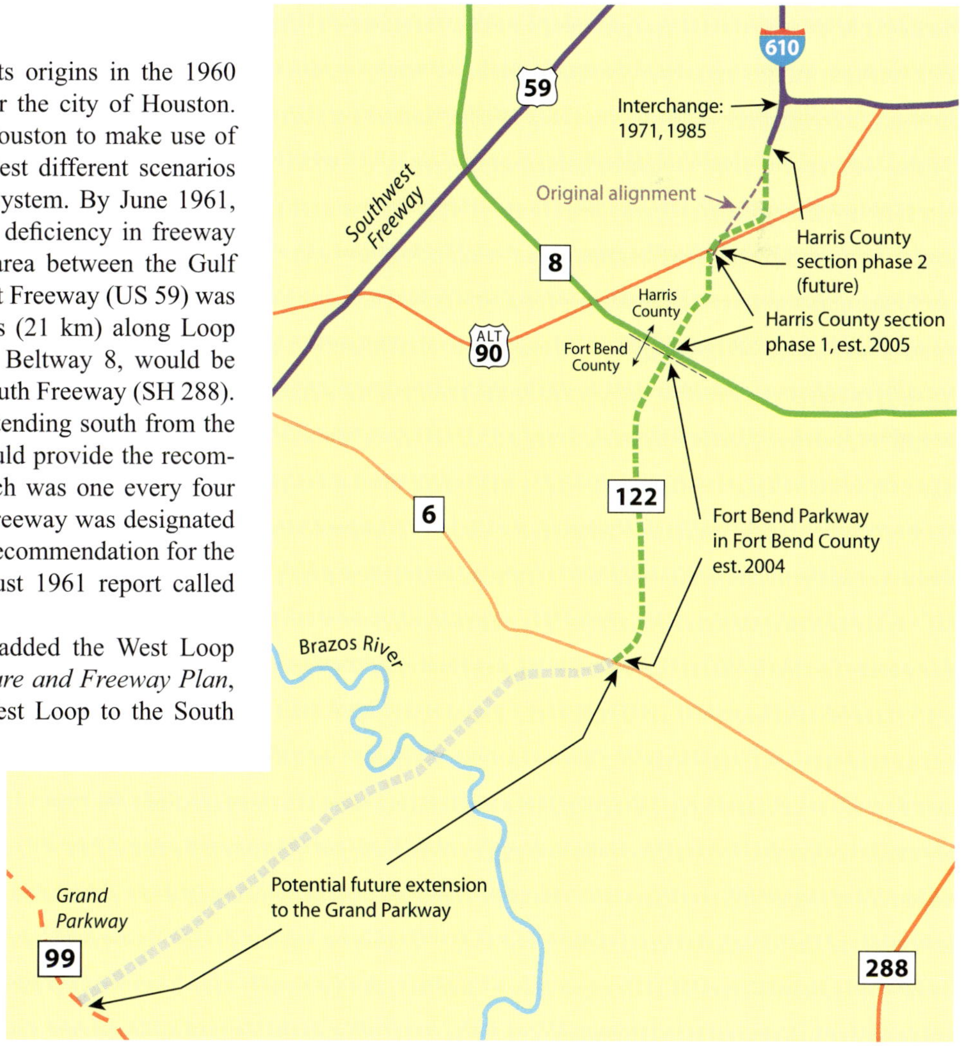

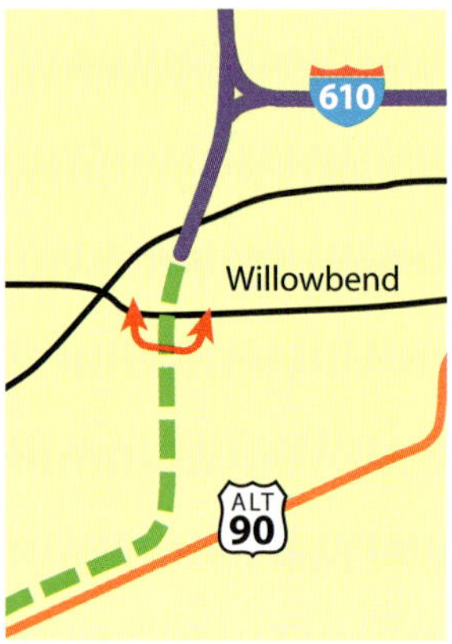

A long-standing problem: This view looks north along the Fort Bend Parkway corridor near the planned terminus at Loop 610, shown in the distance. The route of the tollway will follow Post Oak Road in this section. Urbanization of this section of the proposed freeway corridor was one of the main reasons for the abandonment of plans to construct the freeway in the mid-1970s. The original right-of-way plans for this section, approved by the city of Houston in 1963, showed a narrow 160-foot (49 m) freeway corridor due to the development that had already occurred. Early concepts for the freeway proposed an elevated structure in this area. In 1968 standards for freeway right-of-way required a wider corridor and the city of Houston did not have the financial resources to acquire the needed property. In 1975 the city of Houston ended its efforts to build the freeway. Plans for the tollway in 2003 squeeze the four toll lanes into the corridor with a minimal amount of right-of-way acquisition. (Photo: November 2002)

defining the freeway corridor and proposed the reestablishment of a new right-of-way plan in the urbanized area from Loop 610 to a point about 1.5 miles (2.4 km) to the south. Houston City Council decided to retain the original right-of-way plan established in 1963 and deferred the matter indefinitely on June 25, 1969. Houston City Council was not yet ready to give up on the freeway.

As the city of Houston struggled to keep the West Loop Extension alive, it became clear that no help would be forthcoming from TxDOT. In April 1969, TxDOT approved the addition of the Harrisburg Freeway (the SH 225 extension into downtown) and Beltway 8 to the state highway system, but once again failed to accept the West Loop Extension. Since diminished funding would start to take its toll on TxDOT's construction plans shortly afterward, 1969 was probably the last good chance for inclusion in the state highway system. The city of Houston continued its efforts for TxDOT adoption in 1970, but to no avail.[144]

By early 1975 TxDOT informed the city of Houston that it had "no plans in the foreseeable future to design or construct the freeway." On January 22, 1975, the city of Houston department heads recommended the removal of building restrictions in the right-of-way corridor for the West Loop Extension. This time the right-of-way plans were officially repealed, effectively ending the city of Houston's efforts to get the freeway built. The West Loop Extension appeared to be dead.[145]

The Second Life

While the city of Houston had given up on the West Loop Extension, Fort Bend County just southwest of Houston was ready to step forward and try to make the project happen. In 1984, an update of the Fort Bend County *Thoroughfare Development Plan* identified the need for a freeway-type facility to serve northeast Fort Bend County and connect the region to Houston. Fort Bend County Commissioners Court officially resurrected the moribund freeway in 1985. A committee was formed and a preliminary corridor analysis was undertaken. In January 1986 the committee presented their findings to the Texas Transportation Commission and the commission authorized a detailed study of the corridor. In August 1987 the study concluded that there was demand for the facility and recommended the Hillcroft Avenue corridor for the freeway alignment. The study also recommended terminating the freeway at Beltway 8 rather than Loop 610 because the West Loop could not accommodate the

> "Please be advised that this project has been cancelled."
>
> City of Houston Planning Director Roscoe H. Jones in a letter to Gousha/Chek-chart map company, July 3, 1979

additional traffic load that the freeway would bring. In the 1980s any entity requesting a new freeway from TxDOT was expected to perform all preliminary work, including right-of-way acquisition and environmental clearance, on its own, leaving only the construction to TxDOT. Fort Bend County offered to perform all the preliminary work, and in March 1988 the Texas Transportation Commission officially designated the Fort Bend Parkway as part of the state highway system. While this was great news to Fort Bend County officials, they surely didn't realize they were just beginning a long and costly campaign to get the facility built, one that would require millions of dollars in studies and relentless persistence to overcome the anti-freeway mechanisms built into environmental regulations.[146]

In October 1989, Fort Bend County created a special taxing entity called the Fort Bend Parkway Road District No. 1 to pay for the studies and land acquisition. On February 27, 1990, the Texas Transportation Commission formally entered into an agreement with Fort Bend County authorities, designating the highway as State Highway 122 and stipulating that the state would construct the frontage roads when the county had completed its part of the deal. Fort Bend County officials were optimistic that the frontage roads would be open by 1995.[147]

In 1991 a comprehensive new federal transportation bill, the Intermodal Surface Transportation Efficiency Act, or ISTEA, was signed into law by President George H. W. Bush. ISTEA included a range of new regulations that imposed some significant new hurdles to the Fort Bend Parkway. In April 1993, Fort Bend County was informed that the environmental assessment for the project was not sufficient, and a full environmental impact statement (EIS) would be required. Authorities then began work to extend the environmental assessment into an EIS. Imposing a greater obstacle for the project was a new set of regulations issued by the Federal Highway Regional Administration in July 1993 relating to single-occupant vehicle justification analysis. Now Fort Bend County would have to justify its new lanes for single-occupant vehicles, and at the time methods and techniques for performing the analysis were still being formulated and developed.

ISTEA also required a Major Investment Study (MIS) of the project, so the project consultants went back to work to combine the MIS with the EIS. And then came the noise analysis and more study of alternatives to the freeway. There was also more work to ensure that the regulatory standards applied to the freeway were consistent with standards that had been applied to the Grand

Tollway corridor: This view looks south along the Fort Bend Parkway corridor with Beltway 8 in the foreground. Right-of-way in this section was acquired before the cancellation of the freeway in 1975. As this photo shows, the freeway corridor is a low-density area and the tollway south of Beltway 8 will cause very little disruption. (Photo: November 2002)

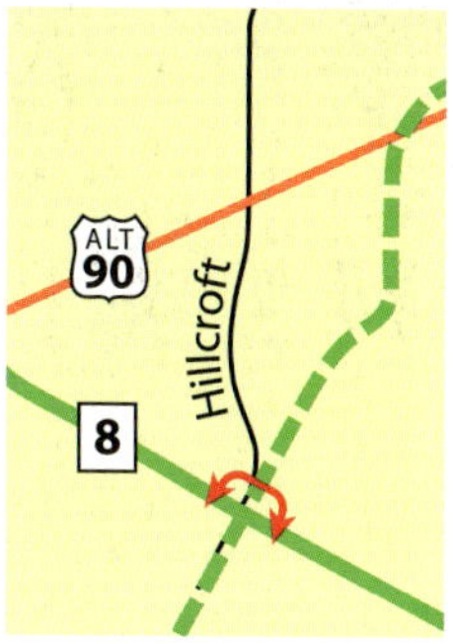

Parkway, another proposed facility that was under study and intersected the Fort Bend Parkway.

In December 1996 Fort Bend County interests began to feel that trying to obtain environmental approval was a futile battle due to the constantly changing regulatory environment. But supporters of the project still believed that with enough persistence, eventually the environmental and compliance issues would be overcome. However, there was a bigger issue lurking: funding. TxDOT had placed a low priority on the project, ensuring a long wait for funds if the project could secure approval. It became apparent that the only way to get the facility built in any reasonable time frame would be as a tollway. As an added bonus, if the project could be done without federal funds, many of the onerous federal regulations would no longer apply.[148]

Moving into the Tollway Era

Whereas the 1980s was the decade of helping the state build new freeways through private transportation corporations, the 1990s was the decade in which the toll

> "It's a history of frustration. It's a history of the finish line constantly being moved a little further away."
>
> Bob Randolph, attorney representing the Fort Bend Parkway interests, in May 1998

road moved to the forefront of Houston transportation planning. The Fort Bend County Toll Road Authority was formed in 1997, and it launched a preliminary feasibility study for a toll road. The results were not good. The consultant concluded that tolls alone could not pay for the cost of the project. However, Fort Bend County officials felt the assumptions used in the study were overly conservative, so a second review was initiated. In 1998 a revision of the study using data more specific to the area that would be served by the tollway provided a favorable result. An engineering study and comprehensive traffic and revenue study were completed in 1999. The tollway south of Beltway 8 was found to be feasible. Around the time of the completion of the studies, the Harris County Toll Road Authority (HCTRA) seriously began to consider building the section of the Fort Bend Parkway within Harris County between Beltway 8 and Loop 610.[149]

Now all that was needed was money. In 2000 Fort Bend County placed a $140 million toll road bond issue on the November ballot. The bonds would be split approximately evenly between the Fort Bend Parkway and the extension of the Westpark Tollway. This vote was perhaps the make-or-break event for the Fort Bend Parkway. After nearly 40 years of effort to get the facility built, it was up to Fort Bend County voters. They approved the bonds with an overwhelming 79% of the vote. Now it appeared that nothing would stand in the way of the Fort Bend Parkway. With the momentum in place, Harris County took steps to move forward with the northern section of the Parkway, authorizing a formal study in November 2000. On April 30, 2002, Harris County Commissioners Court formally approved the plans for the Fort Bend Parkway Tollway section within Harris County. The approved alignment north of US 90A was new, differing from the alignment that had been on planning maps since the 1960s. With HCTRA's agreement to build the northern section, the original plans for the West Loop Extension had been fully restored to Houston's freeway system.

On May 14, 2003, a ground breaking ceremony was held for the construction of the Fort Bend Parkway. It was the culmination of more than 40 years of effort. The persistence, determination, and perseverance had paid off.

42 years in the making: A ground breaking ceremony was held on May 14, 2003, to launch the Fort Bend Parkway Toll Road. It was the culmination of 42 years of effort by Houston-area authorities—the longest struggle for any Houston freeway. And ultimately, the only way it could prevail is as a tollway. In the photo from left to right: Charles Rencher, director, Fort Bend County Toll Road Authority (FBCTRA); Mike Stone, vice chairman, FBCTRA; Bobbie Tallas, secretary, FBCTRA; Jim Condrey, treasurer, FBCTRA; Norm Mason, chairman, FBCTRA; Robert Hebert, Fort Bend County Judge; Mayor Allen Owen, Missouri City; James Thurmond, city manager, Missouri City; Andy Meyers, Fort Bend County Commissioner Precinct 3; Tom Stavinoha, Fort Bend County Commissioner Precinct 1; Grady Prestage, Fort Bend County Commissioner Precinct 2; James Patterson, Fort Bend County Commissioner Precinct 4.

Alvin Freeway, SH 35

The Alvin Freeway is perhaps Houston's only freeway that could be called a survivor. It persisted on Houston's freeway master plan in spite of community protest in the early 1970s, the highway funding crisis of the mid-1970s, and the failure of development and demand to materialize in its corridor. It didn't survive entirely intact, since its freeway status south of Beltway 8 was cancelled. But the Alvin Freeway never died, and in 1999 the first short section opened, 32 years after adoption into the state highway system. It remains somewhat uncertain if or when the full Alvin Freeway will be constructed. However, in the 1990s outlying areas in the Alvin Freeway corridor finally began to see the long-anticipated development that had been predicted in the 1965 study recommending the freeway. The transportation demand resulting from the new suburbanization will probably ensure the survival and ultimate construction of the freeway.

Origins

The Texas Transportation Commission first authorized a study of the SH 35 corridor in May 1964. In July 1965 the study was complete and recommended a new 21-mile (34 km) freeway from central Houston to the city of Alvin. The study cited the impact of the Manned Spacecraft Center (now the Johnson Space Center) and new industrial and residential development in the corridor as key factors in the freeway recommendation. The Alvin Freeway would also provide relief to the chronically congested Gulf Freeway. In November 1967, the Texas Transportation Commission officially adopted the Alvin Freeway into the state highway system and authorized detailed studies and public hearings. This was perhaps somewhat of a surprise, since local officials had unsuccessfully lobbied in previous years to get two other proposed new freeways, the La Porte Freeway Extension and West Loop Extension, adopted into the state highway system.[150]

For the section inside Loop 610, the first public meeting was held on December 5, 1967. There was no controversy over this section, and the route location was approved in 1969.

For the section between Loop 610 and Beltway 8, controversy erupted over plans for the Alvin Freeway to pass through Law Park, a largely undeveloped tract of parkland. Local groups and sympathetic state legislators attempted to pass legislation to force a new alignment study in 1973, but the efforts stalled in the state legislature. The 1974 environmental impact statement reported that the proposed alignment was reviewed by the United States Secretary of Transportation, who "concurred that there were no prudent or feasible alternatives to the proposed location, and that the highway proposed included all possible planning to minimize harm resulting from such use of thc future park site." The plans included depressing the freeway below grade through the park and the addition of noise abatement. The alignment for the Alvin Freeway was approved. Law Park was not developed as it was planned in the early 1970s, and the parkland near the proposed Alvin Freeway was later used for a large police station.[151]

South of Beltway 8, the Alvin Freeway was initially aligned on the east side of the city of Pearland. Controversy over the eastern alignment first arose at a public hearing on December 7, 1967. In 1968 and 1969, Brazoria County and the city of Pearland passed resolutions endorsing an alignment to the west of Pearland. In 1970 the Texas Transportation Commission agreed to a western alignment, but also authorized additional public hearings before an official determination was made. The final environmental impact statement, completed in 1976, recommended an alignment west of Pearland. However, the alignment controversy would become academic as the freeway south of Beltway 8 would later be cancelled.[152]

While the Alvin Freeway had successfully worked its way through the regulatory process, it did not fare so well during the 1970s highway funding crisis. A reevaluation of

Stubby freeway: This view looks south along the Alvin Freeway corridor with the Gulf Freeway (IH 45) in the foreground. The short section of freeway and approximately 1.3 miles (2.1 km) of frontage roads were completed in 1999, 32 years after official approval of the freeway in 1967. As of 2003, there are no imminent plans to extend the freeway main lanes or frontage roads further south. A comprehensive study of the corridor is scheduled to begin in 2003 to determine the future of the freeway. The Alvin Freeway has also been listed as a potential toll road candidate. (Photo: May 2002)

priorities in 1976 caused the Alvin Freeway to be placed on long-term hold. Due to the relatively small amount of development in the corridor, there really was no need to proceed with the freeway in the short- or medium-term time horizons. The reconstruction of the Gulf Freeway near downtown in the mid-1980s included provisions for the ultimate construction of the Alvin Freeway, including long elevated structures along the Gulf Freeway serving as the distribution system into downtown. Starting around 1980, TxDOT held regular meetings with the University of Houston, Texas Southern University, and the city of Houston to define changes to the street system to accommodate the Alvin Freeway, particularly in the area near the Gulf Freeway. By 1985 these plans were finalized. The final environmental impact statement for the section from the Gulf Freeway to Bellfort Street, just south of Loop 610, was approved in the summer of 1992.[153]

In July 1996 construction on the Alvin Freeway finally began. In 1998 the new facility was designated as Spur 5. A very short section of main lanes and 1.3 miles (2.1 km) of frontage roads were completed in 1999. The entire Highway 35 corridor from the Gulf Freeway to Angleton, 40 miles (64 km) south, is the subject of a comprehensive corridor study expected to be underway by 2004. That study will determine the future of the Alvin Freeway—if and when it will be constructed. The planned facility south of Beltway 8 will be reevaluated and could be restored to full freeway status. The Alvin Freeway corridor has also been identified as a candidate for a potential future toll facility by the Harris County Toll Road Authority.

Red Bluff Freeway

The Red Bluff Freeway can trace its beginning to NASA's decision to locate the Manned Spacecraft Center (now the Johnson Space Center) in Houston, which was announced on September 19, 1961. A comprehensive transportation study had been completed just prior to the NASA announcement, and its results had been adopted in September 1961. Because of the Manned Spacecraft Center and Clear Lake City, the huge real estate development planned for the area, the study results for the southeast region needed to be reviewed and updated. Harris County Judge Bill Elliot requested the special study which was performed by the Houston City Planning Commission. The final report, *A Study of Thoroughfare Development in the Southeast Area of Metropolitan Houston and Harris County*, was presented to Harris County on August 13, 1963.

The study recommended numerous modifications to the metropolitan transportation plan, including the addition of a new freeway, the Red Bluff Freeway, which would follow the alignment of the existing two-lane Red Bluff Road. The study recommended a phased construc-

Still semirural: This view looks southeast along Red Bluff Road. The frontage roads for the proposed Red Bluff Freeway were complete by 1969, but land development in the corridor did not occur as anticipated. In 2003 Red Bluff Road remained largely semirural, traversing through forests and pastures. Construction of the freeway main lanes on the 300-foot-wide (91 m) corridor is not planned as of 2003. (Photo: September 2002)

tion, with frontage roads only being constructed initially. Harris County was solely responsible for the project. In 1964 a map showing the right-of-way requirements for the 300-foot-wide (91 m) corridor from Fairmont Parkway to Kirby Road was developed. In January 1966 voters approved a county bond issue which included funds for the Red Bluff Freeway. A final right-of-way map was completed in late 1966, and work on the right-of-way acquisition and frontage road construction moved at full speed in 1967 and 1968. By the end of 1968 the frontage roads were complete from Fairmont Parkway to Kirby Road. Fairmont Parkway between Red Bluff Road and Beltway 8 East was also planned as an ultimate freeway facility, and its frontage roads were completed around the same time.[154]

They Built It and Nobody Came

In 1969 there were no immediate plans to construct the main freeway lanes on Red Bluff Road. And it turned out to be a good thing. Usually with freeways, the adage "build it and they will come" is a rule that never fails. But for some reason, the cars, traffic, and development simply did not materialize in the Red Bluff corridor. In 1977 Red Bluff Road was handling only 14,000 vehicles per day at its busiest point south of Fairmont Parkway. The 1963 planning report had projected 62,000 vehicles per day by 1980. In 1978 efforts were underway to improve the southernmost section of Red Bluff Road between Kirby Road and SH 146. This short section turned out to be major undertaking due to the long water crossing at Taylor Bayou, right-of-way acquisition complications, and an unexpectedly high cost. Only one side of the frontage roads was constructed.[155]

Red Bluff Road remained almost entirely devoid of development through the 1980s and 1990s. As of 2003, the corridor still has a rural feel to it as it traverses through forests and pastures. But slowly, the laws of traffic growth and development are beginning to fill the vacuum on the Red Bluff corridor, with limited residential development taking place near the south end. In June 2001, the Harris County Toll Road Authority adopted a list of potential future projects which included future study of the Red Bluff and Fairmont Parkway corridors. For 35 years motorists on the Red Bluff frontage roads have seen the wide median and wondered, "Will the main lanes ever be built?" Only time will tell.

Other Freeways

State Highway 146

SH 146 extends from Liberty, northeast of Houston, to Texas City, southeast of Houston near Galveston. The section from Baytown to La Porte is a full freeway and includes the Fred Hartman Bridge ship channel crossing. The earliest studies to upgrade SH 146 to a full freeway appear to have occurred in 1955 when the Texas Transportation Commission authorized location surveys for a 300-foot (91 m) corridor extending four miles (6.4 km) south of SH 225, the La Porte Freeway. In September 1959 the city of Baytown, Harris County, and TxDOT agreed to acquire right-of-way for a loop freeway through Baytown, extending from West Main Street near the present-day Hartman Bridge to Ferry Road east of Baytown. The section was approved as an access-controlled freeway in February 1961. The Baytown Loop was originally designated as Loop 201. Loop 201 was absorbed into SH 146 in 1996.[156]

Although planning for a freeway had begun in the 1950s, the SH 146 corridor remained a hodgepodge of highway sections built to various standards up until the 1990s. The SH 146 ship channel crossing opened as a two-lane tunnel on September 22, 1953, but planning for a bridge meeting freeway standards did not begin until 1977. Between 1970 and 1972, expressway and frontage road sections were built south of the La Porte Freeway (SH 225) to Red Bluff Road. Between 1978 and 1981, a section in Baytown between the Houston Ship Channel and Spur 330 was upgraded, with a segment meeting expressway standards and another section a frontage road facility. In the 1990s the freeway era for SH 146 finally arrived. On September 27, 1995, the eight-lane Fred Hartman Ship Channel Bridge was dedicated. In 1996 the freeway from the bridge northward through Baytown was completed, and in 1999 and 2000 freeway sections south of the La Porte Freeway were opened. Also in 2000, reconstruction of the interchange at the La Porte Freeway was completed. The original interchange, opened in 1953, was the first three-level interchange on the state highway system in Texas.

In 2001 a study to determine the future of the non-freeway section of SH 146 south of La Porte was completed. The study recommended upgrading the highway to a full freeway from Fairmont Parkway to Red Bluff Road. From Red Bluff Road southward to FM 518, the construction of express lanes alongside the existing highway was recommended. These express lanes will allow motorists to bypass all intersections in the segment, providing nearly the same service level as a full freeway. South of FM 518, SH 146 will be upgraded to an arterial-type highway with grade separations at major intersections. In December 2002 the Texas Transportation Commission authorized the purchase of the railroad right-of-way that parallels SH 146 for use in the expanded highway facility.[157]

SH 146 holds the distinction of being one of only three freeways in the Houston region to be named after an individual, the others being the Sam Houston Tollway and the Emmet Lowry Expressway in Texas City. The freeway

The Lanier Freeway: Planned since the 1950s, the SH 146 freeway main lanes through Baytown were completed in 1996. This view looks west along the freeway with the Exxon complex in the background. (Photo: July 2002)

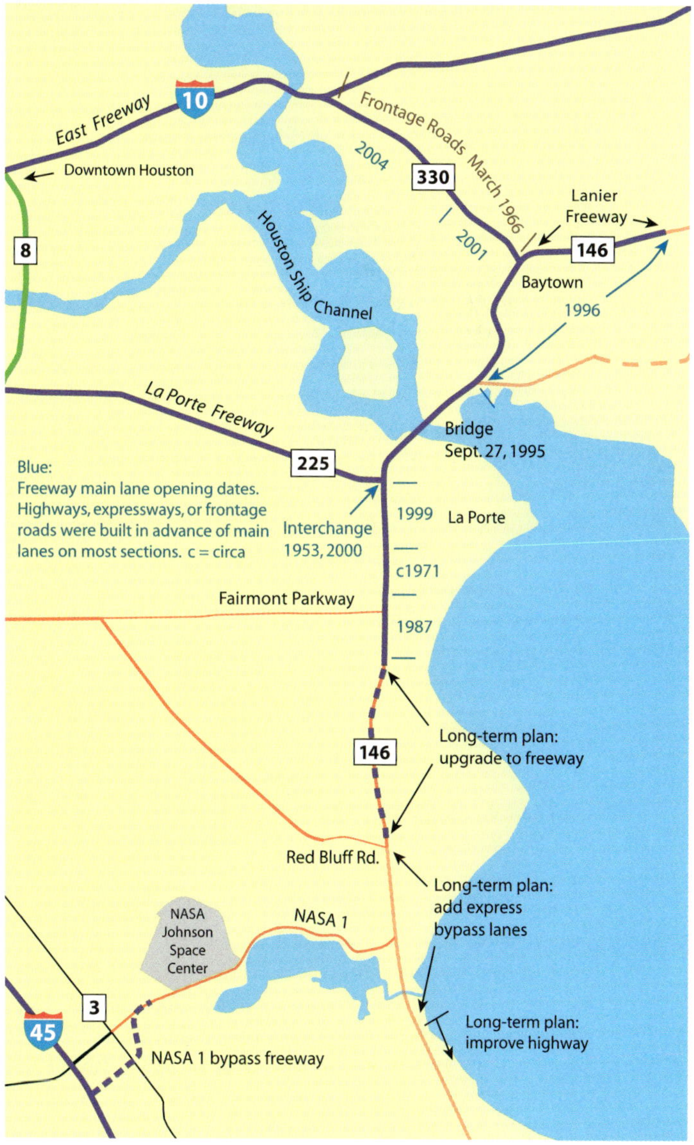

section through Baytown is named the Robert Lanier Freeway in honor of the native Baytonian who became chairman of the Texas Transportation Commission and mayor of Houston.

Spur 330

Spur 330 was authorized by the Texas Transportation Commission in October 1957 and was designated as a controlled-access freeway in January 1961. After completion of the frontage roads in March 1966, Spur 330 went into a long period of dormancy. In November 1996, the city of Baytown made a presentation to the Texas Transportation Commission to request construction of the freeway main lanes. Work on the first section of main lanes was underway in 1998 and was completed in 2001. A contract for the final section of Spur 330 main lanes was awarded in March 2002, with completion scheduled for 2004. The final work to complete Spur 330 consists of the remaining ramp connections at the SH 146 interchange. That work will probably occur sometime between 2005 and 2015.[158]

NASA 1 Bypass Freeway

When the National Aeronautics and Space Administration (NASA) announced that Houston would be the location of the new Manned Spacecraft Center (now the Johnson Space Center) on September 19, 1961, local officials revised transportation plans to improve the roads, highways, and freeways in the NASA area to accommodate the planned development. Somehow a key section of the most vital road serving the NASA campus—NASA

The long wait is over: In July 2002, work was in progress to construct the final section of main lanes for the Spur 330 Freeway. Thirty-eight years elapsed between the completion of the frontage roads and the main lanes.

Road 1—was left behind in the plans. NASA Road 1 was previously Farm to Market (FM) Road 528. A section of FM 528 through the city of Webster between the Gulf Freeway and SH 3 had a very narrow right-of-way, wide enough for only four traffic lanes with no median. The narrow right-of-way and the substantial number of businesses along the roadway complicated efforts to expand NASA 1 for more than 30 years.

In the late 1980s it appeared that construction would move forward on a planned expansion which would have transformed the narrow section of NASA 1 into a six-lane arterial street with a grade separation at SH 3. However, the expansion would have displaced 21 businesses along the highway, and the city of Webster was required to pay a share of the right-of-way acquisition costs amounting to \$1.7 million. Webster officials did not want the wide arterial street through their city and balked at paying the right-of-way costs. Webster held an election for bonds to pay for the right-of-way acquisition costs on May 2, 1992. The bond proposition failed by 16 votes, and the entire NASA 1 improvement project was put on hold.[159]

In 1994 the Webster City Council approved a plan to build a freeway bypass south of the urbanized area of the city. The proposed four-lane freeway bypass was 2.7 miles (4.4 km) long and did not have frontage roads. The revised project worked its way through the public hearing and approval process over the next few years and then had to wait in line for funding. Construction on the \$40 million bypass is scheduled to begin in 2004. So approximately 45 years after the original announcement of the NASA space center, the principal road serving NASA will finally be brought into the modern era. It was a long wait, but the end result will be a new freeway for Houston.[160]

Memorial Parkway

A freeway along the banks of Buffalo Bayou west of downtown was one of the early freeway corridors contemplated for Houston in the city of Houston's 1942 *Major Street Plan*. As the freeway era began to take shape after World War II, it became clear that the parkland along Buffalo Bayou was not suitable for a full Houston freeway with main lanes and frontage roads. Furthermore, Memorial Park and the elite River Oaks neighborhood blocked the path just a few miles west of downtown, preventing a potential freeway from connecting to the rest of the Houston freeway network. When Houston's freeway master plan emerged in 1953, there was no Memorial Drive freeway. But Memorial would get a limited-access facility suitable for its environment—the only true parkway in Houston.

Plans for a transportation corridor along Memorial Drive date back to the 1930s, when Memorial Drive was envisioned as the alignment of SH 73 west of downtown and the western gateway to the city of Houston. In 1937, the route for Memorial Drive had been graded from Shepherd Drive to Piney Point, seven miles (11 km) to the west, allowing motorists to "travel over it in good weather." Most of the route was shell surfaced. Plans were being developed to extend Memorial from Shepherd Drive into downtown Houston. That section would ultimately become a limited-access parkway. The first section of Memorial Drive Parkway, from Shepherd to Waugh Drive,

Memorial Parkway: Memorial Drive was identified as a potential freeway corridor in the city of Houston's 1942 *Major Street Plan*, but did not become part of Houston's freeway system when the overall plan was formulated in 1953. Instead, Memorial became a limited-access parkway serving local traffic. (Photo: May 2003)

opened July 8, 1955. It included the large cloverleaf interchange at Waugh. On January 30, 1956, the section from Waugh to downtown was opened. Memorial Drive was a limited-access facility, but it was designed for low speeds with conventional curbs and no emergency shoulders. The three-level interchange at Shepherd Drive was opened on May 2, 1958. The limited-access section of Memorial Drive was completed on January 28, 1960, when the elevated lanes connecting into downtown Houston were opened. Memorial soon became Houston's busiest thoroughfare, carrying 46,640 vehicles per day at Waugh Drive in 1960. The importance of Memorial Drive as a transportation corridor diminished with the opening of the Katy Freeway in December 1968. Relieved of the traffic load, Memorial could return to its originally intended role as a scenic parkway.[161]

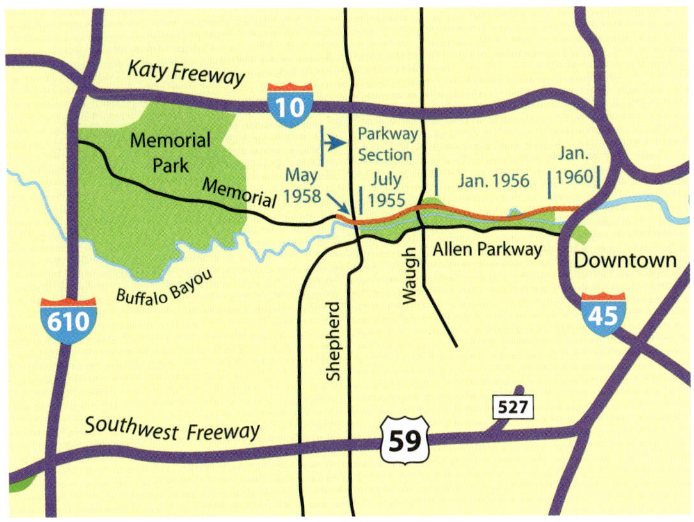

The West Loop: One of the most impressive instances of an "edge city" has risen along the West Loop. This view looks north along the West Loop with the Southwest Freeway interchange in the foreground. (Photo: September 2002)

The Loops

In his 1992 book, *Edge City: Life on the New Frontier*, Joel Garreau documents the rise of new urban centers on the fringes of the nation's cities. These edge cities included not only the traditional elements of suburbia—homes and shopping malls—but also the means of creating wealth—jobs. Garreau identifies 123 edge cities around the nation but makes no effort to rate or rank them by any criteria. After all, the edge city is a somewhat amorphous entity, evolving to meet the needs of the new information economy. But Garreau leaves us with a few hints that perhaps there is one edge city that towers over all the rest, both literally and figuratively, an edge city that stuns the sense of sight by the immense scale of its signature structure. That edge city is the Uptown Houston district located on the West Loop. Just off the West Loop is the 64-floor, 899-foot-tall (274 m) Williams Tower, an architecturally distinctive structure that is made even more striking by its dominating presence over an otherwise impressive landscape of mid-rise office towers. It was the loop freeway, of course, that propelled Uptown Houston to become one of the most impressive edge cities in the United States.

The circumferential freeway bypass has become a standard fixture in the freeway systems of many cities, both in the United States and Europe. But Houston wouldn't settle for just one loop, or even two. By 1965 plans were in the works for three major circumferential loops around Houston, and if the freeway miniloop enclosing downtown is included, there are actually four. Houston is the ultimate implementation of the loop and radial freeway system, with a nearly complete second loop in 2003 and the wheels in motion for the construction of the third loop. Perhaps only two other major cities in the United States, Dallas and San Antonio, hold out any hope of fully completing second freeway loops around their cities. If in fact Houston's third loop, the Grand Parkway, is completed as planned, Houston will be in a class by itself. Almost surely, no other city will ever be able to achieve Houston's loop freeway accomplishment.

The West Loop parking lot: The West Loop is one of Houston's most congested freeways. This view looks north toward Memorial Park. (Photo: James Lyle, TTI, June 2001)

The Loop, Interstate 610

Loop 610 is more than just a freeway. It has come to define a lifestyle and state of mind, not just a geographic section of the city. The "inner looper" is more of an urban person, someone who likes to be close to the arts, universities, events, parks, and entertainment. Many inner loopers seek out the variety, disorder, and nonconformity of Houston's older neighborhoods. Others seek out some of Houston's most affluent and exclusive neighborhoods. The inner looper often lives close-in as a means of avoiding freeways. More than any other geographic group in Houston, the inner looper is likely to be anti-freeway.

While it serves as an informal boundary between central Houston and the rest of the city, Loop 610 is also a vital transportation artery—the most important freeway in the functioning of Houston's loop and radial system. As the focus point for much of the congestion on Houston's freeway system, the West Loop is also the biggest source of dysfunction. The loop ties together many seemingly disparate sections of Houston: edge cities, neighborhoods from the exclusive to the decayed, parks, stadiums, and industry. Along its path, the loop offers an abundance of interchanges, the ship channel bridge, and some impressive freeway sections.

Origins

The need for a bypass loop around Houston was first identified as early as 1931, when Harris County officials were proposing bypass routes to divert traffic from the city center. The early concept proposed using existing streets for the bypass. No progress was made during the 1930s, but efforts to build a loop came back to life in September 1940 when the Houston Chamber of Commerce Highway Committee formed a special subcommittee to study potential bypass alignments. Various benefits were cited, including reduction of traffic inside the city, reduction of accidents, and diversion of trucks, but the possible need for national defense deployments would quickly become the driving factor in moving plans for the loop forward.[1]

In March 1941 the first report recommending a bypass loop around Houston was issued. *Preliminary Study for a Primary Defense Need of Houston and Vicinity—A Bypass or Loop Thoroughfare*, published by the Houston City Planning Commission, explained how military officials had contemplated the logistics of large troop movements through Houston to protect the Houston Ship Channel and its associated industries. Initially authorities had planned to block off city streets to move military convoys through the heart of the city. However, it quickly became clear that this would be very disruptive, and by April 1941 the *Houston Chronicle* was reporting that "military authorities have all but demanded the construction of a belt highway." The study also included a map that indicated the key "war industries" around Houston, such as Dow Chemical in Freeport, Houston Ship Building Corporation along the ship channel, and Consolidated Steel in Beaumont. The contemplated route followed existing and planned arterial streets around the city and was envisioned as a highway or large urban street rather than a freeway. The alignment of the northern half of the proposed loop ultimately became the Loop 610 freeway. The southern half of the proposed loop followed the arterial streets Bellaire Boulevard, Old Spanish Trail, and Wayside. On May 3, 1941, a $5.4 million Harris County bond issue allocating $1,028,354 for the Defense Loop was approved by Harris County voters with 71% of the vote.[2]

Loop 610	
Previous designation	Loop 137
Designated as freeway	1954 (north, west, south) 1960 (east)
First freeway section open	1952 (La Porte cutoff) 1960 (Loop 610)
Freeway complete	September 22, 1975
Reconstruction	Intermittent since 1975
Max traffic volume, 2001	290,000 vehicles per day
Future construction	Reconstruction of West Loop (underway in 2003)

The bombing of Pearl Harbor on December 7, 1941, increased the urgency of constructing the Defense Loop. On June 23, 1942, the Texas Transportation Commission formally adopted the north section of the Loop from IH 10 West (then US 90) on the west side of Houston to IH 10 East (then SH 73) on the east side Houston, calling it the "Loop on US 90." An agreement approved on November 16, 1943, called for the city of Houston to provide a 150-foot (46 m) corridor for the highway. The loop was designated as Loop 137. However, real progress on a true freeway loop would have to wait until after the war.[3]

Putting the Freeway Loop on the Map

As progress on the loop highway crept along in the early 1950s, authorities were formulating a master plan for Houston's freeway system which included a full freeway loop. In July 1953 a Houston delegation appeared before the Texas Transportation Commission in Austin

to request adoption of the proposed new freeway routes into the state highway system. The loop included in the plan followed the previously approved Loop 137 bypass route on the north side of the city and showed the South Loop following the alignment of Holmes Road (see map on page 13). The South Loop was informally called the Holmes Road Freeway during this period. In late 1953 the Texas Transportation Commission officially adopted the spoke freeways requested by Houston officials but did not adopt the loop as a freeway.[4]

The Holmes Road Freeway turned out to be very short-lived. On December 29, 1953, Houston Planning Director Ralph Ellifrit submitted a proposal to realign the South Loop to its present location, which is north of Holmes Road for most of its alignment, citing the availability of open land in the proposed corridor and the complications that would be caused by the railroad along Holmes Road. By the summer of 1954, Houston's overall master plan had evolved to the near-final version and included Ellifrit's route for the South Loop.[5]

A delegation of local officials appeared before the Texas Transportation Commission on September 28, 1954, to request state adoption of the West Loop and South Loop as freeways. In October 1954, the commission officially designated the West and South Loops into the state highway system and approved upgrading the previously approved North Loop to a full freeway. With that designation, the only missing link was the East Loop from the La Porte Highway (SH 255) to the East Freeway (IH 10 East). For the rest of the 1950s Houston authorities would focus their efforts on that section.[6]

Closing the Loop

In December 1955, in a letter to TxDOT head Dewitt Greer, city of Houston planning officials were sounding an alarm about the need to preserve right-of-way for the

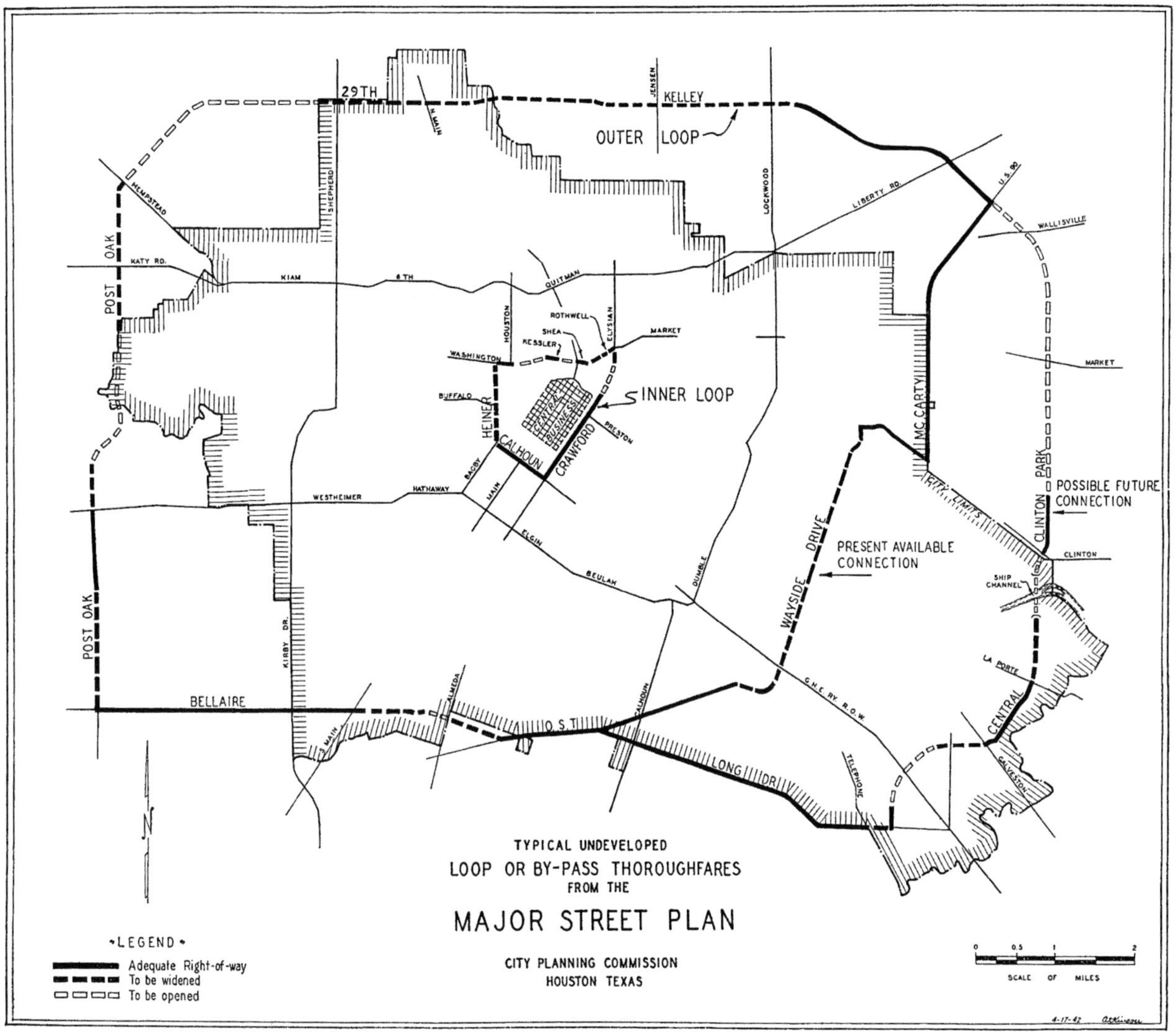

Original loop plan: Although the idea of a loop for Houston had been around since the early 1930s, the first real progress toward the actual designation of a loop occurred in 1941 when the Houston Planning Commission released a report titled *Preliminary Study for a Primary Defense Need of Houston and Vicinity—A Bypass or Loop Thoroughfare*. The above map from the city of Houston's 1942 *Major Street Plan* shows the proposed alignment of the bypass loop. The loop was envisioned as a highway or major arterial street, and for most of its alignment it followed existing routes. The first new construction for the loop would not be completed until 1950.

East Loop corridor, citing a "critical situation with respect to planning and protecting the right-of-way." The alignment of the East Loop had already been shifted twice because of plant expansions in the proposed freeway path. When the Federal-Aid Highway Act of 1956 authorized and funded the construction of the Interstate Highway System, the previously approved C-shaped Loop 137 bypass route was adopted into the interstate system, but the East Loop and ship channel bridge were once again passed over. On April 24, 1958, the Texas Transportation Commission agreed to continue to perform surveys and studies for the corridor and to continue to seek acceptance of the route into the interstate system. The commission did not, however, fully adopt the section into the state highway system.

Texas had a limited amount of mileage it could designate into the federal Interstate Highway System, and there was not enough available mileage to meet all needs in Texas. Dewitt Greer left missing links in Houston and other Texas cities, thinking that the U.S. Bureau of Roads would automatically add the missing sections out of obvious necessity. However, Greer's plan backfired when the Bureau of Roads specified that IH 37 between San Antonio and Corpus Christi would receive the additional interstate mileage available in Texas. In 1960 the Houston City Planning Commission led a new effort to get the East Loop adopted into the state highway system, where its cost would be shared by TxDOT and Harris County. A delegation from Houston appeared before the Texas Transportation Commission August 22, 1960, to make the request. Two days later, the East Loop was finally adopted into the state highway system, allowing local officials to protect and acquire the needed right-of-way. Loop 610 would now truly be a loop.[7]

From countryside to freeway-side: This view shows the Pin Oak Stables with its annual charity horse show in progress in the late 1940s. Post Oak Road runs along the upper part of the photograph, and the future alignment of the West Loop freeway is indicated by the dashed lines. At the upper left corner, the alignment of the future Southwest Freeway is shown. The Pin Oak Charity Horse Show was held from 1945 through 1992. In its prime years, from the late 1940s to the 1960s, it was one of the leading horse shows in the United States and one of the most prestigious social events in Houston. Its parties were attended by Houston's elite as well as film stars and wealthy riding enthusiasts from around the country. The horse show was promoted by James Abercrombie, founder of Cameron Iron Works, and Leopold Meyer, whose family owned the now-defunct Meyer Brothers apparel retail chain in Houston. The land was developed for retail use starting in the late 1980s. The site of the horse show stadium became a parking lot for a large retail store which was later converted to educational use.[8] (Photo: HMRC MSS 67-1074)

Adoption into the Interstate Highway System was still pursued by local officials since interstate status would provide 90% federal funding for the freeway and costly bridge. Finally, during the week of September 10, 1962, the United States Bureau of Roads approved the East Loop as part of the Interstate Highway System.[9]

Building the Freeway

The early work on the loop highway, called Loop 137 at the time, focused on the northeast section between the Eastex Freeway and the East Freeway. The first section, from the East Freeway to Lockwood, was completed in 1950, and the rest of the section to the Eastex Freeway opened on February 26, 1954. The loop highway was generally constructed on a 150-foot-wide (46 m) right-of-way with four highway lanes.

In 1954 the north, west, and south sections of the loop were officially designated as freeways. The 150-foot-wide highway corridor on the North Loop was expanded to a minimum of 300 feet (91 m), and in 1956 a new alignment was approved for a section of the North Loop at the Eastex Freeway interchange. The first full freeway section of Loop 610 had actually opened in December 1952 as the La Porte cutoff on the Gulf Freeway. That section of the

Southwest Freeway interchange construction: These views show the West Loop main lanes under construction at the Southwest Freeway interchange in May 1961. Post Oak Road snaked its way through the construction zone, as shown in the aerial view at right, which looks north. Soon after these photos were taken, this section of Post Oak Road was permanently closed and replaced by the West Loop. (Photo: upper, HMRC RGD6-952; right: TxDOT)

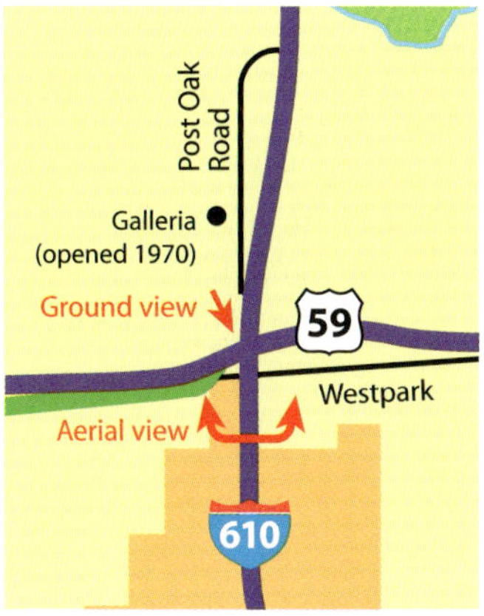

South Loop corridor: Aerial photographs from the 1950s and 1960s show an abundance of drive-in cinemas on the Houston landscape. The 1959 Houston telephone directory listed 18 drive-in cinemas, including the King Center Twin Drive-In shown in the 1960 photo above. No drive-in cinemas in Houston were displaced by freeways, but the South Loop just missed the King Center Twin Drive-In, skirting its edge. The South Loop at this location, indicated by the dashed lines, opened in January 1969. (Photo: The Positive Image)

La Porte Freeway was absorbed into the loop. The initial section to be constructed as part of the loop itself was a 1.3-mile (2.1 km) segment of the North Loop east of the North Freeway. That section opened in 1960. Over the next 15 years the loop would open section by section. The final ribbon for Loop 610 was cut on September 22, 1975, when a 3.3-mile (5.3 km) segment in northeast Houston was opened.[10]

Building the Freeway through Bellaire

The history of Bellaire began in 1908 when a 9,449-acre tract known as the William Marsh Rice Ranch was purchased for development. The developer subdivided the eastern section of the property into small tracts and called it Westmoreland Farms. In a 1909 brochure, Westmoreland Farms was promoted as a "suburban agricultural opportunity" where the country life could be enjoyed within easy reach of the city. A streetcar line operated by the Houston Electric Company began service by the beginning of 1910 to connect the community to Houston. The community was incorporated as a city on June 24, 1918, and it grew slowly prior to World War II. Bellaire had about 330 homes and 1,124 residents in 1940.[11]

After World War II, Bellaire became a classic postwar suburban boom town. In 1950 Bellaire had 3,186 homes. The housing construction boom continued with 600 to 700 homes being constructed each year in 1950, 1951, and 1952. By 1955 Bellaire had been largely built out, and less than 100 homes were constructed. In August 1955 Bellaire had 5,897 homes. Since Houston had annexed all the land around the city, Bellaire became an island city and further growth was not possible. But construction in Bellaire was definitely not over. Transportation officials were working on plans for the West Loop freeway.

In 1941 Houston's loop was designated to pass through Bellaire on South Post Oak Road. But it was planned as a major arterial street, not a freeway. In 1953 local officials revealed plans to turn the loop into a full freeway, and TxDOT officially adopted the loop into the Houston freeway system in October 1954. The freeway would follow Post Oak Road through Bellaire, splitting the city almost exactly in half. By December 1954 a group of Bellaire residents began an effort to stop the freeway.

Trying to Stop a Freeway in the 1950s

Trying to stop a freeway in the 1950s was a daunting and perhaps impossible task. The legal tools for opposing freeways were not available at that time. As an added difficulty for those opposing the freeway, highway engineers sought the most direct, efficient, and least costly routes for freeways. The possibility of curving the freeway around Bellaire to minimize impacts was not considered an acceptable practice at the time since it would have caused an awkward, curving alignment.

The anti-freeway group first protested against the West Loop at a Bellaire City Council meeting on December 6, 1954. At the time, the freeway was informally called the Post Oak Freeway. The *Bellaire Texan* newspaper reported that "city council hid behind the 'need for more facts and figures,' giving nothing more than a 'don't worry about it' brush off to 50 anti-Post Oak Freeway citizens who petitioned the council for a definite stand against any such construction through Bellaire." The mayor tried to assure a skeptical audience that "the entire council is on the side of Bellaire, not Houston." When the protest was reported in the local Houston news, the manager of Houston's Public Works Department wrote a letter to the Harris County Judge, stating, "It looks to me like this is a

Clearing the way: Right-of-way clearance through Bellaire occurred from 1959 to 1961. Completed in March 1968, the freeway main lanes were built in the foreground of this photo. The blurred vehicle in the background is travelling on Post Oak Road, which was a two-lane road. (Photo: *Houston Chronicle*, January 11, 1960)

very dangerous situation."[12]

The opposition then started a petition campaign to force the Bellaire City Council to enact an ordinance that would prevent the expenditure of any city of Bellaire funds for the purchase of right-of-way. Under the terms of TxDOT's adoption of the West Loop freeway route, Bellaire was expected to pay for the freeway right-of-way through the city. The Bellaire City Council complied with the petitioners' request without officially tabulating the petition results by formally notifying the Texas Transportation Commission on April 4, 1955, that the city of Bellaire "respectfully and officially" declined to provide right-of-way for the project. In its letter, Bellaire cited its island status within the city of Houston and its inability to increase its tax base via annexation. The council felt it had complied with the request of the petitioners. However, the petitioners then realized that the city council action would not preclude another entity from purchasing the right-of-way through Bellaire, and they asked city council to take a definite stand against any freeway construction. Harris County would step forward to take responsibility for acquiring the right-of-way, and Bellaire City Council would not actively oppose the freeway.[13]

In November 1955 the proposed right-of-way map for the freeway corridor was released. In the initial plan, right-of-way was to be acquired almost entirely on the east side of Post Oak Road. An article in the November 16, 1955, *Bellaire Texan* presented numerous citizen comments on the freeway, including those of supporters and others who accepted the new freeway as inevitable. Former mayor and prominent resident Abe Zindler, whose estate was immediately adjacent to the freeway route, stated, "If they need it, and that's the best route, then we can't stop progress." In early February 1956, Harris County Commissioners Court adjusted the freeway corridor to lie to the west of Post Oak Road in the northern part of Bellaire, mainly to avoid a Catholic high school. Commissioners Court then proceeded to approve the alignment. The new alignment would displace 190 homes in Bellaire, a clearance corridor that was generally four houses wide along South Post Oak Road.[14]

Only one formality remained: a public hearing on March 29, 1956. About 75 citizens of Bellaire attended the hearing at Harris County Commissioners Court. However, the opponents realized that the hearing was largely a formality. "The freeway is a foregone conclusion. We're

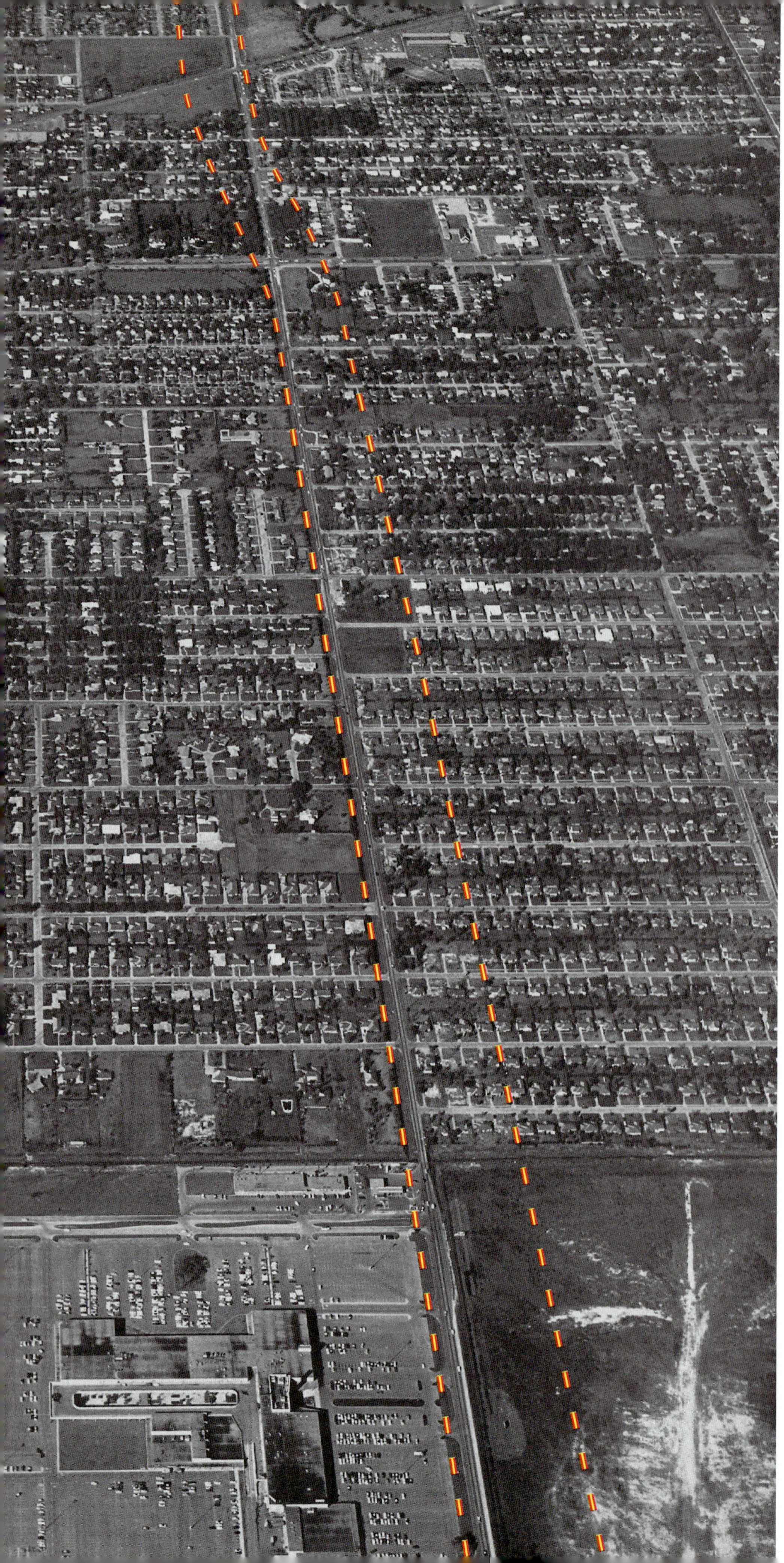

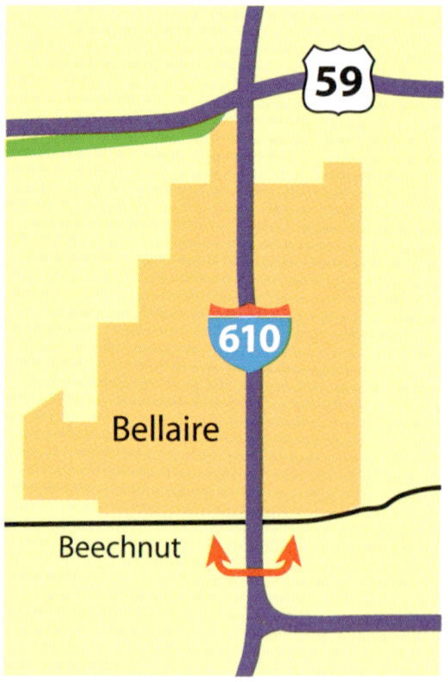

Pre-freeway Bellaire: This view looks north over Bellaire in 1960, just after right-of-way clearance for the West Loop had begun. The West Loop is aligned along Post Oak Road, the two-lane road running through the center of the photograph. The freeway alignment is indicated by the dashed lines. The clearance corridor was approximately four houses wide along Post Oak Road. (Photo: The Positive Image)

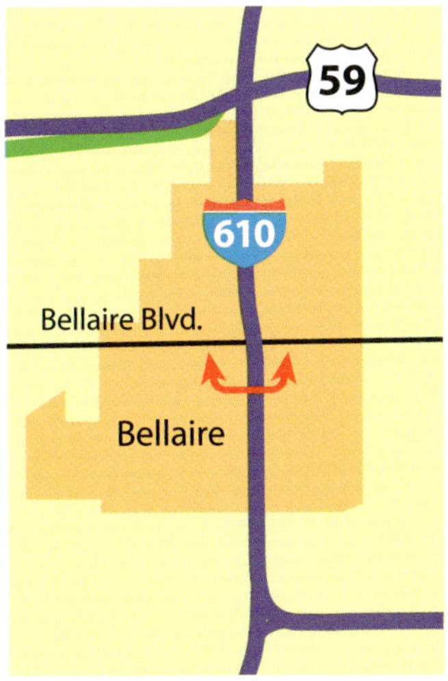

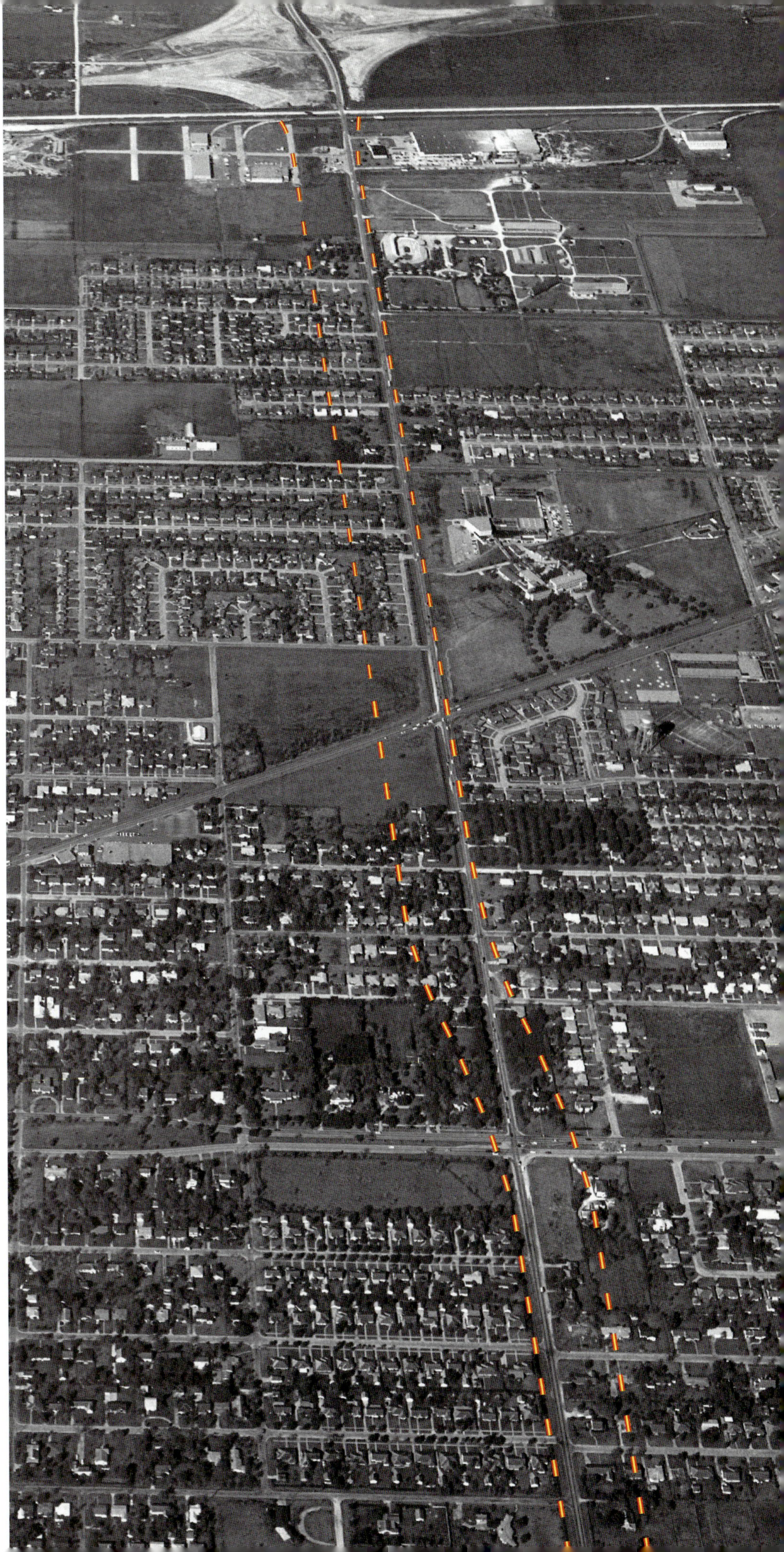

This view, also from 1960, looks north over the north section of Bellaire. At the top of the photo, construction is just beginning on the Southwest Freeway interchange. (Photo: The Positive Image)

Why is this man smiling? In 1957 he was included in *Fortune Magazine*'s list of the 76 wealthiest persons in the United States. Freeways also made him smile. He owned a lot of land along Houston's freeways, including about 90% of the land along the West Loop between Westheimer and Memorial Park. Land along Houston's freeways is particularly valuable because the freeway frontage roads enable commercial development. Who is he? R. E. "Bob" Smith (1894-1973), oilman, rancher, and land speculator.[15] (Photo: HMRC RG-D5-4983)

here only to get the gripes off our chest," remarked one protester. The commissioners unanimously approved the freeway alignment. The West Loop through Bellaire was now a done deal, and the opposition dispersed. When the highway department held a public hearing for the freeway design in September 1957, Houston Urban Project Office head A. C. Kyser reported, "No opposition to the route was voiced and we spent about an hour briefing the people and answering questions that applied to their specific property." At the meeting the freeway corridor was widened to 350 feet (106 m). Right-of-way clearance began in 1959 and was complete by 1961.[16]

Although the freeway opponents may not have realized it, there was a very powerful person who had a strong interest in the Post Oak alignment, probably making it a near certainty: oilman, rancher, and millionaire R. E. "Bob" Smith. Not only was Smith wealthy and well connected, but he was also a strong political backer of Houston Mayor Roy Hofheinz. Smith owned a large amount of land along the Post Oak Road corridor north of Bellaire. In fact, he and his associates owned about 90% of the land along the freeway corridor between Westheimer and Memorial Park, so he stood to reap a huge financial gain from the construction of the West Loop. This did not go unnoticed by Houston City Council. In December 1954, Houston City Council was informed of a new plan to move the alignment of the West Loop slightly east of Post Oak Road for a section between the Southwest and Katy Freeways. The realignment shifted the freeway to vacant land, saving approximately $400,000 in right-of-way costs, approximately 2.28 million in 2003 dollars. The realignment would cause the West Loop to bisect Smith's property, so his land holdings would have freeway frontage on both sides of the freeway. *The Houston Chronicle* reported that "a majority [of council members] indicated opposition on the argument that the routing would increase vacant land owned by Robert E. Smith, political backer of the mayor." The alignment through Smith's property would be adopted, and Smith would reap a financial windfall when the West Loop commercial boom subsequently took place.[17]

The West Loop story has a happy ending for the residents of Bellaire. If ever there is a need for evidence that freeways do not destroy neighborhoods, one needs to look no further than Bellaire. In spite of the West Loop, or perhaps because of it, Bellaire has become one of the most desirable residential areas of Houston. Starting in the 1980s, Bellaire became known as a "tear-down" area, where older homes were torn down and replaced with large and expensive custom homes. Many of the new homes were constructed immediately alongside the West Loop. In 2001, the median home price in Bellaire was $143 per square foot ($1,539 per square meter), far above the Houston average of $66 per square foot ($710 per square meter) and the Harris County average of $60 per square foot ($646 per square meter). New homes in Bellaire typically sell for $500,000 to more than $1 million. Some Bellaire residents may feel that their city would be even better without the freeway, but the convenience and transportation access provided by the West Loop is certainly something that should not be discounted.[18]

In 1997, 43 years after the original controversy over the routing of Loop 610 through Bellaire, TxDOT called for a new series of public hearings to discuss improvements to the West Loop, including the section through Bellaire. The project was a "no capacity added" project, and proposed improvements were very modest, consisting mostly of improvements at entrance and exit ramps. Nevertheless, substantial opposition developed, especially over a plan to extend the West Loop frontage roads underneath the Southwest Freeway just north of Bellaire. It was almost as if the clock had been turned back, and TxDOT officials once again felt the fury that had been released in 1954 when the freeway was originally planned. Some in Bellaire were demanding that the freeway be depressed below grade, but flooding concerns quickly scuttled that idea. The reconstruction of Loop 610 in Bellaire was underway by 1999 with only a few modifications from the original plan. The lack of added capacity, however, would ensure that the freeway would not be able to meet future demand.

Freeway disaster: A cloud of ammonia gas engulfs the West Loop-Southwest Freeway interchange on May 11, 1976, moments after a speeding tanker truck fully loaded with ammonia crashed through the guardrail on a connector ramp and fell to the Southwest Freeway below. Five people were killed, about 50 were hospitalized, and about 150 received treatment. This photo was taken by photographer Carroll Grevemberg about one minute after the accident from the thirteenth floor of an office tower in the Galleria complex. Grevemberg, an audio-video designer at the Transco Company, heard the explosion and then heard someone say, "I wish I had a camera." Grevemberg grabbed his camera, loaded some film, and was able to capture the gas cloud as it expanded over the interchange and then dissipated during the next five minutes. (Photo: Carroll Grevemberg, Grevy Photography, New Orleans)

Disaster

May 11, 1976, 11:18 A.M.: A tanker truck loaded with ammonia speeds northward through Bellaire on the West Loop toward the Southwest Freeway interchange and exits to make the connection to the southbound Southwest Freeway. As the truck rounds the curve, its speed is too great and it cannot stay on the roadway. The truck crashes through the guardrail and falls to the Southwest Freeway main lanes one level below. The resulting explosion unleashes a cloud of ammonia gas, engulfing the interchange. Four people are killed immediately, three by asphyxiation. Others attempting to flee the scene by foot collapse before they can escape the gas cloud. Fifty people are hospitalized, and another 150 are treated and released. The death toll reaches five in the following days.

In terms of loss of life, it ranks as a tragic accident, although certainly not as deadly as other highway accidents over the years, particularly those involving buses. But in terms of extraordinary events and sheer drama, it is the most horrific and memorable accident in the history of Houston's freeway system. Survivors told stories of fleeing the gas cloud and barely making it out alive. Others who had collapsed in the gas cloud were dragged out by good samaritans. In an adjacent neighborhood, foliage looked like it had been hit with a hard freeze—even though it was May. The freeway interchange structure was badly damaged, necessitating the closure of freeway lanes for the rest of May to rebuild a pier structure for the West Loop overpass.[19]

Edge City Extraordinaire

Large business centers generally come in two varieties: those that existed before the freeway era, and those that arose as a result of freeways. The Uptown Houston district along the West Loop provides one of the most dramatic transformations of suburban, freeway-side acreage

(Opposite page) **Transformation:** This view shows the West Loop/Uptown Houston corridor in 1960 just as the first freeway construction is beginning. In the lower part of the photo, land preparation is underway for the West Loop-Southwest Freeway interchange. The dashed lines from top to bottom indicate the route of the West Loop. The dashed lines from left to right indicate the route of the Southwest Freeway. The West Loop main lanes between the Southwest Freeway (US 59) and the Katy Freeway (IH 10) were complete in November 1966, and the opening of the Galleria shopping mall in November 1970 launched the West Loop commercial construction boom. For a modern view from the same perspective, see the chapter introductory photo on page 272. (Photo: The Positive Image)

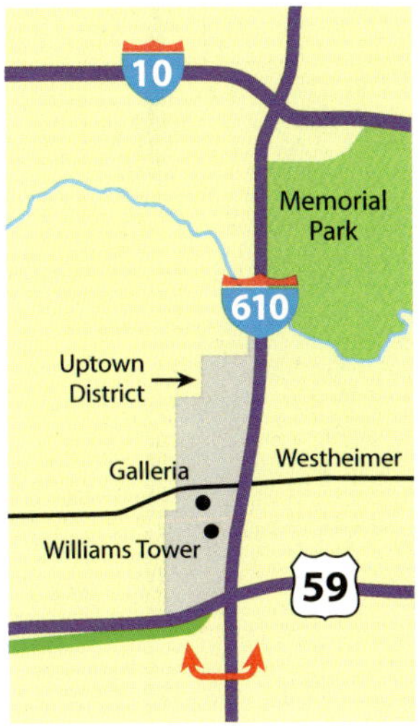

into a major business center. In 1960, the area that would become Uptown Houston was mostly vacant land with a scattering of strip shopping centers, some residential development, a drive-in cinema and a television station. But change was coming. Work was just beginning on the interchange at the West Loop and Southwest Freeway, and by 1961 work was underway on the West Loop frontage roads and main lanes. The first section of freeway main lanes in the Uptown Houston district opened in June 1964, and the entire West Loop was complete in 1968.

The seeds for the West Loop commercial boom were sown by developer Gerald Hines, who began work on the Galleria shopping center in the late 1960s at the corner of Post Oak and Westheimer, just west of the West Loop. The Galleria was a new concept for Houston, featuring a three-level shopping mall with a central ice rink and a glass canopy roof. Office buildings and hotels were integrated into the Galleria complex. The mall opened on November 16, 1970, and became influential in the development of high-end, mixed-use malls across the United States.[20]

The Uptown Houston district boomed along with Houston during the 1970s and early 1980s. An impressive collection of mid-rise office buildings rose along the West Loop. It became one of the most impressive instances of the edge city, a term popularized by author Joel Garreau in his 1992 book, *Edge City: Life on the New Frontier*. The crowning achievement of Uptown Houston was the construction of the landmark 899-foot-tall (274 m) Williams Tower (known as the Transco Tower until 1999) by Gerald Hines Interests in 1983. At the time, it was believed to be the world's tallest skyscraper outside of a central business district. The Williams Tower has a dominating presence on the landscape and dwarfs the nearby structures which are generally in the 20- to 40-floor range. The Williams Tower was the product of a unique era in Houston, a period when energy companies were flush with cash and sought impressive, monumental structures to project their power. Large-scale office construction in Uptown Houston came to an end with the collapse of energy prices and the meltdown of Houston's economy in the mid-to-late 1980s. In 2001 the West Loop commercial district had 23.8 million square feet of office space. Downtown Houston, in comparison, had 39.3 million square feet. In the late 1990s a mini-boom of mid-rise residential tower construction, typically 30 floors tall, got underway. Uptown Houston has accumulated a surprisingly large concentration of high-rise residential structures for a low-density city such as Houston. Many of those high-rise residents are looking to avoid the traffic congestion on the West Loop, no doubt.[21]

A calm, quiet intersection: This view, looking northeast at the Westheimer intersection, shows the first section of the West Loop main lanes to open in the present-day Uptown Houston district. At the time of the photo in December 1964, the freeway main lanes stopped north of Westheimer and the land along the frontage roads was vacant. Within a few years, large-scale commercial development of the land along the West Loop would begin and the Westheimer-West Loop intersection would go on to become one of the most chronically congested in Houston. (Photo: TxDOT)

Uptown Houston: The mid-rise office towers of the Uptown Houston district are dwarfed by the 64-floor, 899-foot (274 m) Williams Tower. Reconstruction of this section of the West Loop began in 2003. Although it is a "no capacity added" project, the reconstruction will substantially improve merging operations and interchange performance. (Photo: May 2002)

The Traffic Boom and the Planning Bust

The commercial boom along the West Loop brought dramatic increases in traffic. In 1968, the year the main lanes of the West Loop were completed, the peak traffic volume was 90,600 vehicles per day. In 1971, just one year after the opening of the Galleria shopping center, peak traffic volume had grown to 146,200 vehicles per day, making the West Loop Houston's busiest freeway. Traffic would continue to grow dramatically through the 1970s and the West Loop would hold the title of Houston's busiest freeway until 1991, when it was eclipsed by the North Freeway. More than any other freeway in Houston, the West Loop sustains severe traffic congestion in both directions during rush hour.

Houston's freeway expansion program became very active in the 1980s. In 1989 work was underway to widen the section of the West Loop between the Katy and Northwest Freeways to 12 and 14 main lanes. The project was complete by the end of 1992. In 1991 the time had finally arrived to formulate a plan to relieve the chronically congested section of the West Loop between the Southwest and Katy Freeways. Big problems required big solutions. The plans for the West Loop proposed a major expansion of the freeway to meet the transportation needs of the area. TxDOT called a public meeting in November 1991 to present the developing plans to the public.

Several options were presented, but attention focused on the leading proposal, the "collector-distributor" option.

It retained the 8-lane freeway in the center, adding 4-lane collector facilities on each side of the freeway to provide 8 more freeway lanes. Frontage roads were generally increased to 4 lanes in each direction. Overall, the proposal generally had 16 freeway and 24 traffic lanes, with exact lane counts varying from point to point. Right-of-way acquisition was minimal since high-rise structures lined both sides of the corridor. The proposal did not include elevated structures and kept all lanes at ground level.

As soon as the meeting started, however, it became clear that this would not be a typical freeway hearing. It became more like an anti-freeway rally. Before public comment even began, one public official after another stepped up to the podium to bash the proposed collector-distributor plan. Most vocal were Houston council members Jim Greenwood, an architect and mass transit advocate, and Sheila Jackson-Lee, who promoted inner-city and minority interests. Greenwood, Jackson-Lee, and others called for a greater emphasis on mass transit. The large crowd at the hearing was unusually receptive to the anti-freeway speakers.

Possibly the most lethal opposition came from the Park People, an organization which worked to improve park resources in Houston. The West Loop cuts through the western edge of Memorial Park, a wooded area with approximately 1,500 acres of parkland. Three and a half acres of Memorial Park were needed to accommodate the widened freeway. The Park People were not willing to accept the loss of a single square foot of parkland for the freeway, and Jackson-Lee summed up the Park People's sentiment when she stated, "In this city, any loss of park space cannot be tolerated." Realistically, users of the 1,500-acre Memorial Park would never notice the loss of 3.5 acres along the western edge of the park, and many park users would have benefited from the improved access to the park. But in situations such as highway project development, reason often does not prevail and emotions can take over.[22]

Also in November 1991, Houston elected Bob Lanier to become mayor of Houston. If anyone could save plans to expand the West Loop, it was Bob Lanier. As chairman of the Texas Transportation Commission during the 1980s, Lanier was a strong advocate for highway construction and had been instrumental in dramatically increasing TxDOT funding. But Lanier was no longer just a highway advocate. He was now a politician. He had to balance the various issues facing him as he entered office, and highway construction was just one issue among many. Lanier gave lip service to the expansion plan, but he was not willing to use up valuable political capital to save it. And it would have taken a lot of capital to keep the expansion plans alive.[23]

How could this happen in Houston? Several factors had converged to cause the anti-freeway outburst. Vocal anti-freeway activists were on Houston City Council. The Houston business community, so often a key supporter of freeways, was remarkably silent. The West Loop business community seemed to be missing in action. Bob Lanier stood by, unable or unwilling keep the plans on track. The press depicted the expansion as a "24-lane" freeway when in fact the freeway section was much smaller, and the potential benefits of the project were not adequately reported.

Was Houston going to succumb to the forces that have devastated transportation planning in other cities? Was this the end of Houston as a forward-looking freeway metropolis? Fortunately for Houston's freeways, the answer was no. It was, in effect, more like a certain alignment of planets had occurred at that particular moment in time in Houston freeway planning. Everything that could have gone wrong in the process did go wrong, so the West Loop expansion died.

By the late 1990s the pavement on the West Loop was crumbling and action was needed. In 1997 and 1998 meetings were held to reach a compromise plan to rebuild the freeway as a "no-capacity-added" project. The final plan included the addition of new merging lanes at entrance and exit points, better lane balance, major modifications to the interchange with the Southwest Freeway, and total reconstruction of the interchange at the Katy Freeway in conjunction with the Katy Freeway expansion. Work began on the southernmost section of the West Loop in 1999, and the final contract valued at $262 million for the work near the Katy Freeway was awarded in July 2003.

The failure of plans to expand the West Loop will have a long-term negative impact on the performance of Houston's freeway system. The importance of the West Loop as a critical link in Houston's freeway system will increase in the future as other freeway projects move forward. In the near future, the Katy Freeway expansion, the Fort Bend Parkway, and the Westpark Tollway will all feed more traffic onto the West Loop. Longer term, a planned expansion of the Northwest Freeway and a potential future tollway along Hempstead Road will bring even more traffic. A long-term planning map published by the Harris County Toll Road Authority in 2001 shows a potential toll road corridor along the Union Pacific railroad which parallels the West Loop about half a mile (0.8 km) to the east. However, the railroad passes through Memorial Park and near high-income neighborhoods, and any effort to construct the route is certain to be highly controversial. The proposal appeared to be dead on arrival when it first received wide publicity in June 2003. But it doesn't take a prophet to conclude that transportation demand will overwhelm the West Loop. At some point, the toll road proposal or an alternative plan—perhaps elevated lanes—will need to be given consideration. Houston motorists can only hope for a more favorable alignment of the planets the next time traffic relief plans are put on the table.[24]

Home for the Dome

The alignment of the South Loop was officially approved by the Texas Transportation Commission in October 1954. At the time, the prairies south of Houston were a quiet area. There was little residential or commercial development in progress, and the area was semirural.

(Above) **The replacement:** This view looks northwest over the Astrodome and its replacement, the $449 million, retractable-roof Reliant Stadium. The Houston Texans football team played its first game in the stadium on August 24, 2002. (Photo: The Positive Image)

(Opposite page) **Building the Dome:** The Astrodome was the world's first fully enclosed and air-conditioned stadium when it opened on April 9, 1965. The upper photo, looking north, shows the beginnings of the Astrodome in 1962. Excavation for the Dome was nearing completion and construction of the South Loop frontage roads had just begun in the lower part of the photo. The frontage roads opened in June 1963. The lower photo, looking southeast, shows the structural shell of the Dome substantially complete in 1964. The Astrodome was renovated and its seating capacity increased in 1989, but the stadium could not meet modern standards. The Houston Oilers football team played its last game in the Astrodome on December 15, 1996, before moving to Nashville, Tennessee. The Houston Astros baseball team played its last game in the Astrodome on October 9,1999, before moving to the new downtown baseball stadium. When the NFL awarded Houston a new football franchise on October 6, 1999, plans moved forward to build a state-of-the-art, retractable-roof football stadium that would also be used by the Houston Livestock Show and Rodeo, the Astrodome's only remaining major tenant. (Photos: upper, The Positive Image; lower, Houston Photographic and Architectural Foundation Trust)

However, Harris County Judge Roy Hofheinz would soon start contemplating about the future of professional sports in Houston, and his vision would find a home on the South Loop.

In the late 1950s local authorities began discussions for a new sports center for the Houston area. In 1958 a stadium location study was completed. Various locations throughout Houston were considered, and a site on the South Loop was rated as number one by both the city of Houston and TxDOT. The accessibility provided by the planned South Loop Freeway and nearby South Freeway was the principal strength of the site, and the land was readily available for sale by millionaire oilman R. E. "Bob" Smith and others.[25]

The construction of a stadium along a freeway was nothing unusual. But this stadium would be different. Judge Hofheinz wanted a futuristic facility to gain worldwide recognition for Houston. His new stadium would be the world's first fully air-conditioned domed stadium. The otherwise nondescript segment of the South Loop would be distinguished by this first-of-its-kind structure.

Harris County voters approved $20 million in revenue bonds on July 26, 1958, for the domed stadium. However, the revenue bonds depended on future revenue to be generated by the facility, and the lack of history of revenue from stadiums made it difficult for Harris County to sell the bonds. The domed stadium moved beyond the talking stage on January 31, 1961, when voters agreed to replace the revenue bonds with $22 million in general obligation bonds, which were backed by general tax revenue. When the lowest bid for the construction of the dome came in at $19,440,000 (excluding excavation, which had previously been completed for $738,000), it became necessary

Key dates in the history of Loop 610	
Early 1930s	Loop or bypass routes are first proposed.
1941	The first formal study for a bypass loop is completed.
1942	The North Loop is adopted into the state highway system as a highway.
1954	The West and South Loops are officially adopted into the state highway system as freeways. The North Loop is upgraded to freeway status.
1960	The first freeway section of the Loop opens. The East Loop is adopted into the state highway system.
1962	Houston's first four-level interchange is completed at the West Loop-Southwest Freeway intersection.
1965	The Astrodome is opened on the South Loop on April 9.
1970	The Galleria shopping center opens near the West Loop on November 16, launching the construction boom along the West Loop.
1973	The ship channel bridge opens on March 2.
1975	The loop is completed on September 22.
1983	The 899-foot-tall (274 m) Williams Tower is completed near the West Loop.
1992	Plans to expand the West Loop are cancelled due to opposition.
2007	Scheduled completion of reconstruction of the West Loop, including a full rebuild of the Katy Freeway interchange.

to raise more funds with a supplementary bond issue of $9.6 million to pay for all project costs and site work. The bonds were approved by Harris County voters on December 22, 1962, bringing the total bond funds to $31.6 million. For comparison, the $31.6 million translates to approximately 154 million in 2003 dollars. Still, that seems like a bargain by modern standards. The replacement for the Astrodome, the retractable-roof Reliant Stadium completed in 2002, cost $449 million.[26]

Ground was broken for the domed stadium on January 3, 1962. The new stadium was officially dedicated on April 9, 1965, when the Houston Astros baseball team played an exhibition game against the New York Yankees.

Even before the first game, a problem had arisen. The first time the Houston Astros ventured onto the new field for practice on April 7, they discovered that the glare from the roof, with its clear plastic panels, made it extremely difficult to catch fly balls. Sunglasses didn't help since the area of glare coming from the roof was so large. A quick solution was needed to make daytime baseball possible in the Astrodome. The week of April 19, crews began painting the clear roof panels with off-white paint to make the panels opaque rather than transparent. The grass in the stadium was already struggling before the roof was painted, and paint further reduced available sunlight by 25-40%. Artificial turf had recently been developed and successfully installed at a private school in Providence, Rhode Island, in 1964. The solution was obvious: the Astrodome would become the first major sports venue in the United States to use artificial turf. The first major league baseball game played on Astroturf took place on April 8, 1966. Astroturf would soon become widely used in sports stadiums across the United States.[27]

The South Loop was still a work in progress when the Astrodome was completed in 1965. Frontage roads for the freeway were in place near the Astrodome, but to the east and west, the South Loop didn't exist at all. On May 16, 1969, a 2.5-mile (4.0 km) section of the South Loop main lanes at the Astrodome was opened, completing all of the South Loop except for a short segment west of the Gulf Freeway interchange. Aside from the Astrodome, the South Loop didn't get any glamorous or distinctive development. During the 1970s the South Loop became a favorite location for the large, boxy warehouses of furniture retailers. The furniture retailers would frequently go out of business or change names, leaving vacant warehouses along the freeway. By the end of the 1970s the furniture warehouse era had largely come and gone.

On March 2, 2002, the Houston Livestock Show and Rodeo ended its 37-year run in the Astrodome with a farewell concert attended by former President George H. W. Bush and featuring a star-studded list of country music performers. A record-setting crowd estimated at 70,200 packed the Dome for what was expected to be its final major event. It was a bittersweet night for Houston. The Dome, once touted as the "Eighth Wonder of the World," had propelled Houston to international prominence and had been a source of civic pride. Now, it was just another passé relic of the 1960s. The Dome faced an uncertain future in 2003 as Harris County officials searched for an economic use for the facility. Local preservationists expressed hope that the Dome could be spared from demolition.[28]

The South Loop has a more certain future than the Astrodome. The section of the South Loop serving the Astrodome and its replacement, Reliant Stadium, was expanded to 10 main lanes in 1994, providing ample traffic-carrying capacity. In the long term, there is the possibility of a large new stack interchange at the intersection with the proposed Alvin Freeway. The future of the Alvin Freeway will be determined by a study expected to be underway by 2004.

The Future of the Loop

The reconstruction of the West Loop will continue until approximately 2007. As of 2003, there are no major studies planned for the north, east, and south sections of Loop 610. For the intermediate future Loop 610 will likely not see major changes. Increasing traffic congestion on the North Loop may eventually prompt some work on that segment, however.

South Freeway interchange: This wide-angle view looks east along the South Loop at the sprawling South Freeway interchange. This interchange was completed in 1978 and opened to traffic in 1981. (Photo: September 2002)

East Freeway interchange: This view looks north along the East Loop at the East Freeway interchange. This interchange was completed in 1976. (Photo: November 2002)

Sam Houston Parkway/Tollway, Beltway 8

Even before the alignment of Loop 610 had been finalized, the city of Houston was formulating plans for a second loop. It was a remarkable act of vision and foresight to recognize the future importance of loop highways in today's predominant suburb-to-suburb transportation patterns. However, the first wave of freeway construction in Houston from the 1950s to the 1970s came and went with very little progress on the Beltway.

The age of the Houston Beltway arrived with the second wave of Houston freeway construction, which started in the 1980s. By 1996 the entire loop had been constructed in some form—freeway, tollway, or frontage road. Completion of the South Belt main lanes in 1997 left only one segment without main lanes. The phenomenal success of the Sam Houston Tollway, the toll main lanes of Beltway 8, even brought traffic congestion to the western and northern sections of the tollway.

Although the main lanes of Beltway 8 are, for the most part, not very interesting, the Beltway has one of the nation's most impressive collections of modern four- and five-level freeway-to-freeway stack interchanges. The Beltway is also unusual in that its tollway sections have continuous toll-free frontage roads.

Origins

Beltway 8 had its origins in a 1952 report by the City of Houston Planning Department, *Proposed Location for An Outer Belt Drive for Metropolitan Houston*. The report was prepared as a basis for fixing a location for a minimum 120-foot-wide (37 m) thoroughfare located four to five miles (6 to 8 km) beyond the city limits, which were generally located near Loop 610 at the time. The 1952 annual report of the City of Houston Planning Commission went on to explain, "This report was prepared in view of the imminent development of much of the area through which the thoroughfare would need to pass and because there will be a great need for such a thoroughfare in the future as the population of the urban area spreads." At the time the report did not envision the Outer Belt as a freeway. Ralph Ellifrit, city of Houston planning director, was the individual most responsible for the birth of the Outer Belt.

In September 1954, based on the recommendation of Ellifrit, the corridor width was increased to 150 feet (46 m) to accommodate a larger arterial highway. Also in 1954, the City of Houston Planning Department performed detailed studies to fix the location of the Outer Belt. Nearly all of the alignment defined in that period would become the ultimate route of today's Beltway 8.[29]

In 1960 Harris County stepped forward and took the leading role in the development of the Outer Belt. Harris County Judge Roy Hofheinz appeared before the Houston Planning Commission in March to discuss the merits of changing the Outer Belt to a full freeway on a minimum 300-foot-wide (91 m) right-of-way. On July 11, 1960, Harris County Commissioner's Court voted to increase the corridor right-of-way width to 300 feet and grant it full freeway status. Harris County would now be in charge of building the 87-mile (139 km) freeway loop. Getting into the freeway-building business was a big undertaking for Harris County and would ultimately be more than it could manage on its own. As of April 1960, a 300-foot-wide right-of-way corridor had been obtained for only 8 miles (13 km). Seven miles (11 km) had 150–250 feet (46–76 m) set aside, and 15 miles (24 km) had 120 feet (36 m) or less set aside. For the remaining 57 miles (91 km), no right-of-way had been acquired. However, only 1 mile (1.6 km) was listed as passing through a built-up area with "damage to buildings." The projected cost of the Beltway was $150 million, still a relatively low cost after adjusting to 750 million in 2003 dollars.[30]

Beltway 8 (Sam Houston Parkway/Tollway)	
Designated as freeway	1960
First freeway section open	1970 (overpasses only) 1982 (toll bridge)
Freeway/tollway complete	Scheduled 2007
Reconstruction	Intermittent pavement repair only
Max traffic volume, 2001	233,000 vehicles per day
Future construction	Main toll lanes of northeast segment; expansion of west and north tollways

(Opposite page) **The Southwest Freeway interchange:** Beltway 8 gently curves through the interchange, completed in 1997. The Beltway 8/Sam Houston Tollway has an impressive collection of modern, multilevel stack interchanges. (Photo: May 2003)

An Early Patchwork

Starting in the 1950s, the city of Houston and Harris County began to acquire right-of-way and build short sections of roadway on the Outer Belt alignment as real estate developers donated land and as funding permitted. In 1958, one of the early sections to open was a 2.5-mile (4 km) section of the East Beltway built as a two-lane roadway north of today's Business Route 90 (Beaumont Highway). That section was made possible by a land donation from a real estate developer and was constructed to provide access to the development. By the late 1960s, short sections of frontage roads, or in some cases only one side of the frontage roads, were open in southeast Houston near the Gulf Freeway and in west Houston south of the Katy Freeway. Harris County constructed a section of frontage roads in Pasadena in 1971. These short sections were generally not very useful even to local traffic because of their short lengths and lack of connectivity.[31]

One trouble spot for the Outer Belt was already developing in 1961. The western Outer Belt was aligned to go through the center of the municipality of Jersey Village northwest of Houston. Jersey Village was just outside the extra-territorial jurisdiction of the Houston Planning Commission, so the commission had no power to approve or reject development plans and could not protect the freeway right-of-way from development. Houston Planning Director Ralph Ellifrit raised an alarm to TxDOT that the freeway right-of-way was at risk of being developed. The right-of-way was not protected, and plans for residential development in Jersey Village proceeded in the path of the proposed Outer Belt.[32]

The only rumbling of community opposition in the

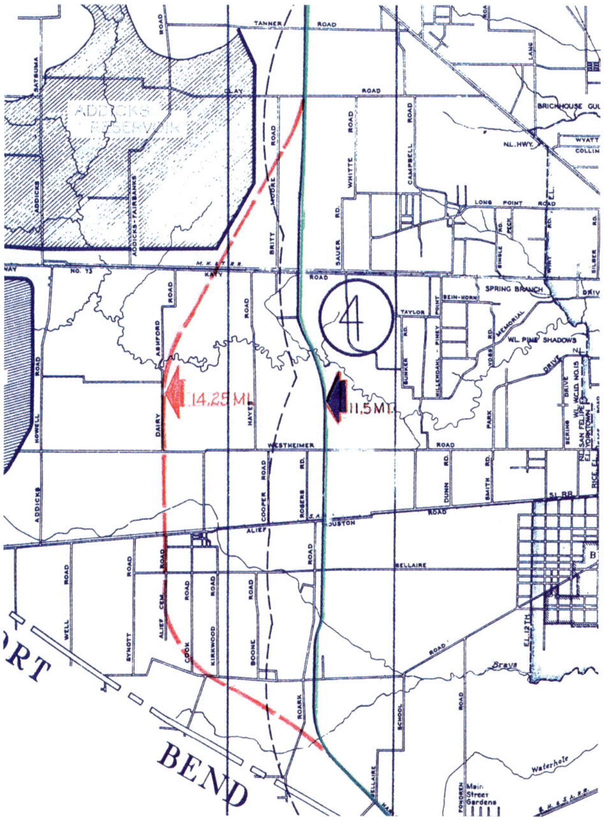

Controversy: Beltway 8 West through the Memorial Bend subdivision was one of the two controversial sections of the Beltway. In 1962 residents of Memorial Bend attempted to have the Beltway realigned three miles west to follow the present-day Dairy Ashford Road, but the increased cost of the longer alignment resulted in its rejection by the Houston Planning Commission on June 19, 1962. The map at left, from the original Outer Belt location studies conducted circa 1954, shows the approximate alignment (in red) that was requested by Memorial Bend residents. As part of the preparation of the environmental impact statement for the West Belt in 1975, a study to determine the feasibility of alternative routes was completed. The only possible alternative route for the West Belt was determined to be SH 6, six miles to the west. The SH 6 route was infeasible due to cost and other impacts. This section of the Beltway, which opened in 1988, has a highly unusual short section of a two-way frontage road, visible on the right side of the photo. (Photo: May 2002; map, city of Houston records)

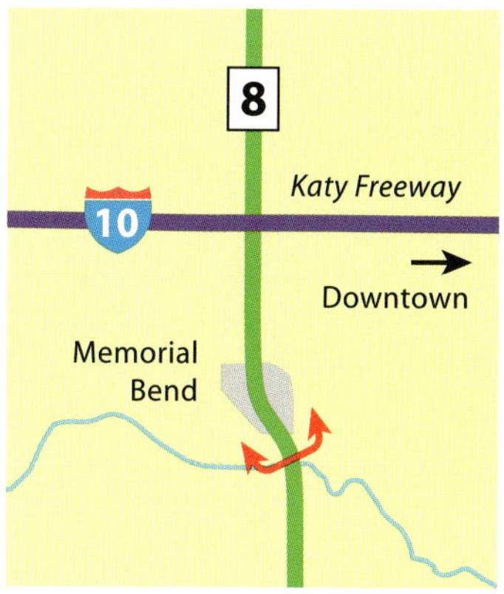

early development of the Outer Belt occurred in 1962 when residents of the Memorial Bend subdivision in west Houston objected to plans to align the Outer Belt through their neighborhood. The neighborhood requested a realignment of the freeway at a meeting of the City Planning Commission on June 5, 1962. The proposed realignment would have shifted the Outer Belt three miles (5 km) west to Dairy Ashford Road. The request was denied on June 19, 1962, primarily because of the substantially increased cost of routing the freeway westward. The West Belt remained on its originally planned route.[33]

The first section of the Outer Belt to be constructed as anything resembling a freeway was the North Belt near Bush Intercontinental Airport. The original plan for the airport was described in an October 1961 engineering report titled *Plan of Development, Jetero Intercontinental Airport*. In its section about roadway access, the report stated, "It is recommended strongly that the North Belt Drive between US 59 and Interstate 45 and its connecting link to the airport be constructed as soon as possible." Progress on this section was delayed due to voter rejection of a county bond referendum in 1963, but a successful bond referendum in January 1966 authorized $14.8 million in funding for the Outer Belt. The North Belt frontage roads were completed in February 1970, shortly after the opening of the airport on June 8, 1969.

"We do not anticipate any improvements by the State on Beltway 8 in the next 20-year period."

TxDOT Houston district head Omer Poorman to Houston mayor Fred Hofheinz, November 12, 1976

The construction of the North Belt made Harris County realize it had bitten off more than it could chew with the full Outer Belt. In June 1967 Harris County was already requesting assistance from TxDOT for the construction of the North Belt-North Freeway interchange. The cost of just the right-of-way for the 87.5-mile (140 km) Outer Belt would probably be more than the county could bear. Responding to requests from local officials, the Texas Transportation Commission officially adopted the Outer Belt into the state highway system on March 7, 1969. In July 1969, the Outer Belt was officially designated as Beltway 8.[34]

Starting in 1972, TxDOT began engineering and environmental studies for Beltway 8. Numerous public hearings for all sections of the Beltway were held in 1975 as part of the preparation of the environmental impact statement. The schematics presented at the 1975 meetings showed a minimum of eight freeway main lanes with a 28-foot-wide (8.5 m) central median and a right-of-way corridor that had a minimum width of 420 feet (128 m). The corridor width would later be downsized to a typical width of about 300 feet (91 m) due to funding shortfalls, a decision which may ultimately come back to haunt Houston's highway planners. For the western segment that included the Memorial Bend subdivision, a special study titled *Investigation of a Possible Relocation of West Beltway 8* was undertaken by an interdisciplinary team. It considered ways to avoid the Memorial Bend subdivision, but because west Houston had become so heavily urbanized by that time, the only other possible alignment was six miles (10 km) to the west on SH 6. The SH 6 route was ruled infeasible due to greatly increased cost, as well as legal and administrative issues. Although opposition in Memorial Bend lingered, the route of the freeway through the neighborhood was effectively finalized in 1975. Another community affected by the Beltway, Jersey Village, would continue to be a flash point of controversy through the 1970s.[35]

Dark Days

The adoption of Beltway 8 into the state highway system in 1969 seemed to be good news for the freeway, but it came at a time when TxDOT was about to descend into a financial crisis that drastically curtailed its ability to construct new freeways and made large, costly projects like Beltway 8 impossible. Starting in the early 1970s, highway construction inflation spiraled out of control and transportation funding was stagnant or shrinking. Houston's boom was driving up property values, making right-of-way acquisition costly. In September 1975, referring to Beltway 8, the chairman of the Texas Transportation Commission Reagan Houston III stated, "Our funds have diminished and our expenses are escalating which leaves little left for new construction."[36]

As TxDOT was backing away from new projects and seemed willing to let Beltway 8 die from neglect, responsibility for saving the Beltway shifted back to Harris County. TxDOT couldn't afford to build it. Harris County couldn't afford to build it. The only remaining option was a toll road. In May 1975 Harris County asked the Texas Turnpike Authority to study the west and northwest sections of the Beltway as a potential tollway. The Turnpike Authority concluded that the route was financially infeasible as a tollway. Harris County Judge Jon Lindsay was starting to doubt that the Beltway would ever be built, and was considering reallocating funds set aside for right-of-way acquisition.[37]

Plans for Beltway 8 reached a low point in August 1976 when a comprehensive study of TxDOT's highway construction program conducted by the McKinsey consulting firm developed two possible scenarios for the future of Houston's freeways, one with expected funding levels and one with an increased level of funding. The Beltway was not included in either plan. Based on the results of this study, the head of the Houston district of TxDOT, Omer Poorman, conveyed the following grim message to Houston Mayor Fred Hofheinz in a letter dated November 12, 1976, "We do not anticipate any improvements by the State on Beltway 8 in the next 20-year period."[38]

But Houston was not going the let the Beltway die. As

The West Belt: This view looks north along the West Belt at the Westchase commercial district. This section of tollway opened in 1988. (Photo: May 2002)

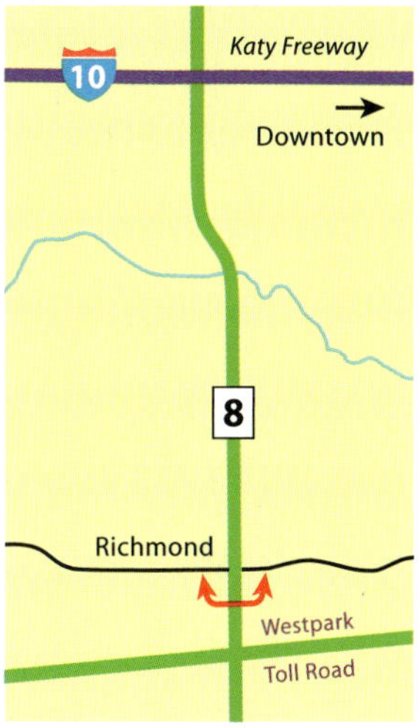

with many freeway projects in Houston's history, the real estate and land development community stepped forward to lobby for the Beltway, forming an organization called the Transportation Development Group which focused on getting the Beltway built. At the urging of the Transportation Development Group, Harris County authorized another tollway study in 1976. In 1977 there was more bad news. The Texas Turnpike Authority once again concluded that the west and northwest sections of the Beltway were infeasible as a tollway but kept the project within its consideration for future study.[39]

The Tide Turns

The prospects for Beltway 8 began to improve by late 1977. TxDOT received additional funding from the Texas Legislature in 1977, allowing it to formulate a new plan of highway priorities. The new plan released in December 1977 restored frontage roads for a key section of Beltway 8 West between the Northwest and Southwest Freeways. Harris County realized that the key task at hand to save the Beltway was to preserve right-of-way in rapidly urbanizing sections of the city, especially west Houston. In 1978 Harris County authorized the use of bond funds for purchasing right-of-way for Beltway 8 West. Also in the summer of 1978, the Texas Turnpike Authority concluded that the Houston Ship Channel Bridge on Beltway 8 East was feasible and issued \$102 million in bonds (approximately 234 million in 2003 dollars) to build the bridge and adjacent sections of tollway. In June 1979 TxDOT began to take a more active role in moving the Beltway 8 frontage roads forward to construction, authorizing its staff to prepare plans and acquire right-of-way in the controversial section through the Memorial Bend neighborhood in west Houston. At a May 20, 1980, public hearing about the Memorial Bend plans, the *Houston Post* reported "unexpected strong public support for construction of the long-delayed Memorial Bend section." TxDOT also began working on plans for frontage roads on other sections of Beltway 8 around Houston.[40]

By 1980 the controversy surrounding the alignment of the Beltway through Jersey Village had been resolved, allowing the Beltway to move forward. Opposition first became vocal in 1973 when the Village Council held hearings where it stated its opposition to plans. As TxDOT continued with the environmental process in

1975, the plans showed the freeway on its original route, right through the middle of Jersey Village on an elevated structure. Two years later in 1977, key players including TxDOT, the city of Houston, and landowners outside of Jersey Village were sticking to plans to build the freeway through Jersey Village. But opposition in Jersey Village continued to build, and in 1977 a bill was introduced in the Texas Legislature to require TxDOT to route the freeway around Jersey Village. The bill did not become law, but TxDOT was forced to respond to the opposition and realigned the Beltway to avoid Jersey Village. Plans to align the Beltway on the east side of Jersey Village didn't make everyone happy. At a public hearing on January 8, 1980, several hundred residents, a "loud delegation" mostly from the subdivision to the east of Jersey Village, turned out to oppose the new alignment. In spite of the opposition, the eastern alignment was adopted. It would be the only major shift in the original planned alignment of the Beltway, but it would be a substantial shift as the Beltway snaked its way around Jersey Village.[41]

Momentum was now on the side of Beltway 8. It would still take the dedicated efforts of a project champion to get the main lanes built. Now that the Beltway had been saved, County Judge Jon Lindsay would take the lead in getting it built.[42]

As early as 1977 Lindsay had begun to realize that Harris County would need to take matters into its own hands if it wanted to see a tollway constructed in the foreseeable future. At the time, he stated that Harris County should be responsible for the Beltway toll road rather than the Texas Turnpike Authority (TTA). However, real progress toward the formation of the Harris County Toll Road Authority would not occur until 1982. By mid-1982 controversy surrounding the proposed Hardy Toll Road had caused the TTA to back away from its plans to build the tollway. Previously the TTA's feasibility studies had rejected the West Belt as a tollway candidate. It appeared that the TTA's Beltway 8 ship channel bridge would be its only project in the Houston area. In August 1982 Lindsay instructed the county attorney to investigate if the county could create its own toll road authority. It turned out that special legislation would be needed. The legal authority for Harris County to form a toll road authority came with Texas Legislature Bill SB970, signed by Governor Mark White in June 1983. Harris County then set a $900 million bond election for September 13, 1983, to launch the Harris County Toll Road Authority (HCTRA). The bonds were passed with 69.7% of the vote. The west and north Beltway 8 main lanes would be built as a tollway.

The Biggest Challenge

As the nation's highway building program built momentum in the 1950s and large-scale construction of the Interstate Highway System was launched in 1956, the people who would build the highways joined the national effort. One such person was Carol Letz, who served in various roles in right-of-way acquisition in the Houston TxDOT office starting in 1957. Letz was involved in nearly all the major right-of-way acquisition events in the history of Houston's freeway system and remains active in her position in 2003.

When Letz was asked to name the most difficult or challenging right-of-away acquisition in her career, without hesitation she responded, "Beltway 8 in west Houston." Really? This section of Beltway 8 included the short controversial section through the Memorial Bend subdivision, but the rest of it was through undeveloped, vacant property. What could be so difficult about buying up vacant land?

It was all a matter of timing. Efforts to acquire right-of-way were underway at the peak of Houston's oil boom in the late 1970s and early 1980s. Land values were increasing at a rate of about 30% per year, so quickly that it was nearly impossible to acquire property using the normal procedures. Typically there was a 60- to 90-day cycle for property appraisal, offer preparation, and obtaining approval from TxDOT headquarters in Austin. In that time period the appraisal would become obsolete due to the rapid escalation in property values. It was a losing battle, and the cost of right-of-way for the freeway went up as every month passed. Another complication was the administrative procedure for acquiring land. Harris County was responsible for acquiring all property and would then be reimbursed by TxDOT for 90% of the cost. Harris County simply did not have the cash on hand to expedite the process. It would make a purchase, wait for the 90% reimbursement to restore its bank account, and then proceed to the next purchase.

Landowners along the western section of the Beltway were seeing their hopes for real estate riches evaporate with the endless delays to the freeway. Just as land developer Frank Sharp had organized a group of landowners in 1957 to donate land for the Southwest Freeway to expedite its construction, landowners along Beltway 8 West formed an organization called the Beltway 8 Group and submitted a proposal to TxDOT in June 1982. The landowners would lock in their property values at existing appraisals that were between 1 and 2.5 years old. In return, TxDOT would agree to begin construction on the frontage roads as soon as all the land could be acquired, which was originally contemplated to be as short as five months. There were 23 parcels of property on the five-mile (8 km) segment of Beltway 8 that needed to be acquired, and the selling price was locked in at $36.8 million. The landowners' concession was estimated to save TxDOT about $17 million.[43]

It was a deal TxDOT couldn't refuse, so it was quickly approved. But there was a problem. Harris County was still responsible for the actual land purchase, and the county didn't have the money to finance the purchases. An overhaul of the land acquisition process was needed. By February 1983 the "re-engineered" process was in place. Harris County would pay its 10% share of the right-of-way cost, then TxDOT would do all the acquisition. As soon as the agreement was finalized, TxDOT's right-of-way acquisition machine went into high gear, acquiring

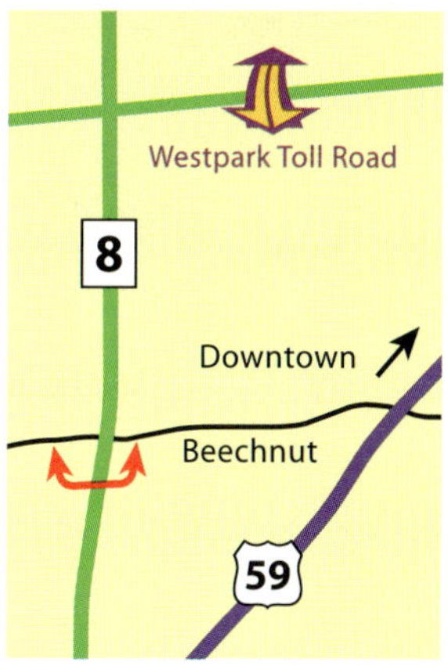

The most difficult right-of-way to acquire in the history of Houston's freeway system: This vacant land in the path of the West Belt between Westheimer and the Southwest Freeway (US 59) looks deceptively easy to acquire, but in fact it was the most challenging right-of-way to obtain in the history of Houston's freeway system. Efforts to acquire this right-of-way were underway at the peak of Houston's energy boom in the late 1970s and early 1980s. Land values were increasing at the rate of about 30% per year, so quickly that standard right-of-way acquisition procedures did not work. Approval of property appraisals typically required 60 to 90 days, but during that time the appraisals would become obsolete. It was a losing battle, and property values were escalating every month.

Landowners who wanted to see the project move forward struck a deal with TxDOT to lock in property values, allowing TxDOT to complete land acquisition. As part of the deal, TxDOT agreed to begin construction of the frontage roads as soon as the right-of-way was acquired. Construction began in 1983 and was completed in December 1985. This project launched the wave of construction that would build nearly the entire beltway by 1996. (Photo: Texas State Library & Archives Commission, June 1982)

Transformation: These views looking east along the North Belt at the North Freeway show the progression of development along the freeway in the Greenspoint area. The upper view on the opposite page shows the construction of the North Belt frontage roads in 1968. The frontage roads were constructed to provide access to Bush Intercontinental Airport, which opened on June 8, 1969. The lower photo on the opposite page was taken circa 1978. Greenspoint Mall opened in 1976 and development of offices, apartments, and retail centers was gaining momentum. The development boom of the Greenspoint area was largely complete by the mid-1980s. The Beltway 8 main lanes east of IH 45 in the Greenspoint area were completed in 1984, and the Sam Houston Tollway connection west of IH 45 was completed in 1990. The Greenspoint area matured in the 1990s and started to succumb to suburban decay. Local business groups worked hard to maintain the area and succeeded in stabilizing it. The suburban development story of Greenspoint culminated with the construction of the stack interchange. The first phase of the North Freeway interchange opened in 1997, and the full interchange was completed in early 2003. The above photo was taken in September 2002 as construction of the interchange was nearing completion. (Photos: opposite upper, The Positive Image; opposite lower, HMRC MSS 287 HH-935-B; above, September 2002)

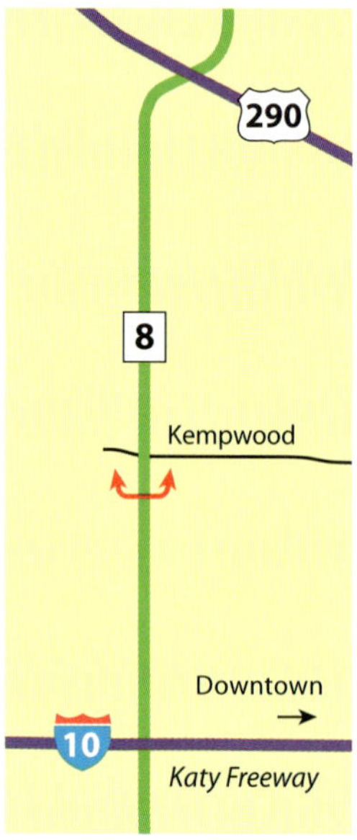

Success: By the late 1990s traffic congestion was a daily occurrence on the West Belt. This view looks north near Kempwood during the afternoon rush hour. The heavy traffic on the Beltway has been a financial windfall for the Harris County Toll Road Authority, enabling it to expand the toll road system. Projects to expand the West and North sections of the Sam Houston Tollway, from the Southwest Freeway to the North Freeway, to eight lanes began in 2002 and are scheduled to be complete by 2005. Expansion work in progress can be seen in the distance in this photo (the light-colored concrete). (Photo: September 2002)

$25 million in property within the next few weeks. It was still a challenging task, and one day before a decision had to be made about whether the first contract could be awarded on the agreed-to date, the right-of-way acquisition job was not complete. On the decision day at 8:15 A.M., Letz received the final right-of-access needed for the project. The job would go to bid, and the construction of Beltway 8 West was soon underway.[44]

The contract award for the short length of frontage roads in west Houston in 1983 marked the beginning of the wave of construction that would build Beltway 8. For 30 years the Beltway had moved forward one inch at a time, and often stood still. But now, TxDOT support and the newly created Harris County Toll Road Authority would push the Beltway into reality over the next 13 years. TxDOT constructed the frontage roads and sections of main lanes on the North Belt and East Belt, as well as most of the interchanges at intersecting freeways. The section of frontage roads included in the Beltway 8 Group funding agreement was opened in December 1985. After that, frontage road sections opened regularly around the Beltway, culminating with the opening of the final section to complete the circular loop in southeast Houston near Hobby Airport in 1996. HCTRA took on the job of building the main lanes as a tollway, with initial work focusing on the West Belt and North Belt.

Sam Houston Tollway, Cash Cow

Proceeding with construction of the Beltway 8 Toll Road was somewhat of a leap of faith for Harris County Judge Jon Lindsay. Revenue studies had rated it as a marginal, risky project and indicated a strong chance the project would lose money. The same study predicted that the Hardy Toll Road would be profitable. But Lindsay pushed forward with the Beltway Toll Road. In July 1985 a ground breaking ceremony was held for the construction of the first segment of the tollway from the Southwest Freeway to the Katy Freeway. In 1986 the Beltway was named the Sam Houston Parkway and was designated as a scenic district to prevent the proliferation of billboards along the frontage roads. The tolled main lanes were designated as the Sam Houston Tollway. The Beltway became the first and only limited-access facility within the city of Houston to be named after an individual—the first president of the Republic of Texas and namesake of the city of Houston. The first segment of the Sam Houston Tollway opened on June 29, 1988. The second section, from the Katy Freeway to the Northwest Freeway, opened in June 1989, and the third section, from the Northwest Freeway to the North Freeway, opened in July 1990. The Sam Houston Tollway was unusual for a tollway in that it had continuous, toll-free frontage roads for its full length.[45]

Even before the first section of the Sam Houston Tollway opened, Harris County officials had substantially lowered traffic and revenue projections. The collapse of Houston's economy in the mid-1980s and the extended recession through the late 1980s had made the project's original traffic projections obsolete. By mid-1989, one year after the opening of the first section of the Sam Houston Tollway and the completion of the Hardy Toll Road, toll revenue was less than 50% of original projections. A revised projection predicted that long-term revenue would be 40% less than the original estimates. During 1989 there was frequent talk of the possible need for a subsidy from Harris County's general tax revenue to meet toll road bond payments. By July 1990, Lindsay was warning that a $20 million annual subsidy from general tax revenue would be needed within two years.[46]

The third section of the Sam Houston Tollway, from the Northwest Freeway to the North Freeway, opened just as Lindsay delivered the financial bad news. Soon, however, the financial picture began to change. The complete, continuous tollway from Southwest Houston to Bush Intercontinental Airport caused a surge in traffic. Just three months after section three opened, Harris County officials were stating that a sharp increase in traffic resulting from the completion of section three had postponed the need for a subsidy for at least five years.[47]

The Sam Houston Tollway traffic boom had begun. Traffic and revenue surged ahead of projections during the 1990s, making the Sam Houston Tollway a cash cow for HCTRA. In the meantime, traffic and revenue on the Hardy Toll Road still lagged behind projections. It turned out that the original traffic projections in 1984, which predicted a successful Hardy Toll Road and a marginal Sam Houston Tollway, had missed the mark on both facilities. But the success of the Sam Houston Tollway allowed HCTRA to easily cover the Hardy Toll Road revenue shortfall.

Key dates in the history of Beltway 8-Sam Houston Tollway

1952	The Outer Belt is first proposed, originally as a major arterial street.
1957	The first roads built on the Outer Belt alignment open.
1960	The Outer Belt is designated as a freeway. Harris County takes ownership of the project.
1969	The Outer Belt is adopted into the state highway system and officially named Beltway 8.
1970	The first substantial section of the Beltway, a section of frontage roads, is completed in conjunction with the 1969 opening of Houston Intercontinental Airport.
1976	The Beltway is near death as the 1970s highway funding crisis makes new freeway construction impossible.
1977	The Beltway is restored to long-term plans.
1982	The Houston Ship Channel toll bridge is opened on May 6.
1983	Harris county voters approve the creation of the Harris County Toll Road Authority (HCTRA).
1985	In December a section of frontage roads opens in southwest Houston, launching the wave of construction that would build the Beltway.
1988	The first main lanes built by HCTRA open.
1989	Houston's first five-level stack interchange is completed at the Katy Freeway.
1994	HCTRA takes ownership of the ship channel bridge from the Texas Turnpike Authority on May 5.
1996	The full Beltway is complete, either as a freeway, tollway, or frontage road.
2005	Scheduled completion of expansion to 8 main lanes on the west and north Sam Houston Tollway.
2007	Scheduled completion of the last remaining section of main lanes in northeast Houston.

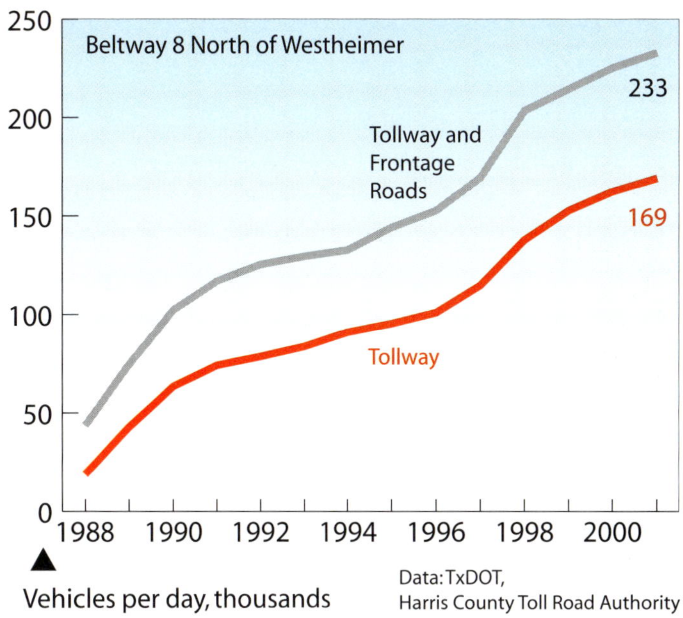

Cash cow: This plot shows traffic volume at the busiest point of the Sam Houston Tollway near Westheimer in west Houston.

Northwest Freeway interchange: This view looks along the southbound Sam Houston Tollway main lanes. This interchange was completed in 1990. (Photo: May 2003)

Construction: This view looks east along the South Belt at the construction zone for the Gulf Freeway interchange in late 1996. (Photo: Williams Brothers Construction Company)

In the early 1990s HCTRA Executive Director Wesley Freise and Lindsay were negotiating with the Texas Turnpike Authority to take over the financially beleaguered Beltway 8 Houston Ship Channel Bridge, which was facing an inevitable default on its bonds in 1996. The bridge transfer to HCTRA occurred on May 5, 1994. As part of the deal, HCTRA received $90 million from TxDOT for the construction of the Sam Houston Tollway between the La Porte Freeway (SH 225) and the Southwest Freeway (US 59), as well as TxDOT's commitment to build interchanges at the Southwest and Gulf Freeways estimated to cost $120 million. The southeast section opened in July 1996, the south section opened in March 1997, and the southwest section opened in May 1997. With those openings and additional progress on freeway sections of Beltway 8 in north and northeast Houston, only one section of Beltway 8 did not have its main lanes in place. That section in northeast Houston, from the Crosby Freeway (US 90) to the Eastex Freeway (US 59), is expected to begin construction in 2005 and be completed in 2007 as a tollway. Traffic congestion on the west and north Sam Houston Tollway prompted HCTRA to move forward with plans to widen the Sam Houston Tollway to eight main lanes from the Southwest Freeway to the North Freeway. The widening projects began in 2002 and are scheduled to be completed in 2005.

The Airport and the Beltway

The 1950s was a decade for big infrastructure dreams to take root. Houston's freeway system was put on the map in the early 1950s, and the Beltway was first proposed in 1952. Another important part of Houston's transportation infrastructure, its major airport, also was taking shape in the 1950s. As initial planning for a major airport took place, one thing became clear: the airport and the Beltway would go hand-in-hand. All potential airport locations were located immediately adjacent to the Beltway, and a site along the North Belt was purchased in 1957 and officially designated for the airport in 1960. The new airport provided the impetus for the construction of the first significant section of Beltway 8 in 1970—the only substantial section in existence for 12 years until the opening of the Beltway 8 ship channel toll bridge in 1982.

The story of Houston's major airport is in many ways a contrast to the story of the development of Houston's freeway network. While local authorities aggressively developed plans for the first-class freeway network in the early 1950s and worked hard to make it happen, the issue of Houston's airport was plagued by indecision and miscues. With the help of local business interests, the wheels were belatedly set into motion in 1957, and after numerous delays the airport finally opened in 1969. Construction and expansion of roads and freeways around the airport played out slowly after the opening of the airport, with significant momentum finally getting underway in the 1980s and major area-wide construction taking place in the 1990s.

History of Bush Intercontinental Airport

Today's William P. Hobby Airport was the city of Houston's first airport, opening in 1937 as Houston Municipal Airport on the site of a private airfield. By the early 1950s it became evident that the airport would not be able to meet Houston's expanding aviation needs. In 1951 the city of Houston contracted for the first study of a second major airport for Houston. The study—known as the Bourne Study—identified three potential airport sites along the present-day Beltway in northwest Houston. However, no action was taken after the release of the report. In the early 1950s the second major airport was still largely in the discussion stage and officials were not ready to commit to the new airport. In the meantime, officials moved forward with the construction of a new terminal at Houston Municipal Airport. The terminal design was declared to be inadequate by Houston Mayor Roy Hofheinz during construction, necessitating changes to correct glaring problems. Unfortunately, the planning errors in the terminal proved to be somewhat prophetic about the future of aviation planning in Houston. In October 1954 the new terminal opened and the airport was renamed Houston International Airport.[48]

Through the mid-1950s the need for a new airport became increasingly urgent, but city officials were slow to take action. In 1957 several events converged to finally move Houston's second airport forward. While the coming of the "jet age" of commercial air service had been anticipated since before 1950, it finally arrived in Houston on May 20, 1957, when a French-built Caravelle jet arrived from Miami while on a demonstration tour through the United States. The medium-range, twin-engine Caravelle had no difficulty with the short 6,565-foot (2,001 m) runway at Houston International, but aviation authorities knew that the runway length would not be adequate for the imminent wave of new jets that would soon arrive from U.S. manufacturers: the Boeing 707, the Douglas DC-8, and the Convair 880. A planned runway extension to 7,300 feet (2,225 m) would be enough to accommodate the new jet aircraft for departures to domestic destinations, but still would not be adequate for a jet departing on an overseas flight. It slowly became clear that Houston had underinvested in its aviation facilities and was falling behind the nation's other major cities in its aviation infrastructure. The *Houston Chronicle* published a series of articles comparing Houston's aviation efforts to those of Dallas. The *Chronicle* concluded that Houston had "stood still" in the preceding years while Dallas had a well-planned program of promoting its aviation activities. Word that Houston is losing ground to its rival Dallas is often enough to spur action.[49]

The Jet Era Arrives and Becomes the Jetero

The most significant event of 1957, however, was the action of a group of Houston businessmen who effectively took matters into their own hands to move the second airport forward while local authorities dawdled. The group of businessmen formed an entity called the

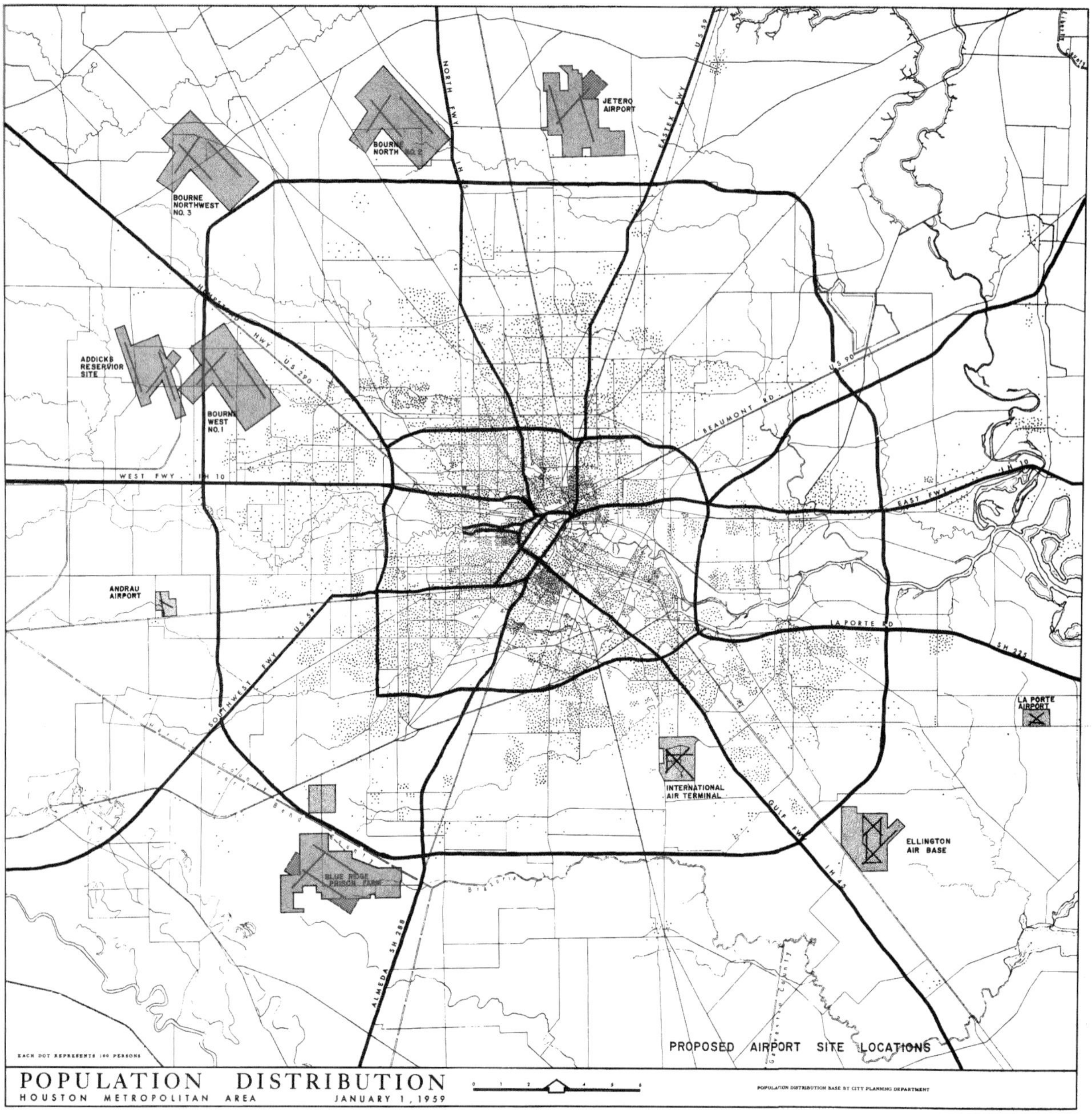

The Beltway and the airport—together from the beginning: This 1959 map from the Houston City Planning Department shows the potential airport sites that were considered. All potential sites were located along the Beltway. The first study for the location for a new airport was conducted in 1951 and identified the three Bourne sites. In 1960 a study delivered the final site recommendation. The "Bourne West #1" location, the preferred site in 1951, was rejected because of high land cost and flight patterns over newly urbanized areas. The "Bourne North #2" site offered no advantages over the Jetero site but would have been more difficult to acquire. The "Bourne Northwest #3" was rejected because of development, including a power plant and high voltage lines. The Addicks Reservoir location was rejected due to the high cost to raise the site above flood level, potential compromise of flood control capability, and flight patterns over urbanized areas. The Blue Ridge prison farm site southwest of Houston was determined to be unsuitable for an airport. The winner: the Jetero site.[50]

Jet Era Ranch Corporation, which purchased a 3,126-acre tract of land 15 miles (24 km) north of downtown Houston for $1,860,938.27. The group held the property for resale to the city of Houston at the original purchase price for use as the site of Houston's new airport. The intended name of the land-holding entity—the Jet Era Ranch Corporation—turned out to be short-lived. A secretary's typographical error transformed the words "Jet Era" into the single word "Jetero" in an early planning document. From that point on, the airport site became known as the Jetero airport site. The name Jetero would persist until 1983 when it was retired as the name of one of the main entrances to the airport.[51]

While the land was now in hand, several formalities

had to be taken care of. In July 1959 Houston voters approved a $50 million bond issue that included $6 million in airport development funds. In April 1960, in response to Mayor Lewis Cutrer's request for a comprehensive evaluation of all potential airport sites, a report was issued titled *Review and Evaluation of Proposed Major Airport Sites to Serve the Houston Area.* This report enumerated the known deficiencies of the existing Houston International Airport and identified the Jetero airport site as the only suitable candidate for the new airport. The legal agreements to officially accept the Jetero airport site and make final payment of principal and interest were completed in June 1960.

The jet era becomes Jetero: The site for Bush Intercontinental Airport was originally purchased by a group of civic-minded Houston businessmen in 1957 to preserve the site until the city of Houston could formulate a plan for a second airport. The holding company for the land was named the Jet Era Ranch Corporation, but a typographical error transformed the words "Jet Era" into "Jetero" and the airport site subsequently became known as the Jetero airport site. Although the name Jetero was no longer used in official planning documents after 1961, the eastern entrance to the airport was named Jetero Boulevard. In 1983, on the recommendation of Houston City Council member Eleanor Tinsley, Jetero Boulevard was renamed Will Clayton Parkway in honor of the cofounder of the Anderson, Clayton & Co. cotton trading firm and undersecretary of state for economic affairs from 1940 to 1948. Tinsley and other political officials felt that the 1950s-sounding name Jetero was not good for Houston's image. However, nostalgia buffs may have been disappointed. The name Jetero recalled an era when jet travel promised to revolutionize travel opportunities—a time when the roar of a jet engine or a condensation trail in the sky inspired awe. (Photo: Houston Airport System)

Planning for the Airport and its Freeways

The first engineering report for the new airport, *Plan of Development, Jetero Intercontinental Airport*, was issued in October 1961. The name of the airport site—"Jetero"—was transferred directly into the new airport name. The 1961 document envisioned a single circular-shaped terminal with several concourses extending from it. In terms of roadways, the document specified access roads that would be required. Most significant was the call for immediate action to construct the nearby section of the Outer Belt, the present-day Beltway 8.

The second major airport planning document was issued in July 1963 and was titled *Volume II, Plan of Development, Terminal Area, Houston Intercontinental Airport.* The "Jetero" designation for the airport was gone, but it would find a second life as the name for one of the main entrance roadways to the airport. This second planning document detailed the comprehensive study that had been undertaken in order to determine the best terminal configuration for the new airport. Planning officials visited all the major U.S. airports and had extensive consultations with airport authorities. Four terminal concepts were selected for detailed study: the mobile lounge, pier, satellite, and unit terminal. The mobile lounge concept, which was in use at Dulles airport near Washington, D.C., featured bus-type vehicles that shuttled passengers between the main terminal and aircraft. The pier concept most closely resembled the original design and featured a central unit for all airline operations with pier structures housing gates radiating from the central unit. The satellite concept also featured a single central unit, but aircraft would be grouped around individual satellite buildings that were connected to the main terminal. The unit terminal concept featured a series of relatively small, stand-alone terminals constructed along a central mall.

The engineers and architects unanimously recommended the unit terminal design, mainly because it best met the requirements of flexibility and expandability. In addition, it would distribute traffic among several terminals rather than one terminal, and it would avoid the sprawl and long passenger walking distances that would eventually occur in a single-terminal design. Although the unit terminal appeared to be the best design at the time, it would prove to be less than optimal for the airline hub and spoke operations that would later dictate airport design. The city of Houston adopted the unit terminal design on September 9, 1963.

The next major engineering report, *Volume III, Plan of Development, Land Use, Houston Intercontinental Airport*, was issued in December 1964. This was the first document to provide details on the two planned entrance roadways, John F. Kennedy Boulevard from the south and Jetero Boulevard on the east. The document specified a 400-foot (122 m) right-of-way to accommodate eventual construction of limited-access main lanes, frontage roads, and mass transit service. Initially, Kennedy Boulevard would be constructed as a four-lane divided roadway and Jetero Boulevard would be constructed as a four-lane divided roadway within the airport property and a two-lane roadway between the airport and the Eastex Freeway.

At this point, all the plans were in place to build the airport. In fact, the initial north-south runway was completed in 1964 before any terminal construction. Terminal construction dragged on two years after the scheduled completion date in 1967 because of labor problems that

Airport construction, August 1964: This view looks north-northwest at the construction site for the airport. The north-south runway was already complete and work was underway on the east-west runway in the upper right of the photo. Land for the terminal complex between the two runways had been cleared. Due to labor problems and project management difficulties experienced by the contractor responsible for the terminal, completion of the airport was delayed two years. Houston Intercontinental Airport began operations on June 8, 1969. (Photo: Houston Airport Authority)

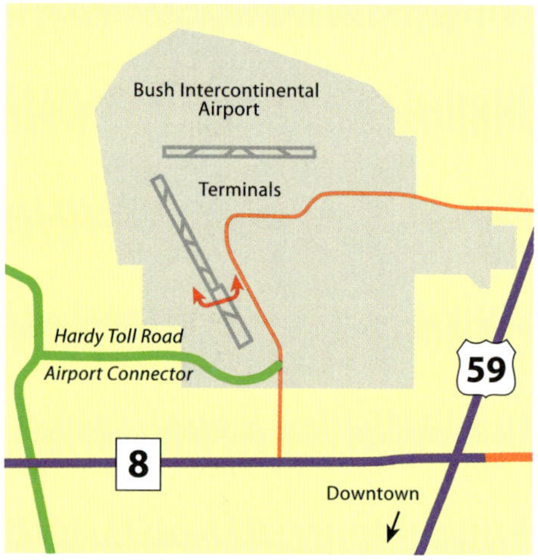

plagued the prime contractor, R. F. Ball Construction. Finally, on Sunday, June 1, 1969, Houston Intercontinental Airport was officially dedicated in a large ceremony featuring an air show, an open house of the entire airport, and the usual political ceremonies. The first regular-service commercial flight landed one week later on June 8 when the airport officially opened for business.

Unfortunately, the pattern of aviation miscues in Houston soon reared its head again. It turned out that the longest runway at Houston Intercontinental, the 9,400-foot-long (2,865 m) east-west runway, was not long enough to allow a fully-loaded wide-body aircraft to take off for long-range flights during warm weather conditions. Air France's nonstop service to Europe incurred a 10,000-pound (4,535 kg) weight penalty due to the short runway. In addition, the runways were not thick enough to handle the weight of newly introduced wide-body aircraft, and the paved area of the taxiways did not extend far enough from taxiway edges to prevent ground erosion due to jet blast. The new "jet era" airport wasn't even capable of handling all jets. To add further insult to injury, Houston's rival 250 miles (400 km) to the north had just begun work on the massive Dallas-Fort Worth International Airport. Dallas-Fort Worth International Airport would become the world's largest airport and would be served by a sprawling freeway network.

The next engineering report, *Volume IV—Plan of Development, Second Stage*, was released in April 1971 and

Original JFK Boulevard connection ramps, 1970-1992: This August 1978 view looks west along the North Belt frontage road at the original connection ramps at John F. Kennedy Boulevard. The ramps were dismantled and replaced in 1992 during the construction of the Beltway main lanes. (Photo: Texas Transportation Institute)

North Belt at JFK Boulevard: This photo shows the same view as the above photo in May 2003.

Key dates in the history of Bush Intercontinental Airport	
1951	The first study of potential airport sites is completed.
1957	Local business leaders purchase the airport site with a land holding-entity called the Jet Era Ranch Corporation.
1960	Voters approve bonds for the airport.
1969	The airport opens on June 8, 1969, after a long delay in completing the terminal.
1981	Terminal C opens, and Continental Airlines begins building its Houston hub operation.
1990	The International Airlines Terminal opens.
1997	The airport is renamed George Bush Intercontinental Airport/Houston, in honor of nation's 41st president who served from 1989 to 1993.
2002	For passenger traffic, the airport is the 8th busiest in the United States and 13th busiest in the world.
2003	The new $225 million Terminal E opens, with 23 gates.

focused on correcting the airport's runway shortcomings as quickly as possible. The 8,000-foot (2,438 m) north-south runway was slated to be lengthened to 12,000 feet (3,657 m). Runways and taxiways would be thickened where necessary. In terms of roadways, the 1971 plan reaffirmed previous plans. The cross section view of the entrance roadways, Kennedy and Jetero Boulevards, now showed a 450-foot-wide (137 m) corridor. The document also included an expanded discussion of mass transit service to the airport in recognition of the increasingly freeway-hostile climate of the early 1970s and greater interest in mass transit. However, it was recognized that a mass transit system was at least 10 to 15 years in the future.

Perhaps the lowest point in the history of Houston Intercontinental Airport—at least psychologically—occurred in the 1970s when Houston was not a hub airport and many travelers were forced to make connections at Dallas-Fort Worth International Airport. For Houstonians, having to connect through Dallas was perhaps the ultimate slap in the face. Attracting a large airline to call Houston its home became a top priority of local officials. The completion of Terminal C in 1981 provided a terminal facility capable of handling a hub operation, and in 1982 Continental Airlines moved its headquarters to Houston. Starting around that time, Continental began building its Houston hub. Finally, Houston Intercontinental Airport had overcome its growing pains and was now positioned to propel itself into the upper tier of U.S. airports.[52]

The 1990s and 2000s saw ongoing improvements to the airport. In 1990, the International Airlines Terminal Building opened. Later in the 1990s, the circular gate facilities at Terminal A, which had a distinctive 1960s feel and seemed to be reminiscent of the 1950s-sounding term "Jetero," were demolished and replaced with linear gate facilities. The circular facilities at Terminal B were retained, however, even as Continental's hub operation expanded into Terminal B. In May 2002, a 6,000 foot (1,829 m) runway was lengthened to 10,000 feet (3048 m) for full jet service, and in 2003 an all-new, 9,400-foot (2,865 m) runway opened. With the addition of the new runway, the airport's land area exceeded 10,000 acres. The first phase of the new Terminal E was dedicated in June 2003.

In 2002, Houston's Bush Intercontinental Airport ranked as the 8th busiest airport in the United States and the 13th busiest airport in the world as measured by total passengers. It was a remarkable level of success for an airport that had been plagued with so many problems during its development and was not ideally situated for airline hub operations. To complement its world-class freeway system, Houston also had an airport that reached world-class status.[53]

The Roads and Freeways

Improvements to the roads and freeways serving Bush Intercontinental Airport were completed around the time of the airport opening in June 1969. The Eastex Freeway from downtown to the airport entrance was upgraded to full freeway status by 1970. The Beltway 8 frontage roads between the North and Eastex Freeways were opened in February 1970. The North Freeway had been completed in 1963. The southern entrance to the airport, John F. Kennedy Boulevard, was a four-lane divided highway, and the eastern entrance to the airport, Will Clayton Parkway, was a four-lane divided highway inside the airport and a two-lane roadway outside the airport. During the 1970s there were almost no freeway or roadway improvements in the vicinity of the airport.

Freeway construction resumed in the mid-1980s. In 1983 work was underway to construct the Beltway 8 main lanes near the airport. The Hardy Toll Road opened in 1987. The 1990s brought a construction boom to airport-area freeways. The Beltway 8 main lanes between the North and Eastex Freeways were completed in bits and pieces, with the final section at the Eastex Freeway opening in December 2002. A four-level interchange at Kennedy Boulevard with two direct connectors was completed in 1992. Major expansion of the North Freeway west of the airport was completed in 1998. Expansion of the Eastex Freeway was completed in 1999. The Hardy Toll Road airport connector opened in January 2000. The five-level stack interchange at Beltway 8 and the North Freeway saw its first phase open in 1997 and was fully completed in early 2003. The first connector ramp at the interchange of the North Belt and Eastex Freeway opened in December 2002, and the second phase, in progress in 2003, will add three more direct connector ramps.

But what about the entrance roadways to the airport? Will they ever become freeways? In the long run, the answer to that question is probably yes, but in the short and intermediate planning horizons neither John F. Kennedy Boulevard nor Will Clayton Parkway will become a freeway. John F. Kennedy Boulevard will be the first to become a full freeway since only one traffic light needs to be eliminated. The lightly-travelled Will Clayton Parkway will probably have to wait a very long time for freeway status.

Will it ever be a freeway? This view looks west along Will Clayton Parkway, the former Jetero Boulevard and one of two entrances to Bush Intercontinental Airport. The roadway is in a frontage road configuration and has a wide right-of-way for its ultimate planned freeway status. However, traffic volumes entering the airport on Will Clayton Parkway are low, so construction of freeway main lanes is in the distant future. (Photo: May 2002)

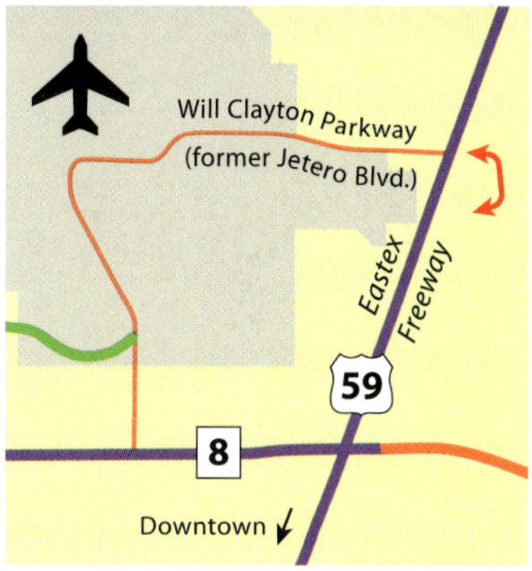

Trying to stay one step ahead of suburbia: This view shows subdivisions encroaching on the only complete section of the Grand Parkway in west Houston. Rapid development along the Grand Parkway corridor particularly northwest of Houston prompted officials to move forward with alignment studies in the late 1990s, defining most of the parkway route around Houston by 2002. This view looks north along the Grand Parkway with the Fry Road intersection in the foreground. (Photo: The Positive Image, April 2000)

The Grand Parkway, SH 99

The name alone is enough to inspire awe. The great railway era had its Grand Central Terminal in New York City. The American West has its Grand Canyon and Grand Tetons. Baseball, tennis, and golf have their grand slams. And as the crowning achievement of its freeway system, Houston would get its Grand Parkway.

It was a freeway befitting of its name: as originally planned, a 130-mile (208 km) mega-loop around Houston that lassoed the two inner loops and the web of spoke freeways. But bringing grand dreams to reality is another matter. The big freeway plans of most major cities died in the 1970s, including Houston's Grand Parkway. Even today, cities that urgently need second outer loops, such as Atlanta and Washington, D.C., have no hope of actually seeing proposed wide outer loops become reality.

But what really makes something grand? In the case of the Grand Parkway, its grandness is its mere existence. The Grand Parkway came back to life in the early 1980s, and its first section was opened in 1994. As of 2003, it appears that most, and probably all, of the Grand Parkway will ultimately be constructed. Building a wide outer loop will be a grand achievement—perhaps something that no other major city will be able to do. If and when the day comes that the Grand Parkway is complete, Houston will be in a class of its own.

Origins

The first discussions of a third loop around Houston began in April 1961 with talk of the "inter-county loop." The Houston City Planning Commission conceived the route and began discussions with surrounding counties. The inter-county loop was contemplated to be a highway or major arterial street about 17 miles from downtown Houston and about 5-6 miles outside the Outer Belt (now Beltway 8). It would have followed existing routes, including Highway 6 and FM 1960 around west and north Houston, and Battleground Road in east Houston. Some new roadways would have been required south of Houston in Galveston and Brazoria Counties.[54]

Serious consideration of a larger freeway loop that evolved into the Grand Parkway began in 1964. In October 1965, plans for the Grand Parkway became public. The Houston City Planning Commission released a draft version of the 1966 *Major Thoroughfare and Freeway Plan* showing the new third loop on the official planning map. The proposed 130-mile (208 km) belt would be 20 to 25 miles (32 to 40 km) away from downtown and would actually be about seven-eighths of a complete loop since it did not include the southeast section from IH 10 East to the southeast terminus at Galveston Bay south of Kemah. The City Planning Commission approved the route and scheduled a public hearing for February 17, 1966. The *Houston Chronicle* enthusiastically endorsed the new route, saying, "No sensible citizen can doubt that this freeway will be needed eventually." The route was adopted into the region's official long-range plan.[55]

As of early 1966, the third loop did not have a name. Since Houston already had a loop and an outer belt, there was no obvious name. Houston Mayor Louie Welch referred to it as the "way out loop" and some news reports called it the "extreme outer belt." The City Planning Commission had more plans for the corridor than just a freeway, however. By October 1966 the full vision of the corridor was made public by the City of Houston Planning Department. The corridor would connect a series of large parks and lakes around the perimeter of Houston. The freeway corridor itself would be very wide, up to thousands of feet wide, to promote its parklike qualities. The freeway and accompanying parks would be a refuge for city dwellers in the future Houston of 1990, which was projected to have a population of 3.1 million. The new freeway loop would be the centerpiece of a plan to provide the growing city with adequate recreation opportunities. The freeway corridor was named the Grand Parkway.[56]

In 1968 the Planning Commission published a map titled *Houston Preliminary General Study Plan for 1990*, showing the Grand Parkway as a park belt around the city of Houston. The concept of the Grand Parkway as a park corridor was short-lived. The lakes proved to be unpopular with landowners and were not seriously considered. A similar map published in 1972 did not include the Grand Parkway or any of its associated parks and lakes. The Grand Parkway freeway was not affected by the demise of plans for parks and preserves, and remained on the official long-range transportation planning maps for the Houston region. The 1970s, however, would not be kind to freeways, especially costly long-range projects such as the Grand Parkway.[57]

Grand Dreams and Harsh Reality

As the financial problems at TxDOT developed into a full-blown crisis in the mid-1970s, local officials were forced to drastically curtail plans for new freeways. The downsizing of freeway plans was detailed in a comprehensive 1976 report which provided a blueprint for sustaining TxDOT in its new era of diminished resources. The report gutted Houston's planned freeway system and

effectively called for the cancellation of Houston's second loop, the Beltway 8, by not recommending any future work on the facility. While Beltway 8 was struggling for its life, the chances of ever building the Grand Parkway diminished.[58]

The Grand Parkway had not been adopted into the state highway system by the Texas Transportation Commission. It was still just a line on the map of the regional long-range transportation plan and was officially under the jurisdiction of the city of Houston. In 1976, just 10 years after it was officially added, the regional planning agency deleted the Grand Parkway from its long-range transportation plan. Only a small section southeast of Houston from SH 146 to near Alvin still remained on the long-range planning map.[59]

Resurrection

In early 1982 two men decided to bring the Grand Parkway back to life. One was a prominent real estate agent in west Houston; the other, an engineer for the consulting firm Turner, Collie & Braden, Inc. It was the beginning of a new era in Houston—an era in which those who benefited the most from new freeways would take the lead in getting the new freeways built. The engineer, Al Knipe, envisioned a lucrative pot of engineering and consulting contracts if the Parkway were built. The real estate agent, Jack Hooper, handled real estate transactions for a prominent family that co-owned the 5,416-acre Cinco Ranch, a planned real estate development which was in the path of the Grand Parkway. Hooper realized the potential to substantially increase property values by constructing the Grand Parkway, so he began to seek donations of land for the freeway. Soon Knipe and Hooper had obtained written

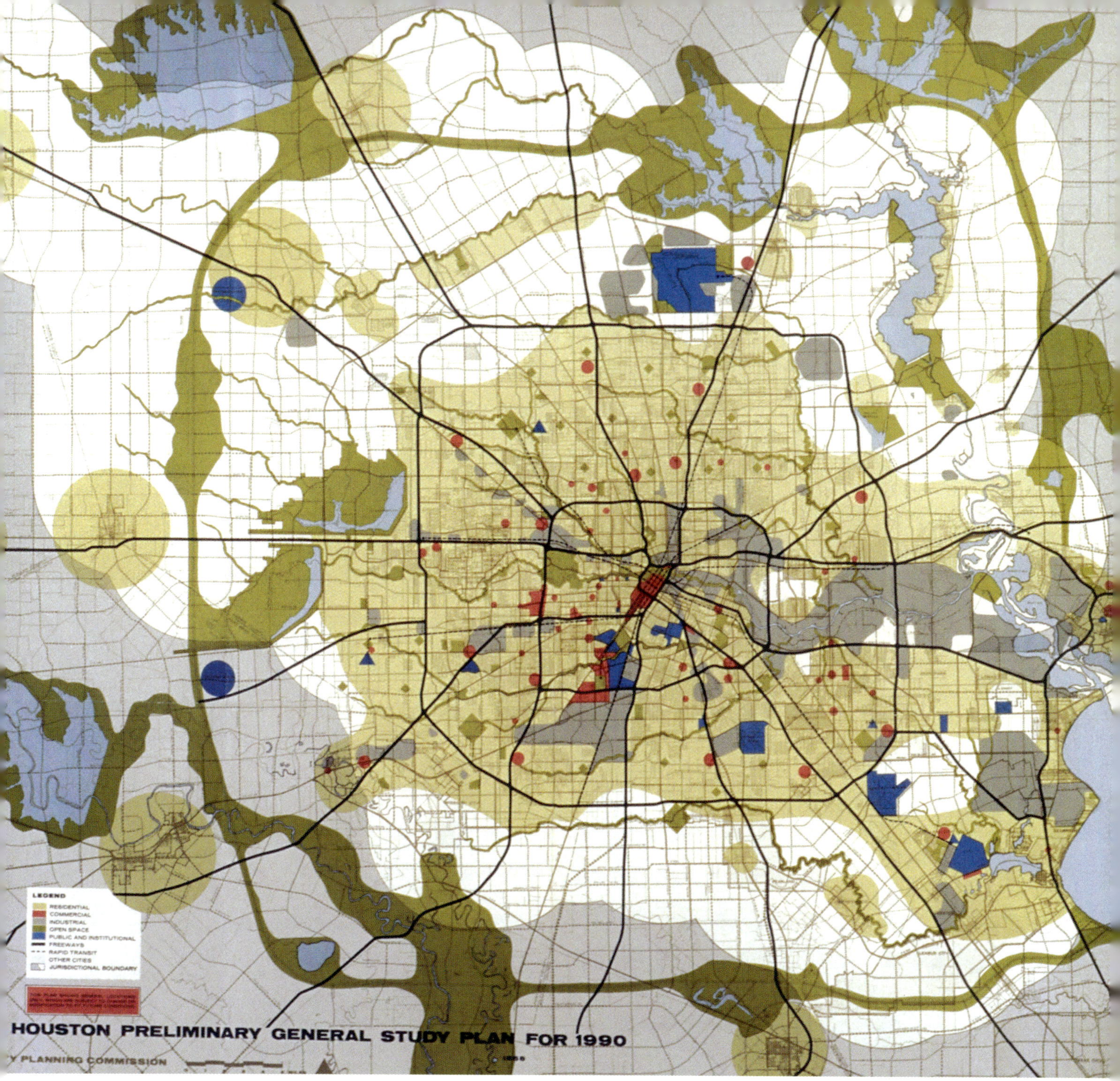

Grand dreams, but only the freeway remains: Plans for the Grand Parkway were first made public in October 1965. The planned freeway was included in the 1966 official regional transportation plan. Also in 1966, the City of Houston Planning Department unveiled plans to transform the parkway into a wide belt of parks and lakes around the periphery of Houston, as shown in this 1968 planning map which depicted the parkway alignment as a wide park strip. The idea of building parks and lakes in association with the Grand Parkway quickly faded, but the Grand Parkway transportation corridor remained.

commitments for land donations for eight miles (13 km) of 300-foot-wide (91 m) freeway right-of-way, verbal commitments for another eight miles, and a list of other landowners who were potentially interested.[60]

Knipe and Hooper soon wondered: Why not build the entire Grand Parkway on donated right-of-way? Based on the favorable response from landowners and with unanimous support from local political officials, it seemed as if it could be done. In addition to the efforts of Knipe and Hooper, county officials in Fort Bend County southwest of Houston and Montgomery County north of Houston were working very successfully to secure right-of-way donations. Knipe and Hooper's efforts received further support from a November 1982 TxDOT study on the future transportation needs of Texas. The report concluded that a third highway loop in Houston would be needed as early as 2002, especially in far west Houston.[61]

By early 1984 local politicians, real estate developers, business organizations, and transportation interests were pushing hard to move the project forward and get the Grand Parkway reinstated to official long-range planning documents. By May 1984 the Grand Parkway was restored to the long-range plan, and in 1985 it was back on the official transportation planning map.[62]

At a special session of the Texas Legislature in June 1984 to address education and transportation, a new state law was passed authorizing the formation of nonprofit transportation corporations. Transportation corporations would accept donations of land and money to be used in developing highway projects. The idea was for the transportation corporation to perform all needed work up to the point of construction, including right-of-way acquisition, environmental studies, and engineering. TxDOT would then pay for the actual construction. The Texas Transportation Commission authorized the formation of the Grand Parkway Association as a transportation corporation on October 25, 1984. In 1985 the commission approved several sections of the Grand Parkway for state-funded construction—if and when the Grand Parkway Association could get those segments ready for construction.

The Grand Parkway Association had an important ally and promoter on the Texas Transportation Commission: Houstonian and commission chairman Bob Lanier. Lanier was pushing for increased private-sector involvement in new highways to stretch state highway construction dollars as far as possible. Lanier was helping the Grand Parkway effort to showcase what he hoped would be the wave of the future. Not only would the local participation save TxDOT a lot of money, but the use of state funds (as opposed to federal funds) to construct the facility would eliminate the need for costly and time-consuming environmental impact studies. Lanier's ownership of a large tract of land along the proposed parkway route in northwest Harris County would later become a point of controversy in his support for the Grand Parkway, however.

The now-official Grand Parkway Association set up an office and lined the walls with maps and aerial photographs. Engineers worked to plot out a route where they could obtain donated right-of-way and avoid existing structures to keep costs low. Progress seemed to be good. Officials in Montgomery County north of Houston reported in February 1985 that they had obtained commitments for 95% of the right-of-way for the section between the North Freeway (IH 45) and the Eastex Freeway (US 59). In trying to identify and obtain a 170-mile-long (272 km) freeway corridor, the Grand Parkway Association was attempting to perform a task that would normally take 10 years or more. In fact, officials had been busy acquiring right-of-way for the Beltway 8 for the previous 30 years, and efforts still weren't complete. Building a huge outer loop in a short time of approximately 10 years was a lofty goal, but it seemed possible at the time.[63]

Not So Fast

It didn't take long for reality to set in. First, it would be impossible to obtain 100% or close to 100% of the right-of-way from donations. While large landowners were typically more than willing to donate the needed land, small landowners and farmers were more difficult to convince. The parkway would disrupt their farming operations and rural lifestyle, and the financial payoff from the Grand Parkway would be sometime in the distant future. Other property owners didn't want their property severed. Inevitably there would be some litigation. A more realistic goal would be to obtain 80-90% of the right-of-way through donations, and as time went on, even that percentage would become impossible.

Second, the cost of preconstruction work turned out to be very high. The Grand Parkway Association needed to obtain financial donations from landowners to pay for preconstruction costs, which were reported at $360,000 per mile in 1985. The collapse of Houston's economy in 1986 and accompanying free fall in property values greatly diminished landholders' enthusiasm for making cash contributions.[64]

Third, the freewheeling business dealings of the Grand Parkway Association came under closer scrutiny in 1986. At issue were potential conflicts of interest because some of the association's directors owned property or had business interests tied to the Parkway's proposed route around Houston. The Texas Transportation Commission adopted new rules in April 1986, prohibiting individuals with real estate interests from serving as directors of private transportation corporations. Three of the five directors of the Grand Parkway Association, including the executive director, were forced to resign.[65]

It became clear that the Grand Parkway Association would need to take the project one step at a time. Around 1987 effort was refocused on the western segment that originally inspired Knipe and Hooper to resurrect the Grand Parkway in 1982.

Building the First Section

In 1987 the Grand Parkway Association worked to finalize the route of the western segment between the Katy Freeway (IH 10) and the Southwest Freeway (US 59), but there was one complication. A landowner just south of the Katy Freeway didn't want the freeway to pass through her property and was threatening litigation to move the alignment to the edge of her property. It was ironic that this landowner was Vivian Smith, widow of oil and real estate magnate R. E. "Bob" Smith. Probably no one else in Houston had used Houston's burgeoning freeway system in the 1950s and 1960s as a vehicle for generating real estate wealth better than Bob Smith. Bob Smith had astutely bought large tracts of land along freeway corridors, especially the West Loop near the Galleria, and cashed in as the freeways and booming city pumped up property values. Regardless of the wealth that freeways had contributed to her estate, Vivian Smith didn't want a freeway through her property. In spite of objections from Smith and others, in March 1988 the Texas Transportation Commission approved the planned alignment of the Grand Parkway through Smith's property. The case would eventually be settled in court years later, after the death of Vivian Smith in 1989.[66]

In the meantime, the Grand Parkway Association was scrambling to come up with the money to fund its end of the deal to provide all land and preconstruction services. By 1989 it became clear that there would be a shortfall. If the Grand Parkway's first segment would be constructed, TxDOT was going to have to pick up a share of the cost that was originally intended to be the responsibility of the Grand Parkway Association. The association was still optimistic that construction could begin in the summer of 1989.

But there was an even bigger hurdle looming: environmental issues. By 1989 there were already disputes about the level of environmental studies needed for the project. The U.S. Fish and Wildlife Service held up approval of the western segment, saying that a comprehensive study of the impact of the entire Grand Parkway was needed. The Army Corps of Engineers sided in favor of the highway and had no objections to its construction. On February 20, 1991, after nearly two years of stalemate, the Department of the Interior refused to support the Fish and Wildlife Service's request for a broader study and authorized the Army Corps of Engineers to issue a permit for filling wetlands. Construction on the 19-mile (30 km) segment was underway by the end of 1991. The west Grand Parkway was officially dedicated on August 30, 1994, in a large ceremony in Sugar Land near the south end of the segment. Only about three miles (5 km) of the segment were constructed to full freeway standards. The rest consisted of main lanes or frontage roads, and was designed to be easily upgradeable to full freeway status with the construction of overpasses.[67]

The first section of the Grand Parkway was built. It wasn't easy, and building future sections wouldn't be any easier.

In the Crosshairs of Environmentalists

The dispute regarding the level of environmental study required for the first segment of the Grand Parkway was not the end of the story. In fact, it was just the beginning. Stopping the Grand Parkway became one of the top priorities of environmental groups, especially the Sierra Club and the Audubon Society. The proposed route of the Grand Parkway traversed through regions with environmental issues far more serious than any other Houston freeway had faced. The segment west of Houston between the Katy and Northwest Freeways crossed the Katy Prairie, a wintering ground for migratory birds. Southwest of Houston, the route crossed the Brazos River bottomland hardwood forest and came near Brazos Bend State Park, a park with a large wildlife population. Northeast of Houston, the route passed near Lake Houston State Park. North of Houston near the community of Spring and southeast of Houston near the community of Sante Fe, residents were vocal that they did not want the freeway nearby. But more than anything else, environmentalists were opposed to the parkway since they believed it would promote urban sprawl in the vast area around Houston. Controversy and the threat of lawsuits continued through the 1990s, with environmentalists asking for more comprehensive environmental impact studies. Ultimately it was determined that each section between radial freeways, called segments of independent utility, could be studied individually. In 2001 environmental groups attempted to use federal air quality regulations to kill the Grand Parkway and other highway projects.[68]

In a 1996 report titled *Road to Ruin*, prepared by the groups Friends of the Earth and Taxpayers for Common Sense, the Grand Parkway was listed among 22 nationwide highway projects that the groups identified as most wasteful and environmentally damaging. The 1999 version of the report listed the Grand Parkway as the fifth worst project in the nation among the 50 that were listed.[69]

Still, there was near-unanimous political support for the construction of the Grand Parkway and particularly strong political support from outlying counties, especially Fort Bend County southwest of Houston. The reason was simple: growth was going to occur with or without the Grand Parkway. The environmentally sensitive areas around Houston would likely be urbanized—it was only a matter of time. There was no mechanism to stop or even slow growth in any region around Houston. Officials could look to any number of cities in the United States that had curtailed freeway construction in the 1970s but still sustained ongoing and unstoppable sprawl. Proactive planning was needed to avoid the transportation quagmires that cities such as Washington, D.C. and Seattle had gotten themselves into. The choice was clear: growth and sprawl with adequate transportation infrastructure, or growth and sprawl with inadequate transportation infrastructure and lower quality of life.

As an added plus for the Grand Parkway, officials with the Grand Parkway Association were touting its parkway-like qualities. Since most of the Grand Parkway would not have frontage roads, it would truly be a parkway in the sense that it would not be lined with commercial establishments and billboards like most other Houston freeways. Local officials painted a particularly grim vision of the future without the Grand Parkway. Suburbanites would be forced onto country roads that would be upgraded to arterial streets. The roads would subsequently be lined with strip malls and signage. Traffic lights would proliferate. The nightmarish stretch of FM 1960 in north Houston between the North and Northwest Freeways, with its endless traffic lights, commercial clutter, and traffic congestion, would become the harsh reality of the future. Environmentalists held steady in their opposition, but it seemed that local interests opposing specific sections of the Grand Parkway recognized the need for the parkway—they just didn't want it in their backyards.

The Grand Parkway opposition never gained critical mass. Opposition by environmental groups was often vocal, but the near-unanimous support of the Houston-area political establishment kept the project on track.

The Grand Parkway Model Falls Apart

In the early 1980s it seemed like a good idea. The Grand Parkway Association would obtain land donations for the right-of-way and pay for all work leading up to

The alternative: This view of FM 1960 north of Houston shows the probable future of suburban transportation in Houston without the Grand Parkway. The farm-to-market road has been engulfed by suburban development and has become a heavily commercialized corridor with nonstop strip shopping centers, signage, and traffic lights. (Photo: March 2003)

construction, making the projects attractive to TxDOT on a cost-benefit basis and expediting construction. The difficulty in completing the first section of the Grand Parkway cast that model into doubt. By the early 1990s any hope of moving the Grand Parkway forward with the originally contemplated model was gone. Numerous factors contributed to the demise of the original Grand Parkway model, but at the top of the list was the high cost of performing environmental studies. The complexity and cost of environmental studies increased substantially in the 1990s due to the Clean Air Act Amendments of 1990 and the Intermodal Surface Transportation Efficiency Act of 1991. In 1991 TxDOT officials decided to perform a full environmental impact statement for each of the 10 unbuilt segments rather than a less comprehensive environmental assessment. Each segment would require millions of dollars in funding and years of study. The growth of Houston and increase in property values also made it more difficult to obtain land donations. The first segment of the Grand Parkway had about 90% of its right-of-way donated, but subsequent sections were not expected to equal that percentage. The long delays in obtaining environmental clearance, uncertainty over the availability of funds for construction, and lack of firm timetables for completing segments made donations less attractive to landowners along the route.[70]

The tremendous difficulty in moving sections of the Grand Parkway to construction became even more evident with the second segment, the I-2 segment east of Houston between the East Freeway (IH 10) and SH 146 in Baytown. The project was an ideal candidate for the original Grand Parkway model since a large landowner (USX Steel at the time) was willing to donate land and contribute to preconstruction costs. Preliminary work began in 1991. Seven years later in August 1998, the environmental impact statement was finally complete and a record of decision providing environmental clearance for the project was issued. The project then had to get in line for state funding. In the meantime, landowners became impatient and allowed offers of land donations to expire. The Grand Parkway Association was originally expected to provide 87.5% of the right-of-way for the segment. In February 2000, the percentage of donated right-of-way was reduced to 80%, and in March 2002 the percentage was further reduced to 60%. A $1.7 million contribution from the new landowner of the USX tract was contingent on construction being underway by September 2003. Contracts estimated at $43 million for the Grand Parkway and $29.4 million for associated work on IH 10 are scheduled to be awarded just prior to the deadline.[71]

Approximately 11 years would lapse between the opening of the first segment of the Grand Parkway and the completion of the second segment. With nine additional unbuilt sections, the timeline for completion of the Parkway was looking to be very long—if it could be done at all.

Building Momentum

By the mid-1990s the Grand Parkway Association evolved into more of a coordination and lobbying entity for the Grand Parkway. The money to keep the project going would have to come from county governments, TxDOT, and toll road agencies.

The poor prospects for obtaining construction funding prompted an effort in 1998 to build the section around northwest and north Houston, from the Katy Freeway (IH 10) to the Eastex Freeway (US 59), as a tollway. In 1999 a consultant concluded that toll revenue would be sufficient to pay for only 25% of the project cost. The Harris County Toll Road Authority offered to build it as a tollway if TxDOT paid for 75% of the cost, but TxDOT rejected the proposal.[72]

In 1998 the wheels were set in motion for substantial progress on the Grand Parkway—a wave of events that will probably lead to construction of numerous sections of the Grand Parkway. A study and public hearing process for the environmental impact statement for segment C southwest of Houston was launched in March 1998. This study considered one of the three most environmentally sensitive sections of the Parkway. The alignment would need to traverse through a region with parks, lakes, a large oil field, the Brazos River bottomland hardwood forest, an observatory, conservation areas, a prison, wildlife populations, and the wide Brazos River flood plain. In October 2000 a recommended preferred alignment which avoided the sensitive areas was approved by the Houston-Galveston Area Council, the regional planning agency that must approve all highway projects. A record of decision providing environmental clearance was expected in 2003. Fort Bend County voters approved a bond issue in 2001 providing $7.7 million for project development and right-of-way acquisition, moving the project development process forward.[73]

After the 1999 study concluding that the northwest and north sections of the parkway were not financially feasible as a tollway, TxDOT and the Harris County Toll Road Authority entered into an agreement to provide $8 million in funding to complete environmental studies. In 2002 the preferred alignment for the corridor was identified. This 52-mile (83 km) stretch included two environmentally-sensitive sections, segment E in west Houston through the Katy prairie and segment G in north Houston through the San Jacinto River bottomland hardwood forest. However, the alignment through urbanized areas just west of the North Freeway (IH 45) proved to be the most controversial. Neighborhood groups threatened litigation to stop the Grand Parkway, but political officials held firm, saying that the Grand Parkway is needed. Ultimately, an alignment acceptable to the neighborhoods was selected, causing a snakelike curve in the route. The draft environmental impact statement was completed in 2003, and a record of decision authorizing construction is expected in 2004.[74]

Key dates in the history of the Grand Parkway

1961	The inter-county loop is first discussed.
1965	Plans for a wide outer freeway loop are revealed.
1966	The Grand Parkway is named, included in the official long-range plan, and proposed to be the centerpiece of a belt of parks around the periphery of Houston.
1976	The Grand Parkway is cancelled.
1984	The Grand Parkway is restored to the official long-range plan. The Grand Parkway Association is formed.
1994	The first section of the Grand Parkway opens west of Houston.
1998	A series of environmental impact and alignment studies begins, defining the parkway route for approximately 103 miles (165 km).
2003	Construction begins on a 10-mile (16 km) section east of Houston. Toll road status becomes increasingly likely for the remaining sections.

In August 2002 a $7.5 million study to define the alignment and complete the environmental impact statement for segment B southeast of Houston began. With the launching of that study, approximately 75% of the Grand Parkway was either built, was about to begin construction, had its alignment defined, or was under study. In 2001 the Harris County Toll Road Authority published a map showing potential future toll road corridors it might consider. The entire Grand Parkway, excluding the completed section west of Houston and the imminent section east of Houston, is shown as a toll road candidate. The northwest and north sections of the Grand Parkway were being restudied for toll road feasibility in 2003. Toll-supported bond financing increasingly appeared to be the only way to build the Grand Parkway in the near future. In April 2003, the Texas Transportation Commission approved a resolution supporting the construction of the remaining sections of the Grand Parkway as tollways, with the Harris County Toll Road Authority taking over the sections in Harris County. The Texas Transportation Commission was essentially telling the Houston area that if it wanted the Grand Parkway built, it would have to be done with local funds. The financial strength of the Harris County Toll Road Authority and the acceptance of tollways in Houston make the financial feasibility of constructing the segments around northwest and north Houston as tollways increasingly likely.[75]

In 2003 the Grand Parkway has achieved greater momentum than at any other time during its 38-year history. Many obstacles remain before construction can begin, and only time will tell if Houston gets its unprecedented third freeway loop. If a wide outer loop can be built in any major American city, Houston is the place where it will happen.

COLOR

Freeway Mass Transit

cowritten with Chuck Fuhs*

One day in the late 1960s veteran engineer Bill Ward was pondering the future of the Houston freeway. Most of the freeways in Houston's central urban system were either complete or under construction. The engineering and design phase for the first wave of freeways was, for the most part, over. It was time to start thinking of what would come next—a new generation of freeways. In particular, the antiquated Gulf Freeway, Houston's first freeway, would need to be modernized.

The idea of dedicated bus or mass transit lanes on freeways had been around since large-scale freeway construction was first contemplated in the 1930s, but the first wave of freeways in the United States, from 1945 to the late 1960s, was designed exclusively for the private automobile. The first true bus lane in the United States would not open until 1969, when a dedicated lane was opened on present-day Interstate 395 in Virginia, just outside Washington, D.C. In Houston, studies of mass transit in the early 1960s had identified express bus service as the most promising option.[1]

Ward believed that the time for the busway had arrived. He envisioned a reversible busway in the center of the modernized Gulf Freeway, separated from traffic by barriers, allowing buses to bypass congestion and providing unimpeded travel for commuters. The busway would extend for more than 19 miles (30 km), connecting directly to transit stations, park-and-ride lots, and employment centers along the freeway.

Ward sketched his ideas for a freeway with the reversible busway. He then embarked on a soft-sell campaign to gain support for the idea. He kept the drawings in his desk and selectively showed them to individuals passing through his office. Slowly, Ward worked to promote the concept of integrating mass transit into Houston's freeways.

The Houston transitway was born. But the process of putting the idea into concrete moved with baby steps through the 1970s. It would take nearly 20 years before Ward's concept would become reality, but by the late 1980s fully integrated transitways were taking shape on most of Houston's radial freeways, and transitways became the backbone of Houston's commuter transit system.

(Opposite page) **Fork in the road:** This view shows the transitway access structure on the North Freeway near FM 1960. (Photo: James Lyle, TTI, June 2001)

* Chuck Fuhs was a manager for Houston's transitway system while employed with the Metropolitan Transit Authority between 1979 and 1985. He worked for many years with TxDOT and Houston area agencies on various highway and transit projects and is currently a transportation consultant with the firm Parsons Brinckerhoff.

Throughout its history Houston has always been a city built by business, for business. And in the world of business, everything eventually comes down to one number—the rate of return on investment, or as some would say, "bang for the buck." In the era of government-subsidized transit systems, mass transit hasn't normally been designed around rates of return. Houston's transit planners have always had their eyes on costly rail systems. But in Houston, the influence of the business mindset seems to extend further than in most other cities. The story of commuter mass transit in Houston has been one of building the best bottom line—getting the most ridership for the minimum investment. Getting improvements to the market quickly. Getting the most bang for the mass transit buck. Houston voters and highly influential members of the business community have steered Houston's transit down this path.

So how do you get the most bang for the mass transit buck in a sprawling, low-density freeway metropolis like Houston? By incorporating mass transit into the freeway system. Houston has developed one of the nation's most extensive systems of dedicated transit lanes on its freeways. Houston's transit lane system was born in response to a crisis, grew through gritty opportunism and the can-do spirit of highway and transit officials, benefited from the demise of competing transit options, and ultimately became the backbone of Houston's commuter transit system. No comparable city has been able to achieve Houston's level of transit and rideshare usage for the investment made.

Since the beginning of Houston's bus lanes in the 1970s, the Houston office of TxDOT has been the most consistent supporter and promoter of the dedicated transit lane system. Houston's commuter transit system is a product of Houston's freeway builders. Officials with Houston's mass transit agency, the Metropolitan Transit Authority, have always envisioned the dedicated bus lanes as an interim solution, to be used until a rail system could be built. But as one rail opportunity after another failed to materialize, the freeway transit lane system grew bigger, better, and more successful. Houston's transit lane system continues to evolve. In 2003 Houston is poised to become a leader in the next generation of freeway-based mass transit: the managed-lane freeway.

The father of Houston's transitways: TxDOT engineer William V. "Bill" Ward (born 1927) first developed the ideas for Houston-style freeway bus lanes in the late 1960s and promoted the concept for Houston's second generation of freeways. Ward was appointed head of the TxDOT Houston Urban Project Office in 1972. The ultimate realization of Houston's transitways included many of the features envisioned by Ward in the late 1960s. (Photo: TxDOT, 1970)

Origins of Express Bus Transit

In the early 1960s the first serious consideration was given to commuter-oriented mass transit in Houston. Because of Houston's similarity to Los Angeles, city officials, including Mayor Louis Cutrer, carefully studied the transit plan that was recommended for Los Angeles in 1960. The Los Angeles proposal included rubber-tired transit cars similar to rail cars, running on elevated fixed guideways positioned in the freeway medians. By 1962 Houston had completed its first comprehensive study of mass transit possibilities. The study ruled out any kind of fixed-rail mass transit, citing Houston's low population density, the increasing use of automobiles, and the emerging freeway network as reasons why fixed rail was unsuitable for Houston. "In none of the sections of the metropolitan area is there a sufficient density of population to warrant development of the fixed type of rapid transit service," the report stated. Houston's population density of 2,600 per square mile was about half of the 5,413 in Los Angeles, and only a small fraction of the 24,387 in New York City, 16,599 in San Francisco, and 15,772 in Chicago. Instead, the report recommended the development of express bus routes on freeways, with supporting feeder systems to take passengers to the express buses.[2]

The idea of buses on freeways just wasn't a concept capable of sparking a surge of interest among local authorities, however. It simply wasn't very glamorous. Various interests had been promoting a futuristic monorail for Houston since the 1950s. Others, including Mayor Louie Welch, were more interested in fixed-rail "rapid transit." Even as freeway construction proceeded at full speed during the 1960s, talk of the need for mass transit continued to intensify. Concepts and ideas for various types of fixed transit would come and go during the 1960s and over the next 30 years, but until 2001 only one would have staying power: the freeway transitway.

Saving the Bus System

When TxDOT engineer Bill Ward sketched his concepts of the Houston busway in the late 1960s, he began TxDOT's long-running support for transitways. These dedicated, barrier-separated lanes for buses and high occupancy vehicles (HOV) are often called HOV lanes. But TxDOT couldn't make it happen alone. Houston's bus service provider would be the other key player, and

Houston's Transitway System	
First dedicated transit lane	North Freeway contraflow lane opened August 28, 1979
First barrier-separated transitway	Katy Freeway opened October 29, 1984
Operational facilities, July 2003 Barrier-separated reversible transitway and elevated two-way transitway "Diamond" lane	 88.2 miles (141 km) 6.6 miles (10.6 km)
Under construction, or complete but not yet operational, July 2003 Barrier-separated "Diamond" lane	 6.1 miles (9.8 km) 5.0 miles (8.0 km)
Daily passenger trips, December 2002	121,090
Daily vehicles, December 2002	39,150

Data: Houston High Occupancy Vehicle Lane Operations Summary, TTI, December 2002

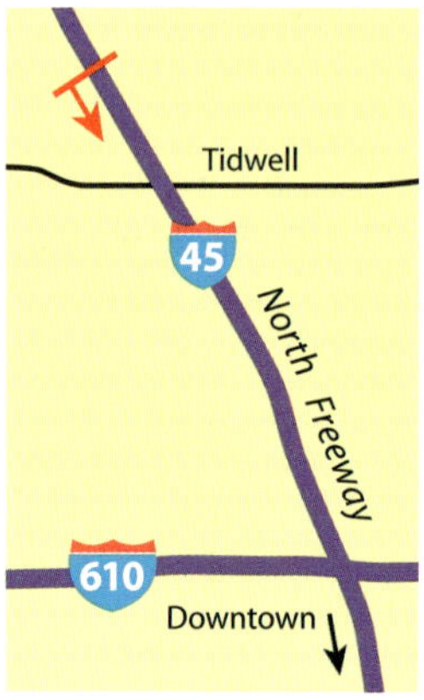

North Freeway contraflow lane in operation, circa 1980: One lane from the off-peak direction was taken and marked off with pylons spaced every 40 feet (12 m). Buses and vanpools travelling in the peak direction had access to the traffic-free contraflow lane. The contraflow lane was implemented as a demonstration project and was operational from August 28, 1979, to November 23, 1984. (Photo: Texas Transportation Institute)

in the 1970s Houston's privately owned bus service, Rapid Transit Lines, was in a dreadful condition. Investment needs in buses and facilities had not been met and maintenance had been neglected. The bus system operated out of a decrepit, turn-of-the-century streetcar maintenance facility called the Milby Bus Barn. The more passengers that Rapid Transit Lines carried, the more money it lost. Ward had to put his plans for the Gulf Freeway transitway on hold due to the lack of a bus operating partner. When TxDOT held public meetings in 1973 for planned improvements to the Gulf Freeway, the plans did not include a transitway. On October 6, 1973, Houston voters rejected the creation of a regional transit authority called HARTA, Houston Area Rapid Transit Authority. The purpose of HARTA was to build a proposed fixed-rail transit system and busway system, including the Gulf Freeway transitway.

The city of Houston had resisted buying out the private bus operator because of the associated costs, but with the defeat of HARTA and the continued deterioration of Rapid Transit Lines, Houston soon had no choice. Negotiations began in 1972, and in April of 1974 the city of Houston closed the deal to purchase Rapid Transit Lines and renamed the bus agency HouTran. While the city's purchase was a step in the right direction, it did not solve the bus service problem overnight. HouTran did not have a funding mechanism other than the city of Houston, and even if substantial funding could be found, it would take years to undo the decades of neglect. The city of Houston brought a commitment to improved bus service, and leading the way was Houston's pro-transit mayor Fred Hofheinz, who entered office in January 1974.[3]

Fast and Cheap

One of Mayor Hofheinz's top priorities was to show that commuter bus service was viable in Houston. However, any kind of major construction would take years to complete. Something fast and cheap to demonstrate the potential of commuter bus service was needed. In November 1974 Hofheinz asked TxDOT to evaluate the feasibility of a contraflow lane. In a contraflow lane system, one lane of freeway traffic from the off-peak direction is "borrowed" and used to provide an extra lane in

Pylon operations: Setting and removing the 1,300 pylons twice each day to designate the contraflow lane was a big job. Operators in a small convoy of trucks manually inserted the pylons into holes drilled in the pavement. For the morning rush hour, pylon installation began at 3:30 A.M. and removal began around 9:00 A.M. For the evening rush hour, pylon placement began at 2:30 P.M. and by 8:00 P.M. the lanes were clear. (Photo: Metropolitan Transit Authority)

the peak direction, with transit vehicles authorized to use the borrowed lane. In Houston's plan, the borrowed lane would be marked off by pylons inserted into holes in the pavement at 40-foot (12 m) intervals. Studies conducted by the Texas Transportation Institute and the city of Houston examined Houston's freeways and the limited number of dedicated transit lanes operating in the United States. Findings indicated that a very high potential existed in the IH 45 North Freeway corridor for a contraflow lane. The corridor had strong directional characteristics—a 65-35% split in the peak/off-peak directions—and already had the city's highest concentration of vanpools. Many of Houston's large employers, particularly the oil companies, had established vanpools in response to the 1970s energy crisis and Houston's traffic congestion. Approximately five other contraflow systems were in operation in North America at the time, including three in the New York-New Jersey area.

The contraflow project and park-and-ride lots would indeed be very inexpensive, approximately $6.8 million in 1978 dollars (roughly 15.6 million in 2003 dollars), but it wouldn't be as fast as everyone had hoped. Three years were required to obtain grants, complete feasibility studies, obtain approvals, and complete the final design. In November 1977 a construction contract was awarded. During the construction phase on August 12, 1978, public transit took a big step forward when Houston voters approved the creation of the Metropolitan Transit Authority (Metro) and a 1% sales tax to support it. Now the previously cash-starved transit agency would have a reliable stream of cash flowing into its coffers. But at its creation, Metro would still need to recover from the underinvestment of the previous private and city of Houston ownership. Houston had one of the oldest bus fleets in the country, and maintenance was still performed at the outmoded Milby Bus Barn. Efforts were undertaken almost immediately to upgrade bus operations, but acquiring new buses and building the required maintenance facilities would take years. Things would get worse before they could get better.

For Metro, 1979 was a disastrous year. Bus service

Opening day banner-busting: On October 29, 1984, Houston's Metropolitan Transit Authority (Metro) began service on the Katy Freeway transitway. And what better way to get Texans excited than with a Friday night football banner-busting. The Katy Freeway transitway was Houston's first barrier-separated transitway. Since this was the second freeway to receive express bus service, Metro proclaimed that it was "coming through again." (Photos: Chuck Fuhs)

collapsed in January when the transit agency ran out of spare parts for its barely functional buses, and nearly half of Metro's routes could not be served. There were between 100 and 200 in-service bus breakdowns on a daily basis. It was easy to find broken-down and abandoned buses on city streets. Newspaper reports ran a daily count of the number of buses that were out of service, and in the dark days of the service crisis, it was common to have more buses out of service than in service.[4]

In the meantime, the North Freeway contraflow project continued to move forward. Because of Metro's inability to provide reliable bus service, it contracted with private bus operators for the contraflow project. Metro secured lease agreements for several park-and-ride lots, leased space for training and deployment of the operating crew, and contracted enforcement to the City of Houston Police Department. On August 28, 1979, the contraflow lane opened for service with a big opening day celebration featuring nearly all local political leaders. With Houston descending into a traffic congestion quagmire, politicians were more than pleased to finally open the lane, which at least provided the perception of action to help ease the traffic crisis. Although the North Freeway contraflow lane was not the first in the United States, its 9.6-mile (15 km) length was the longest such facility in the nation and the only such facility to operate during both peak periods.[5]

The contraflow lane got off to a good start. First day usage was reported to be 1,458 passenger trips with 27 bus trips and 96 vanpool trips. Commuters reported saving up to 20 minutes as the buses and vans on the contraflow lane cruised past the freeway parking lot. After one year, usage had increased to 8,724 passenger trips on 125 bus trips and 412 vanpool trips. In spite of some minor incidents and a collision fatality in April 1980, the facility proved to be safe and reliable. In 1980 the transit lane was extended two miles (3 km) northward in a "concurrent flow" or

The retrofit transitway: The Katy Freeway transitway was incorporated into a pavement repair project and built where the interior shoulders previously existed. Less than 30 months after the Katy Freeway transitway idea was conceived, buses and vanpools began using it on October 29, 1984. Later the lane was opened to high occupancy private automobiles. The major expansion of the Katy Freeway, scheduled for 2003 to 2008, will replace the transitway lane with four managed lanes. (Photo: September 2002)

"diamond lane" configuration by converting the existing median shoulders into traffic lanes. In August 1982, three years after its opening, the contraflow lane was serving 15,600 passenger trips per day. The contraflow lane was a big success.[6]

The contraflow lane had always been a temporary demonstration project and was never intended for long-term operation. By 1983 time was running out for the contraflow project. Traffic congestion was increasing in the off-peak direction due to the loss of a lane for the contraflow service, and TxDOT formally notified Metro that an alternative was needed. Metro was planning a heavy rail system on the nearby Hardy Road corridor, but voter rejection of a $2.35 billion bond issue in June 1983 by a 61-39% margin scuttled the rail plans. Metro turned its attention back to bus lanes and accelerated plans to replace the contraflow lane with a reversible, barrier-separated transitway lane in the median of the freeway. On November 23, 1984, the North Freeway contraflow lane ended its five-year run and was replaced with a the central transitway. The transitway, or a pavement set-aside for a future transitway, was extended northward in conjunction with TxDOT's reconstruction and widening projects, which were completed to the Woodlands in 2003.[7]

From Fast and Cheap to Quick and Dirty

The success of the North Freeway contraflow lane soon had officials turning their attention to another severely congested freeway, IH 10 West, the Katy Freeway. The Katy Freeway west of Loop 610 was slated for a pavement repair project to begin in 1983. In 1982 Metro proposed a plan that would turn the repair project into a new transitway construction project by converting the interior emergency shoulders into a barrier-separated bus lane. Local agencies had developed a close working relationship during the North Freeway contraflow lane construction and operation. This relationship would make it possible to build a new transitway lane in record time.

Once an agreement had been reached, Metro immediately developed plans for the bus lane. Since the project was on a very fast track, there was no time to seek federal funding—the $12 million cost was borne entirely by Metro. The transitway and a complicated flyover structure to provide access to the bus lane were opened on October 29, 1984, less than 30 months after the idea was conceived. This remains the fastest transit guideway facility ever

The wishbone: The wishbone access ramp is one of the distinctive structures on Houston's freeway system, allowing vehicles to make a direct connection between a central transitway and the freeway frontage roads. There are four full wishbone ramps in Houston: two on the North Freeway, one on the Gulf Freeway, and one on the Eastex Freeway. This wishbone ramp is on the North Freeway just south of FM 1960 and was completed in 1998. (Photo: James Lyle, TTI, June 2001)

implemented and was a testament to the enthusiasm and motivation of both Metro and TxDOT to see improvements happen very quickly.

The Katy Freeway transitway was improved and extended in the following years. In December 2002 it was the most heavily used transitway in Houston, serving 29,530 passenger trips and 10,050 vehicles daily. In October 2000, an impressive 1.5-mile (2.4 km) high-level connector ramp providing buses and high occupancy vehicles direct access into downtown was completed. Heavy usage of the transitway on the chronically congested Katy Freeway required raising vehicle occupancy to a minimum of three persons during peak periods in order to keep the transitway traffic free flowing.

The Original Vision Is Realized

The first two transitways were a demonstration project and a project of opportunity. But in 1980 a plan began to take shape to build the Gulf Freeway transitway as a carefully planned facility that integrated all the elements needed to provide transit service effectively: bus services, park-and-ride lots, transit centers, and direct-access ramps to the transitway. It was the realization of Bill Ward's 1960s vision, and a design that would define the next wave of transitways to be built on Houston's freeways. Ward had become head of TxDOT's Houston Urban Project Office in 1972, so he was now in charge of TxDOT's participation in the project. The project was a cooperative effort between TxDOT and Metro, and was entirely funded with state and federal highway funds.[8]

Construction began in 1983, and the first section of the

The half-wishbone: This structure on the Eastex Freeway allows southbound transitway vehicles to exit to the freeway main lanes. (Photo: September 2002)

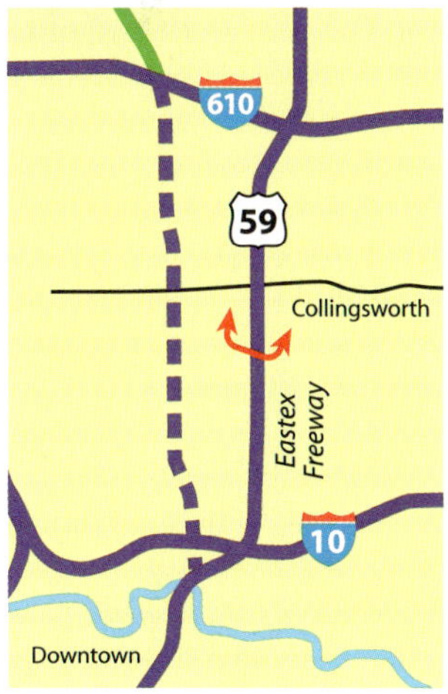

transitway from downtown to the South Loop 610 opened on May 16, 1988. The transitway was pushed southward during the 1990s, culminating with the 1997 opening of a wishbone entrance ramp just south of the Beltway 8 interchange. In August 1999 a plan recommending further southward extension of the transitway facility was finalized. Non-barrier-separated "diamond lanes" are planned from the current transitway terminus to a point nine miles (14 km) south at FM 518.

Regional Transitway Development

Following the June 1983 defeat of Metro's $2.3 billion bond package for a rail system, transitways became the centerpiece of regional transit planning. Many of Metro's staff resources were refocused from rail system planning to the development of transitways. Plans were formulated to add transitways to the Northwest (US 290) and Southwest (US 59) Freeways. Using the same design template as the Gulf Freeway, each of these would be reversible with direct-access ramps to transit facilities located adjacent to the freeway. Both the Northwest and Southwest Freeway projects were funded by Metro, using the Federal Transit Administration's "new start" guideway funds. This funding was rarely applied to facilities for rubber-tired vehicles, and the ability to compete with rail transit for funding showed the new respect transitway solutions had earned.

The Northwest Freeway project was carved from existing median shoulders with an average width of 19.75 feet (6.0 m) between barriers, just wide enough for one vehicle to pass another at reduced speeds if a stall occurs. Metro assisted with design and awarded the construction contracts, with transitway operations starting on August 29, 1988. The most distinctive feature of the Northwest Freeway transitway is its elevated structure and connections along the West Loop. An extension from Little York Road to Highway 6 was opened for service on February 6, 1990.

The Southwest Freeway project was built in conjunction with a major freeway reconstruction project which took place between 1989 and 1992. It became the first

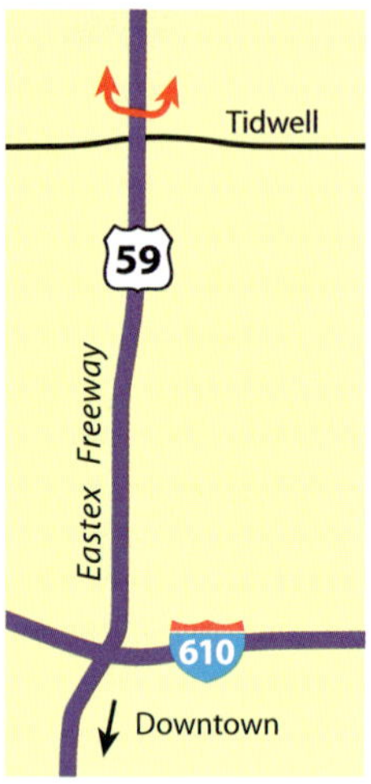

The "Texas T": The T-ramp is the most common structure on Houston's transitway system. In the transportation engineering community, this type of structure is informally called a "Texas T" because of its widespread use in Houston. It connects the transitway to an adjacent transit center or park-and-ride lot, crossing over the freeway main lanes and the frontage road. This T-ramp is located on the Eastex Freeway near Tidwell. (Photo: September 2002)

transitway in the region to be designed so that it could later be converted to two-way operation. This marked a major change in what was then the accepted design practice of serving commute trips in the peak direction only. The extension of the Southwest Freeway transitway into downtown Houston became the system's most controversial project. Original plans called for an elevated structure towering over the existing elevated Southwest Freeway lanes near Montrose, just west of the exit into downtown. Opposition from neighborhoods resulted in a redesign of the project, placing both the 10 regular freeway lanes and the 2 transitway lanes in a new trench that will replace the existing elevated structure. Work on the complicated project began in late 2002. Freeway expansion in Sugar Land, southwest of Houston, was completed in May 2003 and included two non-barrier-separated "diamond" lanes, extending the dedicated transit facility to SH 6.

The final corridor to receive a transitway in the first generation of transitway construction would become the crowning achievement in transitway design. US 59 North, the Eastex Freeway, was slated for a major expansion during the 1990s. A substantial amount of right-of-way acquisition would be required to expand the corridor width to 400 feet (122 m), making available sufficient space for a state-of-the-art transitway. The major freeway expansion began in the early 1990s, with work substantially complete by 1998 and construction in progress in far north Harris County in 2003. The Eastex Freeway transitway opened in March 1999 and features some of the most distinctive transitway structures in Houston, including a wishbone, a half-wishbone, four T-ramps, and an interchange bypass. The transitway will be integrated into the major new downtown interchange at US 59 and IH 10, planned for completion in 2004.

Additional work in the 1990s and 2000s focused on improving access to the transitways, improving connections into downtown Houston, and adding new park-and-ride lots. Throughout the 24-year period of transitway operation, the Texas Transportation Institute has continuously monitored the performance of each facility and the system as a whole. The performance data has helped local agencies make planning and operational decisions that have enhanced the effectiveness of the transitway system.

The dual T-ramp: This structure on the Gulf Freeway provides access to the transitway from both sides of the freeway. (Photo: James Lyle, TTI, June 2001)

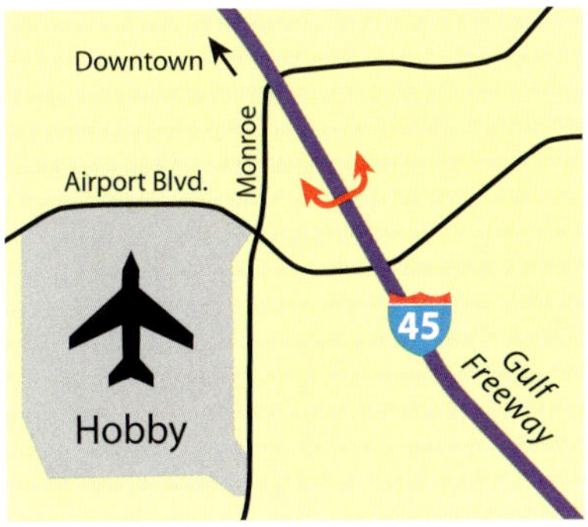

In December 2002 the system had 88.2 miles (141 km) of barrier-separated and dedicated two-way facilities operational and 6.1 miles (9.8 km) under construction, plus an additional 11.6 miles (18.6 km) of diamond lanes operational or about to become operational. The total number of weekday passenger trips was 121,090, including 40,190 via buses and 80,900 via high occupancy vehicles, vanpools, and motorcycles. A total of 39,150 vehicles used the transitways daily.[9]

The Future of the Houston Transitway

In 1998 there was a shift in the political climate in Houston. Mayor Bob Lanier, one of the key architects of Houston's bus and transitway strategy, exited office due to term limitations. He was replaced by Lee Brown, Houston's first black mayor and a strong proponent of building rail transit. Building a light rail line became one of Brown's top priorities, and a route on South Main Street, away from any freeways, was selected for the $325 million project. The project was small enough to be built with available funds, precluding the need for a bond election. Brown pushed the rail line forward as fast as he could, and after a period of litigation that was resolved in March 2001, construction began. After more than 40 years of acrimonious debate on the subject of rail, Houston would get a rail line on Main Street. Perhaps

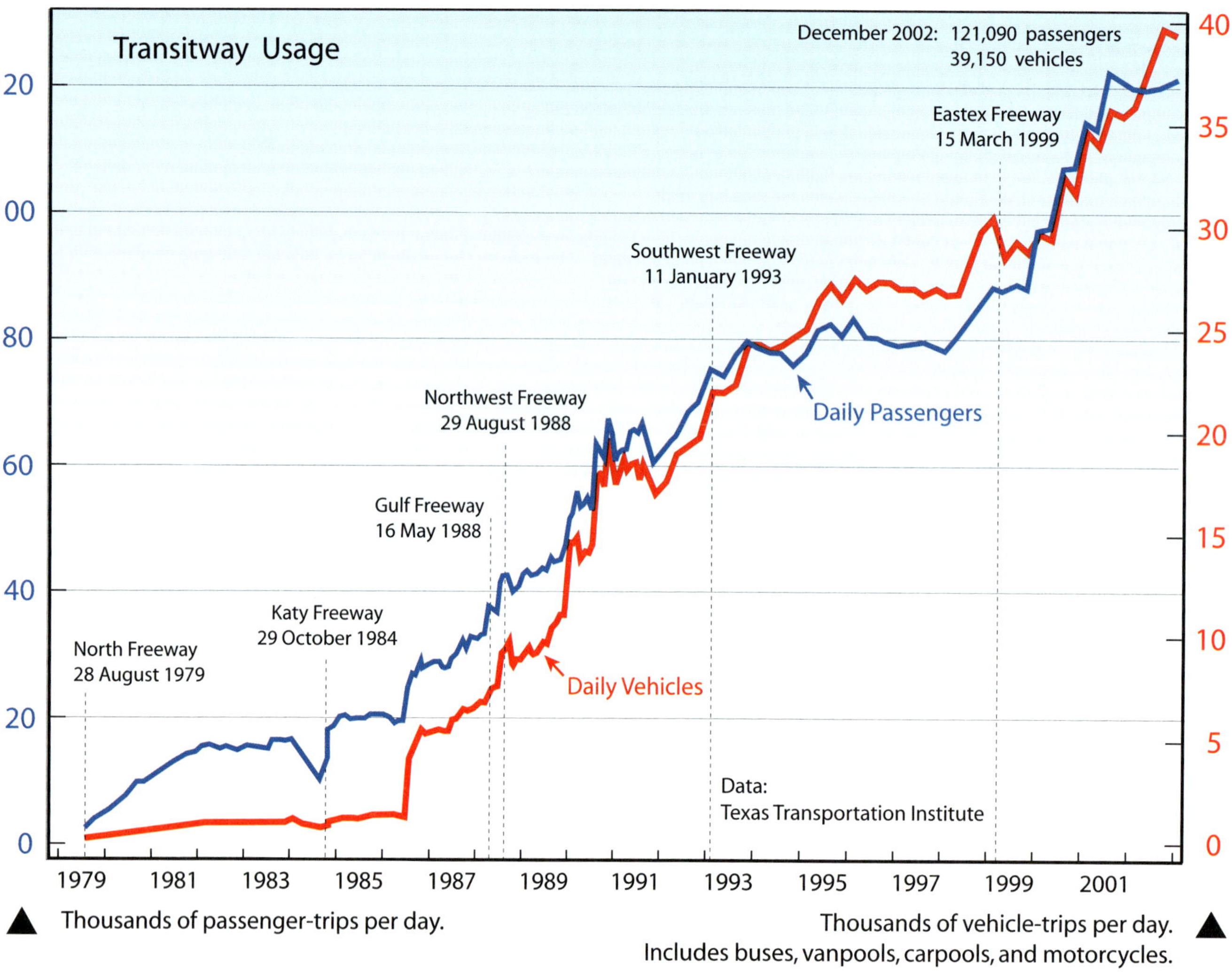

Transitway patronage: Houston transitway vehicle and passenger volume has risen steadily since the opening of the first transit lane in 1979. In early 2003 Houston's system was carrying twice as many weekday passenger trips as Dallas' light rail system.

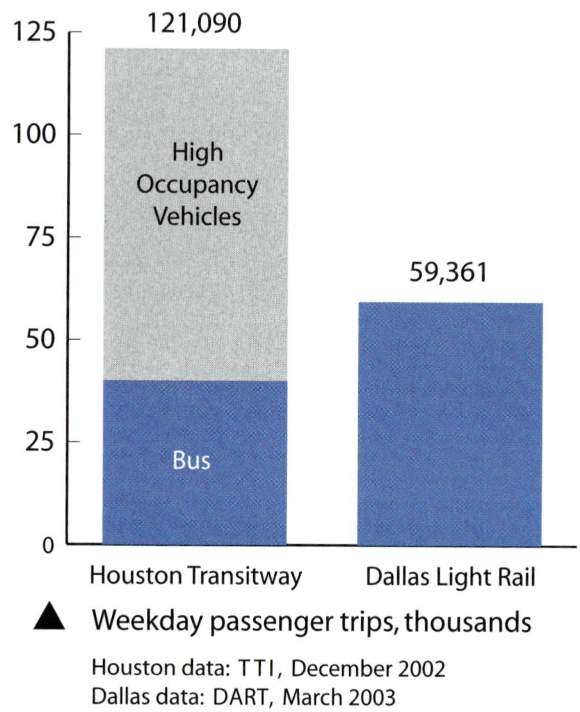

Key dates in Houston's transitway system

1962	A study identifies express bus service as the most promising transit option for Houston.
1974	Local officials initiate efforts to build an express bus lane on the North Freeway.
1979	The North Freeway contraflow lane opens.
1984	The first reversible, barrier-separated transitway opens on the Katy Freeway.
1988	The Gulf Freeway transitway opens. It is the first transitway designed into a comprehensive freeway reconstruction project and includes all needed features for complete transit service.
1999	The most advanced transitway opens on the Eastex Freeway.
2008	Scheduled completion of the Katy Freeway reconstruction including managed-lane transitway.

The elevated bypass: Carpools and buses on the two-way elevated transitway structure along the West Loop North zoom past backed-up traffic on the freeway main lanes. (Photo: September 2002)

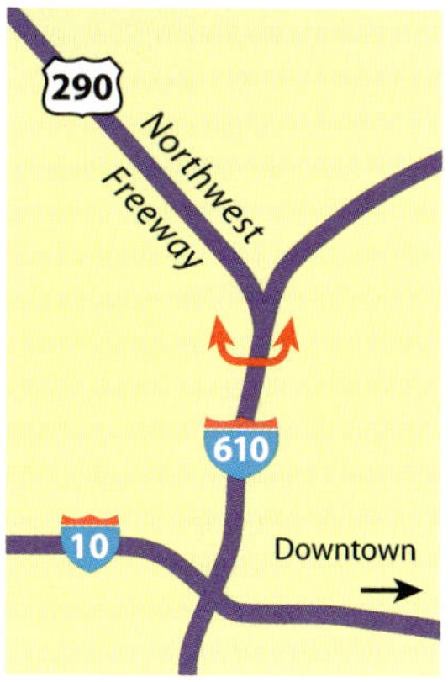

more significant was the shift in Metro's strategy. Brown believed that a city the size of Houston should have a rail system, and he redirected Metro's long-term objective to constructing rail. The performance merits and cost effectiveness of rail transit remained a point of sharp disagreement in Houston, and will surely continue to be an issue in the future. Brown will exit office at the end of 2003 due to term limits, and Metro's long-term strategy could once again be revised. Long-term plans unveiled by Metro in April 2003 included three new light rail lines, but also strengthened the transitway system and included plans to integrate transit into the upcoming generation of new managed lanes and tollways to be built in Houston. The transitway system will continue to be the backbone of Houston's commuter mass transit for the foreseeable future. All existing transitways will be extended in the intermediate or long-term future as suburban freeways are reconstructed and widened.

In the late 1990s a new player and a new concept arrived on the Houston transitway scene with the promise to define the next generation of transportation facilities. The new player was the Harris County Toll Road Authority (HCTRA), which during the 1990s had become one of the most powerful forces on the Houston transportation landscape. The concept that arrived was the idea of managed lanes—freeway lanes that may be used by buses, high occupancy vehicles, or single-occupant vehicles, with single-occupant vehicle access restricted or subject to variable tolls, depending on the time of day. These lanes are also called HOT lanes, or high occupancy toll lanes.

A comprehensive study of the Katy Freeway corridor

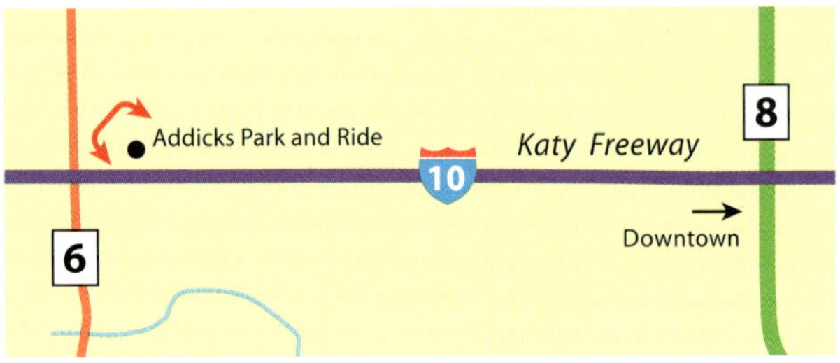

The park-and-ride lot: This park-and-ride lot along IH 10 in west Houston is the largest in the system with 2,441 spaces. The lot is typically near capacity Monday through Thursday. Park-and-ride lots are integrated into all the transitways on Houston's freeways. (Photo: James Lyle, TTI, June 2001)

(Opposite page upper) **The Pretzel:** This structure on the Northwest Freeway, just west of Loop 610, provides access to both freeway frontage roads. (Photo: James Lyle, TTI, June 2001)

(Opposite page lower) **Southwest Freeway:** The Southwest Freeway transitway was constructed in conjunction with the freeway expansion. The first phase of the expansion was completed in 1992, and the transitway opened in January 1993. (Photo: Metropolitan Transit Authority)

163 ELDRIDGE EXP
2201

The interchange bypass: Transitways were added to freeways after the original freeway construction, necessitating structures to get around existing interchanges. A long elevated structure allows transitway vehicles to bypass the IH 10-IH 45 interchange north of downtown. The elevated bypass opened on October 16, 2000. (Photo: May 2002)

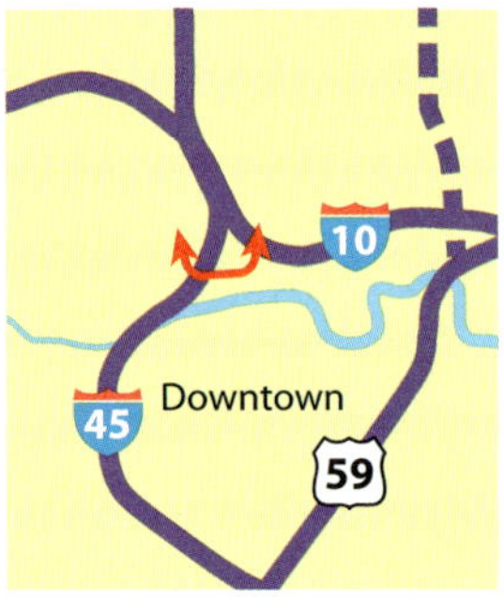

Eastex bypass: This view shows the Eastex Freeway transitway veering away from the freeway main lanes to bypass the interchange at the North Loop. The transitway rejoins the freeway at the upper right corner of the photo. This transitway was completed in 1999. (Photo: September 2002)

Downtown access ramp: This view shows the elevated structure for the two-way transitway ramp along the downtown approach of the Katy Freeway (IH 10). (Photo: May 2002)

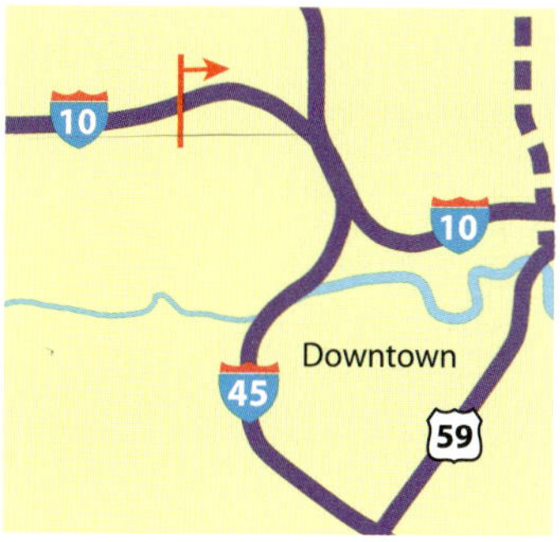

from 1995 to 1997 concluded that the best strategy for improving traffic flow was the creation of four managed lanes, two in each direction, to replace the reversible transitway lane. By making the managed lanes available to buses, high occupancy vehicles, and toll-paying single occupant vehicles, optimal use could be achieved by allowing in the maximum number of vehicles that would still provide free-flow of traffic. In April 2001, local officials announced that HCTRA was considering taking a large role in the project by constructing the four managed lanes as a toll road. In March 2002, the Federal Highway Administration approved the construction of the toll road as part of the massive Katy Freeway expansion project. Finally, on January 30, 2003, the deal was closed when the Texas Transportation Commission officially approved the agreement for HCTRA participation. The can-do spirit that had built Houston's transitway system had once again prevailed, this time with a new player, launching the next generation of transit facilities for Houston. Construction is scheduled for 2003-2008, although the project faces a legal challenge and probable delays resulting from right-of-way acquisition difficulties as of mid-2003.

In September 2002 a second managed lane system for Houston was recommended by the Northwest Freeway major investment study. The recommended managed lane facility has four lanes and will be constructed along Hempstead Highway inside Beltway 8 and along the Northwest Freeway outside Beltway 8. Construction is expected to occur between 2008 and 2015, although the managed-lane facility could be built sooner.

Bridges and Tunnels

Fifty miles inland, on a flat plain drained by small bayous, Houston in its early days did not seem destined to become a city of bridges. There were no rivers to cross and no nearby bays or lakes to block the city's growth. Although Houston was free of impediments, the addition of a man-made barrier would be the event that propelled Houston into the ranks of the nation's largest cities. Dredging of the Houston Ship Channel to a depth of 25 feet (7.6 m) was completed in June 1914, and the channel was officially opened by President Woodrow Wilson on November 10 of that year. The rest, one might say, is history, as the ship channel spurred Houston's industrial boom.

The construction of one great work of infrastructure, the Houston Ship Channel, would ultimately necessitate other construction projects to bridge the man-made divide. Houston would not become a great bridge city on the order of New York City or San Francisco, but would still develop a nice collection of bridges and tunnels to complement its freeway system. In comparison to most cities in the United States, Houston's major bridge crossings are a relatively modern development, with the first high-level bridge span opening in 1973. With newness comes better design and wider spans, but as this history shows, Houston's bridges have all had their share of problems.

The complete history of Houston's bridges, however, predates the construction of the modern Houston Ship Channel. While Houston was still a mosquito-infested outpost on Buffalo Bayou, one of the nation's more prosperous cities was thriving just 50 miles (80 km) to the south—on Galveston Island. If there is anything that can spur bridge construction, it is money. Bridging the mainland to the wealthy port city of Galveston marks the beginning of Houston's bridge history.

Diamonds are the bridge's best friend: The double-diamond towers of Fred Hartman Bridge rise over the Houston Ship Channel, distributing the cables that support the bridge deck. (Photo: TxDOT)

Galveston Causeway

For an island city that has depended on its port and tourism for economic livelihood, the connection to the mainland is a lifeline. In the 19th century, railroads made the vital connection and hurricanes were the biggest threat to the mainland link. As automobile ownership became widespread, the vehicular crossing became the key lifeline, and bridge obsolescence became the biggest threat as traffic volumes rapidly increased. A succession of vehicular crossings that started in 1893 culminated with the present-day structure, completed in 1964. The next chapter in the story of the Galveston Island crossing will unfold starting in 2003, when the existing structures are replaced by an all-new eight-lane, impact-resistant causeway.

Galveston Causeway modern bridge	
Opened	Southbound: July 26, 1961 Northbound: April 29, 1964
Type	Pier and beam
Length	1.6 miles (2.6 km)
Vertical clearance	73 feet (22 m)
Traffic volume, 2001	64,000 vehicles per day
Future construction	Remove existing bridge, build new 8-lane, impact-resistant structure. Construction starts in 2003.

Early Crossings

The settling of Galveston and its use as a port had its origins with Spain and Mexico in the late 1700s and early 1800s, but today's city was begun by real estate investors in 1838, two years after Texas' independence in 1836. After Texas entered the United States in 1845, Galveston's growth was driven by its port, which focused on the cotton trade. Galveston's population, 13,818 in 1870 and 22,248 in 1880, remained relatively small compared to large cities in the eastern United States. St. Louis, Missouri, completed the Eads Bridge over the Mississippi River in 1874, and New York City completed the landmark Brooklyn Bridge in 1884. But for a city the size of Galveston still somewhat on the frontier of America, the Galveston Bay crossings were a big undertaking, especially in consideration of the challenges posed by hurricanes.[1]

As railroad construction gained momentum in Texas in the 1850s, the need for a fixed crossing to the mainland became increasingly urgent. In 1857 the city of Galveston approved $100,000 in bonds for the construction of a railroad bridge. The bridge was opened in 1860, and was seriously damaged by a hurricane in 1867.[2]

The first vehicular crossing of Galveston Bay was opened on November 15, 1893. The 2.14-mile (3.7 km) wagon bridge, as it was called, featured 80-foot (24 m) steel truss spans between concrete piers. By the turn of the century there were three railroad trestles and the wagon bridge connecting Galveston Island to the mainland. Then on Saturday, September 8, 1900, the "1900 hurricane" struck Galveston Island, resulting in the deadliest natural disaster in the history of the United States with an estimated 6,000 deaths. The wagon bridge was destroyed and all three railroad trestles were substantially damaged. The Santa Fe Railroad crossing was in the best condition and became the focus of an intensive repair operation which the *Galveston Daily News* described as the "greatest piece of hustling that has ever been witnessed in railroad construction."[3]

Every foot of the mainland Santa Fe track from the northern end of the Galveston Bay crossing, known as Virginia Point, to Hitchcock, eight miles (13 km) northwest, had been swept away by the storm. Crews struggling to restore the track to Virginia Point worked under extremely difficult conditions, with dead bodies of men and animals strewn along the right-of-way, ground consisting of slush and mud, and a shortage of food and supplies. Eight days later, on Sunday, September 16, the track had been restored to Virginia Point, and all efforts could be focused on the bridge repair. Most of the piling of the bridge itself remained intact, except for an 800-foot (244 m) section which had been torn out by a drifting steamer. Crews initially worked to build track on the existing piling to the point of the 800-foot breach, then pile drivers went to work to close the gap. Just before dawn on Friday,

The wagon bridge: The first vehicular crossing of Galveston Bay was the wagon bridge, completed in 1893. The bridge consisted of 80-foot-long (24 m) steel truss spans on concrete piers. This photo shows two horse-drawn carriages on the bridge. The bridge was destroyed in the 1900 hurricane and was not replaced. (Photo: Rosenberg Library, Galveston)

Construction of the first concrete bridge, circa 1911: After the devastating 1900 storm, the need for a bridge capable of surviving hurricanes was clear. Work on a new concrete structure began in 1909 and was completed in 1912. (Photo: Rosenberg Library G-18221FF2, Galveston)

The Galveston Causeway, circa 1912: The first concrete bridge structure was officially dedicated on May 12, 1912. The causeway had two steam railway tracks, one track with overhead electrical wires for the Galveston-Houston Electric Railway, and a vehicular roadway. (Photo: Rosenberg Library G-18221, Galveston)

September 21, only 13 days after the storm, the first train crossed from the mainland to Galveston Island over the repaired Santa Fe trestle.[4]

After the 1900 storm, the need for an improved railroad bridge and a replacement vehicular crossing was widely recognized, but progress toward a steel or concrete crossing was slow. Finally in 1906 a causeway committee was established by Galveston County officials to move the project for a new crossing forward. In December 1908, an agreement was reached among all bridge users, including the steam railroads, the Galveston-Houston Electric Railway interurban, and Galveston County, which represented automobile interests. In July 1909 a contract was awarded for the construction of the causeway and an official opening celebration was held for the completion on May 25, 1912. The *Galveston Daily News* reported that approximately 30,000 people took part in the celebrations and 1,500 automobiles participated in the opening day parade. The causeway consisted of a drawbridge and 28 arched spans, each 70 feet (21 m) long, with a total bridge length of 2,455 feet (748 m). On both ends of the bridge were approach sections consisting of sand fill between parallel sheet piling rows 154 feet (47 m) apart. The causeway surface was 66 feet (20 m) wide, supporting two steam

Repairing the damage from the 1915 hurricane: The bridge structure itself survived the hurricane with virtually no damage, but the earthfill sections leading to the bridge were washed away. Temporary wooden rail trestles from the shorelines to the bridge were immediately constructed to restore rail service. Permanent repairs extended the concrete bridge structure to replace the earthfill sections. This 1922 photograph shows the repair work nearing completion. Vehicular traffic used a temporary wooden bridge which connected to the unharmed 1912 structure in the distance. (Photo: Rosenberg Library G-18221FF2.12#7, Galveston)

railroad tracks, one electrified track for the Galveston-Houston Electric Railway, and roadway surface suitable for automobiles and trucks.[5]

The new causeway would have a short life, however. Another major hurricane hit Galveston on Sunday, August 15, 1915, inflicting serious damage to the causeway. The concrete arches and drawbridge were "practically uninjured," but the 3,696-foot (1.13 km) northern approach and 2,000 feet (610 m) of the 4,500-foot (1.37 km) southern approach were destroyed, with all the fill sand washed away and only sheetpiling remaining. Once again the Santa Fe Railroad took the lead in restoring the crossing. Plans were made to build wooden trestles to the ends of the concrete structure. Construction on the trestle began with two pile drivers on Saturday, August 21, and by Wednesday five pile drivers were at work. On Thursday afternoon, September 2, just 12 days after construction began, the trestle was completed. Later, a wooden vehicular trestle was built to connect Galveston Island and the mainland with the ends of the concrete structure.[6]

The process of making permanent repairs to the causeway turned out to be lengthy. In May 1917 specifications for the repairs were issued. The existing concrete arch structure would be lengthened on both ends with 28 new arches on the Galveston Island side and 51 new arches on the mainland side. The total bridge length was increased to about 1.6 miles (2.6 km). The roadway on the new structure would have a width of 20 feet, 6 inches (6.25 m). After delays in construction, the rehabilitated bridge was officially dedicated on April 21, 1922. The *Galveston Daily News* reported that approximately 2,000 autos participated in the opening day parade, which began in downtown Galveston and proceeded to the causeway.[7]

Streamlining across the bridge: This photograph, dated 1934, shows a sleek diesel-electric Burlington Zephyr passenger train crossing the causeway. The first Zephyr was introduced into service on May 26, 1934, in Denver, Colorado. The train was known as a "streamliner" for its sleek design and stainless steel cladding. Its revolutionary compact diesel engine allowed it to reach speeds up to 100 miles per hour (160 km/hr). The Zephyr became a national sensation and its streamlined appearance influenced the design of many other products. Still, the Zephyr was competing with a much more powerful force—the developing automobile-oriented society. Big changes were on the way for Galveston Island transportation. The Galveston-Houston Electric Railway, which used the electrified track adjacent to the roadway, would cease operations on October 31, 1936. Work began on a new four-lane vehicular bridge in 1936, and the bridge was opened to traffic on November 30, 1938. Passenger rail service went into decline. (Photo: the Center for American History, University of Texas at Austin, CN 04421)

Entering the Modern Bridge Era

As automobile ownership increased and highways from the mainland improved, the need for a wider bridge exclusively for vehicles became more urgent. Whereas previous crossings had been locally funded, the new vehicular crossing would be financed by TxDOT and the federal Public Works Administration. Work commenced in 1936 and the new causeway was opened to traffic on November 30, 1938. An official dedication ceremony was held on August 15, 1939, when all work was completed. The causeway featured four lanes of traffic without emergency shoulders, and a drawbridge structure at the intracoastal waterway crossing. Total cost after all items were included was approximately $3 million.[8]

By the 1950s, rapidly increasing highway traffic volume and expanding use of the navigation channel for waterborne traffic was putting the bridge under increasing pressure. When an accident closed the bridge for a short time in 1952, a delegation from Galveston first asked the Texas Transportation Commission to authorize the construction of a new causeway which would be wider and have sufficient vertical clearance to eliminate the need for the drawbridge. In 1954 TxDOT authorized the placement of warning signs at both ends of the causeway due to the high accident rate. In August 1956, the Texas Transportation Commission directed the TxDOT Houston office to perform a study of the causeway to determine how needs could be met. The study recommended twin, high-level causeways with a total of six traffic lanes. A new span would be built for southbound traffic, and the existing crossing would have its drawbridge replaced with a high-level span. In October 1957, the commission directed the TxDOT Houston office to proceed with plans for the twin-bridge causeway. Dedication of the first high-level span,

Causeway with drawbridge: The first exclusive vehicular bridge opened on November 30, 1938, and featured a drawbridge at the intracoastal waterway. By the 1950s increasing automobile traffic and frequent bridge openings to accommodate expanding waterborne traffic resulted in long backups at the drawbridge, as shown in the photo at right. Galveston authorities began their efforts to build a high-level crossing in 1952, and in 1956 the Texas Transportation Commission authorized a study which recommended the twin three-lane structures that were completed in 1961 and 1964. (Photos: upper, TxDOT, 1954; lower, TxDOT, circa 1957)

(Left) **Dedication of the first high-level span, July 26, 1961:** This view looks down on the drawbridge of the 1938 highway structure, which would be replaced by a high-level span in the following years. The first high-level span shown in this photo was constructed with lightweight prefabricated concrete beams which deteriorated more rapidly than expected. By the mid-1970s repairs were required, and by the late 1990s replacement of the bridge became necessary. (Photo: TxDOT)

(Below) **Hurricane Carla, 1961:** Just two months after completion of the first new span, Hurricane Carla hit the Texas coast south of Galveston in September 1961. Carla was the most powerful hurricane to hit Texas since the Corpus Christi hurricane of 1919. Sustained winds of 115 mph (184 km/h) were reported south of Galveston in Matagorda, and sustained winds of 88 mph (141 km/h) were reported in Galveston. This photo, taken on September 15, 1961, shows the US 75 approach to the Galveston Causeway and large amounts of debris piled along the roadway. This section was upgraded to a freeway in 1965 and became Interstate 45.[9] (Photo: National Archives 30-N-61-339)

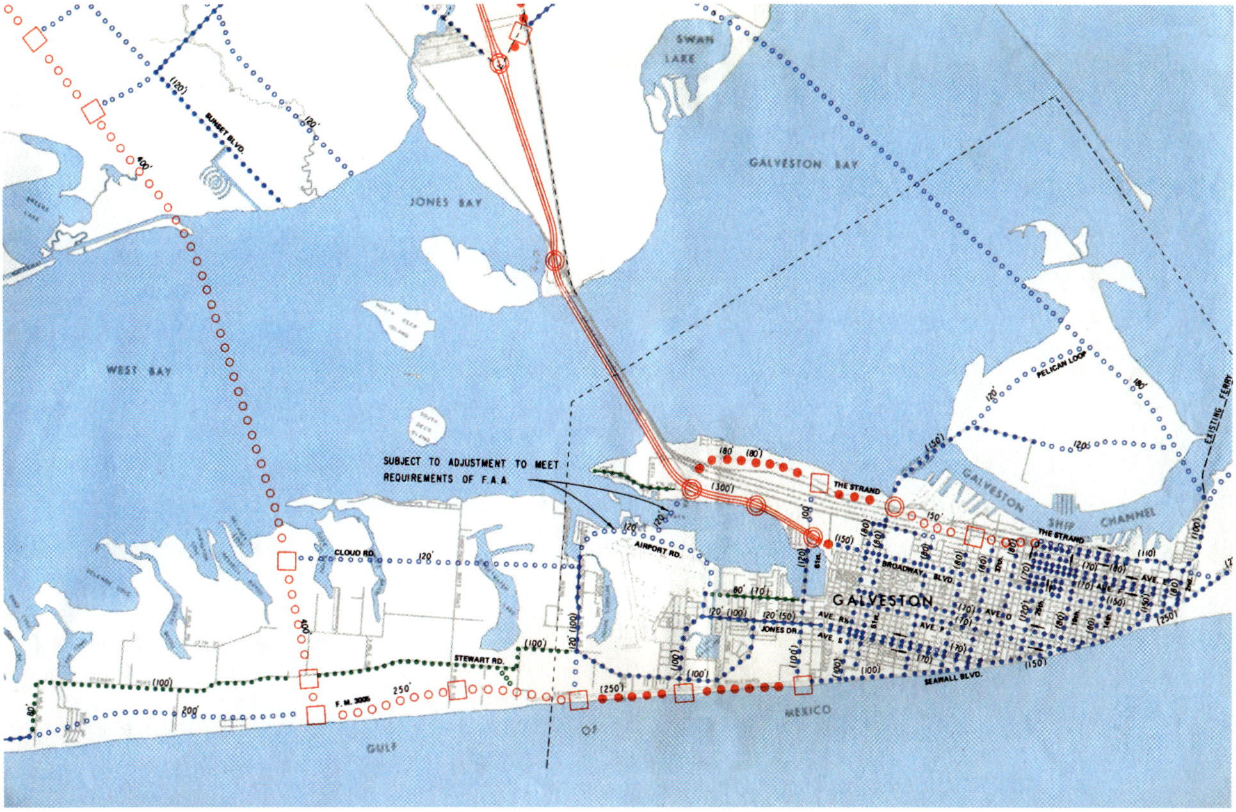

Proposed crossings: The 1969 *Galveston County Transportation Plan* recommended two additional crossings to Galveston Island from the Houston-side mainland. The eastern crossing appears to have faded away very quickly after the plan was published. Efforts to get the western crossing constructed continued until the late 1980s. In 1999 the final nails were put in the coffin of any new crossings to the Houston-side mainland when a study determined that the existing causeway would be reconstructed and expanded. (Source: *Galveston County Transportation Plan, Volume 3*, 1969)

a new 8,400-foot (2.6 km) structure, took place on July 26, 1961. Work then began on modification of the original 1938 vehicular crossing. The entire project was completed on April 29, 1964.[10]

More Bridges?

With the twin-bridge causeway in place, it seemed that Galveston had taken care of its bridge issues. Within a few years, however, proposals for additional crossings appeared. The 1969 *Galveston County Transportation Plan* called for not just one but two new crossings to Galveston Island from the Houston-side mainland, plus a bridge connecting the east end of Galveston Island to Bolivar Peninsula. The so-called Texas City Causeway was proposed to connect Texas City to the northern tip of Pelican Island (which is adjacent to Galveston Island), and a full freeway crossing to west Galveston Island was proposed as part of the West Bay Freeway. Neither of the two proposed crossings moved forward. By the mid-1970s, when the highway funding crisis fully impacted TxDOT, any thought of constructing these new crossings had vanished. In 1973, the more urgently needed crossing to Bolivar Peninsula east of Galveston had been ruled infeasible as a bridge, primarily due to prohibitive cost. In addition, the population increase predicted by the 1969 planning document did not materialize. Without a compelling need, there would be no money forthcoming in the harsh new fiscal climate of the 1970s.[11]

Promoters of the proposed western crossing, however, were not yet ready to give up their dreams. Political leaders and real estate interests envisioned the western crossing opening up rural western Galveston Island to development. In 1986 a preliminary study funded by the city of Galveston indicated there was a need for a second bridge. The proposed bridge began at Interstate 45 on the mainland, just north of the existing causeway, and proceeded in a southwesterly direction across the bay to 8 Mile Road on the island's west end. In 1987 the city of Galveston provided funding to proceed with a preliminary engineering study for the estimated $116 million bridge. In 1989, Galveston officials asked for a study to determine the feasibility of constructing a toll bridge, since there was little hope of ever receiving conventional funding for the bridge, which then had an estimated cost of $145.5 million. The Texas Turnpike Authority studied the crossing and concluded that tolls would generate only 25% of the debt service requirement, so the project was not pursued further. Soon afterwards, the project was dropped due to a low cost-benefit ratio and environmental impacts. Prominent Galvestonian George Mitchell, a real estate developer and leading proponent of the new crossing, looked back on the efforts and stated, "After three years of effort,

Causeway disaster: The Queen Isabella Causeway in south Texas, linking Padre Island to Port Isabel, was struck by an errant barge on September 15, 2001, resulting in the collapse of two bridge sections and eight fatalities. The left photo above, taken on the morning of September 15, shows the center pier still standing. Later in the day it collapsed, shown in the right photo that was taken in the evening. (Photos: TxDOT)

The danger zone: Barge traffic passes underneath the Galveston Causeway at the intracoastal waterway. Barges straying from designated shipping channels and impacting bridge structures have been responsible for six major accidents in the United States since 1980. The collapse of the Sunshine Skyway Bridge in Tampa, Florida, in 1980 prompted the development of standards for impact-resistant bridge structures in the United States. The new Galveston Causeway will conform to the latest standards and will be designed to survive impact from the largest typical barge string. In addition to impact-resistant design, the new causeway will have a much longer main span over the intracoastal waterway, 350 feet (106 m) in comparison to the existing 150 feet (46 m) on the southbound span. (Photo: James Lyle, TTI, June 2001)

The Galveston Causeway, October 2001: This view looks north, with Galveston Island in the foreground and the mainland in the background. The new causeway structure will be built on the same location as the existing causeway. (Photo: The Positive Image)

we knew it was impossible to get done."[12]

In a separate effort, a group of landowners in Brazoria County, on the mainland to the west of Galveston, funded a study of a new facility called the Galveston-Alvin-Pearland Parkway, the GAP Parkway. The GAP Parkway was a resurrection of the cancelled West Bay Freeway. A preliminary engineering report was issued in November 1986, indicating the route was feasible and showing an all-new crossing to the west end of Galveston Island, terminating at the community of Jamaica Beach. After the release of the report, the project failed to gain momentum and soon disappeared from planning maps.[13]

The second crossing was looking increasingly unlikely but was not yet dead. In the 1990s TxDOT initiated additional studies in the region, including the option of improving the existing causeway and adding a new crossing at Offatts Bayou to provide better access to west Galveston Island. The Offatts Bayou crossing proved to be highly controversial in 1994 hearings and was not pursued.[14]

Key dates in the history of the Galveston Causeway

1860	First railroad bridge opens.
1893	First vehicular crossing, the "wagon bridge"
1900	Bridges destroyed by hurricane.
1912	Concrete bridge for railroads and vehicles opens.
1915	Bridge approaches destroyed by hurricane.
1922	Repairs and bridge extension completed.
1938	First dedicated vehicular crossing opens.
1961, 1964	Twin causeway bridges are completed.
2003	Construction begins on a new $136 million causeway with a 148-foot-wide (45 m) deck, 8 traffic lanes, and an impact-resistant structure. Existing bridges will be entirely removed.

Fortress Causeway

In the meantime, age was taking a toll on the existing causeway. As early as 1976, major repairs were required on the newer southbound span to fix the deteriorating bridge footings and pilings. The southbound span, completed in 1961, had been constructed with slender, lightweight prefabricated beams that were not holding up well. An engineering inspection report released in 1997 identified cracking of 50 concrete crossbeams on the span, specified $5 million in immediate repairs, and recommended the formulation of plans to replace the entire causeway.[15]

In 1998 TxDOT undertook a comprehensive study of the Interstate 45 corridor from south Beltway 8 in Houston to Galveston, including the Galveston Causeway. With the recommendations of this study in the summer of 1999, the final nails were put into the coffin of a second bay crossing. The existing causeway was slated to be entirely rebuilt and expanded to eight lanes, providing the only crossing from the Houston-side mainland. Existing arterial streets on Galveston Island would be improved to better integrate local roads with the causeway.[16]

Since the construction of the modern causeway in the early 1960s, design standards for bridge and causeway structures had changed to recognize the risk and minimize the consequences of an all-too-frequent occurrence: barge collisions. Even as the design of the new bridge was underway, two major barge collisions occurred in the United States, resulting in total collapse of the impacted bridge structures and substantial loss of life. On September 15, 2001, a barge collided with the Queen Isabella Causeway near Brownsville in south Texas, collapsing the structure and causing 8 fatalities. On May 26, 2002, a barge struck the IH 40 Arkansas River crossing near Webbers Falls, Oklahoma, collapsing the bridge and causing 14 fatalities.[17]

The development of standards for impact-resistant bridge structures in the United States began after an ocean freighter struck a bridge support of the Sunshine Skyway Bridge in Tampa, Florida, on May 8, 1980, resulting in the collapse of a long-span, high-level bridge. Florida became the leader in research efforts to develop the new bridge standards, and in 1991 the American Association of State and Highway Transportation Officials (AASHTO) published their *Guide Specification and Commentary for Vessel Collision Design of Highway Bridges*. The new Galveston Causeway design was guided by Florida Department of Transportation standards and will be the first in Texas to comply with the AASHTO standards.

The existing causeway will be entirely removed to two feet (0.6 m) below the mud line. In its place a new eight-lane causeway with a 148-foot-wide (45 m) deck will be built. The causeway length will remain at 1.6 miles (2.6 km). The crossing over the intracoastal waterway, where barge traffic passes, will be designed to survive an impact from the largest expected barge without collapsing. Adjacent spans will also be collapse resistant. The main span will be a 350-foot-long (107 m) cast-in-place concrete structure with a 73-foot (23 m) vertical clearance at high tide. A $136 million construction contract was awarded in June 2003.

In a separate study concluded in November 2000, TxDOT identified the most feasible route for a new crossing from Galveston Island to Bolivar Peninsula, east of Galveston. If Galveston ever gets another bridge, it will be the long-studied crossing to Bolivar Peninsula.[18]

SH 146 Ship Channel Crossing Baytown Tunnel and Fred Hartman Bridge

In modern history, bridges and tunnels have been some of the most challenging and dramatic projects undertaken by man. The Brooklyn Bridge. The Golden Gate Bridge. The English Channel Tunnel. These projects often pushed technology, processes, and workers to their limits. The SH 146 ship channel crossing didn't break new engineering barriers during its history, but it did push one thing to the limit: the patience of motorists who used the crossing. Motorists endured years of traffic congestion on the old Baytown Tunnel waiting for construction of the new bridge to begin. When bridge construction finally began in 1987, motorists endured another eight years of waiting before the bridge was completed. On September 30, 1995, the eight-lane cable-stayed span finally opened to traffic. Fittingly for Houston, the old Baytown Tunnel was subsequently sliced up into pieces and removed, its aggregate ground up to be recycled into new freeways.

Origins

After World War II there were no vehicular crossings of the Houston Ship Channel. Harris County recognized the need and approached TxDOT in June 1945 for aid in the development of ship channel crossings. On May 15, 1946, an engineering study was completed. The Texas Transportation Commission stated that TxDOT would not participate in a crossing near Pasadena since it would serve local traffic only. That crossing would become the Washburn Tunnel and would be a Harris County project. The commission agreed to participate in the SH 146 crossing. The TxDOT study had concluded that both a high-level bridge and tunnel were feasible at the SH 146 location. The high-level bridge was estimated to cost \$5,077,000 and the tunnel was estimated to cost \$7,900,000. This was a substantial cost difference. The commission stated, "From the standpoint of motor vehicle traffic movement at minimum cost, it is our opinion that the high-level bridge is the proper solution to the problem." However, the Houston Ship Channel Navigation District and Harris County felt that a high-level bridge "would be detrimental to the

Baytown Tunnel	
Opened	September 22, 1953
Length	4,110 feet (1,253 m)
Removed from service	September 30, 1995
Demolished	1998

Fred Hartman Bridge	
Opened	September 30, 1995
Type	Cable-stayed
Main span length	1,250 feet (381 m)
Main span rank, 2003 cable-stayed bridges	United States: 2 (tie) World: 32
Vertical clearance	175 feet (53 m)
Width	8 traffic lanes Full inner and outer shoulders
Traffic volume, 2001	62,000 vehicles per day
Named after	Fred Hartman, prominent Baytown resident, publisher of the *Baytown Sun* newspaper, promoter of the bridge
Future construction	None planned

Building the Baytown Tunnel: The underwater tunnel sections were fabricated at a steel mill in Orange, Texas. With the ends sealed, the floating tubes were towed 125 miles (200 km) along the intracoastal waterway and the Houston Ship Channel to the tunnel site. (Photo: TxDOT)

Construction of the approach tunnel. (Photo: TxDOT)

Opening day, September 22, 1953. (Photo: TxDOT)

The north portal ventilation structure, 1953-1998: This photograph of the art deco style ventilation structure at the north (Baytown side) portal was taken in 1967, the first year that major traffic backups were reported. It was also the year that local officials began the first serious efforts to replace the tunnel with a bridge. It would be a long road to get the replacement bridge constructed. This structure was demolished during the tunnel removal in 1998. (Photo: TxDOT)

shipping interest using the channel" and preferred the tunnel. In the proposed agreement offered by the Texas Transportation Commission, TxDOT would contribute $5.0 million to the project and the local agencies would be responsible for all additional costs. This proposal was amended on October 15, 1946, fixing the Harris County contribution at $2.9 million and making TxDOT responsible for all remaining costs.[19]

A ground breaking ceremony was held on September 16, 1949. The main underwater sections of the tunnel were formed by 300-foot-long (91 m), 35-foot (10.7 m) diameter steel cylinders fabricated at the U.S. Steel subsidiary in Orange, Texas. The ends of the cylindrical sections were sealed, allowing the tunnel sections to float, and the sections were then towed 125 miles (200 km) on the intracoastal waterway to the tunnel location. The first section arrived in May 1950. The concrete roadways were built into the steel shells, and then each section was sunk separately into the channel. The sections were then joined, the end seals removed, and final interior concrete work completed. The tunnel was opened on September 22, 1953, with a large dedication ceremony. The total project cost was $8,701,138. The two-lane tunnel with its 22-foot-wide (6.7 m) road surface was the first tunnel on the Texas highway system, although Harris County's Washburn Tunnel had opened previously on May 27, 1950. The SH 146 tunnel was 4,110 feet (1252 m) long and its bottom was 89 feet (27 m) below the surface of the water at mean tide. It was named the Baytown Tunnel.[20]

From Tunnel to Bridge

During the 1950s and 1960s the Baytown Tunnel served as the main vehicular crossing of the Houston Ship Channel. The only other crossing was the Washburn Tunnel, 12 miles (19 km) to the west. The expansion of industry along the ship channel after World War II, combined with increased auto ownership and a growing population, soon put the two-lane tunnel under traffic pressure. By 1967 the tunnel was seeing regular traffic slowdowns, especially in the busy summer months when traffic counts exceeded 18,000 vehicles per day. The first efforts to build a bridge began in early 1967 when the Baytown Chamber of Commerce asked the community to join a campaign for a new bridge. Early bridge promoters surely did not anticipate the long road ahead to get the bridge built.[21]

The Texas Transportation Commission authorized preliminary engineering studies of the SH 146 crossing on August 31, 1977, and authorized preparation of detailed plans on November 29, 1982. The final environmental assessment for the project, completed in 1980, recommended an eight-lane structure with a vertical clearance of 175 feet (53 m), a minimum main span of 850 feet (259 m), and a recommended span of at least 1,000 feet (305 m). The bridge would use cable-stayed technology, a technique that was still relatively new in the United States. As of 1980, only one major cable-stayed span had been constructed in the United States—the 750-foot-long (229 m) Pasco-Kennewick span in Washington State. The bridge would be relatively new territory for American

The tease: When the above photograph was taken in March 1990, motorists passing through the tunnel were surely excited to see the towers rising over the Houston Ship Channel. The excitement would turn into frustration as progress came to a near-halt after the tower construction. The towers would stand idle for years as the contractor struggled to obtain steel that met specifications and quality standards. In the meantime, huge traffic jams would develop during rush hour, and local citizens and politicians didn't hesitate to vent their frustration to TxDOT. Construction started to gain momentum in 1994 when the steel fabrication problems were finally solved. The August 1994 photo below shows the bridge deck on the south tower nearly complete. (Photos: TxDOT)

Emerging from the fog: This view shows the main bridge deck nearing completion in 1995. Around the time of this photo, the project schedule was also emerging from a fog, much to the relief of motorists. (Photo: Williams Brothers Construction)

The Fred Hartman Bridge: The bridge was officially dedicated on September 27, 1995, and opened to traffic three days later. The bridge features eight traffic lanes. (Photo: TxDOT)

designers and builders. Many others had been constructed outside the United States, especially in Europe. The final design increased the bridge main span to 1,250 feet (381 m) and featured a distinctive double-diamond design for its two towers, which rose 440 feet (134 m) above the ship channel. Bob Lanier, a native of Baytown, served as chairman of the Texas Transportation Commission from 1983 to 1987. He was able to move the project forward with 100% state funding, an arrangement that opened the door for problems that would plague the project.[22]

Motorists in the Baytown area were delighted when a $91.3 million contract for the new bridge was awarded in December 1986 to a joint venture of Williams Brothers and Traylor Brothers construction companies. Soon, local interests thought, they would be cruising over a beautiful and wide new bridge. Within a year of the contract award Williams Brothers bought out Traylor Brothers' share of the contract, and Williams Brothers became the sole contractor. As the project moved toward construction of the main bridge span, problems began to develop. Over the following years the problems escalated and the construction of the bridge became a project management debacle, delaying the project for years, adding tens of millions of dollars to the cost, and causing design changes that forced reviews of the bridge's quality. The project's problems centered on the steel. Federally funded projects included clauses for the use of American steel, but this project was 100% funded by the state of Texas, so the contractor was free to use steel from any source. Williams Brothers initially planned to use South African steel, but changed that plan because of the controversy surrounding South Africa's apartheid policies. An effort was then made to obtain the steel from Mexico, but the Mexican firm could not deliver. Williams Brothers turned once again to a South African supplier, but the supplier had serious problems fabricating the steel per specifications. News reports painted a picture of chaos at the South African steel plant. Eventually Williams Brothers was forced to turn part of the steel fabrication over to an American firm.[23]

In the meantime, years passed by and the tunnel traffic grew increasingly intolerable. The tunnel itself had serious maintenance problems. By early 1995 local authorities were demanding prompt completion of the bridge. By that time, most of the problems had been worked out and construction was moving forward. Finally, on September 27, 1995, the bridge was dedicated by Governor George W. Bush and named after Fred Hartman, a Baytown publisher and one of the key promoters of the bridge. Hartman died in 1991 at the age of 83 and did not see the completion of the project for which he had worked so hard. The bridge opened to traffic on September 30.[24]

After completion of the bridge, Williams Brothers filed

Tunnel removal, 1998: Because of a planned deepening of the Houston Ship Channel, the tunnel needed to be removed soon after the bridge was opened. Initial plans to cut the tunnel into sections and then transport the sections to an artificial reef area in the Gulf of Mexico did not move forward. In a second round of bidding, the bridge contractor submitted the winning bid with a plan to slice the tunnel into sections and recycle the material. The removal of the Baytown Tunnel appears to be the first removal of an underwater vehicular tunnel in the United States. (Photos: TxDOT)

Fred Hartman, 1908-1991

Fred Hartman was a prominent figure in the newspaper and publishing industry in the Houston area for more than 50 years, serving as publisher of the *Baytown Sun* newspaper from 1950 to 1974. He was a community leader and served in various philanthropic organizations. He worked to promote transportation improvements and was a key force in getting the SH 146 bridge built. Fred Hartman presided over the opening ceremonies of the Baytown Tunnel in 1953, but died in July 1991 before the bridge bearing his name was completed. In an interesting historical curiosity, Hartman was the first boss of Bob Lanier, a native of Baytown who went on to become a leading advocate of transportation improvements in Houston and Texas. In this photo, Fred Hartman participates in the April 18, 1966, dedication of the inner loop section of the East Freeway, the final link connecting Baytown to Houston. Hartman is the third person from the left, wearing a bow tie. Also pictured, from left to right: Eldon Berry, president of the Baytown Chamber of Commerce; Louie Welch, mayor of Houston; Hartman; Earl Caulkins, chairman of the Houston Chamber of Commerce Highway Committee; Herbert Petry, chairman of the Texas Transportation Commission; and V. V. Ramsey, Harris County Commissioner. (Photo: TxDOT)

Key dates in the history of the SH 146 crossing

1946	TxDOT approves tunnel and financial agreements.
1949	Tunnel construction begins.
1953	Baytown Tunnel is dedicated September 22.
1967	Efforts begin to construct a bridge.
1977	TxDOT authorizes engineering studies for the bridge.
1986	Construction contract is awarded.
Early 1990s	Construction practically stops due to steel fabrication problems.
1995	Bridge is dedicated on September 27.
1998	Baytown Tunnel is removed.

a $70 million claim with TxDOT for compensation of costs associated with TxDOT errors. A consultant hired to sort out the claim ruled that Williams Brothers deserved most of the blame, but TxDOT bore some responsibility for the problems. The Texas Transportation Commission awarded Williams Brothers a $23.5 million settlement on February 28, 1996. Including an earlier $3 million change order, the total cost of the bridge was $117.5 million, 29% above the original bid.[25]

The Fred Hartman Bridge features the second longest cable-stayed main span in the United States as of 2003, just behind the 1,300-foot (396 m) Dame Point Bridge in Jacksonville, Florida, and the same length as the Sidney Lanier Bridge in Brunswick, Georgia. It is longer than the well-known Sunshine Skyway Bridge in Tampa, Florida, which has a main span of 1,200 feet (366 m). A surge of bridge construction in Europe and Asia in the 1990s has pushed the Fred Hartman Bridge back to number 32 internationally, with the 2,920-foot (890 m) Tatara Bridge in Japan taking the honor as the world's longest cable-stayed span. The Hartman Bridge features two traffic decks, each with four main lanes and both interior and exterior shoulders. It is a bridge that any city would love to have in its freeway system, and the crown jewel of Houston's bridge collection.

The End of the Tunnel

According to the terms of the original agreements to construct the tunnel, the tunnel and associated buildings would be removed when they no longer carried traffic. Since there were plans to widen and deepen the Houston Ship Channel in the late 1990s, the tunnel removal would proceed promptly.

The tunnel removal appears to be the first and only removal of an underwater vehicular crossing in the United States. The original plan was to cut the tunnel into 350-foot (107 m) sections, float the tunnel sections to the surface, and move them to an artificial reef site in the Gulf of Mexico. After a contract award for the reef conversion, complications arose and the project went out for a second round of bidding. In the second round, the method of tunnel disposal was left up to the contractor. Williams Brothers, the bridge contractor, won with a $17.5 million bid, $4.3 million less than the next lowest bid. Williams Brothers cut the tunnel into sections and recycled the concrete for use in other road construction projects. So rather than being entombed in the Gulf of Mexico, the tunnel material would wind up in another freeway, a fitting end in Houston.[26]

Loop 610 Sidney Sherman Bridge

The Loop 610 ship channel bridge ushered Houston into the era of high-level spans with its opening on March 2, 1973. With a 152-foot-wide (46 m) deck accommodating 10 traffic lanes, it was an impressive structure. Unfortunately, one of its other key dimensions—a vertical clearance of 135 feet (41 m)—was not very impressive. The inevitable major collision with an ocean vessel finally occurred 27 years later, only to be followed by an even worse collision just six months later. The bridge remained intact and was repaired in each case. But in 2001 the bridge may have finally earned a reputation as being a bridge too low.

Sidney Sherman Bridge	
Opened	March 2, 1973
Type	Strutted girder
Main span length	600 feet (183 m)
Vertical clearance	135 feet (41 m)
Width	10 traffic lanes Full inner and outer shoulders
Traffic volume, 2001	122,000 vehicles per day
Named after	Sidney Sherman, colonel in the Republic of Texas Army, promoter of improvements to the Houston Ship Channel
Future construction	None planned

Approach construction: This June 1971 view looking northeast shows the first "V" strut being positioned. (Photo: TxDOT)

Heavy lifting: A steel girder for the approach span is lifted into position in June 1971. (Photo: TxDOT)

Origins

The need for crossings of the Houston Ship Channel was recognized by local authorities after World War II, and efforts were soon underway to get crossings in place. The immediate need would be satisfied by the Washburn Tunnel, opened in 1950, and the Baytown Tunnel, opened in 1953. These two-lane tunnels provided the needed relief, buying time for the first true freeway crossing over the ship channel. The same year as the opening of the Baytown Tunnel, 1953, officials were preparing Houston's freeway master plan—a plan which included a full freeway loop and a crossing of the ship channel. By October 1954, the north, south, and west sections of the loop had been adopted into the state highway system, but not the east section which included the ship channel bridge. Getting the east section adopted into the state highway system and then into the federal Interstate Highway System turned out to be a long, protracted effort for local officials, but by September 1962 the project was fully approved into the interstate system.[27]

By 1962 engineers at TxDOT had already formulated plans for the crossing. In the interest of saving money, TxDOT planned to make the main span as low and as short as possible with a vertical clearance of 135 feet (41 m) and a horizontal clearance between support piers of only 400 feet (122 m). The Navigation District of the Port of Houston objected to the narrow horizontal clearance, citing it as a navigation hazard, and won a victory when the Army Corps of Engineers rejected the proposed design on November 13, 1963. TxDOT then increased the main

Closing the span: This view shows placement of the first main span girder for the bridge on February 19, 1972. As soon as this beam was in position, the low clearance of the bridge became painfully clear to officials at the Port of Houston. (Photo: TxDOT)

span to 600 feet (183 m) by using a strutted design with diagonal supports at each end. In early 1964 the Army Corps of Engineers issued a permit for the bridge. There were concerns that a 135-foot vertical clearance might be inadequate, but with the Army Corps approval, TxDOT had the green light to proceed. The final design for the bridge included a deck 152 feet wide (46 m), accommodating 10 lanes of traffic. The bridge wouldn't be high and its span wouldn't be long, but it would have a very wide deck in comparison to most high-level bridge crossings.[28]

Major work on the bridge was underway by September 1968. On February 19, 1972, the first girder fully spanning the channel was put into place. As soon as that first girder was in place, officials at the Port of Houston realized that the bridge clearance was going to be a problem. In a March 1972 *Houston Post* report, the chief engineer for the Port of Houston, Richard P. Leach, called the bridge's low clearance "a mistake in judgment" and said the bridge would be a navigation hazard. Regarding a possible collision between a ship and the bridge, Leach stated, "I always take the view that if something can happen it will happen sooner or later." Making matters worse were the V-struts at the sides of the span, which provided an even lower clearance for ships veering from the center of the channel. The presiding officer of the Houston Pilots, an association of navigators who guide ships travelling on the channel, lamented, "We protested the vertical clearance ... but nobody listened to us."[29]

Still, TxDOT officials were confident that the clearance would be adequate. And it really was a matter of money. The bridge, as constructed, cost $19 million—approximately 63 million in 2003 dollars. A bridge with a 150-foot (46 m) clearance would have cost an extra $10 million, according to A. C. Kyser, head of the TxDOT Houston Urban Project Office and a key member of the design team for the bridge. Regarding the clearance, Kyser stated, "In our investigation on the 610 bridge we found that no ships were coming into Houston that needed more than a 120 foot clearance." Perhaps TxDOT should have been more concerned. After all, it was their bridge, and if it would be struck by a ship, the liability and repair bill could also be theirs.

Key dates in the history of the Sidney Sherman Bridge

1953	Houston officials propose a full loop around Houston as part of a comprehensive freeway plan.
1962	The East Loop 610 is adopted into the federal Interstate Highway System.
1969	Construction begins.
1973	The bridge is opened March 2.
2000	The first major collision with a vessel occurs.

The 10-lane bridge was completed and opened to traffic on March 2, 1973. The bridge was recognized by the National Society of Professional Engineers as one of the nation's top 10 outstanding engineering achievements for 1972. In 1974 Houston City Council named the bridge the Sidney Sherman Bridge. Sidney Sherman (1805-1873) was a colonel in the Republic of Texas Army and is credited with shouting "Remember the Alamo" during the decisive defeat of the Mexican Army at San Jacinto on April 21, 1836. Sherman was instrumental in early efforts to

Collision: On December 21, 2000, a crane on a cargo ship struck the bridge, causing this large hole in the bridge deck. No vehicles fell through the hole, and there were only three minor injuries. Most of the damage occurred to the bridge substructure underneath the deck. Five months later another major collision caused even more extensive damage to the bridge girders. (Photo: TxDOT)

"They knew it was too low when they built the thing. Who would OK a bridge 135 feet high? I don't know."

Port of Houston pilot quoted in the *Houston Post,* June 23, 1976

make the Houston Ship Channel a navigable waterway.[30]

A Bridge Too Low

It didn't take long for ships to strike the bridge. In fact, it was hit several times during construction, with the bridge structure sustaining minor damage. By 1976 more controversy erupted over the bridge's low clearance. On June 23, 1976, the *Houston Post* reported that at least two ships had hit the bridge, causing minor damage. More significantly, a new class of cargo vessel, a roll-on/roll-off design that allowed cargo to be driven on or off the ship, had to be diverted away from the Port of Houston because it would not fit underneath the bridge. Fingers were pointing at TxDOT. The article quoted a Port of Houston ship pilot who said of TxDOT, "They knew it was too low when they built the thing. Who would OK a bridge 135 feet high? I don't know." The fact was that the Army Corps of Engineers had approved the 135-foot (41 m) clearance, a point that TxDOT emphasized.[31]

The Loop 610 bridge had luck on its side for its first 27 years. There was one minor collision with an oversized vessel in 1988, but the small hole in the bridge deck was repaired within days. The good luck finally ran out on December 21, 2000. A cargo ship's crane struck the bridge, tearing out a large hole in the concrete deck, severely damaging one steel beam underneath the deck and moderately damaging another steel beam. Repair took about six weeks. Three northbound lanes were closed for the entire period, and other major bridge closures were required to make the repair. On May 31, 2001, a freighter's cargo boom struck the bridge, punching another hole in the pavement and causing a bridge girder underneath the deck to sustain extensive damage. Repairs would cost $1.25 million, require two months, and close four of the five northbound lanes. By that time, Houston motorists were hoping for another 27-year run of good luck.[32]

Width is good: The bridge features a deck 152 feet (46 m) wide to accommodate 10 traffic lanes. This view looks south. (Photo: July 2002)

Height is bad: This view looks east along the Houston Ship Channel at the Sidney Sherman Bridge. The 135-foot (41 m) vertical clearance of the bridge caused controversy in the bridge's early years and eventually resulted in collisions with vessels. (Photo: September 2002)

Beltway 8
Sam Houston Tollway Ship Channel Bridge

There's an old saying that "timing is everything." But just as good timing can engender great success, bad timing can send a venture down a road of complications, unmet expectations, or even outright disaster. To state it bluntly: the timing of the Beltway 8 toll bridge was bad. And to further complicate the project's difficulties, a subsequent financial decision was perhaps even worse than all the bad luck that had plagued the bridge.

The project was launched near the peak of Houston's energy boom in 1978 but opened in 1982—just as the devastating collapse of Houston's economy began. Connections to the roadway were poor when it opened and marginal until 1994. Cable-stayed bridge technology, not generally accepted in the United States in the late 1970s, became mainstream within a few years of the design phase and may have offered a more economical bridge design. Compounding the bad luck was a junk-bond refinancing in 1985 that buried the project under a staggering mountain of debt. But the story has a happy, although very expensive, ending, with the financially strong Harris County Toll Road Authority coming to the rescue in 1994.

Sam Houston Tollway Ship Channel Bridge	
Previous name	Jesse Jones Bridge
Opened	May 6, 1982
Type	Concrete box girder
Main span length	750 feet (228 m)
Vertical clearance	175 feet (53 m)
Width	4 traffic lanes, no shoulders
Traffic volume, 2002	28,800 vehicles per day
Future construction	None planned

Origins of the Bridge

The Beltway 8 crossing has it origins in the 1952 city of Houston planning document that recommended a second loop for Houston and designated it as the Outer Belt. In 1960 Harris County took control of the project and designated the Outer Belt as a controlled-access freeway. Initial efforts to construct the Beltway 8 crossing as a toll facility began in the mid-1960s, years before construction even began on Houston's first ship channel bridge, the Loop 610 crossing. Houston officials approached the Texas Turnpike Authority about constructing the bridge, but the Turnpike Authority did not have funds to perform the necessary studies. Bridge advocates then worked to secure the passage of a bill in the Texas Legislature in 1967 that established the Texas Toll Bridge Authority. However, the Texas Toll Bridge Authority effectively duplicated the Texas Turnpike Authority, so Governor John Connally vetoed the bill on June 18, 1967. By 1968 Harris County voters had twice rejected bond funds to construct the bridge, so in 1968 and 1969 Harris County officials focused their effort on building the bridge as a toll facility, funding its own feasibility study. The bridge would not move forward at that time. In 1977 the Texas Turnpike Authority's consulting engineers completed an initial study of the bridge and recommended that a final study be made. In the summer of 1978 final studies indicated the project was feasible, and $102 million in bonds were sold to build the bridge. The bond issue translates to approximately 234 million in 2003 dollars.[33]

The Turnpike Authority's consulting engineers studied numerous designs for the bridge, including a cable-stayed design which was later used on the nearby SH 146 bridge. Starting in the 1980s, cable-stayed bridges became the

design used most frequently for spans of this length. In 1978 only one major cable-stayed span had been constructed in the United States, the 750-foot (229 m) Pasco-Kennewick span in Washington State. The consultant's cost estimate reflected the lack of familiarity among U.S. construction firms with the construction techniques required to build cable-stayed spans and the engineering difficulty in analyzing the structures, since analytical tools were not yet available. Consequently, the cable-stayed design was estimated to be 15% more expensive than the most economical design, the concrete box girder.[34]

The plans for the concrete box girder bridge included a main span of 750 feet (229 m), a vertical clearance of 175 feet (53.3 m), and a total bridge length of 2.0 miles (3.2 km). At the time it was the longest concrete box girder span in the Western Hemisphere and the fifth longest in the world.* Construction was underway by 1979 and progressed well. The bridge, originally named the Jesse Jones Ship Channel Bridge, was opened to traffic on May 6, 1982. Jesse Jones (1874-1956) was probably the most influential individual in the development of Houston into a major city. He played an important role in the 1910 deepening of the ship channel by persuading Houston banks to purchase $1.25 million in Harris County port improvement bonds that previously could not find any buyers.[35]

Pier construction: This 1980 photograph shows the piers for the main span taking shape. The reason for calling this type of bridge a concrete box girder can be seen in this photo. The "box" is the cross section of the bridge's main span, visible at the top of the pier. In this bridge the box actually has the shape of a trapezoid. (Photo: North Texas Turnpike Authority)

The Financial Woes Begin

Traffic volume on the new bridge immediately fell below projections. In 1982, projected traffic volume was 4 million vehicles, but actual traffic volume was only 1.69 million vehicles, 42% of the projection. At the end of 1983 the annual report of the Texas Turnpike Authority first reported on the traffic shortfall, attributing the lack of traffic to the delay in completion of the connections to the bridge. Moody's Investors Service revised the bond rating to "B," which is one level below the rating of the highest rated junk bonds, "Ba."[36]

In 1984, interchanges at the north and south ends of the Beltway 8 segment including the bridge were completed, but traffic increases were not sufficient to generate enough revenue to service the debt. The 1984 annual report of the Texas Turnpike Authority stated that "the lack of adequate revenue stems from the lack of adequate connections to US 90 at the north and IH 45 at the south, combined with a general recession in the Houston ship channel area economy." By May 1985 Merrill Lynch, principal underwriter of the bonds, was warning of a possible default. In July 1985 Governor Mark White sent the Texas Turnpike Authority a letter urging the agency to "take all necessary action to avoid the possible default on these bonds." That's exactly what the agency did, but the price for avoiding default would be very, very high.[37]

The bridge debt was refinanced with a $27.9 million junk bond issue in December 1985. The terms of the junk bonds were indicative of just how risky this proposition

* The Confederation Bridge at Prince Edward Island, Canada, with 43 820-foot (250 m) spans, displaced the Beltway 8 bridge as the longest concrete box girder span in the Western Hemisphere in 1997. After a surge of bridge construction in Europe and Asia in the 1990s, the bridge span no longer ranks in the top 20 internationally. The Stolmasundet Bridge in Norway, with a 988-foot (301 m) span, is the world's longest concrete box girder.

Building a concrete box girder bridge: This photograph taken near the end of 1981 shows the bridge main span about 75% complete. This view also illustrates the construction technique used for a concrete box girder bridge. The "box" is actually the hollow, rectangular cross section of the main bridge span. After the bridge support towers are constructed, the bridge deck is gradually built outwards from the towers in both directions to maintain balance. Each additional section of the bridge is "post-tensioned," meaning that cables are used to put squeezing pressure, or compression, on the concrete. Concrete is strong in compression but weak when it is pulled on (in tension). The two sides meet in the middle, connecting the bridge sections. (Photo: North Texas Turnpike Authority)

A mountain of junk bond debt

The 1985 junk bond refinancing of the toll bridge was financially devastating and practically guaranteed a default or costly bailout when the first junk bond payment came due in 1996. (values in millions of dollars, except interest rates)

Original bond issue, 1978	102
Interest rate	7.54%
Total debt service, interest and principal, 1984	223
Junk bond financing, December 1985	27.9
Interest rate	12.625%
Junk bond total debt service	332
Junk bond minimum liability, based on earliest bond refinance date of July 1, 2002	176
Total debt service, interest and principal, end of 1985	522

was. The interest rate was 12.625%, and no payments of interest or principal would be made for 10 years. The junk bond refinancing bought the authority time, but amounted to a huge bet that traffic and toll revenue would dramatically increase in the future.

The long-term financial implications of the junk bonds were staggering. With no payments for 10 years, the $27.9 million would grow to $95.6 million by 1996 when interest payments would begin. Over its total life through the year 2020, the $27.9 million junk bond issue would require $332 million in debt service payments. Prior to the junk bond financing, the 1984 annual report of the Texas Turnpike Authority listed total debt service obligations of the bridge at $223 million. After the junk bond financing in 1985, total obligations increased to $522 million.

But perhaps most financially devastating was the clause in the agreement that prevented repayment until July 1, 2002. This meant that the Texas Turnpike Authority was liable for $175.8 million in payments on the $27.9 million bond issue, with no option to call (refinance) the bonds. Only those closely involved with the junk bond market in 1985 may really know if such unfavorable bond terms were in fact necessary to obtain the financing. Perhaps the decision makers were content to pass this liability on to the next generation of politicians. Someone else would be in charge when the cards would be turned over on this huge bet.

Authorities would not have to wait until 1996 to know if their bet would succeed or fail, however. Toll-paying traffic on the bridge increased nicely over the next few years but at nowhere near the rate required to meet the obligations of the junk bonds. From 1989 to 1993, bridge patronage remained flat at about six million vehicles per year, generating approximately $10 million per year in revenue. This just met the $9.4 million annual obligation on the original bonds, but in 1996 annual payments of $12

Spanning the channel: This view looks east at the Sam Houston Tollway Ship Channel Bridge. (Photo: September 2002)

million per year would begin on the junk bonds. Default was a certainty if a bailout did not occur. But the savior of the Beltway 8 bridge would soon step forward.[38]

Bailout

While the Texas Turnpike Authority's Beltway 8 bridge had turned into a financial disaster, the Harris County Toll Road Authority (HCTRA) was faring much better. Although HCTRA's Hardy Toll Road was also a financial underperformer, the Sam Houston Tollway had been phenomenally successful and had become a cash cow for the agency. HCTRA was formulating plans to expand its tollway network and there would be more bond issues in the future. A default on the Beltway 8 bridge had the potential to tarnish the image of Harris County in the eyes of bondholders, and this would result in higher interest rates and much higher costs in the long run.

Harris County Judge Jon Lindsay took the lead in negotiating the bailout, which turned out to be no easy task. Work on the bailout began in 1992. A ruling from the Texas attorney general's office in 1992 stated that existing state legislation did not sufficiently define if the

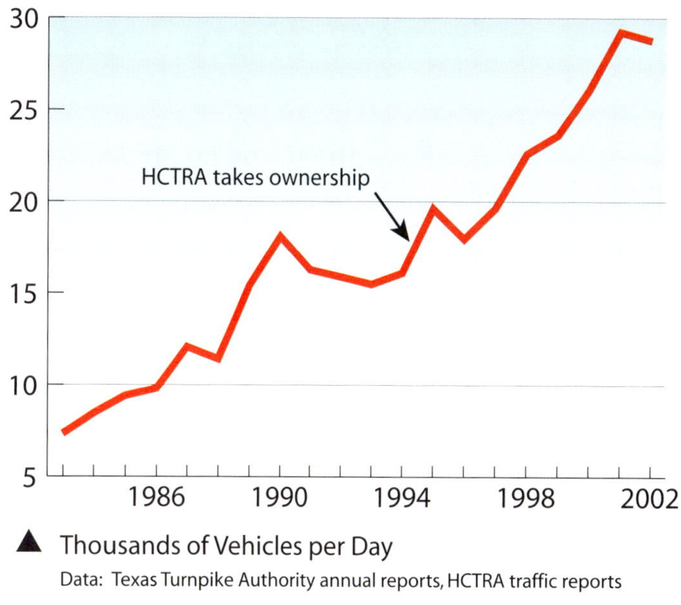

Traffic builds: Since the Harris County Toll Road Authority took ownership of the bridge in 1994, traffic volume has increased from 16,100 vehicles per day to 28,800 vehicles per day in 2002, a slight drop from the 2001 peak of 29,300 vehicles per day.

Key dates in the history of the Beltway 8 bridge

1952	Beltway 8 is first proposed.
1960s	Harris County unsuccessfully tries to construct the bridge.
1977	Studies declare the project feasible as a toll bridge.
1978	$102 million in bonds are sold. Construction begins.
1982	The bridge opens May 6.
1985	Bridge is refinanced with $27.9 million in junk bonds.
1994	On May 5 the Harris County Toll Road Authority takes ownership of the bridge and assumes all debt.
2003	The bridge is near or has reached financial self-sufficiency, based on originally scheduled bond payments.

bridge could be transferred to Harris County. In May 1993 the Texas Legislature approved legislation permitting the transfer. Finally in 1994, everything was in place to make the bailout, and the official transfer occurred on May 5, 1994. Default was avoided, and the bridge became part of the Harris County toll road system. Surely with a big sigh of relief, the Texas Turnpike Authority stated in its 1994 annual report, "The Authority no longer has any financial obligation related to the bridge."[39]

But there was one expensive complication. Under the terms of the 1985 junk bonds, the earliest repayment date was July 1, 2002. The bondholders could have agreed to an earlier repayment, but once they realized that HCTRA had sufficient resources to pay interest through 2002, the bondholders would not compromise on the original terms. This meant an additional eight years of interest at an annual rate of 12.625% would have to be paid, amounting to $12.1 million per year. But HCTRA had no choice—it had to pay.[40]

Why was HCTRA paying $225 million for a bridge that was generating about $10 million per year in revenue? Avoiding a default was one reason. Harris County also managed to negotiate some very attractive incentives for HCTRA. As part of the deal, TxDOT agreed to contribute $90 million towards the construction of the south Sam Houston Tollway between the bridge and the Southwest Freeway, and to spend $120 million on interchanges at the Gulf and Southwest Freeways. The deal sweeteners made it possible to complete nearly all of Beltway 8 as a tollway. In May 1997 the south Sam Houston Tollway was completed. By 2002 bridge traffic volume had increased to 10.5 million vehicles per year, or 28,800 vehicles per day. At an average toll of approximately $2 per vehicle, this translated to a revenue stream of about $20 million per year, bringing the bridge revenue into the range needed to cover bond payments.[41]

The Beltway 8 ship channel bridge was perhaps a bridge that should not have been built. Surely, the timing was wrong when it was completed in 1982. As with any project where there is no obvious answer on whether or not to build, there is risk. But the Houston philosophy of building to meet future growth has always served it well, in spite of this one financial debacle. Because of the decision to build in 1978, Houston has a distinctive structure in its collection of bridges. Compared to the nearby SH 146 cable-stayed bridge, the Beltway 8 bridge is understated and modest. It is a product of its time in history, a window where the concrete box girder structure was the superior design.

The Unbuilt: Galveston-Bolivar Crossing

In Houston, the list of projects that fall into the "unbuilt" category is small and continues to become smaller as time passes. There remains only one large bridge project on that list: the Galveston-Bolivar crossing. But sometime before 2010, there is a good chance that this final major bridge crossing in the Houston area will be scratched off the unbuilt list, and an impressive new high-level span will rise on the east end of Galveston Island. Previous efforts to construct the crossing started in the 1940s but have been unsuccessful, primarily due to the low traffic volume and high cost. But this time things may be different because of advancing bridge construction technology. Cable-stayed designs have become very economical for crossings of this length and height. The advancing technology has unleashed a wave of major bridge construction in Japan, Asia, and Europe. The United States has mostly been on the sidelines for the 1980s and 1990s worldwide bridge construction boom, mainly because most needed crossings in the United States have long been constructed. The forces that have made so many bridges around the world feasible may finally make even the most marginal of projects, such as the Galveston-Bolivar crossing, a reality.

If built, the Galveston-Bolivar crossing will not be a record-setting span. Its main cable-stayed span, recommended to be a minimum of 1,100 feet (335 m), will be among the longest in the United States. The span length will be similar to the Fred Hartman Bridge on SH 146, but will be in the middle of the pack by worldwide standards. The vertical clearance, however, recommended to be 220 feet (67 m), will be among the tallest in the world.

The Galveston-Bolivar crossing has always presented some serious challenges to a fixed crossing. The Houston, Texas City, and Galveston ship channels all pass through the gap between Galveston Island and Bolivar Peninsula, providing heavy ocean vessel traffic. Most significant is the need to bring jack-up offshore drilling rigs into the Galveston Ship Channel. These structures are typically 250–350 feet high (76–107 m) and can exceed 400 feet (122 m) in height on the larger units. Strong currents exist in this area, since the entire Trinity and Galveston Bay waters are connected to the Gulf of Mexico through this location.[42]

Low traffic volume and high bridge cost have always diminished the cost effectiveness of a fixed crossing. In the late 1990s an average of approximately 5,800 vehicles per day used the ferry service. The ferry traffic is highly seasonal, with summer weekend patronage between 9,000 and 10,000 vehicles per day and winter weekday patronage in the 4,000-5,000 range. Ferry service is normally adequate, but during busy summer weekends waiting time can exceed two hours.

History of the Bolivar Crossing

Private operators began $1 toll ferry service in 1929 using the vessels *Galveston* and *Jefferson*. The ferries were purchased by Galveston County at the end of 1929, and were subsequently sold to TxDOT in April 1930. TxDOT took over ferry operation in July 1934 and kept the original vessels in service until 1950 when they were replaced by new ferries.

In 1944 Galveston County officials recognized the

Proposed Galveston-Bolivar crossing	
Design type	Cable-stayed
Main span	At least 1,100 feet (335 m)
Total length	Approx. 2.5 miles (4 km)
Vertical clearance	220 feet (67 m)
Traffic lanes	4
Deck width	85 feet (26 m)
Estimated construction cost (2000)	$211 million

Data: *SH 87 Feasibility Study*, TxDOT, November 2000

Bridge location: This view looks southwest across the location of the proposed Galveston-Bolivar bridge with the city of Galveston in the background. In the foreground is the Bolivar Peninsula ferry landing. The feasible alignment for the bridge leaves Galveston near the strip of land jutting out from the opposite shore, at the upper left corner of the photo. The bridge would land on Bolivar outside the view of this photo. (Photo: James Lyle, TTI, June 2001)

need for improved service between Galveston and Bolivar Peninsula and began efforts to construct a fixed crossing. In July 1944 Galveston County filed an application with the Texas Transportation Commission for aid in the construction of a tunnel. The commission authorized extensive studies, and on May 15, 1946, it submitted three different proposals to Galveston County officials for improvement of the Bolivar crossing. The commission and Galveston County could not agree on terms for the tunnel construction, so on January 22, 1947, the commission submitted a final offer of three options. In option one, TxDOT would provide $5 million in funds for a toll-free tunnel with the remaining funding to be provided by Galveston County. The commission had recently agreed to contribute an estimated $5 million to the construction of the nearby Baytown tunnel, with the Port of Houston contributing the remaining $2.9 million to complete the estimated $7.9 million project, which had a final cost of $8.7 million. For option two, TxDOT would modernize the ferry service at its own expense. In the third option, TxDOT would participate in a toll tunnel, contributing annual funding in the amount of the cost of ferry service. Galveston County did not take action on the proposal, and the offer for state participation in the tunnel was officially rescinded in September 1948. The Texas Transportation Commission then authorized the construction of two new ferry boats, the *Ross Sterling* and *Cone Johnson*, to replace the old *Galveston* and *Jefferson* ferries. Tolls for the ferry service were removed when the new ferries began operating in 1950. A third vessel, the *E. H. Thornton Jr.*, was commissioned into service in early 1959.[43]

In the late 1960s interest in the Bolivar crossing was revived. In response to a request from Galveston County, the Texas Transportation Commission authorized a study in 1968. In November 1968 the study found that a four-lane bridge with a 1,140-foot (347 m) main span and a vertical clearance of 138 feet (42 m) would cost an estimated $30 million—approximately 126 million in 2003 dollars. In July 1969 the commission indicated it could justify an expenditure of $25 million, with Galveston County paying the remaining $5 million. The 1969 *Galveston County Transportation Plan* included the crossing in its long-range recommendation. Concerns over possible bridge interference with maritime industries forced a reevaluation of the proposed bridge clearance in late 1969. In January 1972 the commission authorized a

Some things change, some things remain the same: This photo from circa 1950 shows the ferry landing at Bolivar Peninsula. Ferry facilities have been expanded and improved continuously since TxDOT began the modern ferry operation in 1950. But one thing hasn't changed: teenagers and beach party-seekers go to Bolivar Peninsula looking for action, as these teenagers in the convertible were in 1950. In the 1990s a two-mile stretch of beach at the community of Crystal Beach became known as the "zoo" for its rowdy crowds and party atmosphere.[44] (Photo: TxDOT)

new study of the crossing. The study was completed in September 1973. A fixed crossing was ruled infeasible, mainly because of rapidly escalating costs, and the commission reported that "construction of a bridge or tunnel is not economically feasible at present or in the foreseeable future." The commission authorized improvements to the ferry service, including the purchase of an additional ferry boat. As the highway funding financial crisis escalated during the 1970s, any thought of building a fixed span had just about vanished.[45]

The ferry service operated by TxDOT was enhanced and by the mid-1990s there were six ferry boats, with up to five in service at any given time, and two landings at each end of the ferry route. Still, the ferry became known for lengthy delays during peak travel times, especially during summer weekends when waits often exceeded two hours. Heavy traffic was generated by the teenage and young adult crowds who favored the beaches on Bolivar Peninsula because of the lack of restrictions on alcohol. The ferry service was always susceptible to weather-related disruptions. A report prepared for TxDOT in 1996 by the Texas Transportation Institute identified future capital improvements that would be required for the ferry system, forecasted future traffic growth, and estimated operating costs. The 1996 study also recommended a feasibility study to determine if the ferry system could be replaced by a bridge or tunnel. The SH 87 feasibility study began in 1999 and was completed in November 2000.

The Bridge Gets New Life

The 2000 study looked at all possibilities, including improving the ferry service, building a bridge or tunnel on four possible crossing locations, or doing nothing. The four potential crossing locations included two from the eastern end of Galveston Island, one from Pelican Island (just north of Galveston Island), and one from the mainland at Texas City.

The conclusion: a bridge option is feasible and has a total cost comparable to the cost of needed ferry service improvements. The feasible bridge option was the route from Pelican Island, an alignment which does not obstruct the entrance to the Galveston Ship Channel. The bridge

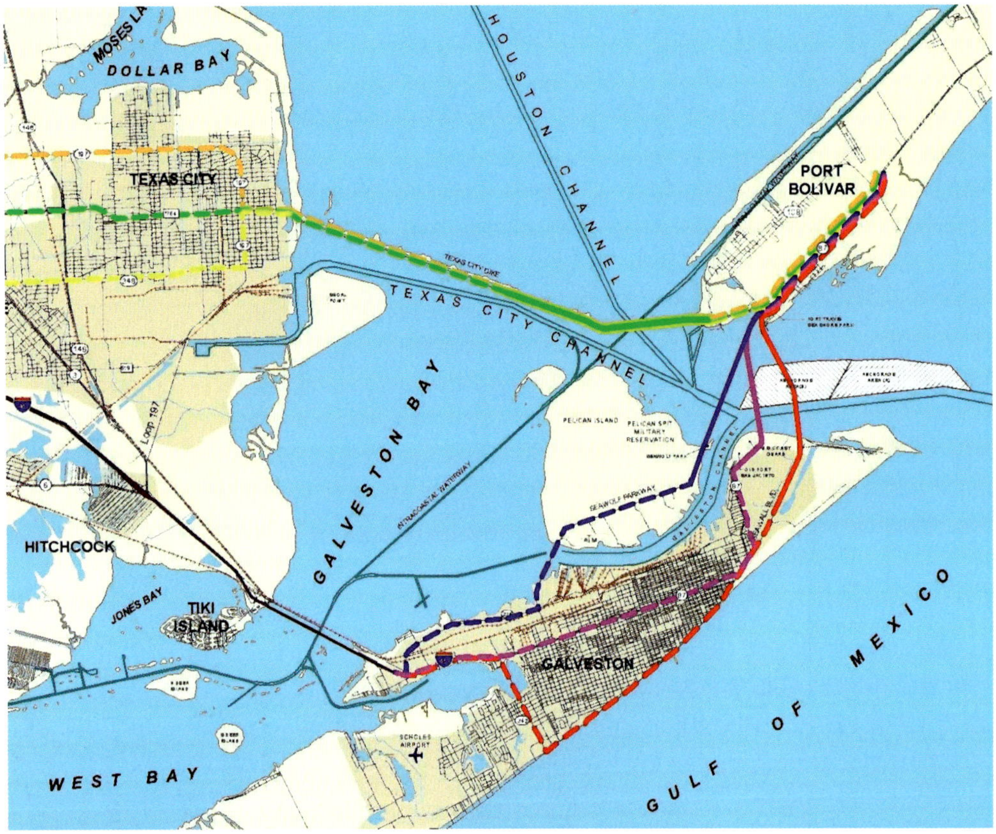

Alternatives studied:
This map shows the crossing routes that were studied in the 1999-2000 study. The red and purple options from the east end of Galveston Island were infeasible because they would have blocked offshore jack-up rigs from entering the Port of Galveston. The routes from the mainland via the Texas City Dike were infeasible due the cost of widening the dike and navigation issues at the crossing location. A bridge crossing on the blue alignment via Pelican Island was determined to be feasible. (Source: *SH 87 Feasibility Study*, TxDOT)

Key dates in the history of the Galveston-Bolivar crossing

1929	Private ferry service is started. Ferries are purchased by Galveston County at the end of the year and then sold to TxDOT in 1930.
1944	Galveston County officials begin efforts to get a tunnel built.
1948	Consideration of a tunnel ends. TxDOT agrees to modernize the ferry service.
1950	New ferries begin toll-free ferry service.
1968	Galveston County officials begin efforts to construct a bridge.
1973	A bridge is declared infeasible. Ferry service is improved.
1990s	Long waiting times for ferry service at peak periods.
2000	A study determines that a bridge crossing is feasible.
2005-2010	Possible construction of new bridge.

cost was estimated to be $211 million, and its total cost, including improvements to access roadways from IH 45, was $243.5 million. The ferry service option needed to meet future needs was estimated to cost $207.6 million. The cost of the tunnel options was in the range of $1.1 to $1.3 billion. The crossings from the eastern end of Galveston Island would have blocked jack-up oil platforms from entering the Galveston Ship Channel, rendering those crossing locations infeasible. The Texas City route would have required a long earthfill project along the Texas City Dike to widen it for a new access road, and the bridge would have been very close to the intersection of the Texas City and Houston Ship Channels, where currents provide hazardous navigation conditions.

The report recommended a cable-stayed main span with a minimum length of 1,100 feet (335 m). For comparison, the nearby SH 146 Fred Hartman Bridge has a main span of 1,250 feet (381 m). The longest cable-stayed bridge in the United States in 2003 has a main span of 1,300 feet (396 m). The longest cable-stayed span in the world in 2003 has a main span of 2,920 feet (890 m), and the world's longest bridge, a suspension bridge, has a main span of 6,532 feet (1991 m). But the proposed bridge will rank among the world's best in terms of vertical clearance. The bridge study carefully investigated the height above water ("air draft") of ships in the world's maritime fleet and attempted to anticipate the future required clearance to ensure the bridge will not restrict waterborne commerce. Perhaps the study authors recalled the controversy surrounding the low, 135-foot (41 m) clearance of the Loop 610 bridge and the limitations it imposed. Within months of completion of the study, there were two major collisions between the Loop 610 bridge and oversized vessels.

The report found that the ocean vessels with the largest air drafts were cruise ships. The tallest vessel identified in the Gulf of Mexico-Caribbean region had an air draft of 228 feet (69 m). An air draft of 208 feet (63 m) seemed to be more typical of the biggest existing and planned cruise ships. The tallest cargo-container vessel identified had an air draft of 180 feet (55 m), and a height of 165 feet (50 m) was determined to be typical of the tallest anticipated future vessels. The recommended vertical clearance for the proposed Bolivar bridge is 220 feet (67 m), which places it in the range of some of the world's tallest and best-known crossings, including the Golden Gate Bridge (232 feet, 71 m), the Verrazano Narrows Bridge in New York City (229 feet, 70 m), the Great Belt Bridge in Denmark (213 feet, 65 m), and the Akashi-Kaikyo Bridge in Japan (220 feet, 67 m). All clearances are for high tide. Even so, in early 2003 the Port of Houston was against a bridge with a 220-foot clearance, citing it as a potential navigation hazard. The proposed 85-foot-wide (26 m) bridge deck will support four traffic lanes and full shoulders. The total length of the proposed bridge is approximately 2.5 miles (4 km).[46]

The report studied the feasibility of using tolls to defray part of the cost of the bridge and concluded that toll financing was a viable option. In June 2001, the Harris County Toll Road Authority (HCTRA) released a map showing potential future toll projects in the Houston area. Included in the map is the Bolivar crossing. In December 2001, the Texas Transportation Commission authorized TxDOT to enter into negotiations with HCTRA for the bridge construction. Legislation passed in the 2003 Texas legislative session resolved legal issues relating to HCTRA's building a project in Galveston County, removing a potential roadblock to the project.[47]

Will the Galveston-Bolivar bridge be built? Only time will tell. The project seems to be closer to moving forward than at any other time in its history. The possibility of industrial development on Pelican Island, the Galveston-side bridge location, will likely influence whether the bridge is built. The potential for an all-new, high-level crossing is something that few cities in the United States can look forward to. The Houston area may get one before 2010.

Making a Jeffersonian crossing: These photos show the *Jefferson*, one of the two ferries which provided service between Galveston and Bolivar Peninsula from 1929 to 1950. The *Jefferson* and *Galveston* were retired in 1950 when the *Ross Sterling* and *Cone Johnson* entered service. (Photos: TxDOT)

16 FT 8 IN

The Freeway Journey

It has taken us to our daily destinations. It has given us the freedom to efficiently go where we want, when we want. It has empowered the transformation of the way we live. It has redefined the American city. The freeway journey has been an amazing trip for America, and an especially remarkable experience for Houston.

Few could have imagined the forces that the freeway would unleash on urban America. Suburbanization. Edge cities. Decentralization. In Houston, these forces have been amplified and accentuated due to the prominence of the freeway. Frontage roads have been most influential in defining the extraordinary role of freeways in Houston. Freeways have also benefited from their high standing in Houston's vision of the future. Freeways have empowered Houston to achieve its goals of growth, prosperity, opportunity, and a seemingly unlimited future. In the post-World War II era, the Houston spirit was built on a foundation of freeways.

The freeway is poised to take us into the future. At this point in the freeway journey, we can pause to reflect. What have been the landmarks on our freeway journey? What have we learned from our 55 years on the freeways? Can a freeway future in 2003 still hold the same promise that it did in 1948?

(Opposite page) **A lifetime of journeys, and the journey of our lifetimes:** Vehicles take a (slow) freeway journey at the Beltway 8-Katy Freeway interchange. While we take journeys on the freeways every day, in a larger sense the freeway has defined the way we live and empowered a lifestyle that has been possible for only the last 40 years. (Photo: James Lyle, TTI, June 2001)

Landmarks on the Journey

As we have seen, freeways have been remarkably influential in nearly every aspect of the development of modern Houston. The freeway has played a large role in nearly all American cities, but in Houston the scope and pervasive presence of freeway-inspired development has elevated the freeway to a seemingly unequalled level of importance. On our freeway journey we have seen the products of the influence of the freeway—the freeway landmarks. Houston's freeway landmarks are among the biggest, most impressive, and most influential in the United States. The landmarks define Houston, but they also exemplify the powerful role of the freeway in shaping the modern American city.

We have seen how freeways have shaped suburbia. Sharpstown was one of the largest suburban residential developments in the United States when it was launched in 1955, and one of the first to be designed around a freeway. Starting with Clear Lake City near the Gulf Freeway in 1962, a wave of large, master-planned suburban communities took root near freeways. The Woodlands, launched in 1974 along the North Freeway, became a model for suburban design.

We have seen how freeways have propelled Houston to the forefront of commercial development. Superb freeway access to downtown spurred a 10-year construction binge that transformed the skyline into one of the more impressive in the United States. One of the nation's most impressive instances of the "edge city" rose along the West Loop, with its Galleria shopping center that served as a model for high-end, mixed-use developments in other cities. Other freeway-based centers of commerce, such as the Katy Freeway Energy Corridor, took root along the loops and radial freeways. The influence of the freeway even led the displacement of an entire neighborhood for the Greenway Plaza development on the Southwest Freeway.

We have seen how freeways have been woven into the fabric of the city, often seamlessly, but sometimes inspiring strong emotion and disagreement. The abrupt and dramatic end of the La Porte Freeway serves as a reminder of the opposition that successfully fought the freeway. The expansive South Freeway had the final word in a neighborhood where the transformation of the inner city first began.

We have seen how the freeways themselves are landmarks. One particular characteristic of Houston's freeways, the freeway frontage road, has been a huge factor in defining Houston's distinct freeway-based urban development pattern. The frontage road urban development pattern exists to a large scale only in Texas, and it reaches its greatest extent in Houston. The frontage road has influenced many distinguishing characteristics of Houston's freeways, including wide corridors, five-level freeway-to-freeway interchanges, and continuous high mast illumination. Houston's freeway-based transitway system has been a pioneer in the cost-effective delivery of mass transit services and has served as a model for other cities.

Lessons Learned on the Journey

There is a much-used political expression that "all politics are local." That expression can be extended to freeways, because if there's one lesson to be learned about freeways, it's that all freeways are local. Local efforts were crucial to the formulation of the original freeway plan, but as time passed, the role of narrowly focused subregions has become more important. In an era of diminished resources, freeway construction has increasingly shifted to constituencies that have political influence, make financial contributions to the project cost, and show near-unanimous support for the project. In Houston there have been many interests who really want freeways, helping propel Houston's freeway system to what it is today.

We have seen that construction of freeways has never been easy—it has been a challenge every step of the way. In the 1950s the need to raise local money to pay for the local share of freeway costs required ongoing and extensive efforts by civic organizations. The 1970s brought the deepest crisis freeway construction has ever faced. The 1980s resurrection required strong political leadership and a broad effort of the business community. Even the seemingly halcyon 1960s was an era of constrained resources. In the modern era, freeways face a daunting maze of regulatory complexity. Houston has demonstrated a very strong perseverance in overcoming challenges and has almost always succeeded.

We have seen that strong political leadership is essential for the construction of transportation infrastructure, freeways in particular. The progress of Houston's freeway system has largely been a reflection of the level of local political leadership. Houston's freeways have been a top priority for some of the most influential political leaders in the Houston area since the end of World War II. We have seen the tides of freeway leadership rise and fall in Houston, but there has nearly always been a rising tide to fill the void left by a falling tide. The absence of a rising tide of freeway leadership in the late 1960s and early 1970s resulted in decline in freeway investment and contributed to the transportation crisis of the late 1970s and early 1980s. Complacency, neglect, or indifference fosters future crises.

We have seen how Houston's freeways have been very resistant to change in certain respects, but very adaptive and progressive in other ways. The basic loop and radial system design formulated in the early 1950s has been perpetuated even though a more gridlike system design would be more optimal. The regional arterial system that feeds the freeways has been neglected. Frontage roads have continuously gained importance in the functioning of Houston's transportation system, providing benefits but also introducing new concerns. Houston's freeways have been adapted to successfully provide mass transit. Freeway engineering has evolved to provide some of the most modern, up-to-date designs in the nation. Safety has always been a top priority of Houston's freeway builders. The net result of clinging to the old in some ways and ag-

Houston vs. Los Angeles

Houston	Los Angeles
Freeways defined the form and development patterns of the city	Sprawl and decentralization characterized the region prior to freeways, primarily due to the Pacific Electric Railway
Widespread frontage roads, the most extensive use in the world	No frontage roads
A transportation system almost entirely focused on freeways	Freeway-focused, but to a lesser extent due to better arterial roads and no freeway frontage roads
Among the national leaders in per-capita lane-miles (major cities)	About 58% as many lane-miles as Houston on a per-capita basis
An aggressive plan for freeway growth and modernization	Modest plans for improvements
Continuously striving for improvement, preparing for the future	Resting on its laurels

gressively advancing in other ways has been a concentration of transportation resources into the freeway, providing larger, more impressive freeways.

We have seen that Houston's demand for freeway capacity is nearly insatiable. Houston's strong growth has surely contributed to the steadily increasing demand for freeway and road capacity, but modern lifestyles and commuting patterns increasingly rely on the freeway. It is difficult, perhaps impossible, for any other form of transportation to compete with the convenience and flexibility of the private automobile, especially in low-density, sprawling cities like Houston. However, we have seen instances where freeways were built and very few cars followed: the Crosby Freeway and the frontage-road-only Red Bluff Road. The South Freeway has also been slow to attract traffic.

The Journey into the Future

In 1948, the freeway was poised to dramatically transform America and shape a new generation of cities, including Houston. In 2003, the freeway doesn't offer any profound changes for the future, but it will continue to provide benefits that are among the most valuable to all of us: personal mobility, housing opportunity, and economic development. There are always efforts to achieve these goals without freeways, particularly among groups opposed to freeways. But in the urban environment of Houston, nothing can deliver the benefits like the freeway can. Competing options, such as transit and high-density housing, will make gains, but the freeway will continue to reign supreme. Perhaps only a major disruptive event—such as a very substantial increase in energy costs—can reduce the demand for freeways. Chances are greater that the advance of technology will produce cars with fewer environmental impacts and less energy consumption. And the cars of the future will still need freeways.

In the larger sense, what is the legacy of the freeway city? Is being mentioned in the same breath with Los Angeles a badge of honor, or a stain of embarrassment? That question has been, and will continue to be, one that arouses strong opinions. But perhaps an oft-cited expression can illuminate the underlying issue: "What's worse than a city the size of Los Angeles with a bunch of freeways? A city the size of Los Angeles without a bunch of freeways."

On September 30, 1948, a crowd of Houstonians gathered on a Gulf Freeway overpass to witness the dedication of Houston's first freeway. Houston's freeway journey began. The journey would build what is perhaps the world's most freeway-focused city. A unique blend of ingredients came together—a mix of factors that could exist in only a few places and combined in the most powerful way in Houston. The Texas commitment to frontage roads. A pro-business, pro-growth philosophy. Great leadership. Driven and visionary individuals. A geographic location and urban environment suited for economical freeway construction. A guiding principle of continuous improvement, that bigger is better. A commitment to freeways that hasn't wavered over time. It is a unique blend of ingredients that produced a great freeway system and a fascinating story. It has been, and continues to be, a remarkable journey.

Southwest Freeway: This view looks west along the Southwest Freeway (US 59) toward the Loop 610 interchange. In 2003 work is in progress to improve and modernize the interchange, originally constructed in 1962. (Photo: May 2003)

Houston Freeway Lane Counts
Interpretation of lane counts:
Main lanes shown first, with transit or toll lanes shown after the dash.
Main lanes are continuous lanes and exclude temporary merging and auxiliary lanes.
H = barrier-separated HOV transitway
D = diamond lane (non-barrier-separated)
B = two-way dedicated HOV transitway
T = toll lanes
(P) = pending; pavement in place for transit lane, but not yet operational
(A) = pavement available, but not in use
6 = 6 main lanes
10-1H = 10 main lanes with 1 barrier-separated HOV transitway
2008: 10-4T = 10 main lanes with 4 toll lanes scheduled for opening in 2008
Indicates lane count transition point
8-1H(P)
10-1H(P)
8-1H(P)
10-1H
8-1H
Hardy Toll Road
Westpark
Future: 8 with potential 4T
Future: 10 with potential 4T
6 (8A)
249
290
59
45
610
10
8
90
225
35
288
6-2D
2006: 8-2D
6-1H
2008: 9-4T
6-1H
2008: 10-4T
12-2B
10-2B
14-2B
12-1H
10-1H
14-1H
2006: 10-2H
6
2005: 8
5
2005: 6
8-2D
4
Future 8-2D
6
Future: 10-2D

Downtown's newest addition: The new Eastex Freeway downtown distribution complex is nearing completion in 2003. (Photo: ©Scott Teven/www.HoustonPhoto.com, May 2003)

Notes

Building the System

1. "Houston's city planners have had a rough time," *Houston Chronicle*, 8 June 1958.

2. Lawrence C. Sandberg Jr., *The Houston North Shore Railway*, Fourth Edition (October 1990); Steven M. Baron, *Houston Electric: the Street Railways of Houston, Texas* (1996).

3. "Mayor talks acquisition of old interurban roadbed with Gray," *Houston Post*, 12 April 1940; "Council authorizes abandonment of four remaining street car lines," *Houston Post*, 13 April 1940; "Holcombe urges Houston-Galveston super highway," *Houston Post-Dispatch*, 15 May 1940.

4. "Street cars to click tune of farewell to Houston early Sunday," *Houston Chronicle*, 7 June 1940; "City preparing to ask bids on removal of rails," *Houston Chronicle*, 15 August 1940.

5. *McGraw Electric Railway Directory*, 1924 (New York: McGraw, 1924); David Brodsly, *L.A. Freeway: An Appreciative Essay* (University of California Press, 1981).

6. Houston City Planning Commission, *The Major Street Plan for Houston and Vicinity, 1942*, Report of the City Planning Commission (Houston, Texas).

7. Brodsly, *L.A. Freeway: An Appreciative Essay*, (University of California Press, 1981), 100; William Kaszynski, *The American Highway* (McFarland & Company, 2000), 130.

8. Texas Transportation Commission Minute Order 20263, 6 October 1943; Texas Transportation Commission Minute Order 21182, 1 May 1945; Texas Transportation Commission Minute Order 22594, 14 October 1946.

9. Map is taken from *Economic Evaluation of the Gulf Freeway* (City of Houston Department of Traffic and Transportation, July 1949).

10. "Modern Expressways Under Way in Houston," *Houston*, November 1947; Texas Transportation Commission Minute Order 19475, 23 June 1942; "Highway Department plans many projects in this area," *Houston Post*, 28 January 1952.

11. Houston Chamber of Commerce, "A Presentation to the Texas Highway Commission: A Need for Expressway Designations in the Houston Urban Area and Harris County," booklet accompanying the presentation made by local officials to the Texas Transportation Commission, July 1953; Texas Transportation Commission Minute Order 34178, 23 September 1953; Texas Transportation Commission Minute Order 35331, 1 December 1953.

12. Ralph Ellifrit to Jim Douglas, head of TxDOT Houston District Office, letter dated 29 December 1953 (city of Houston records); Houston Chamber of Commerce, "Before the Texas Highway Commission, A Plea for Loop Designations in Harris County and the City of Houston," booklet accompanying the Harris County delegation presentation to the Texas Transportation Commission, 28 September 1954; City of Houston Planning Department, map: "Plan for Major Thoroughfares and Freeways, 1955."

13. "Molder of a city," *Houston Chronicle*, 5 January 1947; "City post resigned by Ellifrit," *Houston Post*, 17 December 1963; Houston Metropolitan Research Center, division of the Houston Public Library, audio tape of interview with Ralph Ellifrit, 26 September 1979.

14. "City can't bear cost of freeway right-of-way," *Houston Chronicle*, 18 June 1956; "Freeway growth threatened by 'unconstitutional' fee ruling," *Houston Chronicle*, 19 June 1956; "Court rules freeway tax not valid," *Houston Chronicle*, 2 May 1956.

15. "3 reasons cited for bond failure," *Houston Chronicle*, 4 August 1956; "Bond issue's failure to pass to delay freeways for some time," *Houston Chronicle*, 30 July 1956; "New freeway bond vote planned September 8," *Houston Chronicle*, 2 August 1956; "City partly blamed in freeway program lag," *Houston Chronicle*, 28 January 1958.

16. Houston Metropolitan Research Center, division of the Houston Public Library, audio tape of interview with Ralph Ellifrit, 26 September 1979.

17. "55th Legislature Boosts Highways," *Texas Parade*, June 1957; "What new road plan means to you," *Houston Press*, 17 July 1957.

18. "Master street plan will cost big money," *Houston Chronicle*, 27 September 1957; "Closer planning on expressways pledged," *Houston Chronicle*, 22 October 1957; "Lack of liaison blamed in freeway breakdown," *Houston Chronicle*, 15 November 1957.

19. "$15,000,000 freeway bond vote may be set," *Houston Chronicle*, 1 September 1958; "Tax increase, freeway bonds are defeated," *Houston Chronicle*, 5 November 1958; "County leaders push for bonds," *Houston Chronicle*, 2 December 1958; "Officials are ready to push program for super highways," *Houston Chronicle*, 7 December 1958; "County will speed freeway route buying," *Houston Chronicle*, 1 February 1957.

20. "Federal funds giving big boost to freeway schedule," *Houston Post*, 11 August 1957; "Freeway system cost disclosed," *Houston Chronicle*, 19 July 1957; State of Texas, *Annual Report of the Comptroller of Public Accounts of the State of Texas, 1958*; *Fiscal Year 2003 Historical Tables, Budget of the United States Government* (Washington, D.C.: U.S. Government Printing Office).

21. "Future freeways to have center guard," *Houston Chronicle*, 17 January 1956.

22. "Kyser Did It His Way," *Texas Highways*, July 1972.; "Designer of city's vast freeway system retiring," *Houston Post*, 20 February 1972; "Designer of Houston freeways retires," *Houston Chronicle*, 4 August 1972.

23. "Kyser Did It His Way," *Texas Highways*, July 1972.

24. Carol Letz, TxDOT Houston right-of-way acquisition department, interview by author 29 April 2002; William Ward, head of TxDOT Houston Urban Project Office 1972-1984, interview by author, various dates in 2002.

25. Texas Highway Department Highway Planning Survey, *Houston Central Business District Parking Survey* (1953).

26. Peter C. Papademetriou, *Transportation and Urban Development in Houston, 1830-1980* (Metropolitan Transit Authority, 1982).

27. "100-mph Zone Ahead," *Houston*, November 1960; "City on Wheels," *Houston*, October 1967.

28. City of Houston, *1959 City Planning Commission Annual Report*; "County makes wise move on freeways," *Houston Chronicle*, 11 July 1960; Houston Metropolitan Transportation and Transit Study, *Freeway Phase* (16 August 1961).

29. "Loop circling county proposed," *Houston Chronicle*, 5 October 1965; City of Houston Planning Commission map, "Houston Preliminary Study Plan for 1990."

30. Houston Planning Commission, *A Study of Thoroughfare Development in the Southeast Area of Metropolitan Houston and Harris County* (1963); Texas Transportation Commission Minute Order 57326, 31 March 1966.

31. Texas Transportation Commission Minute Order 60213, 28 November 1967.

32. Texas Transportation Commission Minute Order 62158, 2 April 1969; Texas Transportation Commission Minute Order 62027, 7 March 1969.

33. "Commuters can zip along rebuilt part of Eastex Freeway," *Houston Chronicle*, 16 August 1999; "Names adopted for freeways," *Houston Chronicle*, 23 November 1956.

34. "One freeway renamed, four are studied," *Houston Chronicle*, 29 April 1965.

35. "Beltway to be named Sam," *Northeast News*, 21 October 1986.

36. "John Jones, Jr., heads C. C. highway group," *Houston Chronicle*, 11 February 1953.

37. "Freeways Don't Cost—They Pay," *Houston*, June 1954.

38. "The Gulf Freeway Observes its 500 Millionth Mile of Use," press release prepared by W. J. Van London, head of the Houston Urban Expressway System, 18 November 1954.

39. "Highway Week," *Texas Parade*, January 1952.

40. "Freeway bond drive is launched," *Houston Chronicle*, 27 August 1956; "Freeway bond rally Thursday," *Houston Chronicle*, 28 August 1956; "County to launch work immediately," *Houston Chronicle*, 9 September 1956.

41. "Tax increase, freeway bonds are defeated," *Houston Chronicle*, 5 November 1958; "Civic leaders pledge battle for road bonds," *Houston Chronicle*, 17 November 1958; "Grass-roots bond appeals planned," *Houston Chronicle*, 23 November 1958; "County leaders push for bonds," *Houston Chronicle*, 2 December 1958; "Officials are ready to push program for super highways," *Houston Chronicle*, 7 December 1958.

42. "Road Hearing Rules Draws Fire, Praise," *Engineering News-Record*, 21 November 1968; "Rules altered on new highways," *Houston Post*, 25 October 1968; "AASHO Lashes New Road Rules," *Engineering News-Record*, 12 December 1968.

43. "Kultgen Calls Regulations Serious Threat," *Texas Highways*, January 1969.

44. *America's Highways 1776-1976: A History of the Federal-Aid Program* (U.S. Department of Transportation, 1976), 221-223.

45. "Public Affairs," *Houston*, May 1969; "Public Affairs," *Houston*, December 1969; "Public Affairs," *Houston*, October 1970.

46. "Public Affairs," *Houston*, November 1970.

47. *An Informal History of the Texas Department of Transportation* (Austin: Texas Department of Transportation, 1999), 44; "Highway program in trouble as delays drag," *Houston Post*, 13 February 1972; "Red Tape Tie-Up," *Texas Highways*, September 1971.

48. "The World Moses Built," videocassette of the PBS documentary for the American Experience series (Public Broadcasting System, 1989); David Brodsly, *L.A. Freeway: an Appreciative Essay* (University of California Press, 1981), 47.

49. "The Future of Freeways," *Los Angeles Times*, 8 December 1964; "There's a limit to freeways," *Houston Chronicle*, 12 January 1965; quotation regarding Louie Welch is from an author's interview with a TxDOT official who was speaking off the record.

50. Tom Lewis, *Divided Highways* (New York: Penguin Books, 1997), 220.

51. *Houston Chronicle*, March 8, 1973; *Houston Post*, 19 February 1973.

52. Congressional Research Service Report for Congress, "RL30304: The Federal Excise Tax on Gasoline and the Highway Trust Fund: A Short History," Louis Alan Talley, 29 March 2000.

53. *America's Highways 1776-1976: A History of the Federal-Aid Program* (U.S. Department of Transportation, 1976), 228, 254.

54. HJR 49, 49th session of the Texas Legislature, approved by public vote 5 November 1946; *An Informal History of the Texas Department of Transportation* (Austin: Texas Department of Transportation, 1999), 6, 23; State of Texas Comptroller of Public Accounts, *Sources of Revenue Growth, A History of State Taxes and Fees in Texas*, Publication #96-571 (May 2000).

55. *Engineering News-Record*, 3 July 1976.

56. *Fiscal Year 2003 Historical Tables, Budget of the United States Government* (Washington, D.C.: U.S. Government Printing Office, Washington, D.C., 2002).

57. Texas Comptroller of Public Accounts, *State of Texas Revenue and Expenditure History 1960–2000*, (Austin: State of Texas, November 2000); U.S. Census Bureau Population Estimates; "Costs-tax bind clips roads plans," *Houston Post*, 4 March 1976.

58. *An Informal History of the Texas Department of Transportation* (Austin: Texas Department of Transportation, 1999), 40, 46; "Texas will be out of road funds, committee told," *Houston Post*, 8 November 1975.

59. "State halts most land purchases for road system," *Los Angeles Times*, 24 July 1975; *Los Angeles Times*, 14 September 1975.

60. "Freeway system in L.A. is under fire," *Houston Chronicle*, 13 April 1973; "State freeway building takes new budget jolt," *Los Angeles Times*, 22 November 1975; "State acts to kill 2 freeway plans in southland," *Los Angeles Times*, 23 May 1975; "State drops 3 major road projects," *Los Angeles Times*, 18 July 1975; "Orange County Freeway plan is dropped; fund lack is cited," *Los Angeles Times*, 22 March 1975.

61. "Talk of gasoline tax hike, rationing criticized," *Houston Chronicle*, 25 December 1974; "Ford and gasoline tax hike," *Houston Chronicle*, 24 November 1974.

62. "House rejects new gasoline taxes," *Houston Chronicle*, 12 June 1975.
63. "Gasoline tax hike soundly defeated," *Houston Chronicle*, 4 August 1977; "Senate vote crushes Carter's gasoline fee," *Houston Chronicle*, 7 June 1980.
64. Federal highway bid price indexes, reported by *Engineering News-Record* magazine. Cost indexes appeared in the issues at the end of the second and fourth quarters for the period under consideration.
65. "Angry Senate approves repeal of gas tax hike," *Houston Chronicle*, 15 May 1971.
66. "Welch asking U.S. for help to improve mass transport," *Houston Post*, 11 March 1970.
67. "Welch says city must subsidize or operate mass transit system," *Houston Chronicle*, 28 September 1972.
68. "Fewer Riders, Rising Deficits: Woes of the Transit Industry," *U.S. News and World Report*, 19 January 1970.
69. "The Troubles of Mass Transit," *U.S. News and World Report*, 26 April 1971.
70. "Untangling Big-city Traffic: the Big Push for Mass Transit," *U.S. News and World Report*, 25 May 1970.
71. "Washington Report," *Houston*, September 1963; *Houston*, February 1967; *Houston*, February 1968; "Gas tax for mass transit favored," *Houston Chronicle*, 29 February 1972.
72. *Fiscal Year 2003 Historical Tables, Budget of the United States Government* (Washington, D.C.: U.S. Government Printing Office, 2002); Tom Lewis, *Divided Highways* (New York: Penguin Books, 1997), 221-233.
73. "Plan asks drastic highway cuts here," *Houston Chronicle*, 13 August 1976; *Responding to the Changing Environment—Summary Report* (Austin: Texas State Department of Highways and Public Transportation, July 1976).
74. *Houston Chronicle*, 27 May 1979 (photograph caption).
75. "20 year highway plan," *Houston Chronicle*, 25 September 1977.
76. "MTA, new highway projects don't hold cure, study says," *Houston Post*, 13 May 1979.
77. *U.S. News and World Report*, 7 September 1987.
78. "Houston Gets High on Helicopters," *Newsweek*, 2 August 1982; Hoyle, Tanner & Associates, *Metropolitan Heliport System Plan* (report prepared for the city of Houston, 1985); "Helicopters hunting more area landing places," *Houston Post*, 12 January 1981; "FAA seeks changes in copter ordinance," *Houston Post*, 17 June 1982.
79. "Houston Gets High on Helicopters," *Newsweek*, 2 August 1982.
80. "Helicopter traffic problems outlined," *Houston Post*, 13 November 1981; "Council adopts regulations governing use of helicopters here," *Houston Post*, 30 December 1981; "New set of helicopter regulations imposed," *Houston Post*, 2 December 1983.
81. "Texas will be out of road funds, committee told," *Houston Post*, 8 November 1975; "Road funds spark disagreement," *Houston Post*, 26 February 1976; House Bill 3, 65th Regular Session of Texas Legislature, 1977; "Briscoe signs measure funding state highways," *Houston Post*, 13 April 1977; Texas Comptroller of Public Accounts, *State of Texas Revenue and Expenditure History 1960-2000* (Austin: State of Texas, November 2000).
82. Louis Alan Talley, *RL30304: The Federal Excise Tax on Gasoline and the Highway Trust Fund: A Short History* (Washington, D.C.: Congressional Research Service, 2000); "Even if gasoline tax enacted, funds would fall short for needed highway projects," *Houston Chronicle*, 5 December 1982.
83. "Harris judge favors county to build Beltway toll road," *Houston Post*, 21 April 1977; "West Belt revived," *Houston Chronicle*, 19 May 1978.
84. "Grass roots campaign pays off," *Houston Chronicle*, 14 September 1983; official vote results for Harris County bond election, 13 September 1983 (Harris County Clerk's Office).
85. Roger Hord, interview by author, Houston, Texas, 30 May 2002.
86. "Reaction by officials generally supportive," *Houston Post*, 23 February 1982.
87. "Additional sources of funding to be needed," *Houston Post*, 23 February 1982.
88. "Lanier track record shows a man driven to succeed," *Houston Chronicle*, 13 October 1991.
89. "Two Houstonians don't take positions on board lightly," *Houston Post*, 9 October 1983; "Road revenue is a concrete issue," *Houston Chronicle*, 19 August 2002.
90. "Governor to sign bill on Houston freeway," *Houston Post*, 11 July 1984.
91. Texas Workforce Commission employment data.
92. Michael A. Rosetti and Barbara S. Eversole, *Journey to Work Trends in the United States and its Major Metropolitan Areas, 1960-1990*, publication FHWA-PL-94-012 (Federal Highway Administration, November 1993).
93. Texas Transportation Commission Minute Order 82325, 25 October 1984; Texas Transportation Commission Minute Order 83806, 26 November 1985; transcript of Texas Transportation Commission meeting, November 1998.
94. The following Texas Transportation Commission Minute Orders relate to the Grand Parkway: 82605, 19 December 1984; 83736, 24 October 1985; 83807, 26 November 1985; 85077, 24 November 1986; 85865, 27 May 1987.
95. "Donors to campaigns become contractors on road projects," *Houston Chronicle*, 14 January 1985; "Road funds may face uphill battle," *Houston Chronicle*, 29 October 2001; "Highway plans bring money to politicians," *Houston Chronicle*, 29 August 2002; "Eckels rips big political donations," *Houston Chronicle*, 9 February 1999.
96. Percentage of work done by Williams Brothers Construction is based on analysis of TxDOT-supplied contract data for the Houston region; "King of the road," *Houston Chronicle*, 26 April 1998.
97. Transcript of Texas Transportation Commission, 26 April 2001; transcript of the Texas Transportation Commission, 26 September 2002.
98. Harris County toll road system traffic count statistics, data supplied by Harris County Toll Road Authority.
99. "Local Roads Plan a Challenge to Others," *Houston*, September 1953.
100. *Dallas-Fort Worth Regional Transportation Study* (North Central Texas Council of Governments, July 1967).

101. *Dallas Fort Worth Regional Transportation Study Interim Report 1971* (North Central Texas Council of Governments, May 1972); Arthur Beck, one of the architects of the 1967 revised freeway plan for the Dallas-Fort Worth region, communications with author including letter from Beck dated 27 April 2002.

102. David Brodsly, *L.A. Freeway: An Appreciative Essay* (University of California Press, 1981).

103. *In-Service Experience on Installations of Texas Modular Crash Cushions*, Summary Report 146-2(S) (College Station, Texas: Texas Transportation Institute, Texas A&M University, July 1969); Ted Hirsch, interview by author, 1 July 2002; "A Smashing Success," *Texas Highways*, February 1969; "Barrel cushions get a workout," *Houston Chronicle*, 13 October 1968.

104. "Vehicle Impact Attenuators for Bifurcations," *Texas Highways*, December 1968; "Barrel 'honeycombs' expected to cut fatal crashes," *Houston Chronicle*, 20 September 1968.

105. "Researchers Take Death out of Driving," *Engineering News-Record*, 9 January 1969; "Break-away highway signs saving lives," *Houston Post*, 24 March 1968; *Impact Behavior of Sign Supports-II*, Summary Report 68-2(S), (College Station, Texas: Texas Transportation Institute, Texas A&M University, September 1965); "Freeways' barrels may have saved hundreds of lives," *Houston Chronicle*, 7 December 1980.

106. "The Third Generation Freeway," *Texas Highways*, April 1967.

107. "Future freeways to have center guard," *Houston Chronicle*, 17 January 1956.

108. "The Third Generation Freeway," *Texas Highways*, April 1967.

109. "It's Not Only How High You Place It, It's How You Space it Out," *Street and Highway Lighting*, First Quarter 1969; "High Rise Illumination," *Texas Highways*, October 1966.

110. "Research Lights the Way," *Texas Highways*, October 1965; "Moonlighting in Fort Worth," *Texas Highways*, December 1966.

111. "Use of high-mast interchange lighting increases," *Rural and Urban Roads*, January 1970; "Floods light San Antonio Interchanges," *Street and Highway Lighting*, first quarter 1969; "Texarkana's Tall Towers," *Texas Highways*, July 1968; "Cloverleafs Get High Lights," *Electrical World*, 5 January 1970; "The Sky's the Limit," *Texas Highways*, January 1970.

112. "Blast from the past: company strips, repaints antique moonlight towers," *Austin-American Statesman*, 28 August 1993.

113. Dexter Jones, "Freeway Illumination," text of presentation given at the Southwest Regional Conference of the Illumination Engineering Society (date unavailable).

114. H. Dexter Jones, *Z-Pattern High Mast Freeway Illumination*, Report No. FHWA/TX-80/1+SS22.1, (Texas State Department of Highways and Public Transportation, January 1980).

115. H. Dexter Jones, *Z-Pattern High Mast Freeway Illumination* (see reference 114).

116. International Dark Sky Association web site, www.darksky.org, accessed June 2003.

117. HB 916, Texas Legislature 76th session, 1999.

118. Mike Strech, executive director of the Harris County Toll Road Authority, interview by author, May 2002.

119. "Bond issues prove hit at polls," *Houston Chronicle*, 7 November 2001.

120. James E. McCarty, *Highway Fund Sanctions and Conformity Under the Clean Air Act*, CRS report for Congress (Washington, D.C.: Congressional Research Service, 15 October 1999).

121. Regional Air Quality Planning Committee of the Houston-Galveston Area Council, *Air Quality Reference Guide for the Houston-Galveston Area* (Houston-Galveston Area Council, July 2002).

122. "Road plan conflicts with new state smog levels," *Houston Chronicle*, 18 February 2001; "Air quality demands worry northwest Houston groups," *Houston Chronicle*, 11 April 2001; "Settling suit on emissions prevents cut of road funds," *Houston Chronicle*, 10 May 2001.

123. "Court gives green light to light rail," *Houston Chronicle*, 8 March 2001; "Rail foes halted in their tracks?" *Houston Chronicle*, 8 November 2001.

124. "Rail system is growing but still a minor player," *Los Angeles Times*, 7 January 2003; *Atlanta Regional Commission 2000 Fact Book*, transit ridership statistics, available on the Atlanta Regional Commission web site, www.atlreg.com/index.html, accessed March 2003.

Freeway Metropolis

1. William A. Camfield, *More Than a Constructive Hobby: The Paintings of Frank Freed* (Museum of Fine Arts, Houston and Texas A&M University Press, 1997).

2. William Kaszynski, *The American Highway* (McFarland and Company, 2002); David Brodsly, *L.A. Freeway: An Appreciative Essay* (University of California Press, 1981).

3. "Council authorizes abandonment of four remaining street car lines," *Houston Post*, 13 April 1940; caption for illustration, *Houston Chronicle*, 1 August 1952.

4. Paul R. Tutt, interview by author, 5 August 2002.

5. Kara M. Kockelman and others, "Frontage Roads: An Assessment of Legal Issues, Design Decisions, Costs, Operations, and Land-Development Differences" (paper presented at the Transportation Research Board's Fifth National Access Management Conference, Austin, Texas, 23-26 June 2002); Charles Pinnell and Paul R. Tutt, *Evaluation of Frontage Roads as an Element of Urban Freeway Design* (Texas Highway Department, Departmental Research Report Number 62-3, January 1963).

6. American Association of State Highway Officials, *A Guide for the Application of Frontage Roads on the National System of Interstate and Defense Highways* (Washington, D.C., 1961); Ross D. Netherton, *Control of Highway Access* (Madison, WI: The University of Wisconsin Press, 1963).

7. "Decision to stop building frontage roads draws fire," *Houston Chronicle*, 7 September 2001.

8. Texas Department of Transportation, *An Informal History of the Texas Department of Transportation*, (Texas Department of Transportation, 1999), 33; Dewitt Greer, "Expressways," *Texas Highways*, March 1954.

9. Texas Department of Transportation, transcript of the June 2001 meeting of the Texas Transportation Commission.
10. "Agency backs off its policy on frontage roads," *Houston Chronicle*, 4 June 2002.
11. "2 UH Students propose freeway beautification," *Houston Chronicle*, 16 June 1968; "Try billboard-free freeway, highway engineer urges city," *Houston Post*, 11 July 1968.
12. "Judge will uphold constitutionality of billboard law," *Houston Chronicle*, 11 June 1999.
13. City ordinance 86-2193, 16 December 1986 (establishment of Beltway 8 scenic district).
14. "Signs of the times: billboards move inside city limits," *Houston Chronicle*, 24 April 1994; "New loop fixture is sign of trouble," *Houston Chronicle*, 26 November 1998.
15. "North Loop nearer completion with opening of interchange," *Houston Chronicle*, 17 April 1961.
16. "What Do We Do Next," *Houston*, July 1957.
17. "This Century's Paramount Problem," *Houston*, November 1956.
18. "20 year highway plan," *Houston Chronicle*, 25 September 1977.
19. Sierra Club web site, www.sierraclub.org/sprawl/report98/
20. "Allison's victims, six months later," *Houston Chronicle*, 23 December 2001.
21. "Receding waters reveal new road hazards," *Houston Chronicle*, 12 June 2001.

Downtown Freeways

1. "Van London's roller coaster; maybe double-decker next," *Houston Post*, 5 February 1950; "Urban Expressways for Houston," *Houston*, April 1951; "Double Decker Freeways Due," *Houston Chronicle*, 21 December 1959.
2. "Vast freeway to circle city gets OK," *Houston Chronicle*, 7 April 1953.
3. Ralph Ellifrit to A. C. Kyser, letter dated 30 May 1957 (city of Houston records).
4. Texas Transportation Commission Minute Order 48224, 24 August 1960.
5. City of Houston, *City Planning Commission 1963 Annual Report.*
6. Audiotape of 1979 interview with Ralph Ellifrit, Houston Public Library Metropolitan Research Center.
7. City of Houston planning memo, 6 November 1956 (city of Houston records); "Talks on removal of Union Station have netted city little so far," *Houston Chronicle*, 25 September 1960; "Union Station bottleneck cleared; to let contract," *Houston Chronicle*, 2 November 1960.
8. "Landowners protest over expressway plan delays," *Houston Post*, 31 January 1957; "City may knock out Pierce freeway line," *Houston Chronicle*, 18 April 1957; "Building lines for freeway are revoked," *Houston Chronicle*, 1 May 1957; "Tracts bought for elevated roadway here," *Houston Chronicle*, 15 June 1958; "A new, beautiful freeway link," *Houston Chronicle*, 22 August 1967.
9. "Freeway route change shelved," *Houston Chronicle*, 3 October 1959; Houston Chamber of Commerce president Leon Jaworski to Mayor Lewis Cutrer, letter dated 16 December 1959 (city of Houston records); "City seeks elevated freeway," *Houston Chronicle*, 22 July 1960.
10. Houston-Harris County Transportation Study, *Houston CBD Report* (30 June 1970).
11. Costar Group, *CoStar Office Report, Mid-year 2001, Houston Office Market*; Class A office space is defined by the CoStar report as "a classification used to describe buildings that generally qualify as extremely desirable investment-grade properties and command the highest rents or sale prices compared to other buildings in the same market."
12 "Will 'orphan span' be used?" *Houston Chronicle*, 1 October 1967.
13. "Modern Apartments Replace Schrimpf Alley Slum Area," *Houston*, April 1952.
14. "A new, beautiful freeway link," *Houston Chronicle*, 22 August 1967.

The Spokes

1. Herb Woods, *Galveston-Houston Electric Railway* (Glendale, CA: Interurban Publications, 1976).
2. "Sterling's dream realized – ex-governor proposed Gulf Road in 1930," *Galveston Daily News*, 2 August 1952.
3. Texas Transportation Commission Minute Order 20263, 6 October 1943.
4. "The Houston Expressway," *American City*, November 1948.
5. "Mayor throws switch to light the freeway," *Houston Chronicle*, 1 October 1948.
6. "Judges will view 13,000 names for new expressway," *Houston Chronicle*, 19 November 1948; "It's the Gulf Freeway; the name will bring $100," *Houston Post*, 18 December 1948.
7. "Dream highway slowly assuming shape," *Houston Post*, 25 January 1948.
8. "Dedicated to 7,500,000 Texans, freeway is opened to traffic," *Houston Action*, 9 August 1952; "Freeway to open today," *Houston Post*, 2 August 1952; "Parade of progress continues with Gulf Freeway," *Galveston Daily News*, 3 August 1952; "$20 million plan dooms Gulf Freeway crossovers," *Houston Chronicle*, 29 October 1959; "'Safe' Gulf Freeway," *Houston Chronicle*, 1 March 1953.
9. "The Houston Expressway," *American City*, November 1948.
10. "Houstonian of the Century," *Houston Chronicle*, 29 December 1999.
11. "Gulf Freeway observes it 500 millionth mile of use," 18 November 1954, *Houston Press* vertical file at the Houston Public Library; "66,000 Use Freeway Daily," *Houston*, November 1950.
12. "Houston-Alvin Freeway," *Texas Highways*, January 1966.
13. Photo caption, *Houston Chronicle*, 19 December 1963.
14. "Gulf Freeway TV monitoring research reduces accidents," *Houston Post*, 6 April 1969; "Highway of the future monster of frustration," *Houston Chronicle*, 10 August 1969; TxDOT traffic data.

15. "Institute of Traffic Engineers Award," *Texas Highways*, February 1959; "Gulf Freeway observes its 500 millionth mile of use," 18 November 1954, Houston Press vertical file at the Houston Public Library; "Delay piled on delay on freeway barrier," *Houston Chronicle*, 30 March 1956.

16. "Shoppers get busy as Gulfgate opens," *Houston Post*, 21 September 1956; "City's oldest malls try to shed ragged image," *Houston Chronicle*, 2 April 1991.

17. "Freeway Study," *Texas Highways*, July 1956.

18. "Freeway guinea pig for state, US study," *Houston Post*, 29 June 1950; Texas Highway Department, *Texas Expressways* (February 1954).

19. "Traffic Control for Gulf Freeway," *Texas Highways*, November 1965; "Freeway Metering Revisited," *Texas Highways*, November 1966; "The Omnipotent Eye," *Texas Highways*, January 1967; "Gulf Freeway TV monitoring research reduces accidents," *Houston Post*, 6 April 1969.

20. Tom Wolfe, *The Right Stuff* (1979); communications with Scott Grissom, 2002.

21. *Houston* (advertisement page 23), December 1962.

22. "Traffic-heavy Gulf Freeway to be 8 lanes," *Houston Post*, 20 July 1972; "City seeks new hearing on expansion of freeway," *Houston Chronicle*, 8 March 1973; "Madgelean Bush speaks up, says what she thinks," *Houston Chronicle*, 21 May 1973; *Final Environmental Statement Administrative Action for Interstate 45 between Live Oak Street and Woodridge Street, FHWA-TEX-EIS-72-14-F* (1973) (includes information about opposition groups); "2 citizens groups trying to stop construction of freeways," *Houston Post*, 19 February 1973 (includes quote); "Battle over Gulf Freeway widening flares anew," *Houston Chronicle*, 3 April 1973.

23. "Gulf Freeway, the city's first, will open for the third time," *Houston Chronicle*, 9 May 1988.

24. "Highway of the future monster of frustration," *Houston Chronicle*, 10 August 1969; "Gulf Freeway completion reportedly years away," *Houston Post*, 29 May 1973; "Freeway costing a lot of time," *Houston Post*, 25 March 1976.

25. Texas Department of Transportation, *I-45 South Corridor Major Investment Study Executive Summary* (August 1999).

26. Texas Transportation Commission Minute Order 34718, 23 September 1953.

27. Ralph Ellifrit to Jim Douglas, TxDOT Houston district head, letter dated 29 December 1953 (city of Houston records).

28. "Southwest Freeway will not touch Bellaire," *Bellaire Texan*, 11 January 1956; Texas Transportation Commission Minute Order 48224, 24 August 1960.

29. From the city of Houston files, the suggested material for remarks for Mayor Pro Tem Jones on the occasion of the starting of work on the Southwest Freeway, 22 September 1958, stated the following in regard to cooperation between Sharp and the city of Houston on the Southwest Freeway alignment: "Early in 1954, the first preliminary studies were made for the huge Sharpstown subdivision which the City Planning Commission realized would block off the logical location for the extension of the Southwest Freeway. At this time, in close cooperation with the Highway Department and the developers, the first steps were made to fix the location through this area. Since that time, as section after section of subdivision development has come in, it has been carefully checked with the Highway Department to see that there was no encroachment on the needed right-of-way for the freeway."

30. Mickey Herskowitz, *Sharpstown Revisited: Frank Sharp and a Tale of Dirty Politics in Texas* (Austin, Texas: Eakin Press, 1994); "Add big tract to Sharpstown," *Houston Press*, 12 March 1955; "City of 15,000 homes planned," *Houston Chronicle*, 10 July 1954; "Blooming Prairie," *Newsweek*, 21 March 1955; "U.S. housing chief applauds Frank Sharp for vision, faith," *Houston Post*, 14 March 1955; "Sharpstown jammed for birthday blowout," *Houston Press*, 19 March 1956.

31. "$30 million community center planned by Sharp," *Houston Post*, 12 March 1955.

32. "All of first section of Sharpstown sold," *Houston Press*, 13 April 1955; "Pidgeon, Starlet at Sharpstown today," *Houston Post*, 18 March 1956.

33. "9-lane Southwest Freeway is forecast for Houston," *Houston Chronicle*, 9 August 1955; Ralph Ellifrit to A. C. Kyser, head of the TxDOT Houston Urban Project Office, letter dated 19 September 1955 (city of Houston records).

34. "County speeds up purchases of right-of-way," *Houston Chronicle*, 1 February 1957; Eugene Maier to Mayor Oscar Holcombe, letter dated 11 July 1957, (city of Houston records); "No freeways selected at top-brass meeting," *Houston Chronicle*, 12 July 1957.

35. "Delayed project revived," *Houston Chronicle*, 11 August 1957; "Long delay on new Fort Bend Road seen," *Houston Chronicle*, 1 September 1957.

36. "Two freeway projects in next year's plans," *Houston Chronicle*, 1 March 1958; "S.W. Freeway feeder road pacts to be let," *Houston Chronicle*, 6 March 1958.

37. "County buys last parcel for Southwest Freeway," *Houston Chronicle*, 28 September 1958.

38. "Three miles of freeway are opened," *Houston Post*, 27 July 1961; "SW Freeway service roads—Ten miles of 'em—open Friday," *Houston Press*, 21 December 1961; ."The Pretzel," *Texas Highways*, March 1958.

39. "Drive-in Banking?" *Houston*, December 1956.

40. "Engineers say SW Freeway faces paralysis," *Houston Chronicle*, 19 December 1968.

41. Housing displacements determined by TxDOT right-of-way map; "Greenway Plaza plans expansion," *Houston Chronicle*, 14 July 1968.

42. Ray Miller, *Ray Miller's Houston* (Houston: Cordovan Press, 1982).

43. "Engineers say SW Freeway faces paralysis," *Houston Chronicle*, 19 December 1968; "Help on way to ease SW Freeway frenzy," *Houston Chronicle*, 27 August 1972.

44. "Kyser Did It His Way," *Texas Highways*, July 1972.

45. The account of the Jack Caesar events is taken almost entirely from the testimony provided in the documentary film, *This is Our Home, It is Not For Sale* (1987, Jonathan Schwartz); Martha Terrill, interview by author, March 2002; *Houston Chronicle*, 13 September 1987; "Riverside Terrace is battling HHA," *Houston Chronicle*, 30 November 1966; "Riverside: where time has changed the life-style," *Houston Chronicle*, 18 April 1971.

46. Texas Transportation Commission Minute Order 34718, 23 September 1953.

47. "Ellifrit urges shifting of South Freeway," *Houston Post*, 4 March 1959.

48. Texas Transportation Commission Minute Order 46038, 22 May 1959; city of Houston, *City of Houston Planning Annual Report, 1959*; "Protests few at South Freeway hearing," *Houston Post*, 15 August 1963.

49. Transcript of the public hearing held 27 January 1966 for the South Freeway (city of Houston records).

50. "A strange wake at the death of a neighborhood," *Houston Chronicle*, 24 May 1970.

51. "Gaps and dead ends dramatize stoppage due to money crunch," *Los Angeles Times*, 14 September 1975.

52. "La Porte still has hopes of being a port," *Houston Post Dispatch*, 5 November 1927 (reprinted in the *Bayshore Sun*, November 1990); "The way it was," *Bayshore Sun*, 25 November 1990; "The way it was," *Bayshore Sun*, 2 December 1990.

53. *Texas Highways*, January 1954.

54. "La Porte is gay at opening of concrete road," *Houston Post Dispatch*, 6 November 1927; *La Porte Chronicle*, 28 January 1926; "The World's Largest Corporations," *Fortune*, 8 July 2002.

55. Texas Transportation Commission Minute Order 21182, 1 May 1945; "Another Expressway for Houston," *Texas Parade*, December 1952; Texas Transportation Commission Minute Order 34178, 23 September 1953.

56. *Texas Highways*, January 1954; David Brodsly, *L.A. Freeway: an Appreciative Essay* (University of California Press, 1981), 119.

57. Gus H. Haycock director, City of Houston Parks and Recreation Department, to Mayor Lewis Cutrer, letter dated 24 March 1958 (city of Houston records); Charles A. Easterling, senior assistant city attorney, to Gus H. Haycock, director, City of Houston Parks and Recreation Department, letter dated 7 April 1958 (city of Houston records).

58. Ralph S. Ellifrit, Director, City of Houston Planning, to W. Kyle Chapman, Harris County Precinct 1 Commissioner, letter dated June 23, 1958 (city of Houston records); Texas Transportation Commission Minute Order 48050, 25 July 1960; "Judge rejects Milby Park donor's bid to reclaim title to ground," *Houston Chronicle*, 22 August 1961.

59. "Funds provided to move track," *Houston Chronicle*, 29 August 1958; "Six lane Pasadena Freeway is nearer," *Houston Chronicle*, 30 October 1958; Texas Transportation Commission Minute Order 46694, 23 November 1959; Texas Transportation Commission Minute Order 59274, 1 June 1967.

60. Houston Metropolitan Transportation and Transit Study, *Freeway Phase* (16 August 1961); "Would ease Gulf Traffic," *Houston Chronicle*, 12 July 1961; "Additional study wanted on freeway revisions," *Houston Chronicle*, 13 July 1961; Houston Planning Commission (chairman M. E. Walter) to Houston mayor and City Council, letter dated 3 August 1961 (city of Houston records).

61. State Highway Engineer D. C. Greer to Eugene Maier, letter dated 30 January 1962 (city of Houston records); "$39.1 million Harris freeway jobs urged," *Houston Post*, 1 April 1964; "Expressway plan in draft stage," *Houston Post*, 11 November 1964.

62. Richard Holgin, interviews by author, 25 March 2002 and 28 March 2002.

63. Texas Transportation Commission Minute Order 62158, 2 April 1969.

64. "Residents on right-of-way safe for 3-4 years," *Houston Post*, 10 March 1970.

65. Bill Ward and William McClure to A. C. Kyser, head of the TxDOT Houston Urban Project Office, "Report on a citizens' open meeting on May 28, 1970 at Ripley House regarding proposed SH 225 Freeway," memorandum dated 4 June 1970 (city of Houston records).

66. "La Porte Freeway design team named," *Houston Chronicle*, 19 July 1970; minutes of first meeting of the Highway 225 interdisciplinary design team, 9 September 1970 (city of Houston records).

67. H. Dexter Jones, interview by author, 22 April 2002.

68. Harrisburg Freeway Study progress report for 1970, conferred with memo dated 8 January 1971 (city of Houston records).

69. "Welch proposes tollway study," *Houston Post*, 3 February 1971; "East end toll road plan meets protest," *Houston Post*, 4 February 1971; "Toll freeway plan opposition mounts," *Houston Chronicle*, 7 February 1971; "Mancuso urging toll study plan rejection," *Houston Chronicle*, 8 February 1971; "Council rejects toll road study," *Houston Chronicle*, 10 February 1971.

70. "Highway department standing behind plans for Harrisburg Freeway," *Houston Chronicle*, 21 February 1971.

71. "Eckhardt: put roads near jobs," *Houston Post*, 15 April 1971.

72. Harrisburg Freeway study team progress report newsletter, summer 1972 (city of Houston records); "Highway Department tells tentative route of Harrisburg Freeway," *Houston Chronicle*, 11 January 1973.

73. *Draft Environmental Impact Statement, administrative action for construction of a controlled access freeway State Highway 225, from U.S. 59 to Lawndale Avenue in Houston, Harris County, Texas*, FHWA-TEX-EIS-73-08-9 (May 1973); "HGAC panel approved Harrisburg Freeway," *Houston Chronicle*, 13 August 1973.

74. "Harrisburg freeway study called inadequate by EPA," *Houston Chronicle*, 1 September 1973.

75. Roscoe H. Jones, director, Houston City Planning Department, to Mayor Fred Hofheinz, city of Houston inter-office correspondence dated 11 June 1974 (city of Houston records).

76. William V. Ward, manager of the TxDOT Houston Urban Project Office, to B. L. DeBerry, TxDOT statewide head, letter dated 3 September 1976 (city of Houston records); William V. Ward, interviews by author, various dates in 2002.

77. Texas Transportation Institute, *Evaluation of Residents' Attitudes and Expectations of a Planned Freeway* (April 1975); Harrisburg Freeway Study Team, Texas Highway Department, *Harrisburg Freeway Volume 1 Route Report* (August 1976).

78. "La Porte Freeway; Texas 225 is shortest of freeways; East End protest kept it out of downtown," *Houston Chronicle*, 9 August 1999.

79. Texas Transportation Commission Minute Order 16701, 26 September 1939; documents relating to the Katy Freeway in the city of Houston files; Texas Transportation Commission Minute Order 18793, 17 December 1941.

80. Houston TxDOT district head W. J. Van London to City of Houston Planning Director Ralph Ellifrit, letter dated 1 July 1940 (city of Houston files).

81. Texas Transportation Commission Minute Order 19475, 23 June 1942.

82. "Kyser Did It His Way," *Texas Highways*, July 1972; "Pre-stressed 'Grape Arbor' spans I.H. 10," *Texas Highways*, October 1966.

83. Texas Transportation Institute, *Experiences and Opinions of Residents along Elevated, Depressed, and On-grade Freeway Sections in Houston*, Research Report 148-1, Study 2-1-71-148 (1971).

84. Texas Transportation Institute, *Houston High Occupancy Vehicle Lane Operational Summary* (September 2002); TxDOT traffic maps.

85. "Elevated traffic lanes urged by West Houston Association," *Houston Post*, 31 March 1985; "$1.1 billion to upgrade freeway called for," *Houston Post*, 4 April 1986.

86. "New plans call for 16 lanes on Katy Freeway," *Houston Post*, 31 December 1992.

87. "Katy motorists conclude: build more lanes quickly," *Houston Chronicle*, 17 November 1995.

88. "Latest Katy Freeway plan more flexible," *Houston Chronicle*, 15 November 2000; "24 lanes urged for Katy Freeway," *Houston Chronicle*, 5 April 2001; "Wide approval greets I-10 plan," *Houston Chronicle*, 13 April 2001.

89. Texas Transportation Commission Minute Order 109147, 30 January 2003; "Expenses running $244 million over estimate," *Houston Chronicle*, 30 July 2003; "Katy Freeway project cost raises concern," *Houston Chronicle*, 31 July 2003.

90. "North Freeway undisputed workhorse of freeways," *Houston Chronicle*, 19 July 1999.

91. Texas Transportation Commission Minute Order 21182, 1 May 1945; Texas Transportation Commission Minute Order 22594, 14 October 1946; Texas Transportation Commission Minute Order 28299, 21 March 1950; Texas Transportation Commission Minute Order 26325, 25 February 1949; Texas Transportation Commission Minute Order 32278, 25 June 1952.

92. William R. McCasland, *Evaluation of the First Year of Operation, I-45 Contraflow Lane, Houston*, Report FHWA/TX-81/6+205-9 (Texas Transportation Institute, January 1981).

93. "Contraflow has successful debut," *Houston Post*, 29 August 1979; "MTA to mark 3rd year of I-45 Contraflow lane," *Houston Post*, 28 August 1982.

94. Texas State Department of Highways and Public Transportation, *Environmental Assessment, Interstate Highway 45 from Beltway 8 North in Harris County to Loop 336 North in the City of Conroe* (March 1986); "Newest North Freeway mess is just around the corner," *Houston Chronicle*, 6 June 1985.

95. "Double-decker I-45?" *Houston Post*, 21 June 1983.

96. "Perfect example of smart growth," *Houston Chronicle*, 20 March 2000.

97. "Woodlands developer looks well down road," *Houston Chronicle*, 17 November 2002.

98. "Goodyear's blimp will float away," *Houston Chronicle*, 11 December 1991; "Bye bye blimp base," *Houston Chronicle*, 18 March 1994.

99. "Let's stick to compass-point names," *Houston Chronicle*, 4 May 1965.

100. Texas Transportation Commission Minute Order 21182, 1 May 1945; "Building lines to be fixed on expressways," *Houston Chronicle*, 3 November 1948; "New Freeway Master Plan May be Revised," *Houston*, April 1952; "Vast freeway to circle city gets OK," *Houston Chronicle*, 7 April 1953; Texas Transportation Commission Minute Order 34969, 27 October 1953.

101. "Traffic circle again mishap champ," *Houston Post*, 22 March 1967; "Dangerous crossing to exist 3 more years," *Houston Post*, 30 June 1969.

102. "Bids on Northside Expressway soon," *Houston*, November 1951.

103. "Kingwood life is centered on its villages," *Houston Chronicle*, 15 June 1997.

104. Texas State Department of Highways and Public Transportation, *Environmental Assessment, United States Highway 59 from Interstate Highway 45 in Downtown Houston to Proposed Cleveland Bypass* (July 1985); "Churches at mercy of Eastex plan," *Houston Post*, 17 November 1985; "The growing pains of Highway 59," *Houston Business Journal*, 1 December 1986; "Plans for US 59 lamented, lauded," *Houston Chronicle*, 19 January 1991.

105. "King of the Road," *Houston Chronicle*, 26 April 1998.

106. Texas Transportation Commission Minute Order 22594, 14 October 1946.

107. "Long-Talked Freeway Still 2 Years Away," *Houston Chronicle*, approx. July-Sept. 1957; Texas Transportation Commission Minute Order 34971, 27 October 1953; *Houston Chronicle*, 14 September 1956; Texas Transportation Commission Minute Order 50411, 30 November 1961.

108. "Houston Freeway Open," *Texas Highways*, June 1966.

109. "I-10 reconstruction work lags by year," *Houston Post*, 19 April 1980; "House panel assails I-10 East condition," *Houston Chronicle*, 19 April 1980; "I-10 construction headaches to last at least until 1980," *Houston Post*, 26 May 1977.

110. "U.S. 290 drivers facing changes," *Houston Chronicle*, 5 July 1999.

111. Texas Transportation Commission Minute Order 35331, 1 December 1953; "Commission slates another freeway," *Houston Chronicle*, 1 December 1953.

112. "Hearing set on U.S. 290 construction," *Houston Chronicle*, 15 March 1959.

113. Wilbur Smith and Associates, *Initial Feasibility Assessment Houston Tollways*, prepared for the Texas Turnpike Authority (New Haven, Connecticut: October 1979).

114. "Politicians aim to scotch Hardy Street toll study, plan to boost rail system," *Houston Post*, 19 February 1980; "TMO opposes toll road," *Houston Post*, 16 July 1980.

115. "TTA moves to study feasibility of toll road," *Houston Chronicle*, 30 July 1980; "Commissioners Court now asks toll road study," *Houston Chronicle*, 26 August 1980.

116. "Northside residents need to note hard-driving toll road controversy," (editorial by Richard Vara), *Houston Post*, 28 August 1983.

117. "Tollway plan draws mixed reaction," *Houston Chronicle*, 2 April 1982; Howard Needles Tamman Bergendoff (HNTB), *Preliminary Engineering Feasibility Study, Hardy Tollway* (December 1982).

118. "Mayor favors Metro's rail system over 'competing' toll road," *Houston Chronicle*, 10 November 1982; "Hardy toll project nearing dead end," *Houston Post*, 10 November 1982.

119. "Hardy Tollroad opposition meets with Governor," *The News North Harris County*, 4 May 1983; "Electric rates, teacher pay among White priorities," *Houston Post*, 29 April 1983.

120. "City backs toll project for Hardy," *Houston Post*, 25 August 1983; letter from Houston Board of Realtors president Ray Baxter to its membership, 8 August 1983.

121. "Tollways will need tax money," *Houston Chronicle*, 6 July 1990.

122. "Revenue from toll roads 40% less than expected," *Houston Chronicle*, 21 July 1989; "Expectations of toll road lag," *Houston Chronicle*, 27 July 1998.

123. Howard Needles Tamman Bergendoff (HNTB), *Preliminary Engineering Feasibility Study, Hardy Tollway* (December 1982); "Hardy Toll Road extension is eyed into Montgomery County to Loop 336," *Houston Chronicle*, 29 May 2003.

124. "Lindsay: Hardy plan is flattened," *Houston Chronicle*, 28 September 1994; "Residents reluctantly leave homes," *Houston Chronicle*, 22 July 2003; "Hardy Toll Road extension is eyed into Montgomery County to Loop 336," *Houston Chronicle*, 29 May 2003.

125. Texas Highway Department, *Non-interstate Arterial Highway U.S. 90, Houston to Beaumont* (October 1961).

126. Texas Transportation Commission Minute Order 57326, 31 March 1966; Texas Transportation Commission Minute Order 58563, 28 October 1966; Texas Transportation Commission Minute Order 65454, 4 November 1971; "Speed Urged in Purchase of Freeway Right-of-Way," *Houston Chronicle*, 17 November 1972.

127. Texas Transportation Commission Minute Order 69824, 31 March 1975.

128. "Crosby Freeway will be quick route to Lake Houston town," *Houston Chronicle*, 5 January 1988; "Crosby to celebrate town-to-city connection," *Houston Chronicle*, 22 February 1991.

129. "City's concerted effort won Compaq expansion," *Houston Chronicle*, 17 July 1988; "Lanier reaped profits after promoting roads," *Houston Chronicle*, 26 September 1991.

130. "Compaq to reveal local plans," *Houston Chronicle*, 13 July 1988.

131. Texas Transportation Commission Minute Order 87493, 28 June 1988; Texas Transportation Commission Minute Order 88009, 28 June 1988.

132. "Texas 249 soon to be above it all," *Houston Chronicle*, 8 July 1997.

133. Houston Chamber of Commerce Highway Committee, "A Master Finance Plan for Houston and Harris County" (March 1954).

134. Houston Chamber of Commerce Highway Committee, "A Master Finance Plan for Houston and Harris County" (March 1954).

135. "Westpark toll road plan re-emerges," *Houston Chronicle*, 2 July 1993.

136. Wilbur Smith and Associates, *Initial Feasibility Assessment Houston Tollways*, prepared for the Texas Turnpike Authority (New Haven, Connecticut, October 1979); "Westpark toll road?" *Houston Post*, 18 October 1979.

137. "MTA bond plan soundly defeated," *Houston Post*, 12 June 1983.

138. "Lanier picks two for Metro," *Houston Chronicle*, 1 January 1992; "Metro may be forced to tap cash reserves," *Houston Chronicle*, 27 March 1992.

139. "Metro considers new tollway by Westpark," *Houston Chronicle*, 22 February 1992; "Metro approves record land purchase," *Houston Chronicle*, 25 December 1992; "Metro acquires 158 miles of right-of-way," *Houston Chronicle*, 1 January 1993; "Metro, Southern Pacific sue each other over land deal," *Houston Chronicle*, 30 March 1994; "Metro, Southern Pacific settle lawsuits," *Houston Chronicle*, 15 April 1997.

140. "Lindsay: County should oversee Westpark Tollway," *Houston Chronicle*, 8 July 1993; "West Houston may get new HOV lane," *Houston Chronicle*, 17 November 1995; "Metro to study Westpark Tollway plan," *Houston Chronicle*, 22 August 1996; "Backers of a Westpark Tollway oppose planned Metro transit lane," *Houston Chronicle*, 24 January 1998.

141. "Battle line drawn for Westpark corridor," *Houston Chronicle*, 23 September 1998.

142. M.E. Walter, Houston Planning Commission, to Houston mayor and city council, letter dated 3 August 1961 (city of Houston records); *1960 Traffic and Transportation Study for City of Houston*, analysis of study included in *Final Environmental Statement for Interstate 45 between Live Oak Street and Woodridge Street*, FHWA-TEX-EIS-72-14F, 1973; Houston Metropolitan Transportation and Transit Study, *Freeway Phase* (August 1961).

143. Houston City Council motion 63-2517 and Houston Planning Department drawing CNC-109; Houston City Planning Department Assistant Director Ronald A. Heiser to Louis R. Koehn, assistant director, City of Houston Public Works, letter dated 13 May 1970, including a listing of events regarding the West Loop Extension (city of Houston records).

144. City of Houston resolution 70-74 dated 27 May 1970.

145. E. B. Cape, City of Houston Director of Public Works, to Houston mayor Fred Hofheinz, letter dated 22 January 1975 (city of Houston records).

146. Lichliter Jameson & Associates, "Post Oak/Fort Bend Freeway Corridor, Project Brief" (October 1985); State Department of Highways and Public Transportation, *Fort Bend Freeway Feasibility Study* (August 1987).

147. "Fort Bend Parkway plan gets green light," *Houston Chronicle*, 6 March 1990; "Freeway a step closer to reality," *Houston Post*, 6 March 1990.

148. Transcript of Texas Transportation Commission meeting, 28 May 1998.

149. "Fort Bend Parkway resurfaces," *Houston Chronicle*, 4 May 1997; "Fort Bend not giving up on tollway," *Houston Chronicle*, 23 July 1997.

150. Texas Transportation Commission Minute Order 54363, 5 May 1964; "$55 million Houston-Alvin Freeway system sought," *Houston Chronicle*, 18 September 1965; Texas Department of Transportation, *Final Environmental Impact Statement, Proposed relocation of state highway, FHWA-TEX-EIS-75-10-F 35* (1975); "Houston-Alvin Freeway," *Texas Highways*, January 1966; "Roads will never meet demand; Alvin Freeway will relieve Gulf congestion," *Houston Chronicle*, 12 August 1969; Texas Transportation Commission Minute Order 60213, 28 November 1967.

151. "Freeway fight gains support," *Houston Post*, 1 April 1973; "Freeway plan abandoned," *Houston Post*, 18 May 1973.

152. Texas Transportation Commission Minute Order 63415, 23 February 1970.

153. Texas Department of Transportation, *Final Environmental Impact Statement, State Highway 35 from IH 45 South to Bellfort, FHWA-TEX-82-02-F* (1992); "Councilman to seek delay on vote to close 2 roads," *Houston Post*, 21 August 1985.

154. Harris County engineering department to E.E. Swenke, chief of right-of-way department, letter dated 25 May 1964 (Harris County record storage); Fernando Williams at Williams and Crawford to Richard P. Doss, Harris County Engineer, letter dated 11 October 1966 (Harris County record storage); Letter to W. Kyle Chapman, Harris County Commissioner Precinct 1, dated 29 November 1966 (Harris County record storage).

155. Richard Doss, County Engineer, to Ms. Nancy Gehman, General Superintendent Precinct 1, letter dated 21 January 1977 (Harris County record storage).

156. Texas Transportation Commission Minute Order 38785, 20 September 1955; Texas Transportation Commission Minute Order 49098, 22 February 1961; Texas Transportation Commission Minute Order 106787, 28 March 1996.

157. Texas Department of Transportation, *State Highway 146 Corridor Major Investment Study* (July 2002).

158. Texas Transportation Commission Minute Order 42810, 30 October 1957; Texas Transportation Commission Minute Order 48991 31 January 1961; Transcript of the Texas Transportation Commission, 28 June 2001.

159. State Department of Highways and Public Transportation, *Environmental Assessment NASA Road 1 from IH 45 to SH 146* (February 1988); "Some along NASA Road 1 left marooned," *Houston Chronicle*, 25 May 1992.

160 "Webster agrees on bypass," *Houston Chronicle*, 16 January 1994; Texas Department of Transportation, *Supplemental Environmental Assessment, NASA Road 1 Webster Bypass from IH 45S to 0.6 km east of FM 270* (May 1995).

161. "Beautiful Memorial Drive along Buffalo Bayou banks taking shape," *Houston Chronicle*, 14 March 1937; "New freeway gets little use," *Houston Post*, 9 July 1955; "New freeway speeds cars to old bottleneck," *Houston Post*, 31 January 1956; "Underpass opens Friday at Memorial, Shepherd," *Houston Post*, 27 April 1958; "Memorial Drive extension to be opened," *Houston Post*, 28 January 1960.

The Loops

1. "Proposed north and west loops around city," *Houston Post Dispatch*, 27 December 1931; "Road loops around city are talked," *Houston Chronicle*, 23 or 28 September 1940.

2. "Defense belt roads planned in county's proposed bond issue," *Houston Chronicle*, 13 April 1941.

3. Texas Transportation Commission Minute Order 19475, 23 June 1942; Texas Transportation Commission Minute Order 20301, 16 November 1943.

4. "Council sets up new freeway," *Houston Chronicle*, 19 November 1953.

5. Ralph Ellifrit to Texas Highway Department Houston District Engineer Jim Douglas, letter dated 29 December 1953 (city of Houston records).

6. "Before the Highway Commission, State of Texas, a Plea for Loop Designations in Harris County and the City of Houston," brochure presented to the Texas Highway Commission, 28 September 1954; Texas Transportation Commission Minute Order 36815, 13 October 1954.

7. Ralph Ellifrit, J. M. Nagle, and Eugene Maier (city of Houston department heads) to Dewitt C. Greer, letter dated 1 December 1955 (city of Houston records); Texas Transportation Commission Minute Order 43815, 24 April 1958; "State agrees to finish loop around city," *Houston Chronicle*, 24 April 1958; "East Loop imperiled by State's delay," *Houston Chronicle*, 30 May 1960; "Group will ask loop designation," *Houston Chronicle*, 14 August 1960; Texas Transportation Commission Minute Order 48223, 24 August 1960; "Cutrer hails east loop designation, tells plans," *Houston Chronicle*, 27 August 1960.

8. "Pin Oak turns 50," *Houston Chronicle*, 14 April 1995; "Sun sets on Pin Oak Stables, but it rises on retail sites," *Houston Chronicle*, 3 November 1996.

9. "East loop won't be ready soon," *Houston Chronicle*, 16 September 1962.

10. "Another Expressway for Houston," *Texas Parade*, December 1952; Texas Transportation Commission Minute Order 40812, 26 October 1956; "Long-running serial has ended," *Houston Chronicle*, 21 September 1975; Houston Galveston Regional Transportation Study Newsletter, October 1975.

11. "Rice acreage now city of good homes," *Houston Chronicle*, 16 February 1941; "Westmoreland Farms," sales brochure issued by the South End Land Company, 1909, in the collection of the Houston Public Library Houston Metropolitan Research Center; "History of Bellaire," document in the George Fuermann special collection, University of Houston.

12. "Council Brushes Off Petitioners," *Bellaire Texan*, 9 December 1954; James M. Richards, manager City of Houston Public Works Department, to Harris County Judge Bob Casey, letter dated December 8, 1954 (city of Houston records).

13. "North loop extension puzzle to Bellaire," *Houston Chronicle*, 1 April 1955; "Won't spend $$ on PO Freeway," *Bellaire Texan*, 7 April 1955.

14. "Bellairites outspoken on P.O. Freeway," *Bellaire Texan*, 16 November 1955; "County Okays P.O. Freeway," *Bellaire Texan*, 8 February 1956; "Lose 1.25 million to P.O. Freeway," *Bellaire Texan*, 14 March 1956.

15. "Houston men listed with nation's richest," *Houston Chronicle*, 23 or 28 October 1957.

16. "County sets final P.O. Freeway lines," *Bellaire Texan*, 4 April 1956; "Widen P.O. right of way 50 feet," *Bellaire Texan*, 18 September 1957.

17. Right-of-way maps for Interstate 610, TxDOT Houston office; "Loop freeway gets tough punch," *Houston Chronicle*, 23 December 1954.

18. "Houston market review 2002," *Houston Chronicle*, 7 April 2002.

19. "Hazardous trucks review is urged," *Houston Chronicle*, 12 May 1976; "Ammonia truck crash blamed on high speed," *Houston Chronicle*, 13 May 1976.

20. "20 candles afire on Galleria ice," *Houston Chronicle*, 11 November 1990.

21. The Costar Group, *The Costar Office report, mid-year 2001, Houston Office Market.*

22. "Loop 610 expansion veers into roadblock," *Houston Chronicle*, 27 November 1991.

23. "Chances of wider loop: 'very slim'," *Houston Chronicle*, 13 March 1992.

24. "Mayoral candidates quick to denounce tollway through park," *Houston Chronicle*, 11 June 2003.

25. Associated Sports Center Architects, *Preliminary Architectural Program, A Sports Center for Harris County*, document in the collection of the Houston Public Library Houston Metropolitan Research Center.

26. "Additional $9.6 million asked by county commissioners to construct domed stadium," press release by the Tax Research Association of Houston and Harris County, dated 14 December 1962, in the collection of Houston Public Library Houston Metropolitan Research Center; "Dream comes true as stadium turns out better than imagined," *Houston Chronicle*, 18 August 2002; "The Dome, problems in plastic," *Houston Post*, 8 April 1965; "Fans paying for Reliant's extras," *Houston Chronicle*, 24 June 2002.

27. "History of Synthetic Turf," Southwest Recreational Industries, Inc. web site, www.astroturf.com/historyofsyntheticturf.htm, accessed June 2003; "Fielders find dome flies hard to follow," *Houston Post*, 8 April 1965; "Paint works on glare in dome," *Houston Post*, 21 April 1965.

28. "Rodeo bids farewell to 37 year home with Texas-sized salute," *Houston Chronicle*, 3 March 2002.

29. Houston Planning Director Ralph Ellifrit to W. Kyle Chapman, City Planning Commission, letter dated September 21, 1954 (city of Houston records); city of Houston records, document providing a chronology of events in the history of Beltway 8 up to 1962.

30. "Outer loop as freeway urged," *Houston Chronicle*, 19 March 1960; "County makes wise move on freeways," *Houston Chronicle*, 11 July 1960; document in city of Houston records with Beltway 8 right-of-way status, dated 11 April 1960; "$150,000,000 outer belt to be pushed in 1961," *Houston Chronicle*, 4 December 1960.

31. "Streets Will Grow in the Grass," *Houston*, November 1957.

32. City of Houston Planning Director Ralph Ellifrit to Wiley Charmichael, TxDOT Houston District head, letter dated 6 July 1961 (city of Houston records).

33. Minutes of the City Planning Commission, 5 June 1962; minutes of the City Planning Commission, 19 June 1962.

34. Texas Transportation Commission Minute Order 59380, 1 June 1967; Texas Transportation Commission Minute Order 62067, 7 March 1969; Texas Transportation Commission Minute Order 62581, 31 July 1969.

35. "Beltway 8 Begins," *Houston*, April 1972; *Texas Highway Department District 12 Interdisciplinary Team Report on Investigation of a Possible Relocation of West Beltway 8*, signed by Richard J. Kabat, 20 January 1975.

36. "Beltway's cost dims its future," *Houston Post*, 23 September 1975.

37. "Study sought for toll road," *Houston Post*, 30 May 1975; "County asking state unit to study Beltway 8 plan," *Houston Chronicle*, 30 May 1975; "Beltway 8 future dim, but project isn't dead yet," *Houston Chronicle*, 24 April 1977.

38. State Department of Highways and Public Transportation, *Responding to the Changing Environment—Summary Report*, July 1976; Omer Poorman, head of Houston TxDOT office, to Houston Mayor Fred Hofheinz, letter dated 12 November 1976 (city of Houston records).

39. "Toll study resurrects West Belt," *Houston Post*, 16 September 1976; Texas Turnpike Authority, 1977 annual report.

40. "Beltway 8 completion not due till turn of century," *Houston Post*, 24 December 1977; "West Belt revived," *Houston Chronicle*, 19 May 1978; "Plans for Beltway 8 segment authorized," *Houston Chronicle*, 21 June 1979; "West Beltway 8 project draws support at hearing," *Houston Post*, 21 May 1980.

41. "One freeway route is enough, say Jersey Village residents," *Houston Chronicle*, 18 July 1973; "Proposed route of Beltway 8," *Houston Chronicle*, 18 March 1977; "Jersey Village mayor asks Senate Committee to reroute new outer loop," *Houston Chronicle*, 17 March 1977; "Jersey Village residents oppose Beltway 8 plan," *Houston Post*, 9 January 1980.

42. "Harris County Judge favors county to build toll road," *Houston Post*, 21 April 1977.

43. M. G. Goode, representative of landowners, to Jim Short, director of the Beltway 8 Group, letter dated 1 July 1982 (city of Houston records).

44. Carol Letz, interview by author, 6 May 2002.

45. "County's toll road system a freeway success story," *Houston Chronicle*, 30 August 1999; "As toll roads open, officials look for drivers," *Houston Chronicle*, 27 June 1988.

46. "Revenue from toll roads 40% less than expected," *Houston Chronicle*, 21 July 1989; "Revenues from 2 toll roads rise 7%," *Houston Chronicle*, 28 September 1989; "Tollways will need tax money, $20 million yearly subsidy expected," *Houston Chronicle*, 6 July 1990.

47. "County paints rosy tollway picture," *Houston Chronicle*, 18 September 1990.

48. Houston Airport System web site, www.houstonairportsystem.org; "Does Houston need new airport now?" *Houston Chronicle*, 20 May 1953; "Mayor's blood boils—Hofheinz says airport terminal inadequate," *Houston Post*, 15 July 1953; "Shape of future in new Houston airport," *Houston Chronicle*, 31 October 1954.

49. "Houston should speed up and plan better airport facilities," *Houston Chronicle*, 11 January 1955; "First jet liner here has room to spare," *Houston Post*, 21 May 1957; "Jetliners can use present strip—but not on a hot summer day," *Houston Chronicle*, 22 May 1957; "Way cleared for jet runways here," *Houston Chronicle*, 16 October 1958; "Houston loses ground in aviation while Dallas goes steadily ahead," *Houston Chronicle*, 25 August 1957.

50. City of Houston, *Review and Evaluation of Proposed Major Airport Sites to Serve the Houston Area* (April 1960).

51. Joseph A. Foster, director of City of Houston Aviation Department, to Paul H. Boatman, acting chief of Federal Aviation Agency in Fort Worth, letter dated 5 May 1960.

52. "As Continental wheels, deals, hub is vital spoke," *Houston Chronicle*, 8 December 1996.

53. Airports Council International site, www.airports.org, data last updated 23 October 2002.

54. "Intercounty loop drive pact sought," *Houston Chronicle*, 26 April 1961; "City planners are aiming for loop 17 miles from heart of downtown," *Houston Chronicle*, 7 May 1961.

55. "A dream is always the first step," *Houston Chronicle*, 21 October 1968; "Loop circling county proposed," *Houston Chronicle*, 5 October 1965; "Freeway plan due hearing by city panel," *Houston Chronicle*, 26 January 1966; "Start planning 'way out loop' now," *Houston Chronicle*, 30 January 1966.

56. "Houston 1990: parkways to 'manicured wilderness'," *Houston Chronicle*, 23 October 1966; "City of 3.1 million seen," *Houston Chronicle*, 7 May 1967.

57. Houston Planning Commission, "Houston Year 2000 Map," published 1972.

58. State Department of Highways and Public Transportation, *Responding to the Changing Environment—Summary Report*, July 1976.

59. Houston-Galveston Regional Transportation Study newsletter, January 1977.

60. "Grand Parkway moves at fast clip," *Houston Chronicle*, 25 August 1985.

61. "Houston to need 3rd highway loop in 20 years, study says," *Houston Post*, 13 November 1982.

62. "Groups trying to revive proposal for 160-mile Grand Parkway," *Houston Chronicle*, 14 March 1984; "Officials tout outer loop around Houston," *Houston Chronicle*, 15 March 1984.

62. HB 125, 1984 special session of the Texas Legislature; Texas Transportation Commission Minute Order 82325, 25 October 1984; Texas Transportation Commission Minute Order 82685, 28 January 1985; Texas Transportation Commission Minute Order 83736, 24 October 1985.

63. "Most right of way for Grand Parkway lined up: officials," *Houston Post*, 9 February 1985.

64. "'Mega-loop' off the drawing board," Houston *Chronicle*, 14 May 1989.

65. "New policy forces Mischer to resign from loop board," *Houston Chronicle*, 2 April 1986.

66. "Commission approves parkway at Peek road," *Houston Chronicle*, 30 March 1988.

67. "State gets OK to start on Grand Parkway," *Houston Chronicle*, 27 February 1991.

68. "Environmentalists take new tack in Parkway lawsuit," *Houston Chronicle*, 13 May 1993; "Air quality demands worry northwest Houston groups," *Houston Chronicle*, 11 April 2001; "Local groups challenging clean air suit," *Houston Chronicle*, 18 April 2001.

69. "Activists' report opposes Grand Parkway," *Houston Chronicle*, 13 September 1996; Road to Ruin web site, www.taxpayer.net/TCS/RoadRuin/.

70. "Timetable for Grand Parkway/ Environmental Studies will lengthen the construction period," *Houston Chronicle*, 27 June 1991.

71. "Taxpayers may have to fund parkway design," *Houston Chronicle*, 2 December 1997; transcript of Texas Transportation Commission meeting, 24 February 2000; transcript of Texas Transportation Commission meeting, 28 March 2002.

72. "Grand Parkway backers lobby for new tollway," *Houston Chronicle*, 2 February 1998; "Parkway toll is too steep for Texas," *Houston Chronicle*, 3 April 1999.

73. "Controversial Parkway route wins approval," *Houston Chronicle*, 21 October 2000.

74. Transcript of Texas Transportation Commission meeting, 27 May 1999; "Grand Parkway foes seeking to raise money for legal battle," *Houston Chronicle*, 2 August 2001.

75. Texas Transportation Minute Order 109226, 24 April 2003.

Freeway Mass Transit

1. David Brodsly, *L.A. Freeway: An Appreciative Essay* (University of California Press, 1981), 108; "Transit study to urge rapid freeway service," *Houston Chronicle*, 21 January 1962; "Study turns thumbs down on monorail," *Houston Chronicle*, 27 August 1963.

2. "Cutrer studies Los Angeles mass transit plan," *Houston Chronicle*, 11 December 1960; "Study turns thumbs down on monorail," *Houston Chronicle*, 27 August 1963; "Public transit is given limited role in study," *Houston Post*, 28 August 1963.

3. "Welch says city must subsidize or operate mass transit system," *Houston Chronicle*, 28 September 1972; "Fund use OK'd to buy bus system," *Houston Chronicle*, 8 November 1972.

4. "Gridlock!" *Time*, 12 September 1988; "Not out of woods yet," *Houston Post*, 19 January 1980; "MTA pressed for solutions to problems," *Houston Post*, 12 August 1979.

5. "Contraflow has successful debut," *Houston Post*, 29 August 1979; William R. McCasland, *Evaluation of the First Year of Operation, I-45 Contraflow Lane, Houston*, Report FHWA/TX-81/6+205-9 (Texas Transportation Institute, January 1981).

6. *Houston Post*, caption for photograph of accident on contraflow lane, page 1, 4 September 1981; "MTA to mark 3rd year of I-45 contraflow lane," *Houston Post*, 28 August 1982.

7. "MTA bond plan soundly defeated," *Houston Post*, 12 June 1983; "I-45 contraflow ends after 5 years," *Houston Post*, 24 November 1984.

8. "Reconstructed Gulf Freeway opens today—for the third time," *Houston Chronicle*, 9 May 1988.

9. *Houston High Occupancy Vehicle Lane Operations Summary* (Texas Transportation Institute, December 2002).

Bridges and Tunnels

1. Handbook of Texas Online, subject: Galveston.

2. "From island to mainland: via canoe, skiff, barge, wagon bridge, causeway," *Galveston Daily News*, 10 February 1957.

3. Clarence Ousley, *Galveston in 1900* (Atlanta: William C. Chase); Charles W. Hayes, *History of the Island and the City of Galveston* (Jenkins Garret Press, 1974); "Railroad situation," *Galveston Daily News*, 17 September 1900; "In touch with the world," *Galveston Daily News*, 22 September 1900.

4. "The bridge restored," *Galveston Daily News*, 21 September 1900.

5. "About steel bridges," *Galveston Daily News*, 2 October 1900; "Build the causeway," *Galveston Daily News*, 5 April 1906; program of the Galveston Causeway celebration, 25 May 1912, in the collection of the Rosenberg Library, Galveston; "From island to mainland: via canoe, skiff, barge, wagon bridge, causeway," *Galveston Daily News*, 10 February 1957.

6. "Railroad trestle built to link isle, mainland after 1915 storm," *Galveston Tribune*, 3 September 1915; *General Specifications for Repair of Approaches Destroyed by Storm August 16-17, 1915*, formal engineering specification in the collection of the Rosenberg Library, Galveston.

7. "Program arranged for opening of causeway," *Galveston Daily News*, 26 February 1922; "Causeway is dedicated to uses of commerce," *Galveston Daily News*, 22 April 1922.

8. "From island to mainland: via canoe, skiff, barge, wagon bridge, causeway," *Galveston Daily News*, 10 February 1957.

9. National Oceanic and Atmospheric Administration web site, hurricane history, Carla.

10. "Gulf Coast Ceremonies," *Texas Highways*, September 1961; Texas Transportation Commission Minute Order 37172, 30 November 1954; Texas Transportation Commission Minute Order 40569, 17 August 1956; Texas Transportation Commission Minute Order 42827, 31 October 1957.

11. Texas Transportation Commission Minute Order 67824, 5 September 1973.

12. "Two new causeways to Galveston pushed," *Houston Chronicle*, 4 May 1986; "Causeway's initial costs OK'd," *Houston Chronicle*, 22 May 1987; "Toll status for new bay crossing studied," *Houston Chronicle*, 31 January 1989; Texas Turnpike Authority annual report 1989; "Bridging a Gap: Galveston sharply split over how to make the city more accessible," *Houston Chronicle*, 31 July 1994.

13. Municipal Engineering Company Inc. and Turner Collie Braden Inc., *Galveston-Alvin-Pearland Parkway Feasibility Study* (November 1986).

14. "Offatts Bayou crossing bridge over troubled waters," *Houston Chronicle*, 5 August 1994.

15. Texas Transportation Commission Minute Order 71108, 31 March 1976; "Causeway needs to be replaced," *Houston Chronicle*, 4 November 1997.

16. Texas Department of Transportation, *IH 45 South Corridor Major Investment Study, Executive Summary* (August 1999).

17. "Barge bridge accidents are recalled," *Houston Chronicle*, 27 May 2002.

18. Texas Department of Transportation, *SH 87 Feasibility Study*, November 2000.

19. Texas Transportation Commission Minute Order 22051, 15 May 1946; Texas Transportation Commission Minute Order 22653, 15 October 1946.

20. "Baytown-La Porte tunnel dedicated," *Houston*, October 1949; "The highway tunnel that went a'sailing," *Texas Parade*, July 1950; "3000 see opening of tunnel," *Houston Chronicle*, 23 September 1953; "Our Largest Highway Projects," *Texas Highways Magazine*, April 1954.

21. "In the Public Eye," *Texas Highways*, March 1967; *Texas Highways*, July 1968, 29.

22. Texas Transportation Commission Minute Order 73281, 31 August 1977; Texas Transportation Commission Minute Order 80246, 29 November 1982; Texas Department of Transportation, *Final Environmental Assessment, Route 146 from the Route 146/Route 225 interchange to the Route 146/Route 201 Interchange in Harris County, Texas*, Federal Highway Administration project number F-839, control number 389-12, IPE-379.

23. "Critics test metal of bridge," *Houston Chronicle*, 26 June 1992.

24. "DOT chief vows rapid bridge completion," *Houston Chronicle*, 28 January 1995; "Go ahead, cross that bridge," *Houston Chronicle*, 24 September 1995.

25. "Tale of a troubled bridge over water," *Houston Chronicle*, 26 April 1998.

26. "Baytown Tunnel's reef plan sunk," *Houston Chronicle*, 7 September 1997.

27. "East Loop won't be ready soon," *Houston Chronicle*, 16 September 1962.

28. "Bridge over channel is rejected," *Houston Post*, 14 November 1963; "Permit to be granted for ship channel span," *Houston Post*, 12 December 1963; "$32 million bridge to span ship channel," *Houston Chronicle*, 16 February 1964.

29. "Workers close gap in first bridge over ship channel," *Houston Post*, 20 February 1972; "First ship channel bridge called navigational hazard," *Houston Post*, 5 March 1972.

30. "Remember, bridge has a name," *Houston Chronicle*, 14 February 1991; "Council seeks to bridge conflict over snafu," *Houston Chronicle*, 15 February 1991.

31. "Channel bridge opened," *Houston Post*, 3 March 1973; "Troubled bridge over waters," *Houston Post*, 23 June 1976; "Highway officials miffed about Post bridge story," *Houston Post*, approximately 30 June 1976.

32. "Bridge's puncture wound," *Houston Chronicle*, 30 August 1988; "Ship slams into 610 bridge," *Houston Chronicle*, 22 December 2000; "Traffic flows again on channel bridge," *Houston Chronicle*, 29 January 2001; "Ship's boom hits Loop 610 bridge," *Houston Chronicle*, 1 June 2001; "Bridge to reopen Thursday," *Houston Chronicle*, 31 July 2001.

33. "Channel bridge, road included," *Houston Chronicle*, 7 April 1963; "Ship channel's toll bridge is vetoed," *Houston Chronicle*, 19 June 1967; "County seeks high-rise toll bridge over channel," *Houston Chronicle*, 24 November 1968; "County backs toll bridge over channel," *Houston Post*, 20 December 1968; "Study for channel toll bridge slated," *Houston Post*, 21 January 1969; "Toll bridge feasibility study set," *Houston Chronicle*, 20 February 1969.

34. Jim Griffin, North Texas Turnpike Authority, communications with author, various dates in 2002.

35. "Houston: A Historical Profile," *Houston*, November 1960.

36. "Beltway 8 toll bridge revenues less than expected," *Houston Post*, 25 June 1983.

37. Jesse Sublett, *History of the Texas Turnpike Authority* (Texas Turnpike Authority and its successor, the North Texas Turnpike Authority, 1993).

38. Traffic and revenue data from the Texas Turnpike Authority annual reports.

39. "Takeover plan for toll bridge hits obstacle," *Houston Chronicle*, 19 September 1992; Senate Bill 242, 73rd Texas Legislature, 1993.

40. "Junk bond holders cross over to profits," *Houston Chronicle*, 13 June 1994.

41. "County will finally get $225 million bridge for 'free'," *Houston Chronicle*, 10 March 1994; Harris County Toll Road Authority traffic reports.

42. Texas Department of Transportation, *SH 87 Feasibility Study* (November 2000).

43. "Galveston-Bolivar Ferries," *Texas Highways*, September 1956; Texas Transportation Commission Minute Order 22947, 22 January 1947; Texas Transportation Commission Minute Order 25416, 29 September 1948; "A Bottle was Broken—A Ribbon was Cut," *Texas Highways*, March 1959.

44. "Beach crowds on Bolivar described as powder keg," *Houston Chronicle*, 7 June 1994.

45. Texas Transportation Commission Minute Order 65823, 31 January 1972; Texas Transportation Commission Minute Order 67824, 5 September 1973.

46. Most bridge vertical clearance values are from the Texas Department of Transportation *SH 87 Feasibility Study*, performed by URS Consultants, November 2000. Others are from official bridge web sites.

47. SB 716, 78th Texas Legislature, 2003.

About the Author

Erik Slotboom was born and raised in Sharpstown, Houston's prototypical freeway suburbia of the 1950s and 1960s. After graduating from Sharpstown High School in 1985, Erik received a B.S. in mechanical engineering from Texas A&M University and an M.S. in mechanical engineering from the University of Texas at Austin. He worked as a project engineer in Houston's energy industry for seven years and in 1998 joined the high-tech boom and became a software developer in Austin. Erik started the web site TexasFreeway.com in 2000 and was surprised by the widespread interest in Texas' freeways, prompting him to consider expanding the web site into a book. The high-tech bust in 2001 gave Erik the opportunity to fulfill his long-contemplated goal of writing a book about Houston's freeways.

Freeway, cables, and blue sky: This view shows the deck of the Fred Hartman Bridge (SH 146). (Photo: November 2000)

Index

I

J

K

L

V

W

Z

Ordering and additional information

www.HoustonFreeways.com

photography	Most photography by the author was with a Canon EOS Rebel 2000 and Sigma 28-105 lens, using Fuji Reala film
scanning	Film (modern photos): Nikon Super Coolscan 4000 ED Prints (historical photos): Epson 2450 flatbed TxDOT historical photos professionally scanned with Scitex equipment
image processing	Adobe Photoshop 7; Jasc Paint Shop Pro 7
maps and graphics	Adobe Illustrator 10
layout	Adobe InDesign 2
fonts	Text: Times New Roman Tables, captions, and headings: Arial
printing and binding	C. J. Krehbiel, Cincinnati, Ohio
platesetter (computer-to-plate)	Creo Trendsetter 4557
press	Text: Mitsubishi Diamond-32 four-unit web press Cover: Mitsubishi 3FR sheetfed press
paper	MeadWestvaco Sterling Ultra matte 70 lb